Current Natural Science

Coordinated by Didier BLOCH, Sébastien MARTINET, Thierry PRIEM and Christian NGÔ

Li-ion Batteries

Development and Perspectives

Science Press | edp sciences

EDP Sciences – ISBN(print): 978-2-7598-2555-4 – ISBN(ebook): 978-2-7598-2567-7
DOI: 10.1051/978-2-7598-2555-4

Preface

In order, in particular, to support energy transition, necessary and essential to safeguard our planet, energy storage needs will increase very sharply in the coming decades, whether for stationary applications or for mobility with a global market which should increase from 100 GWh in 2016 to 3 TWh (3000 GWh) in 2030.

Li ion batteries have a number of advantages that help meet these needs. The current challenges are to increase performances (to reach more than 350 Wh.kg^{-1} and 1000 Wh.L^{-1} at cell level) while aiming for increased safety and a cell target cost of around € 80–120 per kWh. The so-called traditional Li ion is now reaching its limits in terms of mass and volume energy densities, which is pushing all scientific and industrial players towards the identification of new technological breakthroughs on new generations of batteries. Particular attention is also fundamental with regard to the sustainability of the solutions proposed by securing supplies, avoiding so-called "critical" materials in terms of environmental impact, using solvent-free processes, but also more generally by considering recycling and full battery life cycle analysis.

This book, by addressing the topic of batteries across the entire value chain from materials to the system, offers readers elements of understanding and reflection allowing everyone to have a better knowledge of the expected assets, but also of the hurdles and issues related to the development of present and new generations of Li-ion or post Li-ion Batteries. The development of these new batteries as a storage solution, beyond being useful for the development of clean energies to support the energy transition, will have a certain environmental and societal impact in the years to come.

Severine JOUANNEAU SI LARBI,
Head of the Electricity and Hydrogen Department
for Transport at CEA/LITEN

DOI: 10.1051/978-2-7598-2555-4.c901
© Science Press, EDP Sciences, 2021

Contents

CHAPTER 4

CHAPTER 5

CHAPTER 6

CHAPTER 7

CHAPTER 8

CHAPTER 9

CHAPTER 10

CHAPTER 11

CHAPTER 12

CHAPTER 13

CHAPTER 14

CHAPTER 15

CHAPTER 16

CHAPTER 17

CHAPTER 18

CHAPTER 19

Chapter 1

Introduction

**Didier Bloch, Sébastien Martinet, Thierry Priem
and Frédéric Le Cras**

After a brief introduction, this first chapter introduces a short history of battery technologies, as well as the operating principles of Li-ion batteries.

It is now clear to everyone that the rapid and drastic greenhouse gases (GHG) emissions is an absolute imperative. This necessarily implies achieving a massive reduction in the consumption of fossil fuels (oil, gas, and coal).

Provided the electricity used to recharge vehicles is decarbonized, the deployment of battery powered electric vehicles could significantly contribute to make this possible, as it shifts the consumption of oil to electricity.

In addition, the deployment of electrified vehicles offers another decisive advantage: as electric motors happen to be much more energy efficient than internal combustion engines, it should make it possible to decrease significantly the overall energy consumption.

This is why the accelerated development of vehicle electrification is a top priority.

In France, for example, in 2019, transport, which is currently essentially oil-based (excluding rail), accounted for about 29% of the country's ~ 150 MTep *final energy consumption*, and contributed to $\sim 39\%$ of the 313 Megatons corresponding GHG emissions (figure 1.1a & b).

Slightly less than half of these emissions come from passenger vehicles alone.

The complete electrification of *e.g.* all French passenger cars and small and medium sized commercial vehicles fleets would therefore make it possible to save 32 Mtoe of oil imports (out of a total of 62 Mtoe, *i.e.* more than 50% of all oil imports), and reduce the country's overall GHG emissions by approximately 25%. Considering the 20–25 years required to completely renew a fleet of vehicles, an ambitious and proactive policy could allow achieving this goal as soon as 2040–2045.

DOI: 10.1051/978-2-7598-2555-4.c001

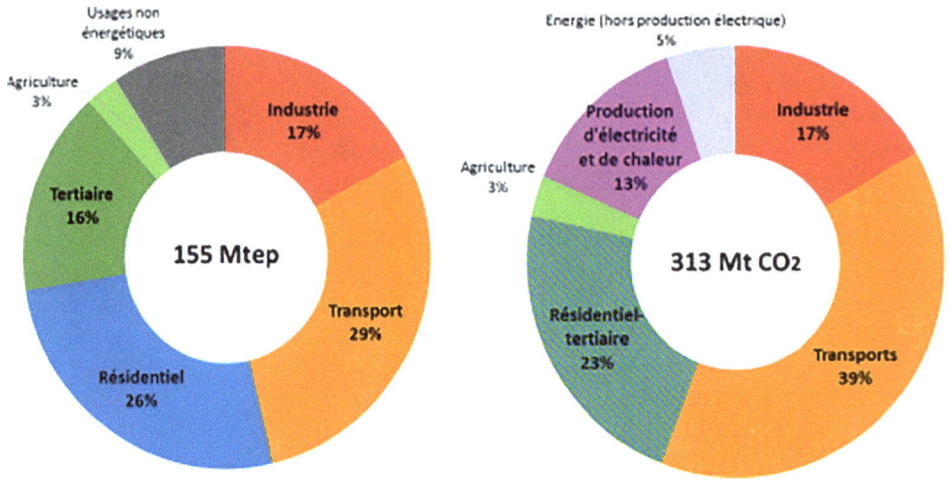

FIG. 1.1 – (a) (left) **Final Energy Consumption by sector, France, 2019**. Transport accounts in France for $\approx 30\%$ of the overall final energy consumed ($\approx$150 MToe), and (b) (right) **GHG emissions corresponding to final energy consumption in France in 2019**. Transport accounts for $\approx 39\%$ of the corresponding overall CO_2 emissions (122 Mt out of 313 Mt). (Note that the 313 Mt CO_2 final-energy-consumption-related GHG emissions correspond to only 70% of the country's global GHG emissions, which take into account additional emissions due to, *e.g.*, industrial or agricultural processes). (source : « Integration de l'electromobilite dans le systeme electrique » RTE-AVERE Report, 15 April, 2019).

In addition, the expected electromobility surge will also spread to other modes of transport, such as, to a certain extent, public and heavy transport, thus multiplying its benefits, and making it possible to achieve this objective earlier than expected.

On the other hand, the deployment of the so-called "renewable" energy (RNE), mainly solar photovoltaic and wind energy, is often being put forward, rightly or wrongly [1], as one of the other possible solutions to help meet the GHG emissions reduction challenge mentioned above. Beyond the fact that the deployment of these electrical energy production devices does not reduce GHG emissions in countries whose electrical energy is *already* decarbonised, these energies are by nature not dispatchable on demand because of their intermittency. Their deployment requires the availability of additional dispatchable electricity production means, as well as massive electricity storage buffers, capable of storing the RNE electricity when it is produced in excess, and delivering it to the grid when there is no wind or sun to meet consumer demand.

A simple calculation shows that these energy storage buffers could be partially made of the storage capacity offered by a fleet of electrified vehicles (which would simultaneously obviate the need to develop and implement specific and costly additional stationary storage means). Let us indeed consider the reasonable assumption of a French fleet of personal electrified vehicles of around 5 million units by 2030 (*i.e.* around 15% of the French personal vehicles fleet). If one considers that approximately

5 kWh are required to recharge every vehicle due to its average, daily, 35 km mileage, and taking advantage of the fact that more than 80% of personal vehicles at parked at any given time of the day, one can see that on board batteries could become a controllable electricity storage buffer system capable of storing approximately 25 GWh (5 million times 5 kWh), that can be stored at any time of the day when solar or wind energy is available. This amount represents approximately one fith of the average daily ENR French production in 2020. It therefore appears on a first hand quite smart and easy to store a significant part of the energy produced by RNE in electrified vehicles, provided the simultaneous deployment of RNE and electrified vehicles is efficiently coordinated. It is even possible to take on a second hand the logic one-step further: the deployment of bi-directional V2G/V2H (Vehicle to Grid/Vehicle to Home) connections could in return allow the vehicles to provide services to the power grid, such as peak power demand shaving, frequency regulation, and substitution of electricity for fossil fuels (oil and gas) whenever possible. Parked vehicles could also be recharged during off-peak hours and be used to directly power homes during peak hours (self consumption). These additional features will reduce the need to build additional power plants generally using fossil fuels, which would otherwise be required to produce the electrical energy in the absence of wind or sun.

Of course, the legitimate question is whether carbon-free electricity power stations will be sufficient to meet the growing electricity demand due to the take-off of the EV's market. To consider, once again, only as an example, the French case, one can reasonably imagine that if the nuclear and hydropower power capacity is *at least* maintained at its present, $\approx$ 450 TWh, 2021 production level, sufficient electricity should be available to meet the 100 TWh additional annual demand of the personal and light/medium sized delivery vehicles' batteries fleets. This assumes, however, a strong hypothesis: batteries should mainly be recharged at night during off-peak hours, and fast charge (>7–11 kW) during day hours should be minimized as much as possible to reduce the risk of grid instability. Additional 2–11 kW recharging could be envisaged, for example, on company's car parks charging stations during low electricity demand periods. Large scale development of fast recharging would undoubtedly require additional power means, and probably a costly reinforcement of the power grid.

Properly coordinated, the concept should lead to a win–win situation for all stakeholders: every citizen will personally contribute to effectively combating climate change, and, finally yet importantly, partially amortize the purchase of the battery in his or her vehicle by selling energy back to the grid. Governments will be able to keep their word and efficiently reduce their GHG emissions, as well as their financial and political dependence on fossil fuels. Car manufacturers will sell as much cars as today. Electricity operators will play with offer and demand to reduce their investments in new power stations and offer additional, profitable services.

The implementation of this vision requires foremost the effective and vigorous coordination of public authorities, which will have to play the role of an orchestra conductor. They will have to foster the implementation of bidirectional recharging infrastructures, which will be a relatively easy task to deal with, but, above all, in close coordination with all industrial players implicated all along the energy value chain, promote the large scale implementation of batteries offering a very long cycle life ($>$2000–5000 charge/discharge cycles), since these batteries will be expected to perform a dual function: mainly for mobility, and also to provide grid services. Such a long cycle life will, by the way, drastically reduce their Total Cost of Ownership (TCO), expressed in €/Wh exchanged, and contribute to a significant reduction of the battery environmental footprint. The expected evolution of the current technical specifications, for which the battery only provides the "mobility" function, therefore, invites to explore new technical options or to improve existing ones, from the material to the complete system.

Lithium-ion batteries will be the catalyst, the pivot of all these expected industrial and societal transformations. Initially developed to replace the nickel–cadmium batteries used at the time to power portable electronic devices, the continuous improvements in their performance and the sharp drop in their production cost now pave the way for their large-scale use as a massive, dispatchable on-demand, electricity storage means. As of today, the battery manufacturer controls a key part of the vehicle's performances and cost.

This is why the very large-scale production of lithium batteries has become a top priority issue, which conditions primarily the future of the automotive industry. As the first one built in Nevada by the Tesla company (Elon Musk) allied with the Japanese Panasonic, "Gigafactories" are now being built all over the world. Each of them mobilizes investments of several billion euros. In the majority of cases, their deployment is based on the technical know-how detained by world leaders, currently Asian (Japan, Korea, and China). As a matter of fact, these countries anticipated and prepared, as early as the beginning of the 20th century, for the major developments in mobility that are taking shape today. China has become the world's largest producer and consumer of lithium batteries in 2020, driven by its domestic market, and by a very proactive political regulation. In Europe, the construction of more than 20 Gigafactories, mainly operated by Asian industrial players, is also underway or planned: Samsung SDI and SK Innovation (Korean) are implementing factories in Hungary; LG (Korean) in Poland; Envision (Chinese) in France; CATL (Chinese) in Germany... [2].

Until then, most European car manufacturers considered batteries to be a simple commodity and expected competition from their Asian suppliers to continue to lower costs and improve performance. However, several factors combined, leading to a gradual awareness in Europe – in 2016 in Germany, mid-2018 in France – of the mandatory need to master the battery manufacturing know-how. As a matter of fact, electrified vehicles are now built "around" the embarked battery, which obviously becomes a critical component. Moreover, the expected growth of the world market will most likely lead to possible issues in battery supply, thus increasing the risk of dependence on manufacturers whose interests are likely to change rapidly depending on the geopolitical context. This awareness led in 2019–2021 to the emergence of European-born consortiums such as those led by Stellantis, Volkswagen, and other car or battery manufacturers such as Northvolt in Sweden, with the perspective to build $>$30 GWh Gigafactories in Sweden, France, Germany or Italy, with the partial support of European and national funding [3].

If Europe wishes to maintain a strong car industry, it must reasonably play both sides of the coin:

- Rebuild a complete industrial sector, across the entire value chain, to manufacture, in the shortest possible time, high-performance and economically competitive batteries, and take a significant share of the fast-growing electric mobility market. As batteries will be manufactured in millions in the near future, it seems unlikely that technologies that are too far from the current state of the art will be used. This is the purpose of the "Present" of lithium-ion batteries presented in this book, to give the reader the most accurate possible idea of the technology used not only today, but in the next 10 years, as long as present-generations Gigafactories will operate.
- Actively encourage the R&D activity necessary for the development of next-generation batteries, since it will be necessary to remain competitive in the long term: this is the purpose of the Post-Li-ion or other types of batteries, presented in this book through dedicated chapters, which will provide the reader with an overview of the options explored to date.

Although the two types of initiatives are closely coupled, the first type of action appears to be more the domain of industrial leadership, and the second type of action more the domain of institutional action. In both cases, the game will involve European industrial and research players, who will be able to coordinate their efforts and select the relevant technical options. They should control and secure the entire battery value chain in a context of intense competition. The manufacture of a battery requires the mastery of a large number of skills in a wide variety of fields, from the extraction of raw materials to their recycling at the end of the battery's life and their reuse in new batteries, *via* the synthesis of active materials, the very large scale manufacture of electrodes, cells, modules, packs, and complete systems including electronic and thermal management.

The avenues for progress are real. From the point of view of the materials used as well as that of the complete on-board system, they depend largely on the functional specifications of the intended application, which has a direct influence on them.

As an example, the trend seems presently to priviledge the development of "all electric" vehicles with >50–70 kWh batteries, in order to offer the longest possible driving range. In this case, materials designed to cycle 500 times (200 000 km with 400 km per cycle) may well meet the demand. But what if the same vehicle must also be used as a grid service provider? Then the material's behaviour must be improved to meet >2000 cycles without major degradation. And what if the battery is supposed to power a Plug-In Electric vehicle (PHEV), which 12–15 kWh, much smaller, and much less material-intensive consuming battery, has on the other hand to cope with much more heavy power discharge rates to power the vehicle, must prove a much larger (>5000–10000 cyles) life expectancy to be recharged every day and offer grid services at the same time? One can see that electrode and electrolyte materials will probably be designed differently in each case. In the end, all doors remain open, in order to adapt materials and batteries to the preferences of the final consumer, the good news being that such design adaptation to meet various needs seem now within reach. Regardless of the option selected, the consortia that will be set up will have to deal with numerous, sometimes contradictory specifications, in order to comply with the final target (reduction of greenhouse gas emissions):

- To ensure user's safety.
- To offer the battery's best possible performance: energy density and power density, of course; but also high cyclability, so as to reduce the cost of ownership per cycle, and enable the vehicle not only to meet mobility needs but also to provide other types of services to the power grid, and reduce the environmental footprint.
- To reduce the presence of sensitive or critical materials to a minimum, and if possible to zero.
- To effectively coordinate all players involved in the entire value chain, including energy producers and operators, manufacturers of electronic materials or components, car manufacturers, recycling industries, etc.
- To manufacture vehicles, especially entry-level and mid-range vehicles, accessible to the greatest number of people. Commercial success will indeed depend on many parameters: social acceptance of the electrified/connected vehicle, changes in consumer purchasing power, renewal rate of the vehicle fleet, availability of recharging systems, and development of alternative mobility solutions.... A great deal of pedagogical work will be necessary in order to answer the legitimate questions of all citizens.

Europe is late, but has the skills and assets needed to catch up with the Asian industry. France and Sweden have, in Europe, key assets in this area, particularly thanks to their decarbonated energy mix. R&D collaborations are already in place and have been established for a long time (European joint projects for example).

Hurdles will be numerous but the European automotive industry survival is at stake and success is the only possible option [3].

1.1   Brief History of Primary and Secondary Batteries

The discovery of the working principle of primary batteries took place in 1800, by Alessandro Volta with the eponymous battery [5] using two different metals, zinc and copper discs, separated by a felt soaked in sodium chloride. It was not until 1859 that Gaston Planté discovered the first rechargeable lead-acid battery [6].

In 2021 battery technologies include only three major families in addition to lead-acid batteries (this latter remaining so far dominant in terms of Wh produced, but not any longer in value): alkaline nickel–cadmium (Ni–Cd), nickel-metal hydride (Ni-MH), and, since 1991, lithium-ion (Li-ion) batteries, which gradually dominate all others.

Two major breakthroughs allowed the large scale commercialization of Li-Ion batteries. In 1980, J. Goodenough *et al.*, discovered $LiCoO_2$ as a high-potential positive electrode material [7]. Shortly after, in 1983, R. Yazami and P. Touzain showed that lithium could intercalate reversibly in low-potential graphite [8]. This made it possible to dispense with the use of metallic lithium, which posed serious safety issues. The combination of these two innovations allowed the development of Li-ion systems and their first commercialization, in 1991, by Sony.

Since 1991, the gravimetric energy density of Li-ion batteries has almost tripled from 100 Wh.kg^{-1} to almost 270 Wh.kg^{-1}. At the same time, their cost has fallen very sharply, thanks in particular to the reduction in manufacturing costs. These improvements, coupled with initiatives of visionary industrial players such as Tesla/Panasonic in USA; LG in Korea; or BYD or CATL in China, made it possible for the electric vehicle market to take off.

Battery technologies are generally compared in terms of their energy density (Wh.kg^{-1} or Wh.L^{-1}) and/or power density (W.kg^{-1}), especially for embedded systems. These performances are usually synthesized by plotting the power density as a function of the mass energy density in a so-called *Ragone* diagram as shown in figure 1.2. For each technology, these performances are described by a beam representing the possibility of modulating the performances according to whether one seeks to favour the energy density or the power density, *via*, for example, the use of thin electrodes to favour the latter. These notions are discussed in chapter 13, which describes the batteries manufacturing processes. It is also worth noting the presence of supercapacitors in this diagram, a technology based on capacitive phenomena and the use of electrochemical double layers. Supercapacitors are covered in chapters 9 and 10.

Of course, the performances of batteries are also measured during use, *i.e.* during charging and discharging cycles, for example. To characterize these performances, one usually reports the evolution of the voltage at the terminals of the cell as a

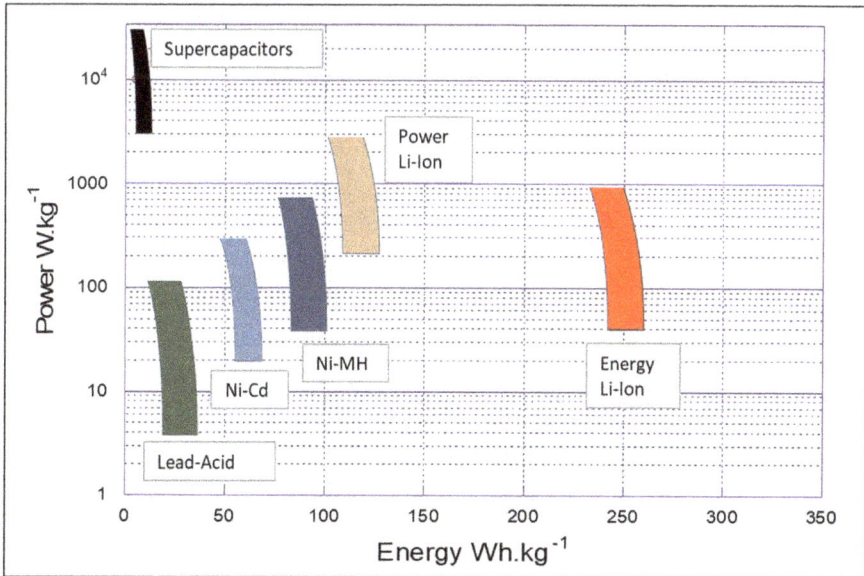

FIG. 1.2 – Ragone diagram of various commercialized battery technologies.

function of the discharged capacity, given in ampere hours (Ah) (1 Ah corresponds to the flow of a current of one ampere over a period of 1 h, *i.e.* 3600 Coulombs), and for various current values. An example is shown in figure 1.3 for the case of a cylindrical "18 650" cell (18 mm diameter and 65 mm height) for currents ranging from 0.2 times the nominal discharge rate (discharge in 5 h) to 2 times the nominal discharge rate (discharge in ½ h). The nominal discharge rate (or C-Rate) is defined as the current that allows the cell capacity to be totally discharged in 1 h. This then allows the use of C/n or nC ratings which represent slower rates, discharge in n h for C/n, or faster, discharge in 1/n h for nC.

In concrete terms, the Li-ion cell shown in the example of figure 1.3 has a nominal capacity of 2040 mAh, *i.e.* is able to withstand a nominal discharge rate C (noted "It" in the figure) and to deliver during one hour a discharge current of 2040 mA. A more rapid discharge rate at 4080 mA, or 2C, lasts almost 30 min, during which time nearly all of the rated capacity is recovered. At slower C/5, or 408 mA, the delivered capacity is close to 2200 mAh. The curves show a significant decrease in the mean voltage when the discharge rate is increased. This is due to an increase in ohmic and/or diffusive losses within the battery cell, which degrades the energy efficiency (energy losses between charge and discharge).

It should be noted that the accumulators (or unit cells) are the elementary cells constituting the batteries in which they are assembled in series and/or in parallel in order to increase the operating voltage (series connection) and/or the on-board capacity (parallel connection). An electronic management system called BMS (Battery Management System) always controls the battery operation and ensures safety. The pack constituting the cell/BMS assembly (see chapter 14) is referred to

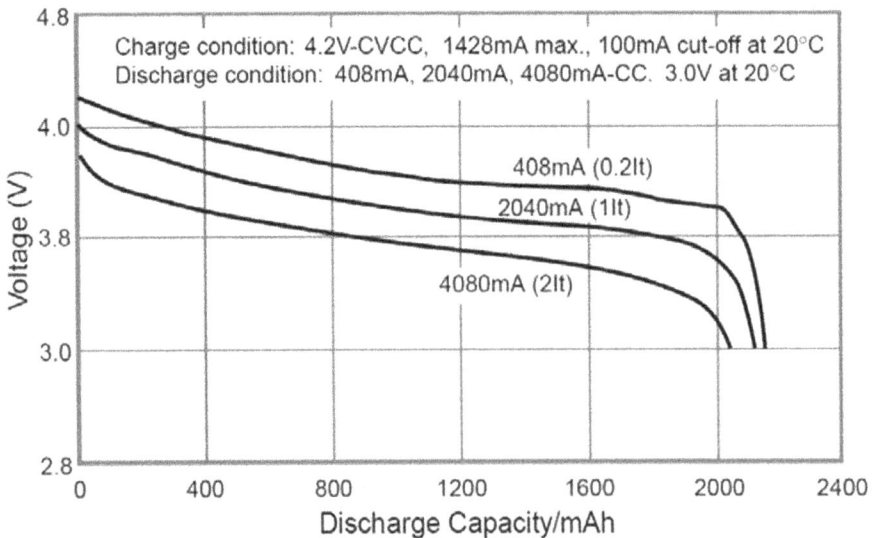

FIG. 1.3 – Discharge curves of a Li-ion cell at various discharge rates: Panasonic CGR18650C [9].

as the "battery pack". It often includes a thermal management system, in order to ensure a uniform temperature between the unit cells, which in turn allows the battery to behave as expected throughout its lifetime.

As mentioned above, after more than 150 years of development, only four main battery technologies have been widely commercialized. Today, Li-ion batteries are taking a growing share of the market thanks to their versatility, performance and cost. At the same time, faced with the challenge posed by the large-scale development of the electric vehicle, and of stationary energy storage, many other options are explored, either to improve the performance of the current Li-ion ("all-solid"...) sector, or to develop alternative solutions – lithium–sulphur, sodium-ion, organic materials... – less demanding in terms of sensitive materials for example.

1.2   General Information on Li-ion Batteries

"Traditional" Li-ion batteries are based on the use of two lithium intercalation materials: a lithiated transition metal oxide at the positive electrode ($LiCoO_2$, NMC, NCA, LFP...) and graphite at the negative electrode. Chapters 2 and 3 introduce new electrode materials which may be used as alternatives to these reference materials. The operating principle is shown in figure 1.4. The battery is manufactured in the discharged state, the positive electrode being the source of lithium, and the graphite of the negative electrode being initially free of lithium ions. During the first charge, lithium ions de-intercalate from the positive electrode, diffuse through the electrolyte and intercalate into the negative graphite electrode. In order to respect the rule of electro-neutrality, the flow of one Li^+ ion in the internal circuit of the accumulator is exactly compensated by the flow of one

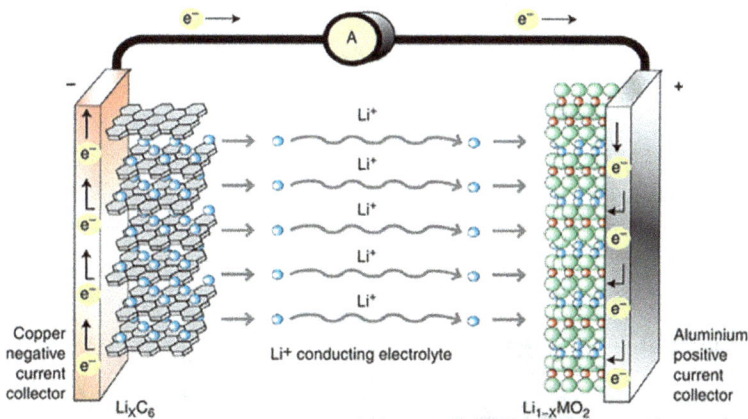

FIG. 1.4 – Working principle of a traditional lithium-ion battery, using a transition metal oxide $LiMO_2$ as positive electrode active material, and graphite as negative electrode active material. From reference [10].

electron in the external electrical circuit. The opposite phenomena occur during discharge.

It should be noted that, during the first charge, a passivation layer, known as SEI (Solid Electrolyte Interphase), is formed on the negative graphite electrode, which ensures the cell's good cycling behaviour. However, it does lead to an irreversible loss of capacity in the order of 5%–10%.

The high voltage difference between the two electrodes – typically 3.6 V nominal voltage – contributes greatly to the high energy density of this battery technology. This compares to 2 V for lead batteries and 1.2 V for alkaline batteries. This is due to the highly reducing nature of lithium metal and the fact that lithium ions are intercalated in graphite at a potential fairly close to the Li^+/Li redox couple potential (−3.04 V *vs.*, SHE, SHE for Standard Hydrogen Electrode), typically on average 100 mV above this value.

Figure 1.5 illustrates the positioning of the various positive and negative materials discussed further in this book. Positive electrode materials (described chapter 2) are mainly positioned at the top left with maximum insertion capacities often well below 300 $mAh.g^{-1}$ (typically 150–190 $mAh.g^{-1}$), while the range of negative electrode materials is wider with three families: conventional carbon-based materials, titanium-based compounds (power-oriented) and materials corresponding to the p-block elements such as silicon with a very high specific capacity (see chapter 3). To promote energy density, it is necessary to select both positive and negative compounds with high specific capacities, but also as far as possible on the scale of potentials.

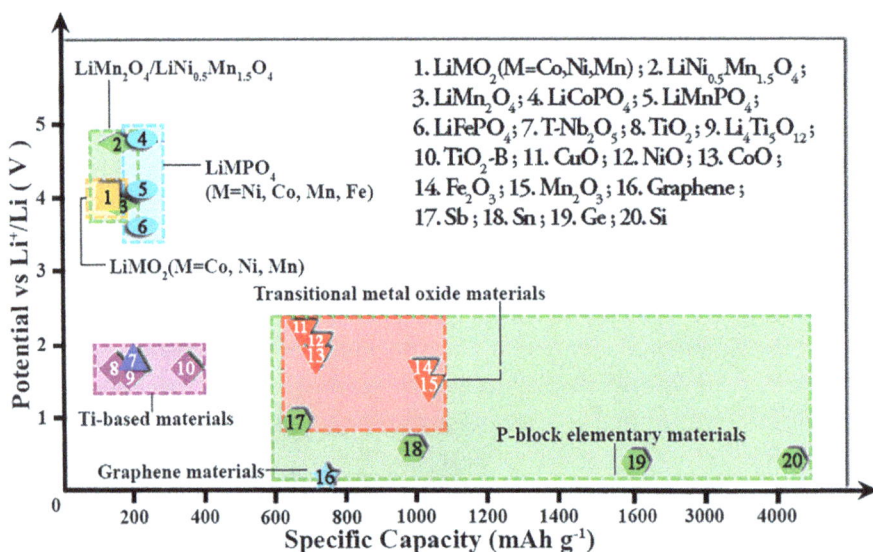

FIG. 1.5 – Li-ion batteries materials mapping. From reference [11].

Bibliography

[1] Linnemann T., Vallana G.S. (2019) Wind energy in Germany and Europe, VGB PowerTech 3 1 2019ho.

[2] https://electrek.co/2019/02/21/lg-vw-battery-cell-supply-ev-gigafactory/.

[3] https://www.greencarcongress.com/2021/07/20210709-stellantis.html. 9 July, 2021 Stellantis to invest more than €30B in electrification and software through 2025.

[4] https://www.transitionsenergies.com/lairbus-des-batteries-est-deja-mort-ne/.

[5] Volta A. (1800) *Philos. Trans.* **2**, 430.

[6] Planté G. (2008) The Storage of electrical energy, Kessinger Publishing, Réimprimé.

[7] Mizushima K., Jones P.C., Wiseman P. J., Goodenough J.B. (1980) Li_xCoO_2 ($0 < x < -1$): a new cathode material for batteries of high energy density, *Mat. Res. Bull.* **15**, 783.

[8] Yazami R., Touzain P. (1983) A reversible graphite-lithium negative electrode for electrochemical generators, *J. Power Sources* **9**, 365.

[9] Panasonic, Lithium-ion Batteries: individual dataSheet CGR18650C, https://www.rosebatteries.com/pdfs/Panasonic%20CGR18650C.pdf.

[10] Zhang Z., Ramadass P. (2012) Lithium-ion Battery Systems and Technology, Encyclopedia of Sustainability Science and Technology, 6122–6149. Springer. DOI https://doi.org/10.1007/978-1-4419-0851-3_663, Print ISBN 978-0-387- 89469-0, Online ISBN 978-1-4419-0851-3.

[11] Tang Y., Zhang Y., Li W., Ma B., Chen X. (2015) *Chem. Soc. Rev.* **44**, 5.

Chapter 2

Positive Electrode Materials for "Lithium-ion" Accumulators

David Peralta, Frédéric Le Cras, Jean-Baptiste Ducros,
Carole Bourbon, Jean-François Colin and Sébastien Patoux

The positive electrode (or cathode) is a key component of a Li-ion cell: it is the most limiting element, for several reasons.

The first of these reasons deals with material's cost: more than 60% of the cost of the electrochemical unit cell is due to the raw materials that make it, and in 2021, the cost of the cathode materials represents 27% of the cell cost (figure 2.1). The cell material's cost share should increase in the next future, with the Gigafactories expected mass production ramp up, which will mechanically reduce production and administrative cost share.

The cathode material market grows rapidly from year to year (410 000 tons in 2021 *vs.* 184 000 tons in 2016) due to the rapid development of the electric vehicle market [1]. However, cobalt, for instance, used in the composition of the cathode active material, is considered as a critical metal: its known resources are relatively limited (it is mined as by-product of nickel and copper extraction) and mainly located in geopolitically unstable countries. This is why compositions with lower cobalt content are currently developed, with a real success. The lithium battery industry remains however exposed to a risk of material's supply shortage. Significant fluctuations in the raw materials cost may be expected in the future, not considering the environmental impact of their extraction conditions.

The second reason deals with the intrinsic performance of the cathode material (energy and power density, electrochemical reversibility, temperature stability...), which largely conditions the behavior of the cell during cycling. The selection of cathode material's blends, and the way in which these blends are implemented, implies a necessary compromise between cost, safety, life expectancy, and energy/power density performance.

DOI: 10.1051/978-2-7598-2555-4.c002

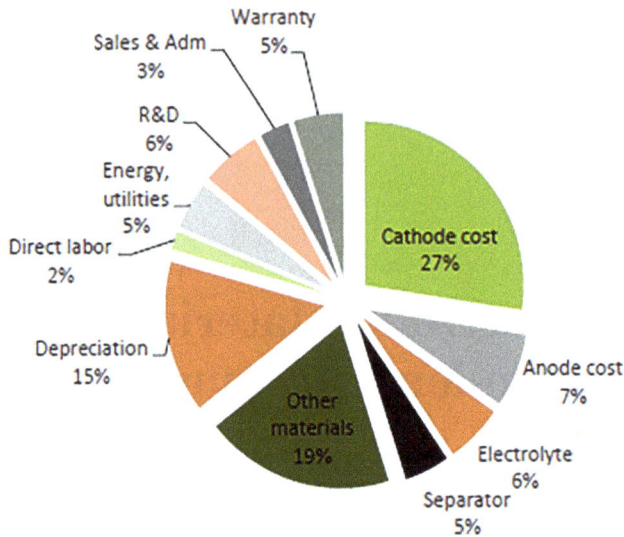

FIG. 2.1 – Li-ion unit cell cost breakdown. Source: Avicenne [1].

Currently, there are very few cathode materials available for lithium-ion battery applications. These compounds belong to three different crystallographic families that give materials special properties: spinels, layered oxides and olivines [2].

The main purpose of this chapter is to provide an overview of these three material families.

2.1   Positive Electrode Materials of "Spinel" Structure

Interest in spinel positive electrode materials was born in the early 1980s with the emphasis on the possibility of chemically and electrochemically extracting lithium from $LiMn_2O_4$ without significantly disrupting the spinel structure of the initial compound [3, 4]. From the outset, this material, free of expensive or toxic elements, has a theoretical specific capacity equal to 148 mAh.g^{-1} at a potential of 4 V *vs.* Li^+/Li. Because the spinel structure is favorable to rapid ionic exchange kinetics, this material family was considered as an alternative to the $LiCoO_2$ layered oxide. With its substituted variants, spinel structure materials were specifically targeted for use in Li-ion power cells.

Description of the Spinel Crystallographic Structure

A number of compounds of the general formula $A[B_2]X_4$ (A and B being cations, X being an anion such as O, S, Se, Te) adopt a crystalline structure isotype of the $MgAl_2O_4$ spinel. This structure is characterized by a pseudo-close-packed anion stacking, and by cations' occupancy of 1/8 of tetrahedral sites and 1/2 of interstitial octahedral sites. In normal distribution, cations A occupy tetrahedral sites, while

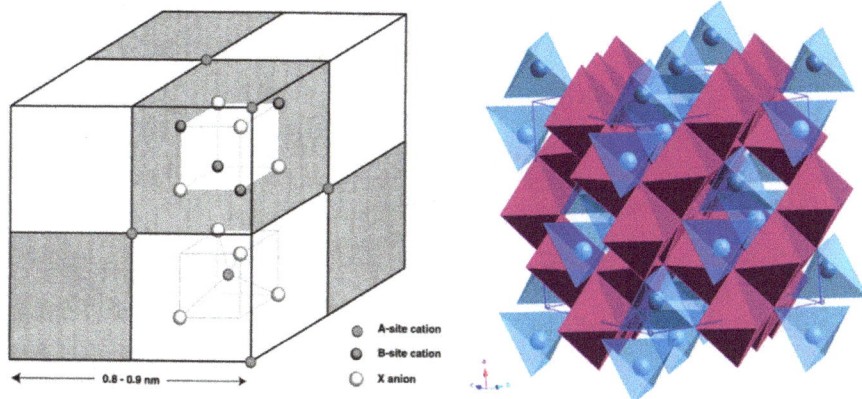

FIG. 2.2 – Left: AB_2X_4 spinel structure consisting of two site types. The mesh parameter of 8–9 Å corresponds approximately to that of oxides [5]. Right: picture of the spinel structure consisting of BX_6 and AX_4 tetrahedrons [6].

cations B occupy octahedral sites of the crystal. However, there may be an inverse B [AB]X_4 distribution in which half of the B cations occupy the tetrahedral sites, while the octahedral sites are occupied by both cation types, and also intermediate distributions $A_xB_{1-x}[A_{1-x}B_{1+x}]X_4$. When the cation distribution on each site type is random, the cubic crystalline structure is indexed in the Fd-3 m space group and includes 8 formula units (figure 2.2).

LiMn$_2$O$_4$ (LMO), a "Low Cost" 4 V Electrode Material for Power Cells

In the case of the LiMn$_2$O$_4$ spinel compound, lithium alone occupies an interstitial network (tetrahedral sites (8a)) delimited by the Mn$_2$O$_4$ skeleton, which consists of octahedral sites (16c), which makes it possible to diffuse Li$^+$ ions within a three-dimensional network of channels. In the case of intercalation materials, such a three-dimensional geometry of the diffusion paths is in principle particularly well suited for applications that require high current densities. At room temperature, LiMn$_2$O$_4$ is an n-type semiconductor with an electronic conductivity of about 10^{-4} S.cm^{-1} [7, 8]. The *gap* value measured on thin-layer samples is in the order of 1 eV [9].

The deintercalation of Li$^+$ from LiMn$_2$O$_4$ occurs primarily at around 4.0 V *vs.* Li/Li$^+$ and is accompanied by the oxidation of Mn^{3+} manganese contained in the Mn^{4+} structure (figure 2.3) according to the equation:

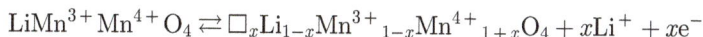

$$\text{LiMn}^{3+}\text{Mn}^{4+}\text{O}_4 \rightleftarrows \square_x\text{Li}_{1-x}\text{Mn}^{3+}_{1-x}\text{Mn}^{4+}_{1+x}\text{O}_4 + x\text{Li}^+ + xe^-$$

Additional Li$^+$ can also be intercalated in LiMn$_2$O$_4$. This reaction results in a potential plateau of around 2.95 V *vs.* Li$^+$/Li typical of a two-phase process

$$\text{LiMn}^{3+}\text{Mn}^{4+}\text{O}_4 \leftrightarrows \square_x\text{Li}_{1-x}\text{Mn}^{3+}_{1-x}\text{Mn}^{4+}_{1+x}\text{O}_4 + x\,\text{Li}^+ + x\,e^-$$

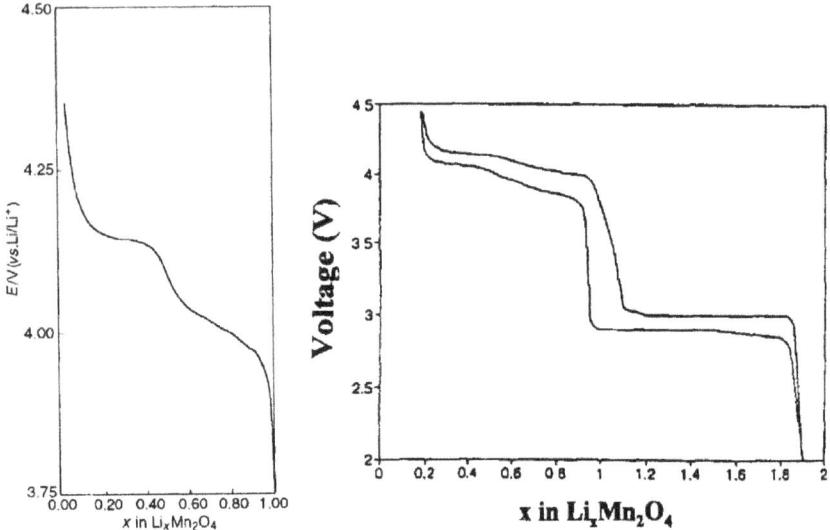

FIG. 2.3 – Left figure: The evolution of the potential of a Lithium/Lithium Electrolyte/$\text{Li}_x\text{Mn}_2\text{O}_4$ cell as a function of the amount x of lithium contained in the $\text{Li}_x\text{Mn}_2\text{O}_4$ cathode material, with $0 < x < 1$. The potential decreases during the cell discharge – the cell functions as an energy source – while the lithium content in the material increases between $x = 0$ (Mn_2O_4 composition) and $x = 1$ (LiMn_2O_4 composition). Right figure: The potential evolution of a Lithium/Lithium Electrolyte/$\text{Li}_x\text{Mn}_2\text{O}_4$ cell, depending on the amount of lithium in the material (charge and discharge of the cell), with $0 < x < 1.8$ [10, 11].

(figure 2.3b). It leads to the formation of a $\text{Li}_2\text{Mn}_2\text{O}_4$ phase with a quadratic cell ($I4_1/\text{amd}$, $\sqrt{2}a = \sqrt{2}b = 7985$ Å, $c = 9.250$ Å) resulting from a significant distortion of the initial cubic cell induced by a cooperative Jahn–Teller effect [12, 13]. Thus, lithium occupies the 16c octahedral sites of the initial cell, making $\text{Li}_2\text{Mn}_2\text{O}_4$ an ordered NaCl compound [14].

At room temperature, LiMn_2O_4-based electrode cycling shows the good reversibility of the lithium intercalation/disintercalation reaction in the high potential '4 Volts' part. However, a progressive loss of capacity is systematically observed, the latter becoming particularly rapid at a higher cycling temperature (55 °C) (figure 2.4). This behavior does not meet requirements of commercial batteries. Numerous studies have attempted to understand the reasons for this and to remedy it. A great number of factors can explain this performance degradation during charge/discharge cycles, among which: formation of the $\text{Li}_2\text{Mn}_2\text{O}_4$ phase on the surface of the particles, generation of Mn^{2+} on the surface of the material (dismutation of Mn^{3+}, formation of Mn_3O_4) [15] followed by dissolution in electrolyte, decomposition of LiPF_6 and organic carbonates electrolytes, non-stoichiometry in oxygen, loss of electron percolation in the electrode...

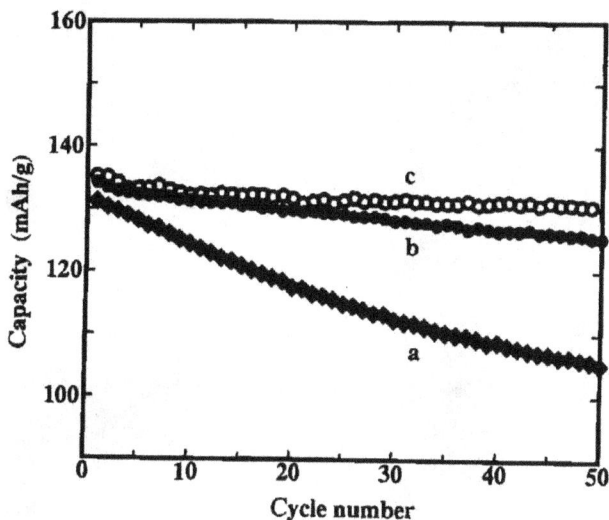

FIG. 2.4 – Specific capacity evolution of $LiMn_2O_4$ cycled at different temperatures between 3.5 and 4.5 V/Li^+/Li in medium 1 M $LiPF_6$ EC : DMC 1 : 2 volume: (a) 55 °C, (b) 25 °C, and (c) 0 °C [16].

The first proposed strategy to limit this capacity loss during cycling is to reduce the Mn^{3+} content in the spinel, to limit the formation of the NaCl-type phase and/or slow down the dissolution of the material. This can be achieved easily by substituting a fraction of manganese with a valence cation of 3 or less and stable in the octahedral environment: $LiM_yMn_{2-y}O_4$ (M = Li^+, Mg^{2+}, Ni^{2+}, Cr^{3+}, Co^{3+}, Al^{3+} among others) [17–20]. As hoped, the electrochemical behavior of these substituted spinels is actually much better, especially for those with an initial average degree of Mn oxidation greater than +3.55 (lithium, magnesium or aluminum doped materials) [21].

At the industrial level, the synthesis of these materials is carried out by thermal treatment at a temperature of 700–900 °C of a mixture based on manganese oxide type γ-MnO_2, lithium carbonate Li_2CO_3 and additives. Commercial powders of spinel materials are generally in the form of pseudo-spherical particles of 10–15 μm in diameter, consisting of crystallites of submicronic size (figure 2.5). The specific capacity obtained is about 105–110 mAh.g^{-1}, corresponding to 430 Wh.kg^{-1} (*vs.* Li^+/Li).

Thanks to their interesting compromise in terms of cost and power density, "LMO" spinel materials have so far been mainly used in applications such as electric portable tools or as blends for EV's applications [23]. Global production of these materials has however stabilized to about 15 600 tons since 2016 and their share in cathode production is expected to decrease in ne next decade, due to the expected domination of the layered oxide compositions (figure 2.6) offering a higher energy density and a better life expectancy.

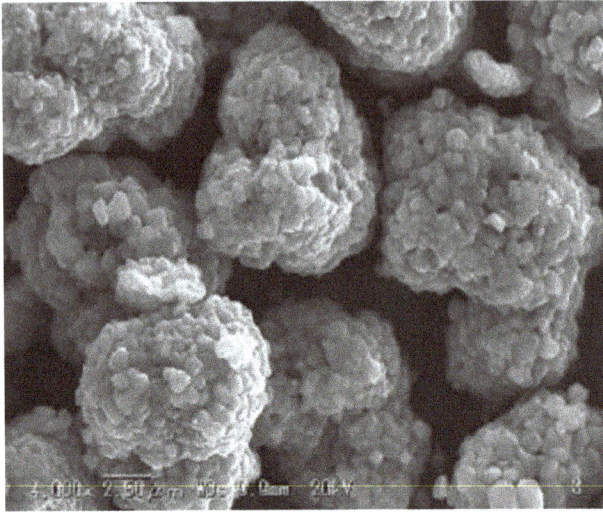

FIG. 2.5 – Morphology of a commercial LMO powder (Nichia) [22].

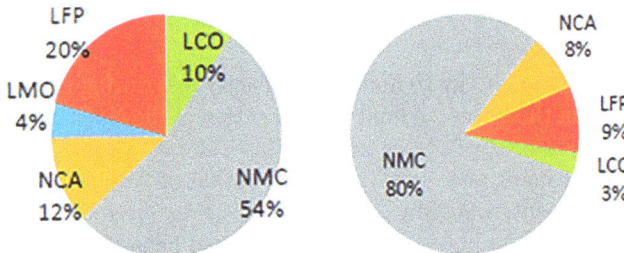

FIG. 2.6 – Cathode active material world's production by composition (left: 2019 : 390 ktons) and 2030 (realistic scenario: 1570 ktons). Nickel-Manganese-Cobalt « NMC »-based compositions will probably be mostly used. Uncertainty remains for LFP composition, free of sensitive materials, which may see a strong resurgence of interest [1].

Manganese Spinels Operating at High Potential – '5 V' Spinels "LNMO"

Work on substituted "LMO" spinels to improve their stability at high temperatures has revealed the presence of processes of de-intercalation/intercalation of lithium at 'high potential', *i.e.* close to 5 V *vs.* Li^+/Li, in particular for "LNMO" spinels of $LiNi_{0.5}Mn_{1.5}O_4$ composition (figure 2.7).

The theoretical specific capacity of this material (corresponding to the exchange of one Li^+ per unit formula) is 147 mAh.g^{-1}. Given the potential of 4.8 V/Li^+/Li at which this step takes place, the theoretical energy density of this material is 705 Wh.kg^{-1}. This very high value (LCO: 625, NMC: 630, NCA: 695 Wh.kg^{-1}),

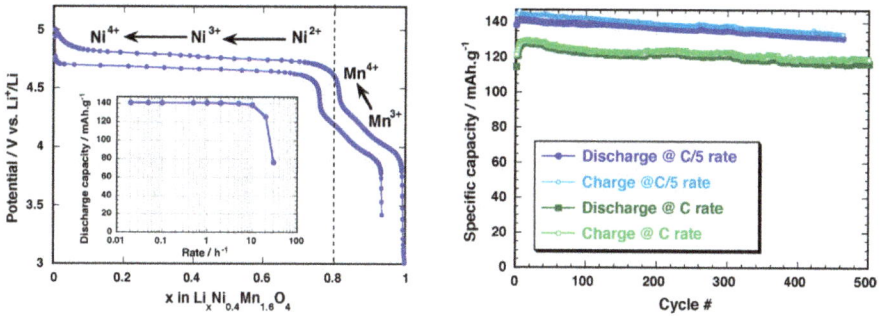

FIG. 2.7 – Left figure: Electrochemical behavior of Li/electrolyte/spinel accumulator LiNi$_{0.4}$Mn$_{1.6}$O$_4$ cycled on the 3–5 V domain, highlighting the successive electrochemical steps at different potentials involving manganese and nickel [24]. Right figure: Evolution of the capacity of a Li/electrolyte/spinel LiNi$_{0.4}$Mn$_{1.6}$O$_4$ cell cycled on the 3–5 V domain in relation to the number of cycles, under two different charge/discharge rates (C-rate: discharge and charge in 1 h; C/5: discharge and charge in 5 h).

combined with a spinel structure favorable to high load/discharge speeds and a 'cheap' composition, should make it a particularly attractive material for applications in the electric vehicle (EV, HEV, PHEV).

However, the ageing processes identified for the "LMO" "4 V" spinels are also observed with the "LNMO" "5 V" spinels. Moreover, due to the particularly high value of the working potential of this electrode at 4.7–4.8 V, *i.e.* significantly above the electrolyte's electrochemical stability domain, the reaction with the electrolyte is exacerbated. In addition, the electrochemical electrode/electrolyte reactivity of the system leads to a rapid self-discharge of the positive electrode, which, while having little effect on the cyclability of the Li/LNMO half-cells, leads to an extremely rapid loss of capacity in Li-ion configuration due to the imbalance of the Li$_x$C$_6$/Li$_{1-y}$Mn$_{0.5}$Mn$_{1.5}$O$_4$ system.

Thus, the introduction of '5 V' electrode materials (including in particular the "LNMO" spinels) into commercial cells remains more than ever an ambitious goal. Many researches are ongoing to assess solutions and eliminate or reduce degradation processes at different levels. These studies aim in particular at limiting the specific surface of the particles; modifying the composition of the spinel material by different doping methods; or modifying the composition of the liquid electrolyte to make it more oxidative-resistant (fluorinated carbonates solvents: FEC, TFEC; sulfonates or nitriles; additives such as tri-allyl phosphate). They also explore other, complementary approches, such as generating a passive film on the material's surface. This type of coating can be carried out by different techniques (sol-gel, ALD or Atomic Layer Deposition,...) of inorganic deposits (carbon, lithiated metal oxides and other lithium compounds). The final objective is to reduce the dissolution of the active material, and act as an artificial "SEI" (Solid Electrolyte Interphase) on the surface of the spinel material and/or the complete electrode. Given that none of these pathways is a 100% effective solution to all issues, a breakthrough combining different approaches is required to lead to a satisfactory behavior of LNMO materials, and counterbalance its deficiencies.

2.2 Positive Electrode Materials with Lithiated Layered Oxide Structure

Generalities

Materials called lithiated lamellar oxides meet the general $LiMO_2$ formula with M that can be an element or a mixture of elements including Co, Ni, Mn and/or Al. MO_6 octahedrons sharing edges are arranged as sheets (figure 2.8). The cross-leaf space is occupied by lithium ions, which can diffuse in two dimensions into the material (thanks to the layered structure).

The best-known layered oxide is without any doubt the formula material $LiCoO_2$ (LCO). This material was the first layered oxide marketed as cathode material, and, despite many years of research, it remains widely used today, especially for the manufacture of batteries used in mobile phone or laptops. The use of LCO as a battery material was first proposed by Goodenough in 1981; the first commercial

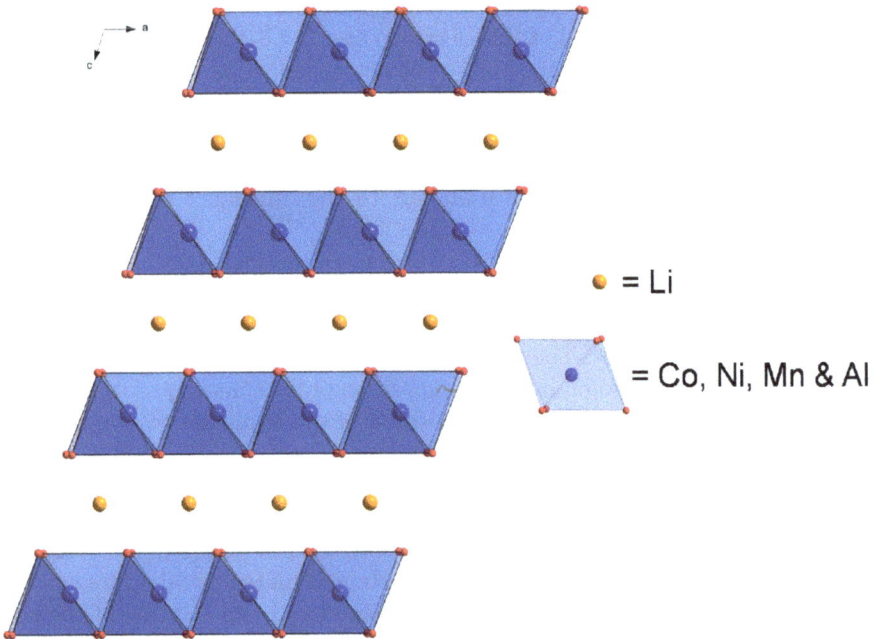

FIG. 2.8 – Structure of a $LiMO_2$ layered oxide with M that can be Ni, Mn, Co, Al or a mixture of these elements.

accumulators using this material were marketed by Sony corporation in 1991 [25–27]. Since then, the major improvements to these materials have been to replace a part of the cobalt with nickel and manganese, or even aluminum [28–31]. These substitutions have increased reversible capacity while reducing toxicity and product cost. The most well-known material developed for this purpose is the $LiNi_{1/3}Mn_{1/3}Co_{1/3}O_2$ formula commonly referred to as NMC 111 because metal elements of MO_6 layers have identical concentrations of nickel, manganese and cobalt. When a cell is cycling between 2.7 and 4.3 V, 0.6 lithium can be extracted, thanks to the electrochemical activity of the Ni^{2+}/Ni^{4+} and Co^{3+}/Co^{4+} electrochemical couples, which is particularly interesting because lithium is extracted from the material in a potential area where the electrolyte is stable. For this reason, the main development concerning layered oxides is to increase nickel content in materials. To do this, two routes are followed: the first is to reduce the level of cobalt and manganese contents in the case of $LiNi_xMn_yCo_zO_2$ (NMC) formula materials; the second is to reduce cobalt content and replace manganese with aluminum to obtain $LiNi_xCo_yAl_zO_2$ (NCA) formula materials. Some of these nickel-rich materials are currently commercialized for "high energy" applications (NMC 532, NMC 622 or NCA for electric vehicles); others are still under improvement (NMC 811 and NMC 9-0.5-0.5). While the increase in nickel content in materials allows extracting more lithium at a constant potential, issues related to safety of use happen [29]. The main benefits of each compound are detailed below in this chapter.

To increase the specific capacity, rather than increase the nickel content, another solution is to increase the lithium content in materials. These materials named "over-lithiated layered oxides" have supplementary lithium in their structures. This lithium can be inserted/disinserted thanks to the electrochemical activity of nickel and oxygen couples. This family of materials would significantly improve the energy density of batteries: it is still under research [32].

This part of the chapter provides a brief overview of the state of the art commercial lithium layered oxide structure materials and those still in search.

The "LCO": $LiCoO_2$

The "LCO" material is mainly composed of cobalt. It is mainly used as battery cathode material for applications in mobile electronics (smartphones, laptops, tablets...). The main suppliers of "LCO" in the world are Umicore, L&F, Pulead and B&M, that share 55% of the world market (figure 2.9).

LCO can crystallize in two structures named "high temperature form" ("HT-LCO") and "low temperature form" ("LT-LCO"). The obtained structure depends mainly on the synthesis temperature (>700 °C to obtain "HT-LCO"). The material used commercially is the "HT-LCO", a layered structure with rhombohedral symmetry (space group $R\bar{3}m$ (figure 2.10). In the "LT-LCO", some of the Co atoms substitute lithium atoms in their sites and vice versa. The result is a material with cubic symmetry (spinel with space group Fd-3 m) [33–35].

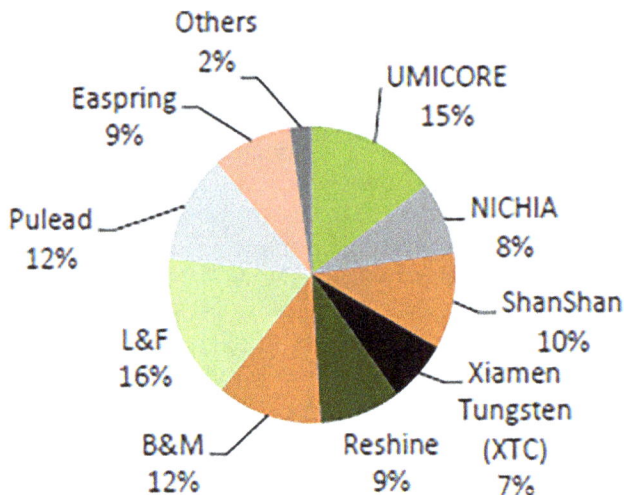

FIG. 2.9 – Main suppliers of LCO materials in the world [1].

The "HT-LCO" structure has a reversible potential of 3.93 V in relation to lithium, while the "LT-LCO" structure suffers from a significant polarization phenomenon (potential of 3.75 V in charge/oxidation versus 3.45 V in reduction/discharge [36, 37]). It is therefore important to take the greatest care to avoid the forming of the LT-LCO phase of spinal structure when synthesizing the material.

The layered structure "LCO" has a hexagonal structure (space group $R\overline{3}m$ with $a = b = 2.816$ Å and $c = 14.08$ Å). The stacking formed by the atoms of Li, Co and O is ABC and the structure can be assimilated as alternating layers of CoO_6 and liO_6 [25, 38, 39].

The insertion and disinsertion of lithium result in several phase modifications [2, 40]. The first modification happens between $0.75 < x < 0.94$. On the corresponding potential tray, two hexagonal phases with different cell parameters coexist ($a = 2.82$ Å, $c = 14.08$ Å and $a = 2.82$ Å, $c = 14.19$ Å^3). Between $0.6 < x < 0.75$, only one hexagonal phase is present. Between $0.56 < x < 0.51$ a monoclinic phase is observed. Between $0.5 < x < 0.35$ a new hexagonal phase is observed. Beyond $x = 0.35$, there are two phases, hexagonal and monoclinical (respectively, $a = 2.82$ Å, $c = 14.26$ Å and $a = 4.91$ Å, $b = 2.82$ Å, $c = 5.02$ Å, $\beta = 111.4°$). The monoclinic phase becomes majority when more lithium is removed (figure 2.11).

Because of these numerous phase changes, $LiCoO_2$ is usually cycled by limiting the extraction of lithium to 0.6 mol of Li per material mole, which represents a theoretical capacity of 140 mAh.g^{-1}. Research is still being conducted to "smooth" these phase transitions to extract more lithium from the material. For small cells, the "LCO" material is often preferred, but for larger batteries, manufacturers favor another lamellar oxide named "NMC".

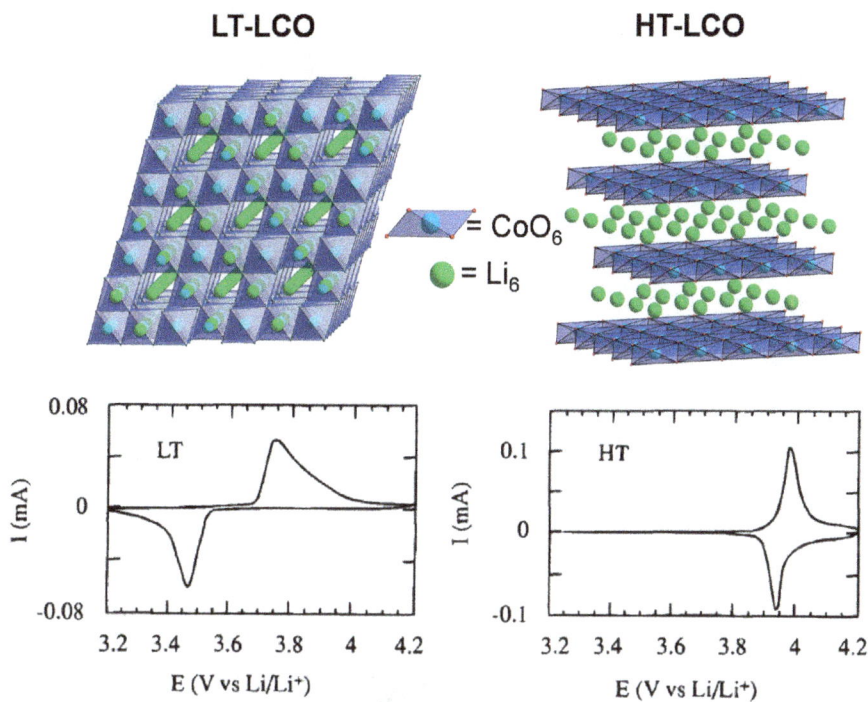

F<small>IG</small>. 2.10 – Associated structures and voltammetries of "LT-LCO" and "HT-LCO" [36]. Potential scanning voltammetry is an electrochemical spectrometer technique that reveals the behavior of active electrode materials: a Lithium/liquid electrolyte/active cathode material cell is submitted to a slow, step-by-step scanning of potential from its equilibrium position. At each stage of the voltage scan, the average current (supplied or consumed by the cell until it reaches its electrochemical equilibrium state) is recorded.

The NMC: $LiNi_xMn_yCo_zO_2$

NMCs consist of a mixture of cobalt, nickel and manganese. The result is a higher energy density and a lower material cost than the LCO material. This family of materials is often used in combination with "LMO" for the manufacture of cathodes for electric vehicles. For example, car manufacturers such as Nissan, Toyota, Mitsubishi and Honda use "NMC" batteries. The main suppliers of materials are Umicore and ShanShan. Battery manufacturers such as LG, Panasonic and Samsung control the production of their proprietary "NMC" material to meet their own needs [1] (figure 2.12).

The formulation $LiNi_{1/3}Mn_{1/3}Co_{1/3}O_2$ (commonly referred as "NMC 111") is proposed by Ohzuku in 2001 [42]. The theoretical capacity of this material is 277 mAh/g when 1 mol of lithium is removed. Ohzuku studies the behavior of the material in cycling and shows that the cell parameters change rapidly beyond $x = 0.6$ (figure 2.13). Beyond this value, the life of the cycling material is

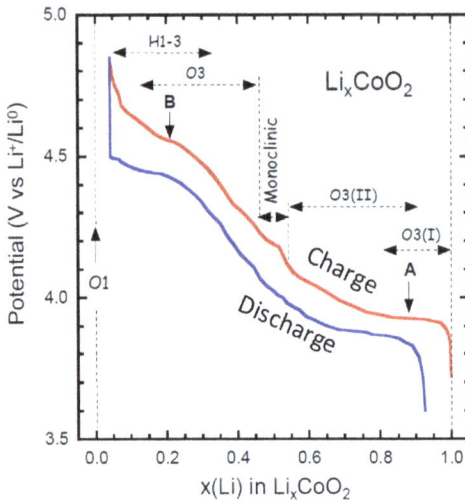

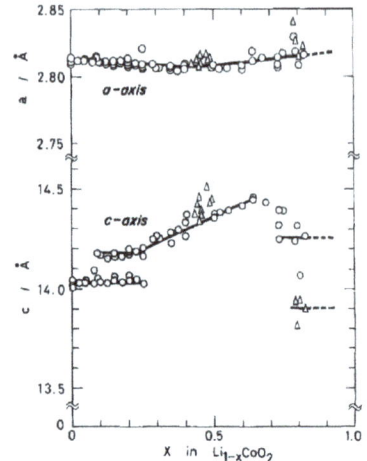

Fig. 5. Lattice parameters of hexagonal unit cell of $Li_{1-x}CoO_2$. Triangles indicate the converted unit cell parameters from monoclinic cell.

FIG. 2.11 – Evolution of the LCO structure in relation to cycling [2, 41].

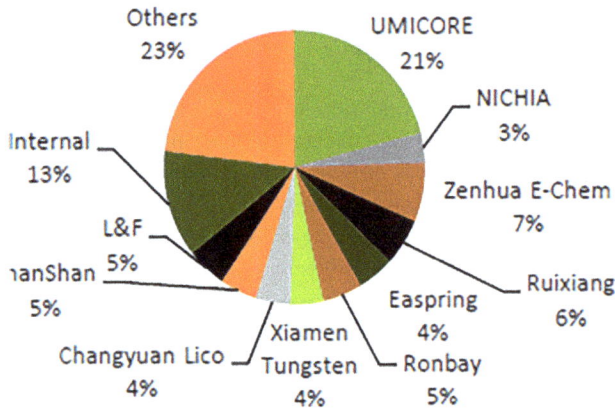

FIG. 2.12 – Leading NMC materials suppliers in 2019 : China represents 65% of the market [1].

significantly affected. Therefore, the cycling point of this material is limited to about 4.3 V and the capacity reached by this material is 160 mAh.g^{-1}.

For several years, there has been a gradual increase in the nickel content of the "NMC". By increasing the amount of nickel, it becomes possible to disinsert and reinsert a greater amount of lithium while remaining in the field of electrolyte stability. Demand for "NMC 111" has strongly reduced during the past 5 years: this composition has been replaced by compositions such as $LiNi_{0.6}Mn_{0.2}Co_{0.2}O_2$ and $LiNi_{0.8}Mn_{0.1}Co_{0.1}O_2$ (respectively, "NMC 622" and "NMC 811"), which make it possible to manufacture batteries with even higher energy density (figure 2.14).

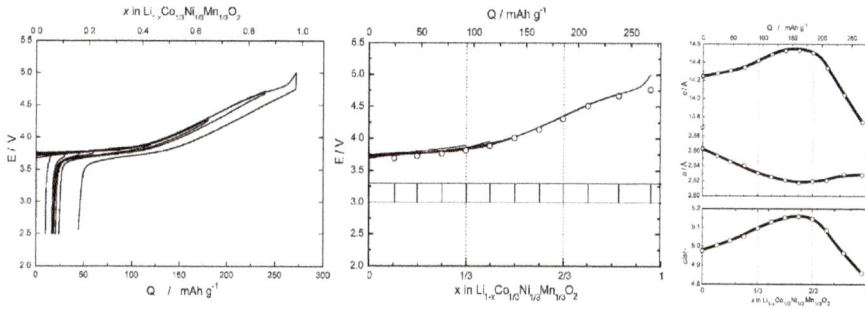

FIG. 2.13 – Electrochemical performance and structural evolution of an NMC 111 based on the amount of lithium dis/insert [43].

These materials with a high nickel content are commonly referred to as Ni-Rich. The theoretical capacity of the materials can be directly related to the amount of nickel in the oxides. The nickel contained in the structure can be completely oxidized from its oxydation state 2–3 to 4+ while it is difficult to oxidize cobalt beyond 3.6 + without oxidizing the oxygen of the structure [32, 44]. It is important to note that in these structures, manganese oxidation state is 4+ and remains inactive in cycling conditions.

We see that it is interesting to increase the density of energy of materials by increasing the ratio of Ni/Co concentration in the material. Another benefit of this substitution is the small amount of manganese contained in the structure. The average potential in the first cycle decreases when the amount of manganese decreases and the nickel content increases, allowing to extract more lithium in the first charge in the same voltage windows [32, 44]. It should be noted that, on the other hand, the discharge potentials are similar regardless of the composition of the material. The capabilities of the manganese-poor "NMCs" are therefore higher.

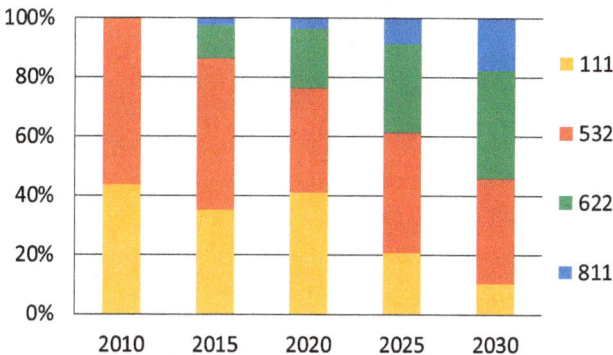

FIG. 2.14 – Evolution of the composition of "NMC" materials over time [1].

Thus, by changing the formulation and substituting more and more cobalt with nickel, it is possible to increase the energy density while reducing the cost of raw materials (cobalt being the most expensive element).

However, the increase in the nickel content of the material unfortunately does not have only advantages: there is a price to pay for it (figure 2.15).

The stability of the material, and thus the safety of its use, is directly related to the amount of nickel in the structure. The more nickel the structure is enriched with, the less thermally stable the material [29]. In addition, the battery cycling behavior is also degrading when the nickel content of the cathode material increases. Battery capacity is decreasing, and the number of charge/discharge cycles that can be made with high nickel-rich materials is currently below the specifications required by car manufacturers. This reduction in cycling capacity is partly explained by the volume expansion of the electrode and also by an evolution of the surface structure of the materials (the latter evolves gradually from a lamellar structure to a rocksalt or spinel structure) [29, 32, 45, 46].

Among the materials called "Ni-Rich", the "NMC 532" and "NMC 622" are now commercial and used in recent generations of electric vehicle batteries. Much of the research effort currently aims at stabilizing the "NMC 811" and "NMC 9-0.5-0.5" materials. As mentioned above, the lack of stability of these materials during cycling can be mainly attributed to an evolution of the material's surface. The solutions envisaged to stabilize these materials imply therefore the modification of their surface [47]. To this end, two approaches are explored: "coating" solutions, or the creation of a "core shell" structure. In the case of a "coating", a thin layer of a stable material is deposited on the surface of the "Ni-rich" particles to stabilize the

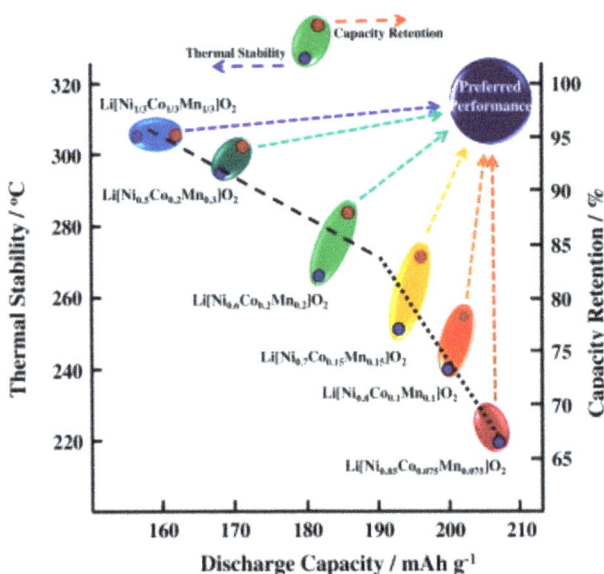

FIG. 2.15 – Advantages and disadvantages of different NMC compositions [29].

electrolyte/material interface. Different types of "coating" are assessed in the literature: fluorinated materials (AlF_3...), glasses, but most of the materials used are oxides (Al_2O_3...) [48–53]. The effectiveness of the coating has been demonstrated on numerous occasions, and coated materials are already being marketed. However, the concept of "coating" can be pushed further by creating a "core–shell" structure. The core shell can be seen as a multilayer composite material [54–59]: a very rich nickel material is fully covered by a thick shell of a more stable material, usually with a high manganese content. The main problem with this solution is the appearance of cracks in the material during prolonged cycling. In fact, nickel-rich or nickel-poor materials do not exhibit the same volume expansion during cycling, resulting in particle stress and rupture. As a result of this limitation, a new concept called "full-gradient concentration" has recently been proposed [60–62]. In this material configuration, the internal and external compositions of the particles are different. The core of the particles is made of nickel-rich material and the surface is rich in manganese, but the material is not multi-layer: the concentration of nickel and manganese changes gradually throughout the particle's thickness. This allows to homogenize the volume expansion and makes it possible to considerably improve the cyclability of the material. The main disadvantage of this concept is the difficulty encountered to perform a synthesis that must tune the material's composition at the nanoscale (figure 2.16).

The "NCA": $LiNi_xCo_yAl_zO_2$

"NCA" materials also belong to the family of layered oxides. In this case, the material is composed of nickel, cobalt and aluminum. This material is used not only in power portable electronics equipment but also in power portable tools requiring a good power/energy compromise. It is also important to note that a large part of the production of NCA material is dedicated to the automotive sector, particularly for

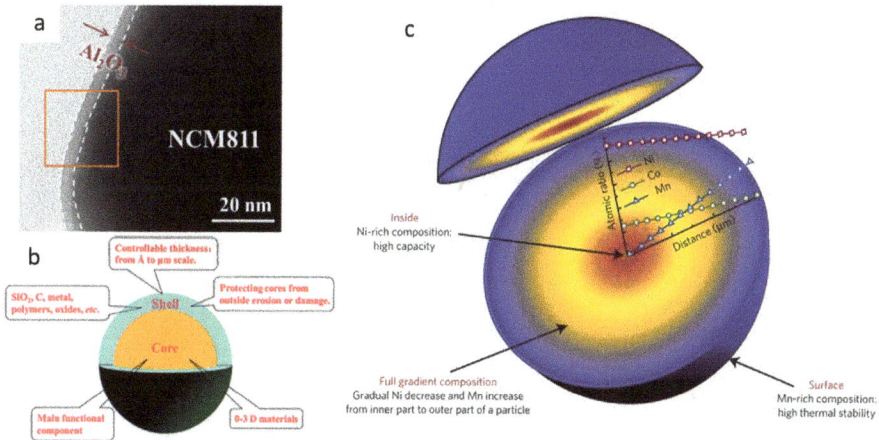

FIG. 2.16 – Comparison of (a) "coating", (b) a "core shell", and (c) a "full-gradient concentration" [48, 54, 60].

the supplying electric cars of the Tesla brand. Sumitomo is the world's leading supplier of "NCA" material because it supplies Panasonic and Tesla companies (figure 2.17).

NCA materials are high nickel layered oxides. The formula material Li $(Ni_{0.8}Co_{0.15}Al_{0.05})O_2$ is currently the only commercial representative of this subfamily of compounds. This material can reach capacities close to 200 mAh.g^{-1} [63–65]. In NCA-type layered oxides, nickel provides high capacity, cobalt improves power performance and limits lithium-metal exchange in the structure, whereas aluminum improves the structural and the thermal stability of the material.

Similarly to NMC, the nickel content is likely to increase in the coming years in order to increase the specific capacity of these materials (figure 2.18).

The issues associated with these materials and the optimization perspectives explored are very similar to those of "NMC" materials: problems related to the modification of the structure of the material during cycling also lead to problems of

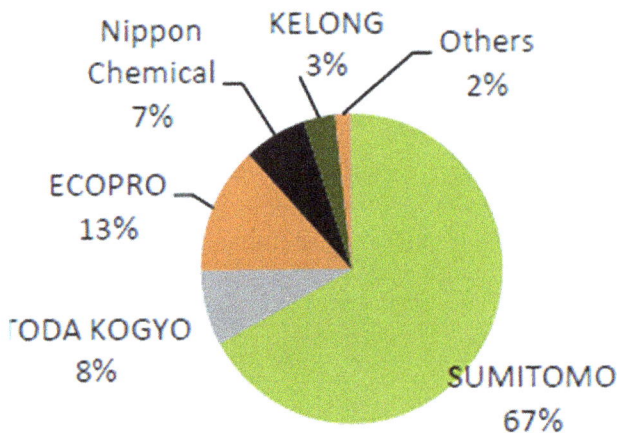

FIG. 2.17 – Leading NCA materials suppliers in 2019 [1].

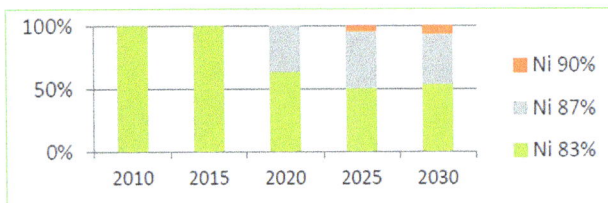

FIG. 2.18 – Expected evolution of the composition of the NCA materials over time [1].

stability in cycling [66]. The "coating" and "core–shell" are also studied to stabilize the material/electrolyte interface [67–71].

The Lithium Rich Layered Oxides

Layered oxides commonly referred to as "Li-rich" are not commercial materials. Indeed, these high capacity materials present many issues that have not yet been resolved. The literature dealing with these materials is extremely dense and we will here only briefly explain the advantages and disadvantages of this material family.

The "lithium rich" layered oxides can be seen as a mixture of $LiMO_2$ and Li_2MnO_3 compositions [72–74]. It is possible to index their structures according to two groups of space, $R\bar{3}m$ if indexed according to the $LiMO_2$ part or $C2/m$ if indexed with the Li_2MnO_3 part [75]. The local structure of the material is controversial among scientists, some consider the structure as a composite material (made up of Li_2MnO_3 independent sheets and $LiMO_2$ sheets), others consider the material as a solid solution (figure 2.19) [76, 77]. This explains why it is possible to write the same formula for a material in two different ways: for example, $Li_{1.2}Ni_{0.2}Mn_{0.6}O_2$ (compact formulation) may also be written in composite form $0.5Li_2MnO_3 \cdot 0.5LiNi_{0.5}Mn_{0.5}O_2$.

In all cases, the $LiMO_2$ and Li_2MnO_3 components provide a compact cubic structure defined by oxygen atoms, where all octahedral sites are filled by alternating lithium, transition or lithium layers, and transition metals. An X-Ray diffraction analysis (XRD) of this material reveals that, with the exception of peaks between 20 and 25°, each peak can be indexed with a hexagonal structure of the type, α-$NaFeO_2$ (space group $R\bar{3}m$) [78]. The peaks between 20 and 25° result from

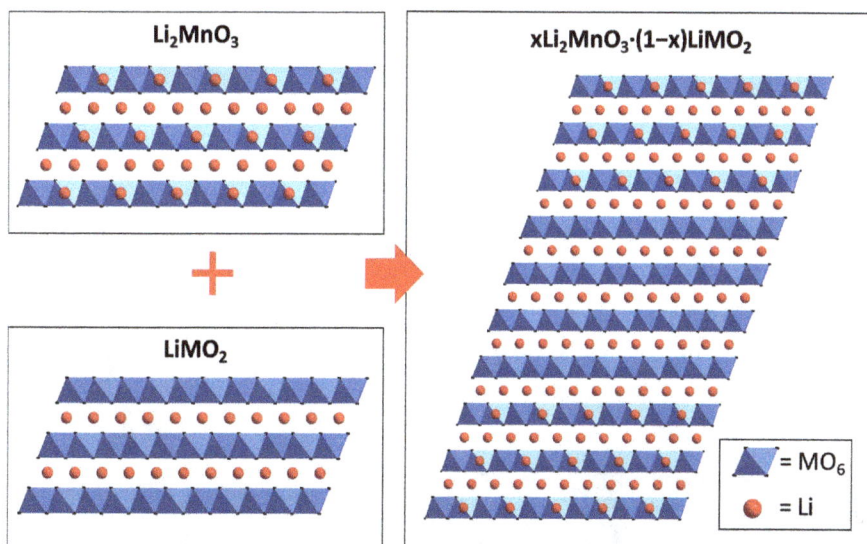

FIG. 2.19 – Structure of a Li-rich layered oxide (010).

an order between lithium and metallic ions in transition metal sheets, giving a Li_2MnO_3 type structure with a C2/m space group (figure 2.20).

The particularity of this material lies in the oxidation mechanisms used to extract lithium from the structure. Typically, the specific capacity achieved with these materials is much higher than the theoretical capacity calculated by Ni^{2+}/Ni^{4+} couple (manganese remains at oxidation level 4 + and is therefore inactive). For example, $Li_{1.2}Ni_{0.2}Mn_{0.6}O_2$, for which the theoretical capacity is 126 mAh.g^{-1}, but the specific reversible experimental capacity is greater than 250 mAh.g^{-1}, can be cited (figure 2.21). This phenomenon has been particularly studied in the literature in order to elucidate the origin of this additional capacity.

During charge, the first part of the curve can be attributed to the oxidation of the transition metals, then a plateau corresponding to the activity of the oxygen

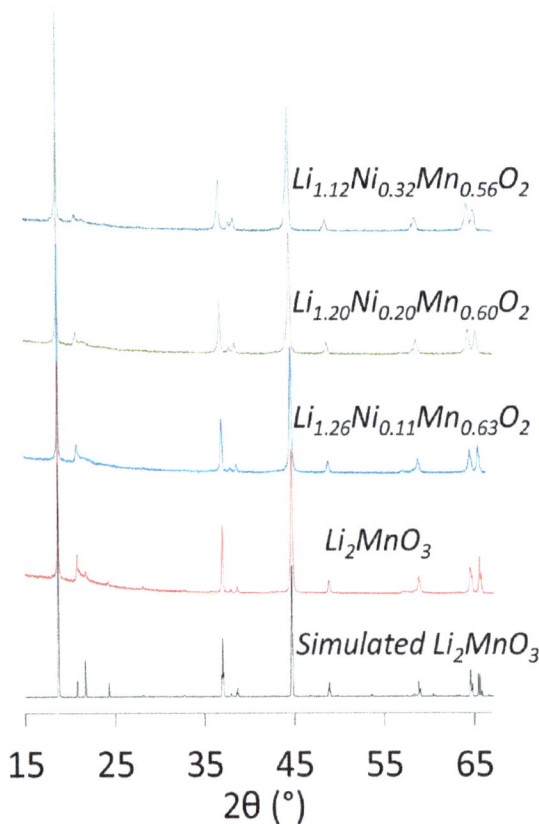

FIG. 2.20 – Example of XRD (X-ray diffraction) analysis of different Lithium rich layered oxides [79].

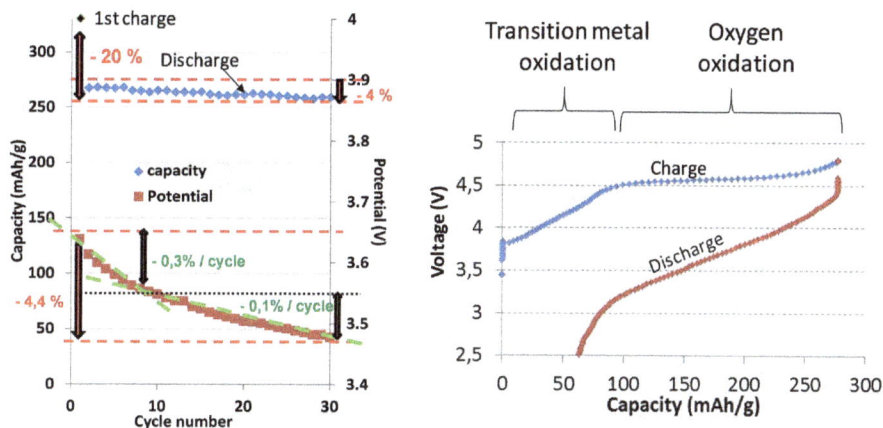

FIG. 2.21 – Highlighting the drawbacks of lithium-rich and the galvanostatic curve of $Li_{1.2}Ni_{0.2}Mn_{0.6}O_2$ [ref: Source CEA].

contained in the structure can be observed [80–83]. It has been shown that during this partially reversible cycle, oxygen is found in the form of peroxides or superperoxides [84]. The result is a very high capacity with an inexpensive material, which justifies the great enthusiasm of many research groups to optimize this material. In return for these advantages, lithium rich layered oxides suffer disadvantages that prevent their commercialization (figure 2.21) [85, 86]. The first and main limitation of these materials is the decay of the average potential during cycling, resulting in a loss of energy density and making it difficult to measure the battery state of charge. Many "coating" and doping agents have been considered, but none of these solutions has so far been able to solve the problem [87, 88]. Phenomena involved in this material are very complex and many studies have focused on their understanding [73, 89]. It is shown that the drop in potential during cycling is not due to a single phenomenon, but to the contribution of various mechanisms that are well summarized by Passerini *et al.* [90]. Peralta *et al.* show that the drop in potential during the first cycles is due to two phenomena occurring at high potential: activation of Li_2MnO_3 sheets and migration of some of the transition metals into the lithium interslab [79]. Currently, the only way to stop material evolution is to cycle below 4.3 V, but unfortunately, under these conditions, the reversible capacity recovered is too low to make this material competitive (figure 2.22). To date, many groups are still working in hope to make this material viable.

Several possible solutions have recently been proposed in the literature to limit voltage decay, for example, the possibility of using O_2 layered phases instead of O_3, or by inserting defects into the structure, or by replacing manganese with ruthenium [84, 91, 92].

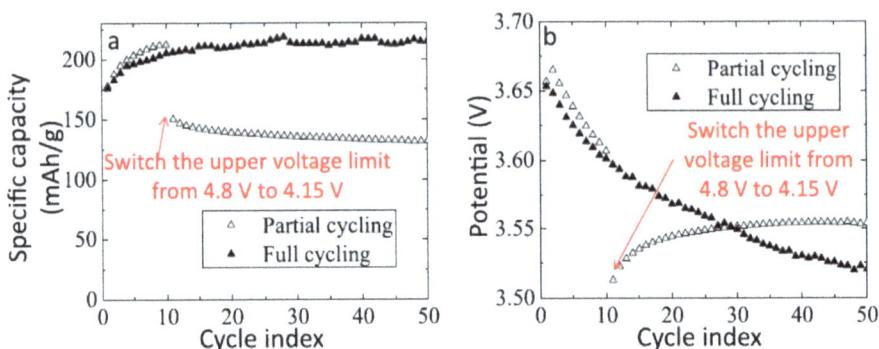

FIG. 2.22 – Highlighting the influence of the high end of charge voltage on the stability of the mean potential of $Li_{1.2}Ni_{0.2}Mn_{0.6}O_2$ [79].

2.3 Positive Electrode Materials with Olivine Structure

Today, lamellar oxides based on "LCO" or "NMC" materials are mainly used as positive electrode materials (cathode) in commercial batteries.

However, these lamellar oxides have limited stability. For example, when they are exposed to a high temperature (around 200 °C) or when they are overcharged in a battery, they are likely to release large amounts of oxygen. The oxidizer thus released into the electrolyte reacts with the electrolyte and feeds a thermal runaway mechanism within the battery, involving safety risks for users.

In addition to mastering manufacturing processes, it is necessary to develop materials that are inherently more stable, and thus safer. Due to its very good thermal stability, the olivine structure $LiFePO_4$ "LFP" material meets this criterion. Furthermore, it does not contain toxic or strategic elements (except lithium) and its production cost can be reduced by increasing the current volume of production.

LiFePO$_4$: Main Advantages and Disadvantages

With an 8% increase expected between 2015 and 2030, demand for $LiFePO_4$ is growing, and may well be underestimated [1]. The supply and demand come mainly from China, for all kinds of applications: power tools, electric buses, xEVs, electric bikes, and stationary applications.

The main advantages of $LiFePO_4$ are its long life expectancy and its thermal stability. The presence of the polyanionic $(PO_4)^{3-}$ group, consisting of highly stable P-O covalent bonds does not generate oxygen even when the material is completely decomposed at high temperatures [93]. The combustion reactions are therefore highly limited. $LiFePO_4$ also allows fast charge/discharge cycles. The high availability and low toxicity of the Fe element make $LiFePO_4$ a material of choice.

The biggest disadvantage of the $LiFePO_4$ material is its limited energy density. Batteries using "LFP" technology offer an energy density of $\sim 120–160$ Wh.kg^{-1}, which remains lower than those produced with "NMC" lamellar oxide materials (~ 250 Wh/kg). Until recently, batteries using $LiFePO_4$ had a better life expectancy (about 5000 cycles) than batteries made from lamellar oxides, which makes them very attractive. However, when implemented under equivalent conditions, the latter are currently experiencing a steady improvement in their life expectancy, in some cases exceeding 5000 cycles one can see here the positive influence of mass production ramp up on manufacturing reproducibility and on the materials reliability. On balance, the technology chosen in the future is likely to allow, for a given cost, the exchange of as much energy as possible throughout its lifetime, for an equal mass of material, in order to give it a decisive advantage from the point of view of life cycle analysis.

A sharp decrease in production costs is expected for $LiFePO_4$ material in the coming years, while lamellar oxides will likely be linked to fluctuations in cobalt price [1]. The cost of producing $LiFePO_4$ is expected to decrease, reaching less than \$10/kg in 2025 [1]. At the same time, the cost of producing lamellar oxides is moving towards a limit that can be estimated at around \$25/kg, regardless of its chemical composition (LCO, NMC, NCA). In addition, the cost of producing lamellar oxides could increase significantly if Co or Ni prices are soaring which will not be the case with $LiFePO_4$, which contains only abundant elements, except lithium, common to all materials. Only $LiMn_2O_4$ (~ 6–7 \$/kg) is cheaper than $LiFePO_4$ [1], but its performances are much less attractive in terms of life span or even energy density. To summarize, LFP materials offer many decisive advantages (no sensitive material, very long life expectancy, safety, low cost), they also allow reducing the pressure on NMC-based materials supply. But they present a (major) drawback : a relatively low energy density. A lot of things will depend on the improvements that will be achieved in the coming years to reduce, if possible, the energy density gap between LFP and NMC-based materials. LFP will anyway most probably used in the coming years for entry level vehicles, if not more. This explains why any forecast is difficult in this area.

State of World Art

Material Synthesis

Numerous synthesis methods are used to prepare $LiFePO_4$, whether in solid-state reaction route or using solvent-based synthesis methods [94].

Solvent-based methods generally allow accurate control of particle morphology. For example, it is possible to obtain spherical secondary particles, thereby increasing the practical density of the material or significantly reducing the size of the particles. However, the ratio of structural defects (in particular the antisite ratio) obtained using a solvent-based synthesis is generally higher than using a solid-state reaction route. The solid-state reaction route is often preferred for its greater versatility, especially since primary particles of less than 50 nm can be obtained when a first stage of high-energy mechanical activation is used [95].

In the solid-state reaction route, a homogeneous mixture of Fe(II), lithium and phosphorus-based precursors is formed by mechanical milling. The resulting powder mixture, sometimes compacted to increase its reactivity, is heated in a furnace under an inert atmosphere, between 600 and 800 °C, for several hours. The powder can be milled/compacted/heated several times in order to improve the homogeneity of the obtained material.

Fe(III) precursors are sometimes preferred as they are air-stable and less expensive than Fe(II) precursors, which can reduce the cost of the synthesis. In this case, reduction of Fe(III) to Fe(II) occurs in the furnace, during the thermal treatment. The reducing atmosphere is obtained via thermal decomposition of the additional carbon source (often a polymer) under inert atmosphere.

Structure and Properties of LiFePO$_4$ Material

LiFePO$_4$ crystallizes in an orthorhombic structure with Pnma space group (n°62). LiFePO$_4$ is described as a "1D" material, which implies that Li$^+$ ion diffusion only occurs in a single direction, in the channels parallel to the b axis (figure 2.23) [2].

It is important to limit structural defects in LiFePO$_4$ to optimize its electrochemical performance. In particular, it is necessary to obtain well-ordered crystallographic structures. Indeed, the higher the antisite ratio (*i.e.* exchange of the position of Li$^+$ ions and Fe^{2+} ions), the more difficult Li$^+$ ion 1D intercalation/extraction is. In addition, for a constant antisite ratio, the long diffusion channel increases the probability of blocking the Li$^+$ 1D insertion channel, and vice versa [97]. In general, antisite ratio comprised between 1% and up to 7%–8% are measured in LiFePO$_4$ [98, 99].

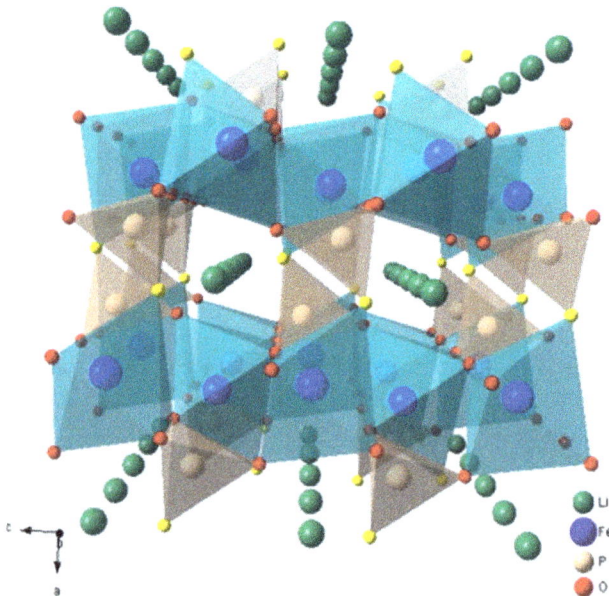

FIG. 2.23 – Olivine structure of LiFePO$_4$ [96].

Due to its low ionic diffusion coefficient ($\sim 10^{-14}$ cm^2/s) and electronic ($\sim 10^{-9}$ S/cm) conductivity, the electrochemical performance of LiFePO$_4$ was very limited in the first experiments [94]. The use of LiFePO$_4$ in applications which require high power has become possible thanks to the use of nanometric particles and the addition of a conductive coating, usually based on carbon [100]. This strategy allows both the increase in the overall electronic conductivity of the material and also a sharp decrease in the diffusion path of Li$^+$ ions in the material. Indeed, the characteristic time (T), which corresponds to the diffusion duration of Li$^+$ ions across a particle's crystallite, is defined by $T = L^2/4\pi D_{\text{Li}}$. In this equation, L represents the diffusion path and D_{Li} represents the diffusion coefficient of Li$^+$ ions in the material. As a result, the characteristic time corresponds to a few hours for micrometric particles ($\sim 10\ \mu$m) whereas it decreases to a few seconds when the particle size is nanometric (~ 20 nm).

However, using both nanometric and carbon-coated particles in the same composite material greatly limits its practical density, usually between 0.5 and 1.5 [101]. Thus, it greatly decreases the energy density of such low density composite material compared to the reference LiFePO$_4$ material with micrometric size particles and no carbon-coating. Different synthesis methods are used to jointly optimize primary particle size, carbon rate and homogeneity, as well as secondary particle morphology, to increase LiFePO$_4$ density [102].

Charge/Discharge Curves

The charge/discharge curve of the material occurs at a constant voltage of around 3.45 V *vs.* Li$^+$/Li (figure 2.24).

The theoretical specific capacity of LiFePO$_4$, corresponding to the reversible extraction of one Li$^+$ ion per unit form, corresponds to 170 mAh.g^{-1}. Therefore, the theoretical specific energy density of LiFePO$_4$ is 586 Wh.kg^{-1}, related to the mass of the active material. The reversible Li$^+$ ion extraction from LiFePO$_4$ is accompanied by a volume change of 6.8% [94].

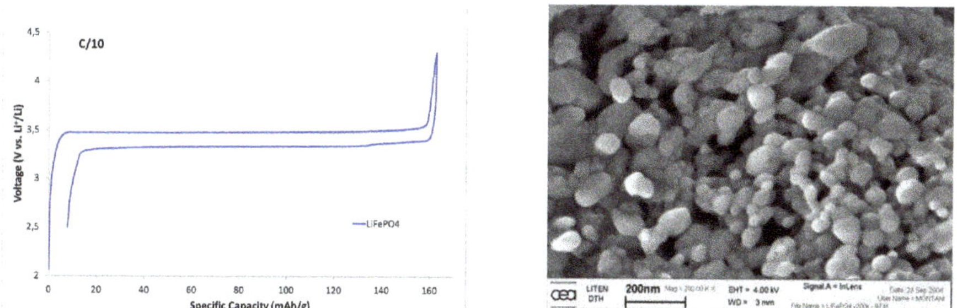

FIG. 2.24 – Charge/discharge curve of LiFePO$_4$ obtained at a constant C/10 rate and SEM image of the corresponding LiFePO$_4$ material [ref: Source CEA].

Increasing the Energy Density

The low energy density of LiFePO$_4$ material is mainly due to the low redox potential of Fe(III)/Fe(II) couple (*i.e.* 3.45 V *vs.* Li$^+$/Li). Substitution of this element by another transition element such as Mn, Co or Ni, significantly increases the battery's operating voltage [103]. Thus, the obtained material is isostructural to LiFePO$_4$ and its theoretical specific capacity is similar to that of LiFePO$_4$. However, the electrochemical activity of Ni(III)/Ni(II) and Co(III)/Co(II) redox couples is greater than the anodic stability of the electrolytes commonly used in Li-ion batteries. Therefore, both LiNiPO$_4$ and LiCoPO$_4$ materials are not usable today as they would require the development of new electrolytes with higher stability at higher voltage.

Interestingly, the operating voltage of LiMnPO$_4$ is around 4.1 V *vs.* Li$^+$/Li and it is compatible with current electrolytes. The theoretical specific energy density of LiMnPO$_4$ is 700 Wh.kg^{-1}, which is around 20% higher than LiFePO$_4$. However, due to its low conductivity (both ionic and electronic) and its more unstable structure, this material is not competitive with the LiFePO$_4$ material, both in terms of capacity and cyclability [104]. After optimizing the morphology of the material, the best performance in the literature is in the order of 85% of the theoretical capacity at low C-rate (C/20) [105]. The Fe/Mn mixed materials, with the general formula LiFe$_{1-x}$Mn$_x$PO$_4$, appear to be a good compromise. The higher the Mn content, the higher the material's average operating voltage – and thus the material's energy density increases. However, adding Mn reduces the mixed material performance very quickly in terms of conductivity, cyclability, power and stability in cycling. The optimal Mn ratio appears to be between 0.6 and 0.8 [106]. Thus, the theoretical gain in terms of energy density is between 11 and 16% for LiFe$_{1-x}$Mn$_x$PO$_4$ (0.6 ≤ x ≤ 0.8) compared to LiFePO$_4$. Today, the commercial supply of the mixed material is beginning to appear, but remains very limited.

The positive electrode, or cathode, is one of the key elements of a Lithium-ion battery. The performance in terms of power, energy, life cycle, and safety of the battery (and thus the complete battery, which consists of an assembly of elementary cells) is directly linked to the nature of the materials selected to implement this electrode. Three families of materials have been used commercially so far. Spinel structure compounds are used for low-cost and power applications, layered oxides are used for "high energy" applications, and olivine structures are favored for applications that require high cyclability and high safety, at the expense of energy density.

The mass marketing of electric vehicles is already driving research in this area. Several lines of research are explored. For example, the use of cobalt, a critical material, may, in the medium term, call into question the production of very large amounts of cathode material. Therefore, very low cobalt cathode materials such as "NMC 811" and "NMC 9–0,5–0,5" are being developed for "high energy" applications. In the short term, highly nickel-rich layered oxides are expected to become widespread, as well as "LFP" materials for entry or mid range vehicles -and may be more-. In the longer term, materials for very high energy applications such as "superlithiated lamellar oxides" or "rocksalt" structures may be used to link with future generations of batteries, such as "all solid", "Lithium-sulfur", "all-polymer" or "Lithium-air" technologies, which are currently being investigated.

References

[1] Avicenne Energy - The Rechargeable Battery Market: Value Chain and Main Trends 2019-2030-IBS -11 March, 2021.

[2] Julien C.M., Mauger A., Zaghib K., Groult H. (2014) Comparative issues of cathode materials for Li-ion batteries, *Inorganics* **2**, 132, doi: https://doi.org/10.3390/inorganics2010132.

[3] Hunter J.C. (1981) Preparation of a new crystal form of manganese dioxide: λ-MnO2, *J. Solid State Chem.* **39**, 142, doi: https://doi.org/10.1016/0022-4596(81)90323-6.

[4] Thackeray M.M., Johnson P.J., de Picciotto L.A., Bruce P.G., Goodenough J.B. (1984) Electrochemical extraction of lithium from $LiMn_2O_4$, *Mater. Res. Bull.* **19**, 179, doi: https://doi.org/10.1016/0025-5408(84)90088-6.

[5] Sickafus K.E., Wills J.M., Grimes N.W. (1999) Structure of spinel, *J. Am. Ceram. Soc.* **82**, 3279, doi: https://doi.org/10.1111/j.1151-2916.1999.tb02241.x.

[6] Birgisson S., Jensen K.M.Ø., Christiansen T.L., von Bülow J.F., Iversen B.B. (2014) In situ powder X-ray diffraction study of the hydro-thermal formation of $LiMn_2O_4$ nanocrystallites, *Dalton Trans.* **43**, 15075, doi: https://doi.org/10.1039/C4DT01307G.

[7] Julien C., Ziolkiewicz S., Lemal M., Massot M. (2001) Synthesis, structure and electrochemistry of $LiMn_2 - yAl_yO_4$ prepared by a wet-chemistry method, *J. Mater. Chem.* 11, 1837, doi: https://doi.org/10.1039/B100030F.

[8] Mandal S., Rojas R.M., Amarilla J.M., Calle P., Kosova N.V., Anufrienko V.F., Rojo J.M. (2002) High temperature co-doped $LiMn_2O_4$-based spinels. Structural, electrical, and electrochemical characterization, *Chem. Mater.* **14**, 1598, doi: https://doi.org/10.1021/cm011219v.

[9] Kushida K., Kuriyama K. (2000) Observation of the crystal-field splitting related to the Mn-3d bands in spinel-$LiMn_2O_4$ films by optical absorption, *Appl. Phys. Lett.* **77**, 4154, doi: https://doi.org/10.1063/1.1336552.

[10] Thackeray M.M. (1997), Manganese oxides for lithium batteries, *Prog. Solid State Chem.* **25**, 1, doi: https://doi.org/10.1016/S0079-6786(97)81003-5.

[11] Kanamura K., Naito H., Yao T., Takehara Z. (1996) Structural change of the $LiMn_2O_4$ spinel structure induced by extraction of lithium, *J. Mater. Chem.* **6**, 33, doi: https://doi.org/10.1039/JM9960600033.

[12] Mosbah A., Verbaere A., Tournoux M. (1983) Phases $Li_xMnO_2\lambda$ rattachees au type spinelle, *Mater. Res. Bull.* **18**, 1375, doi: https://doi.org/10.1016/0025-5408(83)90045-4.

[13] Ohzuku T., Kitagawa M., Hirai T. (1990) Electrochemistry of manganese dioxide in lithium nonaqueous cell III. X-Ray diffractional study on the reduction of spinel-related manganese dioxide, J. Electrochem. Soc. 137, 769, doi: https://doi.org/10.1149/1.2086552.

[14] David W.I.F., Thackeray M.M., De Picciotto L.A., Goodenough J.B. (1987) Structure refinement of the spinel-related phases $Li_2Mn_2O_4$ and $Li_{0.2}Mn_2O_4$, *J. Solid State Chem.* **67**, 316, doi: https://doi.org/10.1016/0022-4596(87)90369-0.

[15] Tang D., Sun Y., Yang Z., Ben L., Gu L., Huang X. (2014) Surface structure evolution of $LiMn_2O_4$ cathode material upon charge/discharge, *Chem. Mater.* **26**, 3535, doi: https://doi.org/10.1021/cm501125e.

[16] Xia Y., Zhou Y., Yoshio M. (1997) Capacity fading on cycling of 4 V Li/LiMn$_2$O$_4$ Cells, *J. Electrochem. Soc.* **144**, 2593, doi: https://doi.org/10.1149/1.1837870.

[17] Gummow R.J., de Kock A., Thackeray M.M. (1994) Improved capacity retention in rechargeable 4 V lithium/lithium-manganese oxide (spinel) cells, *Solid State Ion* **69**, 59, doi: https://doi.org/10.1016/0167-2738(94)90450-2.

[18] Ohzuku T., Kitano S., Iwanaga M., Matsuno H., Ueda A. (1997) Comparative study of Li [Li$_x$Mn$_{2-x}$O$_4$ and LT-LiMnO$_2$ for lithium-ion batteries, *J. Power Sources* **68**, 646, doi: https://doi.org/10.1016/S0378-7753(96)02573-6.

[19] Sánchez L., Tirado J.L. (1997) Synthesis and electrochemical characterization of a new Li-Co-Mn-O spinel phase for rechargeable lithium batteries, *J. Electrochem. Soc.* **144**, 1939, doi: https://doi.org/10.1149/1.1837725.

[20] Robertson A.D., Lu S.H., Howard W.F. (1997) M3+ Modified LiMn$_2$O$_4$ spinel intercalation cathodes II. Electrochemical stabilization by Cr^{3+}, *J. Electrochem. Soc.* **144**, 3505, doi: https://doi.org/10.1149/1.1838041.

[21] Amatucci G.G., Pereira N., Zheng T., Tarascon J.-M. (2001), Failure mechanism and improvement of the elevated temperature cycling of LiMn$_2$O$_4$ compounds through the use of the LiAl$_x$Mn$_{2-x}$O$_{4-z}$F$_z$ solid solution, *J. Electrochem. Soc.* **148**, A171, doi: https://doi.org/10.1149/1.1342168.

[22] Tokunaga T., Yoneda K., JP5549321, 2010.

[23] Kitao H., Fujihara T., Takeda K., Nakanishi N., Nohma T. (2005) High-temperature storage performance of Li-ion batteries using a mixture of Li-Mn spinel and Li-Ni-Co-Mn oxide as a positive electrode material, *Electrochem. Solid-State Lett.* **8**, A87, doi: https://doi.org/10.1149/1.1843792.

[24] Patoux S., Daniel L., Bourbon C., Lignier H., Pagano C., Le Cras F., Jouanneau S., Martinet S. (2009) High voltage spinel oxides for Li-ion batteries: from the material research to the application, *J. Power Sources* **189**, 344, doi: https://doi.org/10.1016/j.jpowsour.2008.08.043.

[25] Mizushima K., Jones P.C., Wiseman P.J., Goodenough J.B. (1981), Li$_x$CoO$_2$ (0<x$\leqslant$1): a new cathode material for batteries of high energy density, *Solid State Ion* **3,** 171, doi: https://doi.org/10.1016/0167-2738(81)90077-1.

[26] Whittingham M.S. (2004) Lithium batteries and cathode materials, *Chem. Rev.* **104**, 4271, doi: https://doi.org/10.1021/cr020731c.

[27] Ozawa K. (1994) Lithium-ion rechargeable batteries with LiCoO$_2$ and carbon electrodes: the LiCoO$_2$/C system, *Solid State Ion* **69**, 212, doi: https://doi.org/10.1016/0167-2738(94)90411-1.

[28] Yabuuchi N., Ohzuku T. (2003) Novel lithium insertion material of LiCo1/3Ni1/3Mn1/3O2 for advanced lithium-ion batteries, *J. Power Sources* **119–121**, 171, doi: https://doi.org/10.1016/S0378-7753(03)00173-3.

[29] Noh H.-J., Youn S., Yoon C.S., Sun Y.-K. (2013) Comparison of the structural and electrochemical properties of layered Li[Ni$_x$Co$_y$Mn$_z$]O$_2$ (x = 1/3, 0.5, 0.6, 0.7, 0.8 and 0.85) cathode material for lithium-ion batteries, J. Power Sources 233, 121, doi: https://doi.org/10.1016/j.jpowsour.2013.01.063.

[30] MacNeil D.D., Lu Z., Dahn J.R. (2002) Structure and electrochemistry of Li[Ni$_x$Co$_{1-2x}$Mn$_x$]O$_2$ (0$\leqslant$x$\leqslant$1/2), *J. Electrochem. Soc.* **149**, A1332, doi: https://doi.org/10.1149/1.1505633.

[31] Rossen E., Jones C.D.W., Dahn J.R. (1992) Structure and electrochemistry of Li$_x$Mn$_y$Ni$_{1-y}$O$_2$, *Solid State Ion* **57**, 311, doi: https://doi.org/10.1016/0167-2738(92)90164-K.

[32] Manthiram A., Knight J.C., Myung S.-T., Oh S.-M., Sun Y.-K. (2016) Nickel-rich and lithium-rich layered oxide cathodes: progress and perspectives, *Adv. Energy Mater.* **6**, 1501010, doi: https://doi.org/10.1002/aenm.201501010.

[33] Li W., Reimers J.N., Dahn J.R. (1994) Lattice-gas-model approach to understanding the structures of lithium transition-metal oxides LiMO$_2$, *Phys. Rev. B.* **49**, 826, doi: https://doi.org/10.1103/PhysRevB.49.826.

[34] Huang W., Frech R. (1996) Vibrational spectroscopic and electrochemical studies of the low and high temperature phases of LiCo$_{1-x}$ M$_x$O$_2$ (M = Ni or Ti), *Solid State Ion* **86–88**, 395, doi: https://doi.org/10.1016/0167-2738(96)00158-0.

[35] Antolini E. (2004) LiCoO$_2$: formation, structure, lithium and oxygen nonstoichiometry, electrochemical behaviour and transport properties, *Solid State Ion* **170**, 159, doi: https://doi.org/10.1016/j.ssi.2004.04.003.

[36] Garcia B., Farcy J., Pereira-Ramos J.P., Baffier N. (1997), Electrochemical properties of low temperature crystallized LiCoO$_2$, *J. Electrochem. Soc.* **144**, 1179, doi: https://doi.org/10.1149/1.1837569.

[37] Kang S.G., Kang S.Y., Ryu K.S., Chang S.H. (1999) Electrochemical and structural properties of HT-LiCoO$_2$ and LT-LiCoO$_2$ prepared by the citrate sol-gel method, *Solid State Ion* **120**, 155, doi: https://doi.org/10.1016/S0167-2738(98)00559-1.

[38] Delmas C., Fouassier C., Hagenmuller P. (1980) Structural classification and properties of the layered oxides, *Phys. BC.* **99**, 81, doi: https://doi.org/10.1016/0378-4363(80)90214-4.

[39] Ménétrier M., Saadoune I., Levasseur S., Delmas C. (1999) The insulator-metal transition upon lithium deintercalation from LiCoO$_2$: electronic properties and 7Li NMR study, *J. Mater. Chem.* **9**, 1135, doi: https://doi.org/10.1039/A900016J.

[40] Yabuuchi N., Kawamoto Y., Hara R., Ishigaki T., Hoshikawa A., Yonemura M., Kamiyama T., Komaba S. (2013) A comparative study of LiCoO$_2$ polymorphs: structural and electrochemical characterization of O2-, O3-, and O4-type phases, *Inorg. Chem.* **52**, 9131, doi: https://doi.org/10.1021/ic4013922.

[41] Ohzuku T., Ueda A. (1994) Solid-state redox reactions of LiCoO$_2$ (R$\bar{3}$m) for 4 volt secondary lithium cells, *J. Electrochem. Soc.* **141**, 2972, doi: https://doi.org/10.1149/1.2059267.

[42] Ohzuku T., Makimura Y. (2001) Layered lithium insertion material of LiCo1/3Ni1/3Mn1/3O2 for lithium-ion batteries, *Chem. Lett.* **30**, 642, doi: https://doi.org/10.1246/cl.2001.642.

[43] Yabuuchi N., Makimura Y., Ohzuku T. (2007) Solid-state chemistry and electrochemistry of LiCo1/3Ni1/3Mn1/3O2 for advanced lithium-ion batteries III. Rechargeable capacity and cycleability, *J. Electrochem. Soc.* **154** A314, doi: https://doi.org/10.1149/1.2455585.

[44] Lee K.-S., Myung S.-T., Amine K., Yashiro H., Sun Y.-K. (2007) Structural and electrochemical properties of layered Li [Ni$_{1-2x}$Co$_x$Mn$_x$] O$_2$ (x = 0.1 – 0.3) positive electrode materials for Li-ion batteries, *J. Electrochem. Soc.* **154**, A971, doi: https://doi.org/10.1149/1.2769831.

[45] Jung R., Metzger M., Maglia F., Stinner C., Gasteiger H.A. (2017) Oxygen release and its effect on the cycling stability of LiNi$_x$Mn$_y$Co$_z$O$_2$ (NMC) cathode materials for Li-ion batteries, *J. Electrochem. Soc.* **164**, A1361, doi: https://doi.org/10.1149/2.0021707jes.

[46] Jung S.-K., Gwon H., Hong J., Park K.-Y., Seo D.-H., Kim H., Hyun J., Yang W., Kang K. (2014) Understanding the degradation mechanisms of LiNi$_{0.5}$Co$_{0.2}$Mn$_{0.3}$O$_2$ cathode material in lithium ion batteries, *Adv. Energy Mater.* **4**, 1300787, doi: https://doi.org/10.1002/aenm.201300787.

[47] Hou P., Yin J., Ding M., Huang J., Xu X. (2017) Surface/interfacial structure and chemistry of high-energy nickel-rich layered oxide cathodes: advances and perspectives, *Small* **13**, 1701802, doi: https://doi.org/10.1002/smll.201701802.

[48] Zhu W., Huang X., Liu T., Xie Z., Wang Y., Tian K., Bu L., Wang H., Gao L., Zhao J. (2019) Ultrathin Al$_2$O$_3$ coating on LiNi$_{0.8}$Co$_{0.1}$Mn$_{0.1}$O$_2$ cathode material for enhanced cycleability at extended voltage ranges, *Coatings* **9**, 92, doi: https://doi.org/10.3390/coatings9020092.

[49] Myung S.-T., Lee K.-S., Yoon C.S., Sun Y.-K., Amine K., Yashiro H. (2010) Effect of AlF3 coating on thermal behavior of chemically delithiated Li$_{0.35}$[Ni1/3Co1/3Mn1/3]O2, *J. Phys. Chem. C.* **114**, 4710, doi: https://doi.org/10.1021/jp9082322.

[50] Peng Z., Deng X., Du K., Hu G., Gao X., Liu Y. (2008) Coating of LiNi1/3Mn1/3Co1/3O2 cathode materials with alumina by solid state reaction at room temperature, *J. Cent. South Univ. Technol.* **15**, 34, doi: https://doi.org/10.1007/s11771-008-0008-9.

[51] Kim H.-S., Kong M., Kim K., Kim I.-J., Gu H.-B. (2007), Effect of carbon coating on LiNi1/3Mn1/3Co1/3O2 cathode material for lithium secondary batteries, *J. Power Sources* **171**, 917, doi: https://doi.org/10.1016/j.jpowsour.2007.06.028.

[52] Yao Y., Liu H., Li G., Peng H., Chen K. (2013) Synthesis and electrochemical performance of phosphate-coated porous $LiNi_{1/3}Co_{1/3}Mn_{1/3}O_2$ cathode material for lithium ion batteries, *Electrochimica Acta* **113**, 340, doi: https://doi.org/10.1016/j.electacta.2013.09.071.

[53] Wang H., Tang A., Huang K., Liu S. (2010) Uniform $AlF3$ thin layer to improve rate capability of $LiNi_{1/3}Co_{1/3}Mn_{1/3}O_2$ material for Li-ion batteries, *Trans. Nonferrous Met. Soc. China* **20**, 803, doi: https://doi.org/10.1016/S1003-6326(09)60217-X.

[54] Su L., Jing Y., Zhou Z. (2011), Li ion battery materials with core–shell nanostructures, *Nanoscale* **3**, 3967, doi: https://doi.org/10.1039/C1NR10550G.

[55] Sun Y.-K., Myung S.-T., Kim M.-H., Prakash J., Amine K. (2005) Synthesis and characterization of $Li[(Ni_{0.8})_{0.8}(Ni_{0.5})_{0.2}]O_2$ with the microscale core−shell structure as the positive electrode material for lithium batteries, *J. Am. Chem. Soc.* **127**, 13411, doi: https://doi.org/10.1021/ja053675g.

[56] Sun Y.-K., Myung S.-T., Park B.-C., Amine K. (2006) Synthesis of spherical nano- to microscale core−shell particles $Li[(Ni_{0.8}Co_{0.1}Mn_{0.1})_{1-x}(Ni_{0.5})_x]O2$ and their applications to lithium batteries, *Chem. Mater.* **18**, 5159, doi: https://doi.org/10.1021/cm061746k.

[57] Cho Y., Lee S., Lee Y., Hong T., Cho J. (2011) Spinel-layered core-shell cathode materials for Li-ion batteries, *Adv. Energy Mater.* **1**, 821, doi: https://doi.org/10.1002/aenm.201100239.

[58] Cho Y., Oh P., Cho J. (2013) A new type of protective surface layer for high-capacity Ni-based cathode materials: nanoscaled surface pillaring layer, *Nano Lett.* **13**, 1145, doi: https://doi.org/10.1021/nl304558t.

[59] Jun D.-W., Yoon C.S., Kim U.-H., Sun Y.-K. (2017) High-energy density core–shell structured $Li[Ni_{0.95}Co_{0.025}Mn_{0.025}]O_2$ cathode for lithium-ion batteries, *Chem. Mater.* **29**, 5048, doi: https://doi.org/10.1021/acs.chemmater.7b01425.

[60] Sun Y.-K., Chen Z., Noh H.-J., Lee D.-J., Jung H.-G., Ren Y., Wang S., Yoon C.S., Myung S.-T., Amine K. (2012) Nanostructured high-energy cathode materials for advanced lithium batteries, *Nat. Mater.* **11**, 942, doi: https://doi.org/10.1038/nmat3435.

[61] Lim B.-B., Myung S.-T., Yoon C.S., Sun Y.-K. (2016) Comparative study of Ni-Rich layered cathodes for rechargeable lithium batteries: $Li[Ni_{0.85}Co_{0.11}Al_{0.04}]O_2$ and Li $[Ni_{0.84}Co_{0.06}Mn_{0.09}Al_{0.01}]O_2$ with two-step full concentration gradients, *ACS Energy Lett.* **1**, 283, doi: https://doi.org/10.1021/acsenergylett.6b00150.

[62] Hou P., Zhang H., Zi Z., Zhang L., Xu X. (2017) Core–shell and concentration-gradient cathodes prepared via co-precipitation reaction for advanced lithium-ion batteries, *J. Mater. Chem. A.* **5**, 4254, doi: https://doi.org/10.1039/C6TA10297B.

[63] Myung S.-T., Cho M.H., Hong H.T., Kang T.H., Kim C.-S. (2005) Electrochemical evaluation of mixed oxide electrode for Li-ion secondary batteries: $Li_{1.1}Mn_{1.9}O_4$ and $LiNi_{0.15}Co_{0.15}Al_{0.05}O_2$, *J. Power Sources* **146**, 222, doi: https://doi.org/10.1016/j.jpowsour.2005.03.031.

[64] Lee K.K., Yoon W.S., Kim K.B., Lee K.Y., Hong S.T. (2001) Characterization of $LiNi_{0.85}Co_{0.10}M_{0.05}O_2$ (M = Al, Fe) as a cathode material for lithium secondary batteries, *J. Power Sources* **97–98**, 308, doi: https://doi.org/10.1016/S0378-7753(01)00516-X.

[65] Yoon S., Lee C.W., Bae Y.S., Hwang I., Park Y.-K., Song J.H. (2009) Method of preparation for particle growth enhancement of $LiNi_{0.8}Co_{0.15}Al_{0.05}O_2$, *Electrochem. Solid-State Lett.* **12**, A211, doi: https://doi.org/10.1149/1.3211133.

[66] Liu C., Qian K., Lei D., Li B., Kang F., He Y.-B. (2018) Deterioration mechanism of $LiNi0.8Co0.15Al0.05O2$ /graphite–SiO_x power batteries under high temperature and discharge cycling conditions, *J. Mater. Chem. A.* **6**, 65, doi: https://doi.org/10.1039/C7TA08703A.

[67] Yoo G.-W., Jang B.-C., Son J.-T. (2015) Novel design of core shell structure by NCA modification on NCM cathode material to enhance capacity and cycle life for lithium secondary battery, *Ceram. Int.* **41**, 1913, doi: https://doi.org/10.1016/j.ceramint.2014.09.077.

[68] Shin J.-W., Son J.-T. (2019) Core-shell-structured $Li[Ni_{0.87}Co_{0.08}Al_{0.05}]O_2$ cathode material for enhanced electrochemical performance and thermal stability of lithium-ion batteries, *J. Korean Phys. Soc.* **74**, 53, doi: https://doi.org/10.3938/jkps.74.53.

[69] Liu W., Tang X., Qin M., Li G., Deng J., Huang X. (2016) FeF3-coated $LiNi_{0.8}Co_{0.15}Al_{0.05}O_2$ cathode materials with improved electrochemical properties, *Mater. Lett.* **185**, 96, doi: https://doi.org/10.1016/j.matlet.2016.08.112.

[70] Dai G., Yu M., Shen F., Cao J., Ni L., Chen Y., Tang Y., Chen Y. (2016) Improved cycling performance of $LiNi_{0.8}Co_{0.15}Al_{0.05}O_2/Al_2O_3$ with core-shell structure synthesized by a heterogeneous nucleation-and-growth process, *Ionics* **22**, 2021, doi: https://doi.org/10.1007/s11581-016-1750-x.

[71] College of Materials Science and Engineering, Hunan University, Changsha 410082, China., Qi H. (2017) Facile fabrication and low-cost coating of $LiNi_{0.8}Co_{0.15}Al_{0.05}O_2$ with enhanced electrochemical performance as cathode materials for lithium-ion batteries, *Int. J. Electrochem. Sci.*, 5836, doi: https://doi.org/10.20964/2017.07.01.

[72] Thackeray M.M., Johnson C.S., Vaughey J.T., Li N., Hackney S.A. (2005) Advances in manganese-oxide 'composite' electrodes for lithium-ion batteries, *J. Mater. Chem.* **15**, 2257, doi: https://doi.org/10.1039/B417616M.

[73] Boulineau A., Simonin L., Colin J.-F., Canévet E., Daniel L., Patoux S. (2012) Evolutions of $Li_{1.2}Mn_{0.61}Ni_{0.18}Mg_{0.01}O_2$ during the initial charge/discharge cycle studied by advanced electron microscopy, *Chem. Mater.* **24**, 3558, doi: https://doi.org/10.1021/cm301140g.

[74] Johnson C.S., Li N., Lefief C., Vaughey J.T., Thackeray M.M. (2008) Synthesis, characterization and electrochemistry of lithium battery electrodes: $xLi_2MnO_3 \cdot (1 - x)$ $LiMn_{0.333}Ni_{0.333}Co_{0.333}O_2$ ($0 \le x \le 0.7$), *Chem. Mater.* **20**, 6095, doi: https://doi.org/10.1021/cm801245r.

[75] Rossouw M., Thackeray M. (1991) Lithium manganese oxides from Li_2MnO_3 for rechargeable lithium battery applications, *Mater. Res. Bull.* **26**, 463, doi: https://doi.org/10.1016/0025-5408(91)90186-P.

[76] Zhonghua Lu, Zhaohui Chen, Dahn* J.R. (2003) Lack of cation clustering in $Li[Ni_xLi_{1/3 -2x/3}Mn_{2/3-x/3}]O_2$ ($0 < x \le 1/2$) and $Li[Cr_xLi_{(1-x)/3}Mn_{(2-2x)/3}]O_2$ ($0 < x < 1$), doi: https://doi.org/10.1021/cm030194s.

[77] Li J., Shunmugasundaram R., Doig R., Dahn J.R. (2015) In situ X-ray diffraction study of layered Li–Ni–Mn–Co oxides: effect of particle size and structural stability of core–shell materials, doi: https://doi.org/10.1021/acs.chemmater.5b03500.

[78] Koga H., Croguennec L., Mannessiez P., Ménétrier M., Weill F., Bourgeois L., Duttine M., Suard E., Delmas C. (2012) $Li_{1.20}Mn_{0.54}Co_{0.13}Ni_{0.13}O_2$ with different particle sizes as attractive positive electrode materials for lithium-ion batteries: insights into their structure, doi: https://doi.org/10.1021/jp301879x.

[79] Peralta D., Colin J.-F., Boulineau A., Simonin L., Fabre F., Bouvet J., Feydi P., Chakir M., Chapuis M., Patoux S. (2015) Role of the composition of lithium-rich layered oxide materials on the voltage decay, *J. Power Sources* **280**, 687, doi: https://doi.org/10.1016/j.jpowsour.2015.01.146.

[80] Koga H., Croguennec L., Ménétrier M., Douhil K., Belin S., Bourgeois L., Suard E., Weill F., Delmas C. (2013) Reversible oxygen participation to the redox processes revealed for $Li_{1.20}Mn_{0.54}Co_{0.13}Ni_{0.13}O_2$, *J. Electrochem. Soc.* **160**, A786, doi: https://doi.org/10.1149/2.038306jes.

[81] Muhammad S., Kim H., Kim Y., Kim D., Song J.H., Yoon J., Park J.-H., Ahn S.-J., Kang S.-H., Thackeray M.M., Yoon W.-S. (2016) Evidence of reversible oxygen participation in anomalously high capacity Li- and Mn-rich cathodes for Li-ion batteries, *Nano Energy* **21**, 172, doi: https://doi.org/10.1016/j.nanoen.2015.12.027.

[82] Koga H., Croguennec L., Ménétrier M., Mannessiez P., Weill F., Delmas C., Different oxygen redox participation for bulk and surface: a possible global explanation for the cycling mechanism of $Li_{1.20}Mn_{0.54}Co_{0.13}Ni_{0.13}O_2$, *J. Power Sources* **236**, 250, doi: https://doi.org/10.1016/j.jpowsour.2013.02.075.

[83] Sathiya M., Ramesha K., Rousse G., Foix D., Gonbeau D., Prakash A.S., Doublet M.L., Hemalatha K., Tarascon J.-M. (2013) High performance $Li_2Ru_{1-y}Mn_yO_3$ ($0.2 \le y \le 0.8$) cathode materials for rechargeable lithium-ion batteries: their understanding, doi: https://doi.org/10.1021/cm400193m.

[84] Sathiya M., Rousse G., Ramesha K., Laisa C.P., Vezin H., Sougrati M.T., Doublet M.-L., Foix D., Gonbeau D., Walker W., Prakash A.S., Ben Hassine M., Dupont L., Tarascon J.-M. (2013) Reversible anionic redox chemistry in high-capacity layered-oxide electrodes, *Nat. Mater.* **12**, 827, doi: https://doi.org/10.1038/nmat3699.

[85] Croy J.R., Kim D., Balasubramanian M., Gallagher K., Kang S.-H., Thackeray M.M. (2012) Countering the voltage decay in high capacity $xLi_2MnO_3 \bullet (1-x)LiMO_2$ electrodes (M=Mn, Ni, Co) for Li+-ion batteries, *J. Electrochem. Soc.* **159**, A781, doi: https://doi.org/ 10.1149/2.080206jes.

[86] Croy J.R., Kang S.-H., Balasubramanian M., Thackeray M.M. (2011) Li_2MnO_3-based composite cathodes for lithium batteries: a novel synthesis approach and new structures, *Electrochem. Commun.* **13**, 1063, doi: https://doi.org/10.1016/j.elecom.2011.06.037.

[87] Bloom I., Trahey L., Abouimrane A., Belharouak I., Zhang X., Wu Q., Lu W., Abraham D. P., Bettge M., Elam J.W., Meng X., Burrell A.K., Ban C., Tenent R., Nanda J., Dudney N. (2014) Effect of interface modifications on voltage fade in $0.5Li_2MnO_3 \cdot 0.5LiNi_{0.375}Mn_{0.375}Co_{0.25}O_2$ cathode materials, *J. Power Sources* **249**, 509, doi: https://doi. org/10.1016/j.jpowsour.2013.10.035.

[88] Wu F., Li N., Su Y., Lu H., Zhang L., An R., Wang Z., Bao L., Chen S. (2012) Can surface modification be more effective to enhance the electrochemical performance of lithium rich materials? *J Mater Chem.* **22**, 1489, doi: https://doi.org/10.1039/C1JM14459F.

[89] Boulineau A., Simonin L., Colin J.-F., Bourbon C., Patoux S. (2013) First evidence of manganese–nickel segregation and densification upon cycling in Li-rich layered oxides for lithium batteries, *Nano Lett.* **13**, 3857, doi: https://doi.org/10.1021/nl4019275.

[90] Wang J., He X., Paillard E., Laszczynski N., Li J., Passerini S. (2016) Lithium- and manganese-rich oxide cathode materials for high-energy lithium ion batteries, *Adv. Energy Mater.* **6**, 1600906, doi: https://doi.org/10.1002/aenm.201600906.

[91] Yabuuchi N., Hara R., Kajiyama M., Kubota K., Ishigaki T., Hoshikawa A., Komaba S. (2014) New O2/P2-type Li-excess layered manganese oxides as promising multi-functional electrode materials for rechargeable Li/Na batteries, *Adv. Energy Mater.* **4**, 1301453, doi: https://doi.org/10.1002/aenm.201301453.

[92] Guo H., Wei Z., Jia K., Qiu B., Yin C., Meng F., Zhang Q., Gu L., Han S., Liu Y., Zhao H., Jiang W., Cui H., Xia Y., Liu Z. (2019) Abundant nanoscale defects to eliminate voltage decay in Li-rich cathode materials, *Energy Storage Mater.* **16**, 220, doi: https://doi.org/10. 1016/j.ensm.2018.05.022.

[93] Doughty D.H., Roth E.P. (2012) A general discussion of Li ion battery safety, *Electrochem. Soc. Interface* **21**, 37, doi: https://doi.org/10.1149/2.F03122if.

[94] Satyavani T.V.S.L., Srinivas Kumar A., Subba Rao P.S.V. (2016) Methods of synthesis and performance improvement of lithium iron phosphate for high rate Li-ion batteries: a review, *Eng. Sci. Technol. Int. J.* **19**, 178, doi: https://doi.org/10.1016/j.jestch.2015.06.002.

[95] Ozan Toprakci, Toprakci A.K., Liwen Ji, Xiangwu Zhang (2010) Fabrication and electrochemical characteristics of $LiFePO_4$ powders for lithium-ion batteries, *KONA Powder Part. J.* **28**, 50.

[96] Milović M., Jugović D., Cvjetićanin N., Uskoković D., Milošević A.S., Popović Z.S., Vukajlović F.R. (2013) Crystal structure analysis and first principle investigation of F doping in $LiFePO_4$, *J. Power Sources* **241**, 70, doi: https://doi.org/10.1016/j.jpowsour.2013.04.109.

[97] Wang J., Sun X. (2012) Understanding and recent development of carbon coating on $LiFePO_4$ cathode materials for lithium-ion batteries, *Energy Environ. Sci.* **5**, 5163, doi: https://doi.org/10.1039/C1EE01263K.

[98] Chung S.-Y., Choi S.-Y., Yamamoto T., Ikuhara Y. (2008) Atomic-scale visualization of antisite defects in $LiFePO_4$, *Phys. Rev. Lett.* **100**, 125502, doi: https://doi.org/10.1103/ PhysRevLett.100.125502.

[99] Yang S., Song Y., Zavalij P.Y., Stanley Whittingham M. (2002) Reactivity, stability and electrochemical behavior of lithium iron phosphates, *Electrochem. Commun.* **4**, 239, doi: https://doi.org/10.1016/S1388-2481(01)00298-3.

[100] Ravet N., Chouinard Y., Magnan J.F., Besner S., Gauthier M., Armand M. (2001) Electroactivity of natural and synthetic triphylite, *J. Power Sources* **97–98**, 503, doi: https://doi.org/10.1016/S0378-7753(01)00727-3.

[101] Yuan L.-X., Wang Z.-H., Zhang W.-X., Hu X.-L., Chen J.-T., Huang Y.-H. (2011) Goodenough J.B., Development and challenges of $LiFePO_4$ cathode material for lithium-ion batteries, *Energy Environ. Sci.* **4**, 269, doi: https://doi.org/10.1039/C0EE00029A.

[102] Liu H., Liu Y., An L., Zhao X., Wang L., Liang G. (2017) High energy density LiFePO4/C cathode material synthesized by wet ball milling combined with spray drying method, *J. Electrochem. Soc.* **164**, A3666, doi: https://doi.org/10.1149/2.0011714jes.

[103] Zaghib K., Mauger A., Groult H., Goodenough J.B., Julien C.M. (2013) Advanced electrodes for high power Li-ion batteries, *Materials* **6**, 1028, doi: https://doi.org/10.3390/ma6031028.

[104] Wang J., Sun X. (2015) Olivine $LiFePO_4$: the remaining challenges for future energy storage, *Energy Environ. Sci.* **8**, 1110, doi: https://doi.org/10.1039/C4EE04016C.

[105] Wang D., Buqa H., Crouzet M., Deghenghi G., Drezen T., Exnar I., Kwon N.-H., Miners J. H., Poletto L., Grätzel M. (2009) High-performance, nano-structured $LiMnPO_4$ synthesized via a polyol method, *J. Power Sources* **189**, 624, doi: https://doi.org/10.1016/j.jpowsour.2008.09.077.

[106] Yamada A., Chung S.-C. (2001) Crystal chemistry of the olivine-type Li (Mn y Fe1 − y) PO 4 and (Mn y Fe1 − y) PO 4 as possible 4 V cathode materials for lithium batteries, *J. Electrochem. Soc.* **148**, A960, doi: https://doi.org/10.1149/1.1385377.

Chapter 3

Negative Electrode Materials

Cédric Haon, Céline Barchasz and Philippe Azaïs

This chapter introduces the various active materials used in the negative electrode (anode) of lithium batteries. The first part deals with possible solutions and associated mechanisms; the main materials used are presented and described. The variants of carbon are presented in the second part with graphite in particular, which represents nearly 90% of the anode market. The third part details silicon as a serious alternative to graphite, while the fourth part deals with the use of lithium metal as an anode material.

3.1 Negative Electrode Materials: Several Solutions

The global market for anode materials for Li-ion batteries was 260 000 tons in 2019 (figure 3.1A) and it has been growing steadily since 2010. Graphite (natural and artificial) represents more than 90% of the market, the remaining part being shared among amorphous carbon, silicon and tin-type materials along with $Li_4Ti_5O_{12}$ ("LTO") (figure 3.1B).

In 2021, graphite remains the preferred choice thanks to its working electrochemical potential close to 0 V *vs.* Li^+/Li and its low cost. The two main selection criteria of an anode material for Li-ion batteries are its electrochemical potential *vs.* Li^+/Li and its specific capacity. The electrochemical potential should be as close as possible to the potential of the Li^+/Li redox couple in order to maximise the voltage difference between the positive and negative electrodes in the full system, and the specific capacity must be as high as possible (which means that the material should be able to store and exchange the highest possible quantity of lithium ions). These two parameters allow to classify anode materials (figure 3.2). There are many alternatives to graphite, such as titanium oxides (LTO), which have

DOI: 10.1051/978-2-7598-2555-4.c003
© Science Press, EDP Sciences, 2021

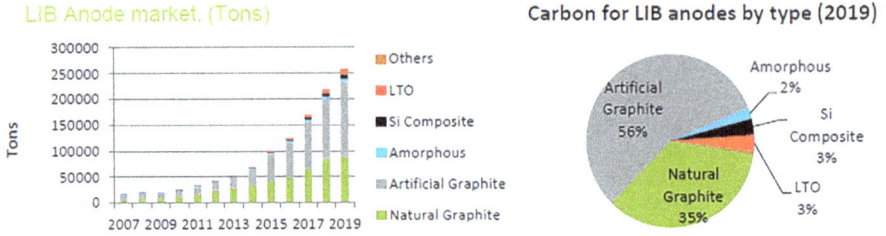

FIG. 3.1 – (A) (left) Li-ion batteries anode market evolution (tons) and (B) (right) Li-ion batteries anode materials by type in 2019 (by weight). Source: Avicenne Energy [1].

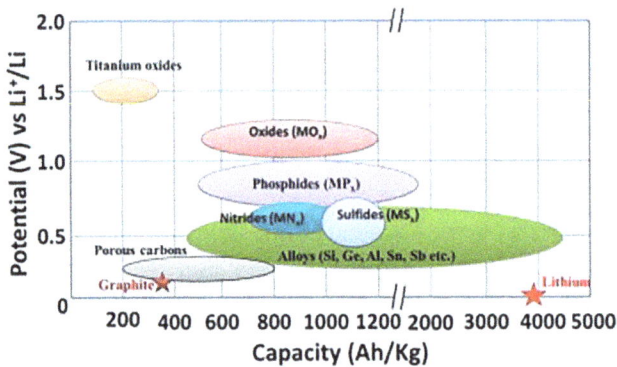

FIG. 3.2 – Potential (*vs.* Li^+/Li) and specific capacity of the main possible anode materials [2].

a higher potential and lower capacity, or alloy materials such as silicon. The ultimate anode remains lithium metal, which is used, for example, in batteries produced by the French company Blue Solutions. Graphite remains so far preferred to lithium metal, as the use of metallic lithium anode imposes specific operating conditions to allow an acceptable cycle life.

Various mechanisms may be at stake between lithium and anode materials: insertion (or intercalation), conversion and alloy formation.

3.1.1 *Insertion-Intercalation*

Graphite is considered as a reference intercalation material. It has a capacity of 372 mAh.g^{-1} and a working potential of about 0.1 V *vs.* Li^+/Li. During intercalation of lithium ions, the crystal structure of the host material is preserved; however, a volume change of the crystal lattice occurs, around 10% for graphite. The intercalation reaction for graphite is the following one:

$$Li^+ + 6C + e^- \leftrightarrow LiC_6$$

The same mechanism is also observed for "LTO" Lithium Titanate material according to the following reaction:

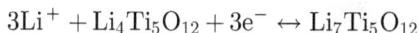

$$3Li^+ + Li_4Ti_5O_{12} + 3e^- \leftrightarrow Li_7Ti_5O_{12}$$

Lithium Titanate has a spinel structure with tunnels in all three spatial dimensions, which allows rapid diffusion of Li^+ ions into the material, even at high charge/discharge rates. During the insertion–disinsertion process, crystalline structure expansion or contraction are minimized. Indeed, the oxidized phase $(Li_4Ti_5O_{12})$ and the reduced phase $(Li_7Ti_5O_{12})$ have a substantially identical volume $(\Delta V = \pm 0.07\%)$. For this reason, "LTO" is considered as a "zero stress" material. The insertion reaction is said to be biphasic (coexistence of oxidized and reduced phases within the electrode) and takes place at a potential of 1.55 V *vs.* Li^+/Li; the theoretical specific capacity related to the formation of the reduced phase is 175 mAh.g^{-1}. As the insertion potential of the lithium ions is above the reduction potential of the electrolyte, no passivation layer is formed at the electrode/electrolyte interface (usual electrolytes do not degrade, unlike in graphite batteries). The voltage and the specific capacity of LTO unit cells are lower than when graphite is used as negative electrode. This penalizes the energy density, however LTO is of interest for its safety, power and lifetime performance. Battery manufacturers such as Leclanché or Toshiba produce batteries with titanium oxide-based anodes.

3.1.2 Conversion

Figure 3.2 also shows the metal oxides, phosphides, nitrides and sulphides for which the reaction with lithium takes place according to a conversion mechanism [3]. For all these families of materials, the conversion reaction is based on the reduction of the metal center and the reversible formation of lithium oxide, lithium phosphide, etc. according to, in the case of the oxide, the reaction:

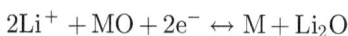

$$2Li^+ + MO + 2e^- \leftrightarrow M + Li_2O$$

Conversion materials are made of metallic elements such as iron, cobalt or chromium for the most studied. They have high specific capacities, between 500 and 1800 mAh.g^{-1}, which can increase the energy density of the cells, and partially counterbalance the loss of energy density due to their potential (between 0.5 and 1.5 V *vs.* Li^+/Li), which is higher than graphite. For now, they are not used in commercial cells because they are hindered by a low coulombic efficiency in the first cycle and by a quite fast capacity fading during cycling.

3.1.3 Alloying

For the last category of anode materials such as tin, silicon, germanium or aluminium, the reaction of lithium with these metals or metalloids results in the formation of an alloy according to the following reaction:

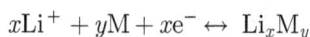

$$xLi^+ + yM + xe^- \leftrightarrow Li_xM_y$$

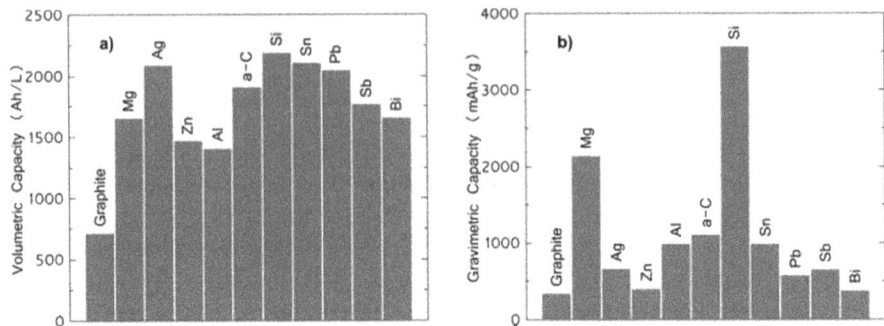

FIG. 3.3 − (a) Calculated volumetric capacity in the fully lithiated state and (b) gravimetric capacity of different alloying elements compared to graphite and amorphous carbon [4].

Figure 3.3 presents the main alloying elements with the associated gravimetric and volumetric capacities, compared to graphite, with silicon in particular having the highest capacities, 3579 mAh.g^{-1} for an operating potential of 0.4 V *vs.* Li$^+$/Li.

It is possible to significantly increase the energy density of the cells thanks to the high capacities these materials can reach. However, the structural transformations associated with the formation of the alloys limit the industrial development of these alternatives despite a lot of R&D work achieved on this topic and in particular on silicon, which is detailed later in this chapter.

3.2 Carbon

3.2.1 Historical Background

Carbon exists in many allotropic forms: graphite, diamond, nanotubes, fullerenes, vitreous carbon, carbon black, activated carbons... [5]. Nevertheless, graphite is the most thermodynamically stable state.

The definitions of the different types of carbon were clarified, in French language, by the "Groupe Français d'Etude des Carbones (GFEC)" in the early 1980s [6–8]. Heavy alkalis (potassium, rubidium and cesium) are intercalated relatively easily into graphite in the MC8 form (M being the alkali metal), as shown by Fredenhagen and Cadenbach [9] as early as 1926. The intercalation of lighter metals is less easy: it is made possible by the very small size of the lithium ion in a rich composition (LiC$_6$, figure 3.4), as demonstrated as early as 1955 by Herold [10] but in phases less rich in sodium[1]. Subsequent work shows that lithium can be thermally inserted into many other structurally imperfect carbonaceous materials [11]. Graphite is then envisaged as early as 1973 by M. Armand as an anode lithium insertion material [12]. In 1976, J.O. Besenhard demonstrates, for all alkalis, the possibility of carrying out the

---

[1]Less thermodynamically favourable than heavy alkalis and less favourable than lithium from a steric point of view.

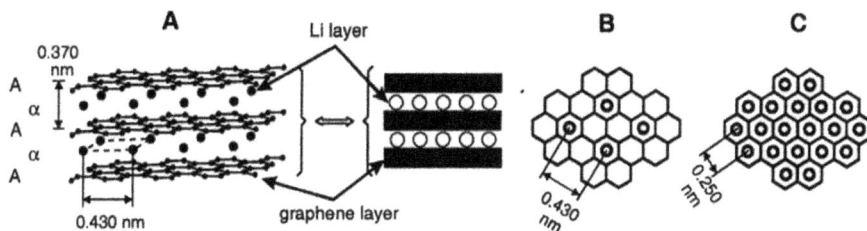

FIG. 3.4 – AA layer sequence and the ordered lithium interlayer of the compound LiC_6. View of LiC_6 (B) and LiC_2 (C) in the plane [17].

intercalation of ions in graphite by electrochemical means using DME (1,2-DiMethoxyEthane) and demonstrates its reversibility [13]. These first trials are continued by R. Yazami and P. Touzain [14]. In 1983, J.O. Besenhard and H. P. Fritz identify the correlation between the stages of graphite intercalation and the values of associated electrochemical potentials [15] (figure 3.5). In the early 80's, one of the problems that remains to be solved is the problem of co-intercalation of the solvent with lithium ion species, leading to a strong volume expansion of the graphite and a very strong irreversibility of the electrochemical reaction. A solution is found, which consists in using carbonate mixtures, other than propylene carbonate (PC) [16]. These mixtures generate a passivation layer (SEI or Solid Electrolyte Interphase) by partial decomposition of these solvents in the presence of a lithium superacid salt (LiPF6).

3.2.2 *Interest*

The interest of graphite, and carbonaceous materials in general, for lithium-ion batteries can be summarized in five points:

- The material is readily available (in its natural state or easily synthesized), making it relatively inexpensive.
- The material is non-toxic and environmentally friendly.
- The full insertion into the carbon structure results in a limited volume expansion (generally <10%) that is mechanically acceptable at the electrochemical cell level.
- The specific capacity is relatively high (theoretically 372 mAh.g^{-1} for the LiC_6 compound) and of the same order of magnitude as those of positive electrode materials (cathodes).
- The potential of the lithium-rich LiC_6 compound is close to that of the lithium metal, which is a decisive advantage for having a high potential difference with the positive electrode[2], and therefore a high energy density.

---

[2]The disadvantage in terms of safety, however, is that under certain charging conditions (high charging speed, cold charging, excessive ageing), a lithium metal deposit ("lithium plating") can form on the surface of the carbon electrode.

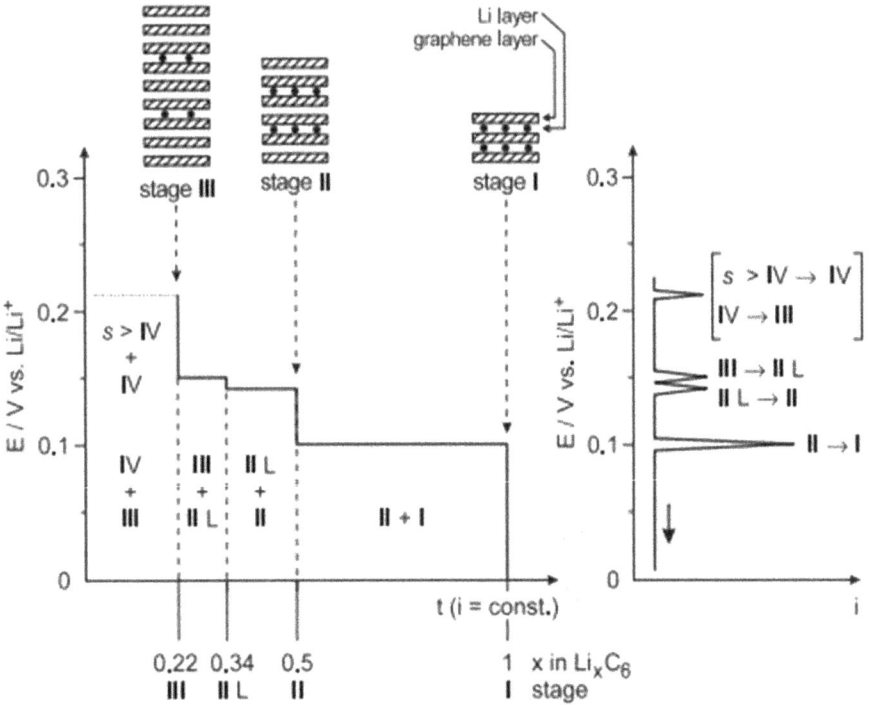

F$_{IG}$. 3.5 – Formation of intercalation stages during the electrochemical intercalation of lithium in graphite. On the right: schematic voltammetry curve corresponding to these stages [18].

Unlike many materials used in electrochemistry, graphite exists in its natural state. It is relatively abundant but very geolocalized (China (>78%), India (7%), Mexico (4%), North Korea, Brazil, Turkey, Sri Lanka, Canada) [19]. If the global carbon market is more than $13 billion [20], the largest share of the market is dedicated to other products. Indeed, there is manufacture of electrodes from synthetic graphite for the production of aluminium and the reduction of steel (electric arc furnaces) as well as the manufacture of carbon fibers and the production of specific parts for high temperature applications (furnace liners, contactors, motor brushes, etc.). The battery market, although growing strongly, requires carbon and graphite in powder and currently represents only a minor part (around 10%) of the total market for industrially produced carbons and graphite. However, this situation will change with the development of electric mobility.

3.2.3 *Relationship between Structural Characteristics and Performance*

The carbon materials used for Li-ion batteries can be classified into two main categories: natural graphite and synthetic graphite/carbon. The production of natural graphite is based on mining, followed by mechanical separation phases, a phase known as "floatation", grinding, spheroidization (aiming at making the particles spherical to reduce irreversible capacity) and a final purification phase [21]. Natural graphite is widely used for markets that do not require a high degree of sustainability ("consumer" markets: mobile phones, tablets, PCs, electric bikes). Synthetic graphite, which is more expensive than natural graphite, is produced by thermal process (pyrolysis) at high temperatures (>1800 °C) from purified oil industry by-products (polymers, resins, and other high-carbon content products). Non-graphitizable synthetic carbons are produced by high-temperature pyrolysis of natural materials (biomass: wood, algae, and sucrose, for example) or synthetic polymers (phenolic resins, epoxy resins, etc. with high oxygen or heteroatom content) [22]. These materials are called "hard carbons" because of their high content of heteroatoms (oxygen, nitrogen) in the precursor, which leads to a significant local disorder preventing graphitization even at high temperatures [23]. These hard carbons should not be considered as so-called amorphous or vitreous carbons: hard carbons have at least an enlarged 002 band present in X-ray diffraction.

Synthetic graphite have the advantage of being more pure (carbon content >99.95%) than natural graphite (generally <99.9%). Graphite crystallizes in two Bravais lattices: hexagonal (most stable form, ABA stacking) and rhombohedral (ABC stacking). The intercalation of lithium in graphite therefore depends on the ratio of hexagonal and rhombohedral crystals [24–26]. Indeed, the stacking defects of the graphene planes affect the storage capacity, and the ratio of the two types of lattices causes disorder at the nanometric scale. The two Bravais lattices are present in various proportions in natural and synthetic graphite [27], unless they have been treated at very high temperatures (>2800 °C) [28]. The main impurity influencing performance is the presence of surface clusters, especially at the edge of the sheet. On these "edges" are concentrated reactive chemical species [29]. The planes ("basal") have very few surface groups due to the lamellar structure of the graphite [30]. This has been demonstrated by chemical treatments [31, 32], or by measuring the active surface area (ASA) [33]. Inorganic impurities (Fe, Si, Al,...) present in natural graphite can potentially have a negative impact on the performance of the Li-ion battery [34, 35].

A lithium-containing passivation layer, commonly called "SEI" for "Solid–Electrolyte Interphase", forms on the surface of the electrode. It is due to the instability of the solvents in the electrolyte below a potential of about 1 V. The composition of the SEI depends on several factors: the morphology of the graphite, its composition and the composition of the electrolyte [36], the geometric surface, the surface chemistry, the conditioning of the accumulator[3] and of course the composition of the electrolyte. This is also the case for non-graphitic carbon [37].

---

[3]The kinetics of the different decomposition reactions (and therefore the degradation products) of the electrolyte necessary for the formation of the "SEI" is linked to the charging and discharging regime of the formation: the composition of the "SEI" puts in competition different reactions leading to different products.

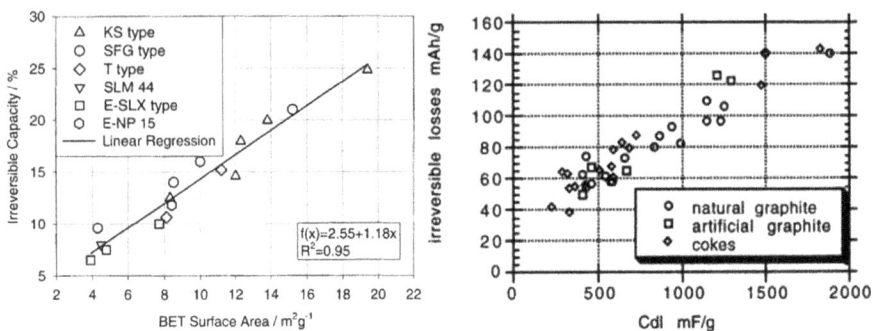

FIG. 3.6 – (Left) Percentage of irreversible capacity at the first charge according to the B.E.T. surface area measured from nitrogen at 77 K for various graphite (synthetic or natural) [52]. The capacity is measured at 10 mA.g^{-1} in the electrolyte EC:DMC (1:1) 1 M LiPF6. (Right) Irreversible capacity on first charge as a function of the double layer capacity (Cdl) measured by impedance spectroscopy. (O) natural graphite, (□) synthetic graphite and (◊) petroleum coke. Electrolyte: LiTFSI 1 M in EC:DMC (1:1) [53].

One of the important parameters to consider for the performance of Li-ion graphite anode is the outer surface of particles[4] (figure 3.6, left) on which the SEI is formed. An electrochemical measurement which allows selecting carbon materials is the identification of the electrochemical double layer value: the irreversible capacity of the first charge increases linearly with this double layer value (figure 3.6, right). A too large area leads to the generation of a large amount of SEI (covering all the particles by partial decomposition of the electrolyte). However, a too small surface area (*e.g.* use of large particles) is not desirable to achieve high power density (too slow diffusion of lithium ions would not allow to benefit from the whole capacity of the particle). This is demonstrated by H. Wang *et al.* by crushing two series of natural graphite by different ways [38]: The "pivot" value seems to be around 5 m^2/g. This trend is confirmed by K. Zaghib *et al.* [39]. To maximize the reversible capacity of the anode material, several strategies are proposed: coating [40–43], chemical grafting, or other surface chemical reaction [44, 45]. One strategy is to protect the graphite (natural or not) with a carbon shell [46–51] to improve cycling behaviour. This path is now favoured by major players as carbon is dedicated to applications requiring very good durability (electrified vehicles, stationary applications).

Many other types of carbon can be used as an anode active material for Li-ion batteries: mono- or multi-wall carbon nanotubes (SWNT, MWNT), carbon nanofibers (CNF), graphene,… However, all these materials have a large external surface area, leading to the formation of a very high irreversible capacity. Moreover, the cost

---

[4]The surface can be determined by nitrogen adsorption at 77 K or by krypton adsorption at room temperature. The surface area (m^2.g^{-1}) is obtained by treatment according to the Brunauer, Emmett and Teller (B.E.T.) equations, as described in Brunauer S., Emmett P.H., Teller E. (1938) Adsorption of gases in multimolecular layers, *J. Am. Chem. Soc.* **60**, 309.

of these materials (energy and financial) is high, which is not compatible with the current economic equation for the targeted applications.

Hard carbon offers interesting performances to improve the power density of the accumulators. With the emergence of light hybridization vehicle technologies (HEV: soft-hybrid, mild-hybrid), there is a need for high power density even at low temperature while very good durability is maintained [54]. However, the work is driven by a strong constraint at the level of the synthesis of these disordered materials at the nanometric scale, in order to strongly limit the irreversible capacity while maintaining a relatively high operating voltage (typically 3.5 V for an NMC/hard carbon system instead of 3.7 V for most NMC/graphite systems). In contrast to the almost superimposable graphite charge/discharge profile, the charge/discharge profile of a hard carbon (and also of a soft carbon that is not fully graphitized) has a hysteresis (figure 3.7). This profile is related to the fact that the path of the ions during charging is not the same as the one during discharge.

3.3 Silicon

3.3.1 (De)lithiation Mechanisms

As described above, the insertion of lithium in silicon takes place *via* the formation of an alloy, the most lithiated phase is $Li_{15}Si_4$ in its crystalline form and corresponds to

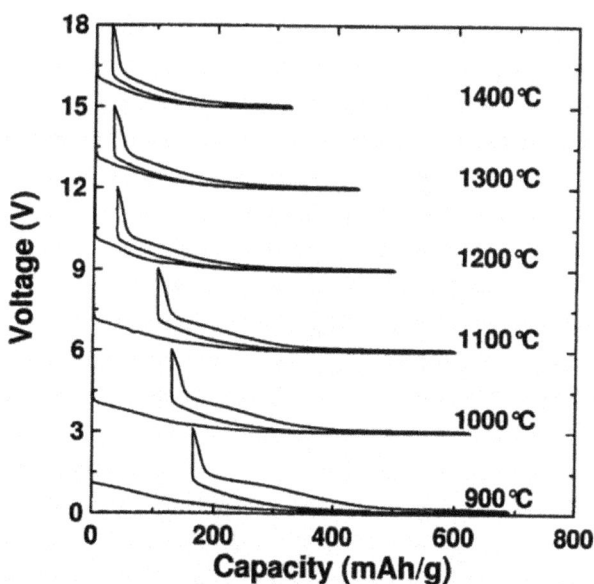

FIG. 3.7 – Charge/discharge profile of hard carbons from the pyrolysis of sucrose at different temperatures under argon [55]. Electrochemical hard carbon/lithium metal cells. Electrolyte: EC:DEC (1:2) 1 M LiPF6.

the capacity of 3579 mAh.g^{-1}. However, the lithiation and delithiation of silicon have different characteristics and happen through amorphous phases. First of all, the first lithiation step differs from the following ones; it is carried out according to a two-phase reaction mechanism at a fixed potential of around 0.1 V with the coexistence of pure silicon (mostly crystalline) and an amorphous Li$_x$Si alloy (figure 3.8 – step I). Numerous studies in the literature have focused on this first step and have shown that the particles have a core–shell structure and the value of x is between 2.2 and 3.5. When the value of 3.5 is reached, the end of the first lithiation continues with a gradual increase in concentration until the composition Li$_{15}$Si$_4$ (figure 3.8 – step II) is reached. The material may here be amorphous or crystalline, depending on a very small variation of the lithium concentration around the value of 3.75 when the potential drops below 50 mV. Then, delithiation depends on whether the Li$_{15}$Si$_4$ phase is crystallized or not. Indeed, when crystallization occurs, the profile of the potential curve shows a plateau (figure 3.8 – step III & VII) which corresponds to a two-phase mechanism with the coexistence of the crystallized phase and of an amorphous phase of composition Li$_2$Si$_5$. Delithiation ends with a solid solution. When the Li$_{15}$Si$_4$ phase remains amorphous, there is no plateau (figure 3.8 – step V). Delithiation is progressive with two pseudo-plateaus at 0.3 and 0.5 V corresponding to the Li$_{2.3}$Si and Li$_{1.7}$Si phases. Whatever the mechanism, at the end of delithiation the silicon is amorphous and the following lithiation is then different from the first lithiation. Two pseudo plateaus are then observed at 0.3 and 0.5 V; they correspond to the Li$_{2.3}$Si and Li$_{3.2}$Si phases (figure 3.8 – step IV & VI).

3.3.2   Degradation Mechanisms

In spite of its interest in increasing energy density, silicon is not widely used in commercial batteries because the life expectancy of silicon-based electrodes is limited due to degradations caused by repeated changes in volume. Indeed, during charge–discharge cycles, the morphological transformations of silicon are significant

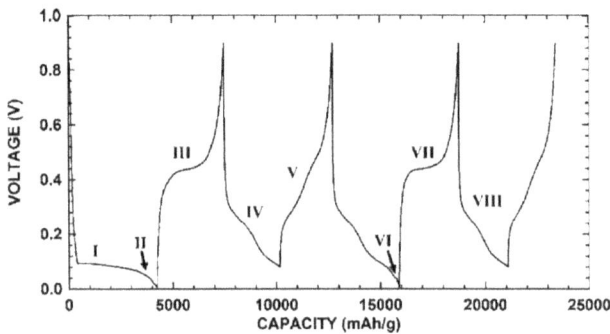

FIG. 3.8 – Potential curve of a cycled crystalline silicon-based electrode to illustrate phase changes [56].

and generate volume changes of up to 300%; and are responsible for the main degradation mechanisms (figure 3.9):

- Mechanical pulverization: the particles do not tolerate repeated volume changes and fracture. This pulverization can lead to decohesion of the particles in the percolating network as well as lithium trapping. This problem can be solved using nanometric particles. For crystalline silicon, the size limit for pulverization is 150 nm and for amorphous silicon 800 nm.
- Electrode delamination: when the polymeric binder, which ensures adhesion to the collector and cohesion between the particles of active material and conductive additive, is no longer able to contain macroscopic deformation, some of the active material may become electrically insulated. Many works in the literature are focused on the use of alternative binders to limit this phenomenon.
- Instability of the passivation layer: at each charge–discharge cycle, the SEI is partially reformed. Thus, the decomposition products can lead to particle trapping or a strong decrease in the porosity of the electrodes. Most importantly, it continuously consumes lithium. The use of fluoroethylene carbonate, which decomposes before the solvents conventionally used, stabilizes the SEI, but another way for improvement is the development of new silicon-based materials; this is the subject of the following paragraph.

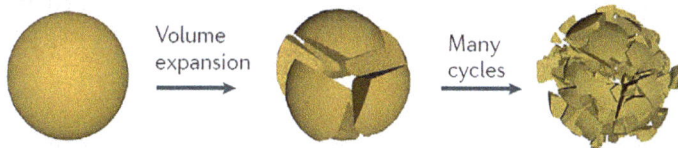

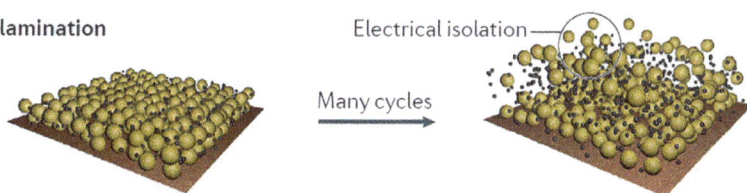

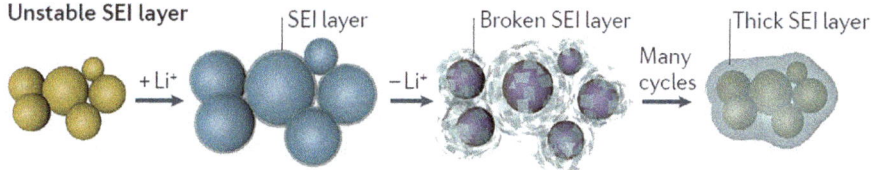

FIG. 3.9 – Main degradation mechanisms of silicon anodes due to volume changes [57].

3.3.3 Material Improvement Approaches

Over the last twenty years, a large number of studies have focused on the development of silicon-based materials. The research work has progressively led to the development of composite materials that contain silicon in a matrix in order to reduce the interface with the electrolyte and limit the impact of volume changes at active material particles and the electrode scale. To this end, various research approaches have been considered:

- SiO_x

  Silicon oxides, which are composed of silicon nano-domains in a SiO_2 matrix according to the "Random Mixture" structural model, are the first to be commercialized in Li-ion cells (*e.g.* Panasonic). In commercial batteries they are combined with graphite because although the reversible capacity is potentially high (1300–1500 mAh.g^{-1}), the irreversible capacity is also high ($\sim 50\%$); the latter is due to structural changes that lead to the formation of stable Li_4SiO_4 and Li_2O phases. Various strategies have been explored to reduce the initial irreversible capacity by combining the oxide with amorphous carbon and/or graphite. There are different methods for preparing these materials: vacuum reduction of SiO_2 with carbon, SiO_2 reduction according to temperature and pressure conditions or, the most common method, by sublimation of an Si + SiO_2 mixture. The main industrial players that have filed patents on the topic are located in Japan (Shinetsu Chemical, Osaka Titanium Technologies, Sumitomo Titanium, Matsushita Electric, Japan Storage Battery and Panasonic) and Korea (Samsung SDI). These patents relate to the synthesis and improvement of these materials.

- Silicon–carbon composites

  A lot of work in the literature is devoted to the development of silicon-carbon composites, where carbon can be in the form of graphite, amorphous carbon, graphene or carbon nanotubes. Carbon, depending on the form used, is interesting for several reasons:

  - Mechanical buffering effect of graphite, which also shows small volume changes ($<10\%$).
  - Good ionic and electronic conductivity.
  - Reduction of electrochemical sintering.
  - Possible impact on the composition of the "SEI".

These materials are obtained by various processes: grinding with graphite, mixing with a carbon precursor and then pyrolysis, combination of graphite and carbon precursor by grinding and then pyrolysis, spray-drying and other syntheses, particularly in the case of the preparation of complex nanostructures. The development of these materials requires the study of many parameters such as the type of silicon used and its dispersion or the choice of the organic precursor of amorphous carbon.

- Alloys:

    Since the early 2000s, silicon-based alloys have been developed for Li-ion applications. The number of publications remains lower than those related to silicon-carbon composites or complex nanostructures, but some studies are regularly published. Between 2000 and 2010, the materials studied were rather binary silicides ($CaSi_2$, $NiSi_2$, Mg_2Si...) prepared by mechanosynthesis with various specific capacities and limited performances; in recent years, multi-element alloys with new synthesis processes such as melt spinning have been developed, with interesting performances, in particular those of Si–Ti–Ni alloys developed by Samsung SDI. At the same time, during the same period, the American company 3 M has been working continuously on the development of this type of material. It seems that the material (which was closed to the industrialisation) is made of an inactive Si–Al–Fe matrix, an active silicon phase and another active phase made of tin and a rare metal. 3M however claimed to have abandoned this activity in 2019.

3.4 Lithium Metal

Lithium is the lightest solid element in the periodic table, with a molar mass of 6.941 g.mol^{-1} and a density of 0.53 g.cm^{-3}, which confers an attractive specific storage capacity of 3860 mAh.g^{-1}. It is the most reducing metallic element, with an electrochemical potential (Li^+/Li couple) of -3.04 V *vs.* SHE, which makes it possible to obtain large potential differences when combined with a well-chosen positive electrode. These advantages make the lithium metal electrode an anode of choice for high energy density post lithium ion technologies. It is estimated (figure 3.10) that the use of a lithium metal anode, in combination with high energy density cathode materials (*e.g.* NMC, sulphur), should surpass the available energy densities of lithium ion (*e.g.* with silicon or graphite) [58].

The first work on the development of lithium batteries was initiated by G.N. Lewis in 1912, while research on the subject accelerated in the 1970s, thanks in particular to the work of S. Whittingham, with the first rechargeable Li-metal battery marketed by Exxon (Li/TiS_2 couple) [59]. However, the commercialization of Li-metal cells was quickly abandoned in favour of Li-ion technology due to the safety problems encountered during use. One of the main Li-metal players at the time, *i.e.* Moli Energy, had to recall more than 1.5 million batteries in 1989 due to problems of explosions and burns from users [60]. The lithium metal electrode is known to lead to the formation of dendrites during charging (lithium deposition), which can lead to short circuit, heating of the battery and thus to safety problems [61].

Due to these safety issues, the development of Li-metal batteries was largely slowed down in the 1990s. Only primary lithium batteries (non-rechargeable), which do not cause problems with dendrites, and "Li-metal Polymer" batteries, developed and marketed by the company BlueSolutions, remained [62]. However, the use of lithium metal as an anode material has motivated a renewed interest since 2015 [63–65] due to the research and development work on "post-lithium-ion" technologies, such as lithium/sulphur, lithium/air, "all-solid" batteries, which are supposed

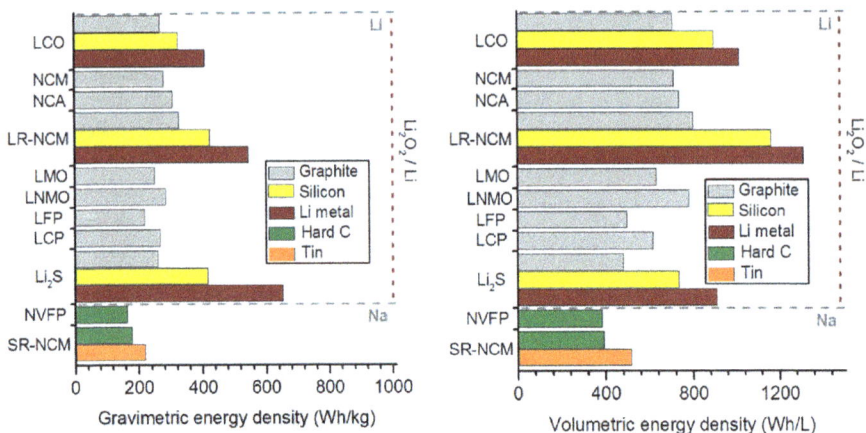

FIG. 3.10 – Estimations of gravimetric (left) and volumetric (right) energy densities at cell scale for different combinations of electrode materials [58].

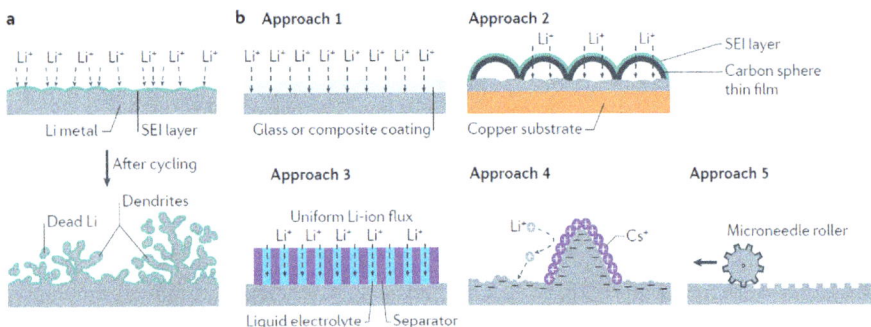

FIG. 3.11 – Failure mechanisms (a) and strategies (b) developed to minimize dendrite formation and improve the lithium-metal electrode interface [57].

to offer advantageous perspectives in terms of energy density. The work focuses in particular on the development of protection strategies (polymers, ceramics, carbon) or texturing of the lithium-metal electrode [66], or on the use of electrolyte additives (figure 3.11) [57].

At this time, graphite remains the negative electrode material of reference. Its relatively high specific capacity and its operating potential close to lithium metal allow energy densities to be maximized. In addition, the material is inexpensive and non-toxic. It therefore remains difficult to replace since it also offers a very good cycle life. Nevertheless, there are alternatives such as silicon-based materials, in particular to increase the energy density (the expected gain being of the order of 20%–30% in gravimetric and volumetric energy density). However, the emergence of silicon-based solutions (silicon-carbon, alloy or SiO_x composites) requires electrode degradation to be limited in order to improve cycle life. Numerous studies on the subject highlight the difficulty of stabilizing the "SEI" and even if very interesting performances are obtained, it is essential to consider the material in the entire system by taking into account the interactions with the binder, the electrolyte and the cathode material as well as the design of the cell and its use. A significant gain in energy density ($>50\%$) may be possible thanks to the development of so-called "post-lithium-ion" technologies, which are still at the R&D level: "all-solid" batteries; lithium/sulphur; lithium/air... For these technologies, the anode of choice remains lithium metal: this will be of renewed interest if a solution is found to the problem of its limited cycle life.

Bibliography

[1] Avicenne Energy – The Rechargeable Battery Market: Value Chain and Main Trends 2019-2030-IBS – 11 March, 2021.

[2] Goriparti S., Miele E., De Angelis F., Di Fabrizio E., Proietti Zaccaria R., Capiglia C. (2014) *J. Power Sources* **257**, 421.

[3] Poizot P., Laruelle S., Grugeon S., Dupont L., Tarascon J.-M. (2000) *Nature* **407**, 496.

[4] Obrovac M.N., Chevrier V.L. (2014) *Chem. Rev.* **114**, 11444.

[5] Lefrant S., Bernier P. (1997) *Le carbone dans tous ses états.* Taylor & Francis, 584 pages, ISBN: 9789056990572.

[6] Marchand A. *et al.* (1984) Comité international pour la caractérisation et la terminologie du carbone, publication de 30 définitions, *Carbon* **22**, 629.

[7] Marchand A. *et al.* (1986) Comité international pour la caractérisation et la terminologie du carbone, publication de 24 définitions, *Carbon* **24**, 657.

[8] Marchand A. *et al.* (1986) Comité international pour la caractérisation et la terminologie du carbone, publication de 14 définitions, *Carbon* **24**, 775.

[9] Fredenhagen K., Cadenbach G. (1926) Die Bindung von Kalium durch Kohlenstoff, *Zeitschrift für anorganische und allgemeine Chemie* **158**, 249.

[10] Hérold A. (1955) *Bull. Soc. Chim.* **187**, 999.

[11] Guérard D., Hérold A. (1975) *Carbon* **13**, 337.

[12] Armand M.B. (1973) "Fast Ion Transport in Solids", Solid State Batteries Devices, Proceedings of the NATO Sponsored Advanced Study Institute, pp. 665–673.

[13] Besenhard J.O. (1976) *Carbon* **14**, 111.

[14] Yazami R., Touzain P. (1983) *J. Power Sources* **9**, 365.

[15] Besenhard J.O., Fritz H.P. (1983) The electrochemistry of black carbons, *Angew. Chem. Int.* **22**, 950.

[16] Dey A.N., Sullivan B.P. (1970) The electrochemical decomposition of propylene carbonate on graphite, *J. Electrochem. Soc.* **117**, 222.

[17] Winter M., Moeller K.-C., Besenhard J.O. (2003) Carbonaceous and graphitic anodes, Chap. 5, *Lithium batteries, science and technology* (G.A. Nazri, G. Pistoia, Eds), pp. 144–194, ISBN: 978-1-4020-7628-2.

[18] Winter M., Besenhard J.O., Spahr M.E., Novak P. (1998) Insertion electrode materials for rechargeable lithium batteries, *Adv. Mater.* **10**, 725.

[19] Source USGS.

[20] "Understanding the Worldwide 'Graphite' Market", TREM12, Critical Material for Energy & Security March 13–14, 2012, Ashbury Carbons.

[21] Shaw S. (Roskill Information Services), "Graphite demand growth: the future of lithium-ion batteries in EVs and HEVs", 37th ECGA General Assembly, avril 2013.

[22] Prem Kumar T., Sri Devi Kumari T., Manuel Stephan A. (2009) Carbonaceous anode materials for lithium-ion batteries – the road ahead [Revue], *J. Indian Inst. Sci.* **89**, 393.

[23] Schiller C., Méring J., Oberlin M. (1969) Etude d'un Carbone Dur (Coke de Saccharose). Effets des Traitements Thermiques, *J. Appl. Cryst.* **1**, 282.

[24] Kohs W., Santner H.J., Hofer F., Schröttner H., Doninger J., Barsukov I., Buqa H., Albering J. H., Möller K.-C., Besenhard J.O., Winter M. (2003) A study on electrolyte interactions with graphite anodes exhibiting structures with various amounts of rhombohedral phase, *J. Power Sources* **119–121**, 528.

[25] Huang H., Liu W., Huang X., Chen L., Kelder E.M., Schoonman J. (1998) Effect of a rhombohedral phase on lithium intercalation capacity in graphite, *Solid State Ionics* **110**, 173.

[26] Simon B., Flandrois S., Guérin K., Fevrier-Bouvier A., Teulat I., Biensan P. (1999) *J. Power Sources* **81–82**, 312.

[27] Shi H., Barker J., Saïdi M.V., Koksbang R. (1996) Structure and lithium intercalation properties of synthetic and natural graphite, *J. Electrochem. Soc.* **143**, 3466.

[28] Spahr M.E., Wilhelm H., Joho F., Panitz J.-C., Wambach J., Novak P., Dupont-Pavlovsky N. (2002) Purely hexagonal graphite and the influence of surface modifications on its electrochemical lithium insertion properties, *J. Electrochem. Soc.* **149**, A960.

[29] Ng S.H., Vix-Guterl C., Bernardo P., Tran N., Ufheil J., Buqa H., Dentzer J., Gadiou R., Spahr M.E., Goers D., Novak P. (2009) Correlations between surface properties of graphite and the first cycle specific charge loss in lithium-ion batteries, *Carbon* **47**, 705.

[30] Anab S.J., Li J., Daniel C., Mohanty D., Nagpure S., Wood III D.L. (2016)The state of understanding of the lithium-ion-battery graphite solid electrolyte interphase (SEI) and its relationship to formation cycling [Revue], *Carbon* **105**, 52.

[31] Suzuki K., Hamada T., Sugiura T. (1999) Effect of graphite surface structure on initial irreversible reaction in graphite anodes, *J. Electrochem. Soc.* **146**, 890.

[32] Peled E., Menachem C., Bar-Tow D., Melman A. (1996) Improved graphite anode for lithium-ion batteries, *J. Electrochem. Soc.* **143**, L4.

[33] Novák P., Ufheil J., Buqa H., Krumeich F., Spahr M.E., Goers D., Wilhelm H., Dentzer J., Gadiou R., Vix-Guterl C. (2007) The importance of the active surface area of graphite materials in the first lithium intercalation, *J. Power Sources* **174**, 1082.

[34] Gallego N.C., Contescu C.I., Meyer III H.M., Howe J.Y., Meisner R.A., Payzant E.A., Lance M.J., Yoon S.Y., Denlinger M., Wood III D.L. (2014) Advanced surface and microstructural characterization of natural graphite anodes for lithium ion batteries, *Carbon* **72**, 393.

[35] Zaghib K., Song X., Guerfi A., Rioux R., Kinoshita K. (2003) Purification process of natural graphite as anode for Li-ion batteries: chemical versus thermal, *J. Power Sources* **119–121**, 8.

[36] Peled E., Golodnitsky D., Ulus A., Yufit V. (2004) *Electrochim. Acta* **50**, 391.

[37] Suzuki K., Hamada T., Sugiura T. (1999) Effect of graphite surface structure on initial irreversible reaction in graphite anodes, *J. Electrochem. Soc.* **146**, 890.

[38] Wang H., Ikeda T., Fukuda K., Yoshio M. (1999) Effect of milling on the electrochemical performance of natural graphite as an anode material for lithium-ion battery, *J. Power Sources* **83**, 141.

[39] Zaghib K., Nadeau G., Kinoshita K. (2000) Effect of graphite particle size on irreversible capacity loss, *J. Electrochem. Soc.* **147**, 2110.

[40] Lux S.F., Placke T., Engelhardt C., Nowak S., Bieker P., Wirth K.-E., Passerini S., Winter M., Meyer H.-W. (2012) Enhanced electrochemical performance of graphite anodes for lithium-ion batteries by dry coating with hydrophobic fumed silica, *J. Electrochem. Soc.* **159**, A1849.

[41] Zhou Y.F., Xie S., Chen C.H. (2005) Pyrolytic polyurea encapsulated natural graphite as anode material for lithium ion batteries, *Electrochim. Acta* **50**, 4728.

[42] Tsumura T., Katanosaka A., Souma I., Ono T., Aihara Y., Kuratomi J., Inagaki M. (2000) Surface modification of natural graphite particles for lithium ion batteries, *Solid State Ionics* **135**, 209.

[43] Zhao H., Ren J., He X., Li J., Jiang C., Wan C. (2008) Modification of natural graphite for lithium ion batteries, *Solid State Sci.*, **10**, 612.

[44] Sethuraman V.A., Hardwick L.J., Srinivasan V., Kostecki R. (2010) Surface structural disordering in graphite upon lithium intercalation/deintercalation, *J. Power Sources* **195**, 3655.

[45] Groult H., Nakajima T., Perrigaud L., Ohzawa Y., Yashiro H., Komaba S., Kumagai N. (2005) Surface-fluorinated graphite anode materials for Li-ion batteries, *J. Fluor. Chem.* **126**, 1111.

[46] Hoshi K., Ohta N., Nagaoka K., Bitoh S., Yamanaka A., Nozaki H., Okuni T., Inagaki M. (2009) Production and advantages of carboncoated graphite for the anode of lithium ion rechargeable batteries, *Tanso* **240**, 213.

[47] Zhang S.S., Xu K., Jow T.R. (2004) Enhanced performance of natural graphite in Li-ion battery by oxalatoborate coating, *J. Power Sources* **129**, 275.

[48] Ohta N., Nagaoka K., Hoshi K., Bitoh S., Inagaki M. (2009) Carbon-coated graphite for anode of lithium ion rechargeable batteries: graphite substrates for carbon coating, *J. Power Sources* **194**, 985.

[49] Wan C., Li H., Wu M., Zhao C. (2009) Spherical natural graphite coated by a thick layer of carbonaceous mesophase for use as an anode material in lithium ion batteries, *J. Appl. Electrochem.* **39**, 1081.

[50] Zou L., Kang F., Zheng Y.-P., Shen W. (2009) Modified natural flake graphite with high cycle performance as anode material in lithium ion batteries, *Electrochim. Acta* **54**, 3930.

[51] Wu Y.P., Rahm E., Holze R. (2003) Carbon anode materials for lithium ion batteries, *J. Power Sources* **114**, 228.

[52] Joho F., Rykart B., Blome A., Novák P., Wilhelm H., Spahr M.E. (2001) Relation between surface properties, pore structure and first-cycle charge loss of graphite as negative electrode in lithium-ion batteries, *J. Power Sources* **97–98**, 78.

[53] Flandrois S., Simon B. (1999) Carbon materials for lithium-ion rechargeable batteries [Revue], *Carbon* **37**, 165.

[54] Fujimoto H. (2010) Development of efficient carbon anode material for a high-power and long-life lithium ion battery, *J. Power Sources* **195**, 5019.

[55] Buiel E. 1998 "Lithium insertion in hard carbon anode materials for Li-ion batteries", Ph.D. thesis, Dalhousie University, Halifax, Nova Scotia (Canada).

[56] Obrovac M.N., Krause L.J. (2007) *J. Electrochem. Soc.* **154**, A103.

[57] Choi J.W., Aurbach D. (2016) *Nat. Rev. Mater.* **1**, 16013.

[58] Berg E.J., Villevieille C., Streich D., Trabesinger S., Novak P. (2015) *J. Electrochem. Soc.* **162**, A2468.

[59] Whittingham M.S. (1976) *Science* **192**, 1126.

[60] Whittingham M.S., "Nanomaterials as anodes for lithium batteries", *Lithium Batteries Discussion 5th*, Arcachon, France, 14 juin 2011.

[61] Orsini F., Du Pasquier A., Beaudoin B., Tarascon J.M., Trentin M., Langenhuizen N., De Beer E., Notten P. (1998) *J. Power Sources* **76**, 19.

[62] https://www.blue-solutions.com/blue-solutions/technologies/batteries-lmp/.

[63] Zhang X.-Q., Cheng X.-B., Zhang Q. (2017) *Adv. Mater. Interfaces*, 1701097.

[64] Zhang J.-G., Xu W., Henderson W.A., Li metal anodes and rechargeable lithium metal batteries, *Springer series in materials science*, ISBN: 978-3-319-44053-8.

[65] Placke T., Kloepsch R., Dühnen S., Winter M. (2017) *J. Solid State Electrochem.* **21**, 1939.

[66] Xu W., Wang J., Ding F., Chen X., Nasybulin E., Zhangad Y., Zhang J.-G. (2014) *Energy Environ. Sci.* **7**, 513.

Chapter 4

Organic Electrode Materials

Philippe Poizot, Saïd Sadki and Thibaut Gutel

This chapter describes a new type of electrode materials for secondary batteries based on organic electroactive compounds which could substitute in a near future some inorganic active materials (see the 2 last chapters). After introducing the most important challenges related to their use in electrochemical devices, we will present the five types of organic active materials before reviewing the main strategies to efficiently implement them into organic batteries.

Introduction

In the context of decreasing mineral resources associated with constantly growing technical and economic constraints, the use of organic electroactive materials presents an interesting alternative to conventional ones in secondary batteries.

Indeed, these compounds could not only offer similar or even better electrochemical performances in comparison with their inorganic counterparts (theoretical energy density until 300 Wh.kg^{-1}, high power density with charge/discharge rate of up to 50 C, figure 4.1a) but could also reduce the environmental footprint and the cost of energy storage systems. Constituted by naturally abundant elements (C, H, O, N and S), organic electrode materials could be synthesized from low cost and potentially biosourced precursors using green chemistry techniques (low power consumption, aqueous synthesis,...) (figure 4.1b) [1].

Moreover, the diversity of organic compounds combined with the versatility of organic chemistry offers the possibility to design a huge number of structures with tunable physico-chemical and electrochemical properties that could be adapted to the requirements of targeted applications. Various electroactive organic moieties also enable to imagine new cell configurations (figure 4.2) based on different insertion mechanisms and playing with cations (inorganic mono- or divalent cations such as Li$^+$, K$^+$, Na$^+$ or Mg^{2+}... but also organic ones like R$_4$N$^+$ or R$_3$P$^+$) charge

DOI: 10.1051/978-2-7598-2555-4.c004

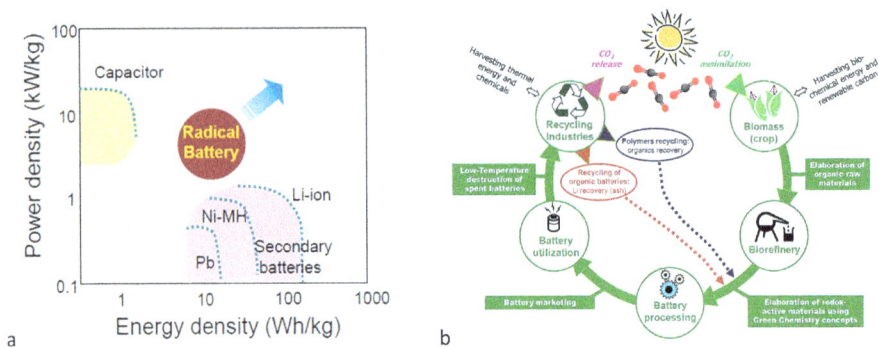

a   b

FIG. 4.1 – (a) Ragone plot showing Organic Radical Batteries (ORB) among various other electrochemical energy storage technologies [2]. (b) Cycle life of a "greener" Li-ion battery taking benefit of redox-active organic electrode materials [1].

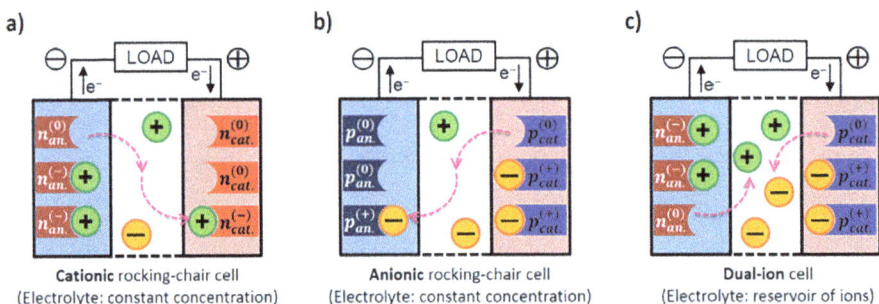

FIG. 4.2 – Various battery configurations: "Cation-ion" (a), "Anion-ion" (b) et "Dual-ion" (c) [6].

balance[1] but also anions[2] (Cl^-, Br^-, PF_6^-, BF_4^-, $TFSI^-$, etc....) or even using both simultaneously in the so-called "dual-ion" cell [4, 5]. In this last case, the electrolyte constitutes an ion reservoir. From these points of view, inorganic electrode materials do not offer such diversity.

Finally, these compounds offer some new technological opportunities in terms of applications: flexible batteries for portable electronics, aesthetical batteries for their integration in urban furniture, biodegradable and/or disposable devices for smart packaging market... and also pave the way for new implementation processes: coating of slurries or solution, printing and also processing techniques coming from plastic industries such as injection molding, extrusion, etc. Two types of organic compounds

---

[1] n-type mechanism (System *B* according Hünig classification) [3].

[2] p-type mechanism (or System *A* according Hünig classification) [3].

have been proved so far promising for electrochemical energy storage using solid-state electrodes: conjugated (conducting) polymers and organic structures that possess some electroactive functions (note that both types could be combined into one material). However, organic electrode materials still do not emerge into commercial products because they suffer from two major limitations: low cycle life usually related to significant solubility into common liquid electrolyte of battery and low electronic conductivity that requires the introduction of high amount of carbon additives (*i.e.* above 15%wt). Finally, we also have to mention that these compounds present very low densities, which limit volumetric energy density.

4.1 Different Types of Organic Electrode Materials

4.1.1 π-Extended System (Conducting Polymers)

Intrinsic semi-conductor polymers are characterized by the presence of extended π-conjugated system (A. J. Heeger, A. G. MacDiarmid, H. Shirakawa, Nobel prizes of Chemistry in 2000). They can be synthesized by chemical or electrochemical (electropolymerization) techniques [7–11]. The main examples are shown in figure 4.3.

Stille and Suzuki's (Nobel prizes of Chemistry in 2010) coupling methods lead to a renewed interest into these materials by offering the possibility to improve their molecular weight and polydispersity index. These polymers could be reversibly oxidized (p-doped) or reduced (n-doped) [12] (for a global overview on conducting polymers, see [3, 11–14]). At doped state, these conjugated polymers are intrinsically electronic conductors. Moreover, they can be used as electrode material for secondary batteries. Works on electronic conducting polymers have been performed with this objective during the last 25 years [13–18]. As the literature is already abundant in this field since 80's, the following part of this chapter describes the most important applications for each type and shows significant progress (composite materials constituted by these polymers are also included).

4.1.1.1 Polypyrrole Derivatives (PPy)

Thanks to its high electronic conductivity and its strong mechanical and chemical stability, polypyrrole is one of the first polymers studied as positive electrode material for secondary batteries. It is not stable at n-doped state and consequently

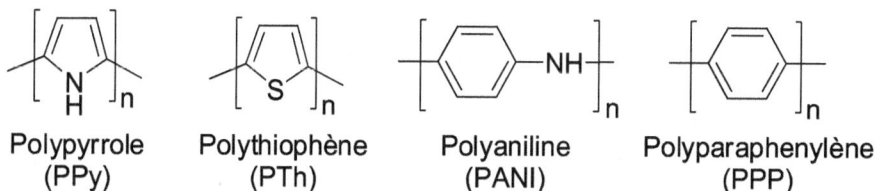

Polypyrrole   Polythiophène   Polyaniline   Polyparaphenylène
(PPy)   (PTh)   (PANI)   (PPP)

FIG. 4.3 – Main types of conducting polymers.

could not be used as negative electrode materials [19, 20]. Many examples of batteries based on PPy have been described in the literature since 80's and show a good cycle life and an interesting theoretical energy density (300 Wh.kg^{-1} with graphite based counter electrodes) [21]. At the end of 80's, VARTA/BASF consortium produced some prototypes constituted by PPy/Li [17]. The cycle life of these systems is around 400 cycles with a loss of capacity to about 10%. The coulombic efficiency is very close to 100% and after 1000 cycles, the specific capacity decreases only 20%.

4.1.1.2 Polythiophene Derivatives (PTh)

Polythiophene (PTh) presents excellent stability at p-doped and neutral state and high electrochemical properties. PTh shows moderate stability at n-doped state and could be used as negative electrode materials. One of the first examples based on electropolymerized PTh leads to some specific capacities of 100 mAh.g^{-1} and a doping ratio of 0.24–0.30 electrons per monomer [22, 23]. Li/PTh system is characterized by several voltage plateaus around 3 V. One of the most important thiophene derivatives is 3,4-ethylenedioxythiophene that could be polymerized chemically or electrochemically. The corresponding polymer presents a high specific capacity as electrode material (about 691 mAh.g^{-1} during the first cycle) [24]. This capacity decreases during the second cycle and becomes stable at about 350 mAh.g^{-1} after 44 cycles.

4.1.1.3 Polyaniline Derivatives (PANI)

Since 1985, the following works done by MacDiarmid (Nobel prize of Chemistry) *et al.* [25], PANI appears as a promising material due to its specific redox (doping) and acid–base properties. Indeed PANI is air-stable and offers high and tunable conductivity [26, 27]. The results show that PANI could be a good candidate as electrode material for batteries. Its discharge capacity could be closed to the theoretical value and a maximum voltage of about 4.2 V *vs.* Li$^+$/Li could be obtained with PANI. Taguchi *et al.* have demonstrated the stability of PANI for more than 500 cycles with a current density of 0.1 mA.cm^{-2} and a specific capacity of 83 mAh.g^{-1} [28]. Recently a new and original lithiated version of this polymer has been synthesized leading to a capacity of 230 mAh.g^{-1} and the possibility to be implemented in thick electrodes [29].

4.1.2 Stable Radical

These compounds in particular nitroxide radical (R-NO$^•$), well-known in the field of magnetism and for ESR and NMR spectroscopy, could be oxidized and reduced (bipolar behavior, figure 4.4a). In the case of 2,2,6,6-tetramethyl-1-piperidinyloxy (TEMPO, figure 4.4), only oxidation process is fully reversible involving a charge compensation by an anion. With a relatively high redox voltage (~ 3.6 V *vs.* Li$^+$/Li), this system also benefits of an ultra-fast electrochemical kinetic well-adapted to high power applications. Unfortunately, the molecular weight of the redox unit

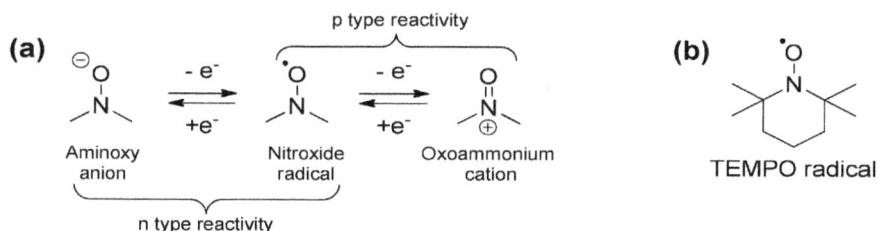

FIG. 4.4 – (a) (bipolar) redox reactivity of nitroxide radical and (b) structure of TEMPO radical.

limits the energy density of so-called "ORB" (Organic Radical Battery) based on this type of positive electrode materials.

At molecular state (non polymeric), these structures are highly soluble into common organic electrolytes and their use requires the implementation of immobilization strategies (see section 4.2).

4.1.3 Organosulfides & Thioethers

At the origin of Li-sulfur technology, the disulfide bond (–S–S–) could be integrated into an organic backbone and consequently used for electrochemical storage according to a bond breaking/formation mechanism associated to the generation of negative charge balanced by a cation (n-type) as shown in figure 4.5.

Although the introduction of an organic skeleton (electrochemically inert in most of the cases) increases the molecular weight of the final compound and consequently decreases the specific capacity in comparison with elemental sulfur, it also offers the opportunity to tune the physico-chemical (solubility into electrolytes) and electrochemical (redox potential and electrochemical kinetics) properties of disulfide based products. These compounds, known as thiol, usually show high specific capacity and stable cycle life, but still suffer from slow electrochemical kinetics and low electronic conductivity [30].

4.1.4 Carbonyl Functions

The reduction of conjugated carbonyl structure (figure 4.6) is usually a reversible phenomenon, well established in molecular electrochemistry.

This n-type electrochemical reaction could be used in Li-ion or Na-ion cells. Quinones, which are well known for their involvement in various biological processes based on electron transfer, represent the most famous examples and offer many

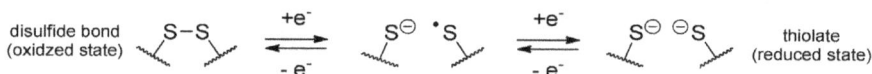

FIG. 4.5 – Redox reactivity of disulfide bond.

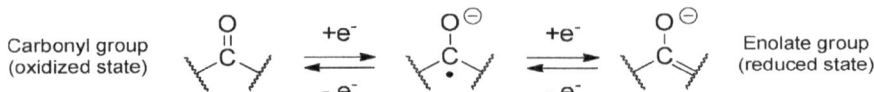

FIG. 4.6 – Redox activity of carbonyl group.

opportunities to (eco-)design electroactive molecules prepared using environmentally friendly syntheses [1, 31–33]. Nowadays, several quinones (in molecular or polymeric forms) have been evaluated as active materials for positive electrodes (charge state of the cell). Their electrochemical activity occurs in 2.5–3 V *vs.* Li$^+$/Li voltage windows, but their redox potential could be increased or decreased playing with inductive/mesomeric effects. Even if only rare examples have been reported, it is also possible to prepare these compounds directly in reduced forms (lithiated or sodiated enolates) which allows the same working configuration used with conventional electrode materials such as LiFePO$_4$ [5, 34, 35]. It should also be mentioned that quinone-like redox structures could be modified in order to directly increase redox voltage such as in tetracyanoquinondimethane (TCNQ) [36]. The carbonyl bond is also involved in imide and carboxylate electroactive functions leading to more cathodic voltage (<2.5 V *vs.* Li$^+$/Li) and offers the possibility to assemble full organic cell in cation-ion configuration [35, 37] in particular in the case of conjugated carboxylates in which redox potentials could be closed to 0.65 V *vs.* Li$^+$/Li [38]. To sum up, the use of redox carbonyl functions offers real advantages (tunable potential, bio-sourcing) for future battery developments.

4.1.5 Aromatic Amines

As in the well-known example of polyaniline, aromatic amines could also present a p-type (anion uptake) reversible electrochemical activity at a voltage around 3.5 V *vs.* Li$^+$/Li (figure 4.7).

Even if many researches have investigated this polymer, it has been only very recently demonstrated that this function could be also electroactive in non-polymeric structures (molecular polyanionic or macrocyclic organic compounds [39–41].

4.2 Implementation Strategies

Most of the research groups actually focus on the development of strategies to prevent the most promising organic redox functions from dissolution and very few works are interested in low electronic conductivity issues. Two main approaches

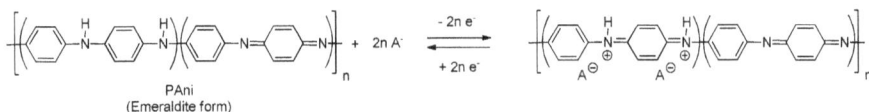

FIG. 4.7 – Redox reaction of PANI.

have been studied in the literature: immobilization of organic or inorganic structures on various substrates and chemical modifications in particular the preparation of multi-charged (organic polyanions) systems leading to insoluble species and intermediates.

4.2.1 Grafting on Inorganic or Organic Support

4.2.1.1 Functionalization of Neutral Polymers

- *Disulfide polymers*: In order to prevent molecular compounds from dissolution, polymeric analogues incorporating an intramolecular disulfide bond have been developed in particular dimercaptothiadiazole (DMcT) and polyamides derivatives [42–44]. They store cations *via* a depolymerization/polymerization mechanism and their intermediates are generally soluble into electrolytes.
- *Nitroxyde polymers*: In collaboration with NEC company, Nishide's team (Waseda University, Japan) since 2002 started some pioneer work about the functionalization of various polymer backbones [14], in particular polymethacrylate, polyether, polynorbornene and polystyrene with nitroxide radical. The performances of these materials are really impressive at high C-rate (50 C–100 C corresponding to charge/discharge in few minutes) even if limited specific capacities are reported. Batteries based on this technology are the most advanced ones and NEC announced their commercialization since 2013, but they have never been put on the market until now [45].
- *Carbonyl based polymers*: Neutral structures based on electroactive carbonyl such as quinones generally offer fast decreasing electrochemical performances along cycling due to their high solubility into conventional liquid electrolytes. Obviously, the first investigation started by focusing on polymeric analogues and quite promising performances has been obtained using polyquinone and polyimides based structures. Many examples are now reported in the literature [31].

4.2.1.2 Functionalization of Conducting Polymers

Grafting redox moieties on conducting polymer backbones which are intrinsically electronic conductors has been also recently considered even if many works have been performed in 80's [46]. The most simple synthetic strategy consists in functionalization of the monomer by redox group before (electro-) polymerizing them. Various redox groups have been grafted on conjugated chain such as TEMPO derivatives, quinones, etc.... The objective is to simultaneously increase the specific capacity and, therefore, the energy of the batteries, the electronic conductivity and the electrochemical stability of electroactive polymers. Indeed the huge interest generated by the discovery of conducting polymers in the 80's was hindered by fast capacity decay and high self-discharge. Grafting redox group is only interesting if redox potentials of the electroactive function match with the electrochemical window in which conducting polymers are p- or n-doped. Some of them can even limit the over-voltage avoiding irreversible oxidation/reduction of the conducting polymer skeletons acting like a redox shuttle. Finally, electronic conducting polymers are generally poorly soluble in the electrolyte for battery.

This part does not pretend to be exhaustive as the nature of the grafted group is largely diverse and because many combinations are possible but will focus on the most relevant examples from literature.

- *Conducting polymers functionalized by TEMPO group*: Aydin *et al.* reported TEMPO grafted polythiophene in position β, β' [47]. The obtained performance of 79 mAh.g^{-1} corresponds to 83% of the theoretical capacity on a voltage window of 2.5–3.8 V. This work shows the possibility to use polythiophene-TEMPO as electrode material, but stays at the prospective stage. The synthesis of EDOT bearing a TEMPO radical has also been recently published by Armand *et al.* [48].

- *Conducting polymers functionalized by disulfide bond*: Amaike *et al.* developed a new pyrrole derivative, bearing an intramolecular cyclic disulfide bonds [49]. The first discharge shows a flat plateau at 2.5 V corresponding to a specific capacity of about 398 mAh.g^{-1} which is very close to the theoretical one. Abruna *et al.* have also functionalized PEDOT with disulfide bonds leading to good results in the case of the implementation of PEDOT-DMcT based positive electrodes [50–52]. A working voltage of 3 V with specific capacities of 130 mAh.g^{-1} was reported. PEDOT skeleton leads to a rigid character keeping thiolate units close enough to rebuild effectively disulfide bond during the charge processes as described in the following figure (figure 4.8).

- *Conducting polymers functionalized by quinones derivatives*: Novak *et al.* synthesized poly(5-amino-1,4-naphthoquinone) and assembled full cell using graphite based negative electrode. The obtained mean voltage was 2.6 V and the specific energy about 100 Wh.kg^{-1}. However, the authors observed a very fast capacity decrease along cycling which needs to be elucidated [53]. Some other electronically conducting and redox materials are described in the review of Holze *et al.* [54].

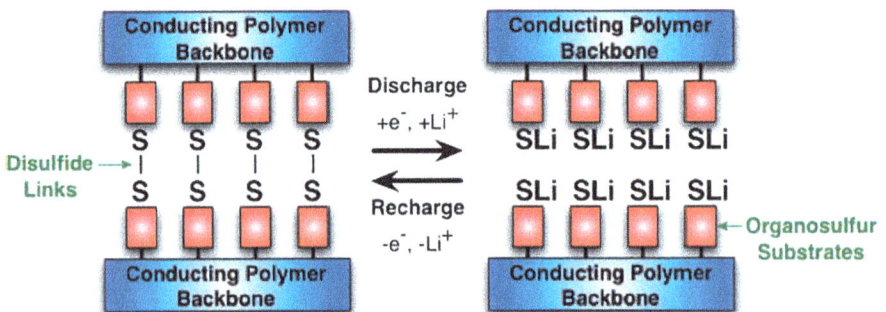

FIG. 4.8 – Schematic view of the concept of conducting polymers crosslinked by disulfide bonds [50].

4.2.1.3 Stabilization by Physisorption

As the dissolution issue is one of the main challenges encountered in the development of Li/S battery, some efficient approaches developed for this technology have also been implemented for organic electrode materials. In particular, physisorption, which uses the intermolecular forces that could intervene with a substrate such as alumina or carbon materials, has been applied to small organic molecules (quinones) presenting high specific capacity and also a strong tendency to dissolve in apolar protic media. With alumina, few studies have shown some clear limitations in the capacity decay along cycling [55, 56]. However, the introduction of alumina into composite electrode constitutes a dead mass. Better approaches consist in using carbon properties in particular new morphologies of carbons with high specific surface (activated carbon, nanotubes and graphene). In addition to bring better electronic percolation inside the composite electrode, some π-stacking interactions could take place with many organic electrode materials presenting an aromatic character [57–59]. However, the total carbon content still stay relatively high ($\sim 40\%$wt) and the electrode preparation process is quite complex.

4.2.2 Polyanionic Salt Formation

Liquid electrolytes used in Li- and Na-ion batteries being mainly polar and aprotic, unfavorable anion solvation is expected (no possible H-bond). Therefore, the use of organic salts (polyanionic organic structures) leads to low intrinsic solubilities [60]. The more permanently negative charges are present on the organic backbone (whatever the oxidation state); the lower the solubility of this compound, and consequently the higher the cycle life. According to this approach, many carbonyl based salts (n-type) have been synthesized in lithiated and sodiated forms: enolates, imidates and carboxylates which cover a large voltage window ($\Delta E > 2$ V) [1, 31, 33]. Several representative examples are shown in figure 4.9. Moreover, organic chemistry offers the opportunity to combine, for example, enolate and carboxylate functions in the same structure opening the possibility to synthesize a unique compound that could simultaneously play the role of positive and negative electrode material and enable to assemble fully organic batteries (group n°6 of the figure 4.9).

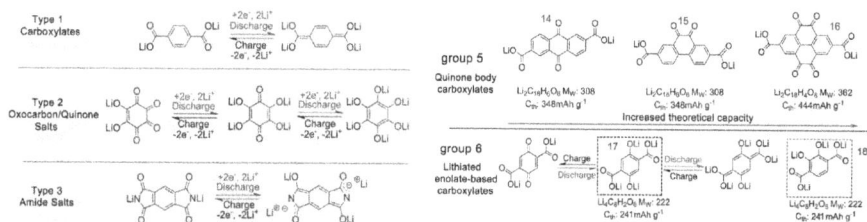

FIG. 4.9 – Representative examples of polyanionic organic structures based on C = O/C–O electroactive functions [33].

Conclusion

The use of conducting polymers as electrode material has been proposed since the 80's and leads to the commercialization of two Li-metal batteries using polymer positive electrodes; but low cycle life and high self-discharge lead to remove these products from the market. Following the introduction of Li-ion technology by Sony in 1990, almost all research activities were focused on transition metal based intercalation materials and organic electrode materials were not considered anymore. Due to some recent environmental and economic considerations, they are back into the game and intensively studied since 10 years even if no product is on the market at this time. Indeed NEC announced the commercialization of ORB in 2013, but it never happens.

Organic electrode materials	Reported specific capacity (mAh.g⁻¹)	Potential (V $vs.$ Li⁺/Li)	Theoretical energy density (Wh/kg$_{AM}$⁻¹)	Reported energy density (Wh.kg$_{MA}$⁻¹)	Cycle life	TRL
Conducting polymers						9
- Polypyrrole	80–100	2–4.5	300	90–390	500	Commercialized by Varta (PPy) and Bridgestone (PANI)
- Polyaniline	100–150	2–4		90–540		
- Polythiophene	50–100	3–4		100		7
Stable radical	100–150	3–4	300–500	10–50	500	Commercialization announced by NEC (3 mAh, 5 kW)
Disulfide	300–400	2–3	400–1400	100–150	–	4
Thioether	200–600	2–3.5	400–1500	–	–	3
Carbonyl functions						
- Quinonoïde	200–600	2–3.5	300–500	50–150	>1000 cycles	5
- Dimide	100–200	2–3	150–250			
- Anhydride	100–200	2–3	150–250			
- Carboxylate	200–300	0.5–1.5				
Aromatic amine	300–400	3–4	400	100	–	3

Various organic redox moieties exist and the diversity of organic chemistry let imagine infinite redox compound possibility but their practical use are still hindered by limited cycle life attributed to dissolution phenomena of active materials along cycling. Various strategies have been proposed in order to prevent these problems, leading to promising materials.

References

[1] Poizot P., Dolhem F. (2011) Clean energy new deal for a sustainable world: from non-CO_2 generating energy sources to greener electrochemical storage devices, *Energy Environ. Sci.* **4**, 2003, https://doi.org/10.1039/c0ee00731e.

[2] Nishide H., Suga T. (2005) Organic radical battery, *Electrochem. Soc. Interface*, 32.

[3] Deuchert K., Hünig S. (1978) Multistage organic redox systems—a general structural principle, *Angew. Chem. Int. Ed. Engl.* **17**, 875, https://doi.org/10.1002/anie.197808753.

[4] Novák P., Müller K., Santhanam K.S.V., Haas O. (1997) Electrochemically active polymers for rechargeable batteries, *Chem. Rev.* **97**, 207, https://doi.org/10.1021/cr941181o.

[5] Gottis S., Barres A.-L., Dolhem F., Poizot P. (2014) Voltage gain in lithiated enolate-based organic cathode materials by isomeric effect, *ACS Appl. Mater. Interfaces* **6**, 10870, https://doi.org/10.1021/am405470p.

[6] Poizot P., Dolhem F., Gaubicher J. (2018) Progress in all-organic rechargeable batteries using cationic and anionic configurations: toward low-cost and greener storage solutions? *Curr. Opin. Electrochem.*, https://doi.org/10.1016/j.coelec.2018.04.003.

[7] Sadki S., Schottland P., Brodie N., Sabouraud G. (2000) The mechanisms of pyrrole electropolymerization, *Chem Soc Rev.* **29**, 283, https://doi.org/10.1039/A807124A.

[8] Moliton A., Hiorns R.C. (2004) Review of electronic and optical properties of semiconducting π-conjugated polymers: applications in optoelectronics, *Polym. Int.* **53**, 1397, https://doi.org/10.1002/pi.1587.

[9] Roncali J. (1992) Conjugated poly(thiophenes): synthesis, functionalization, and applications, *Chem. Rev.* **92**, 711, https://doi.org/10.1021/cr00012a009.

[10] Reynolds J.R. (2007) *Handbook of conducting polymers*.

[11] Heinze J., Frontana-Uribe B.A., Ludwigs S. (2010) Electrochemistry of conducting polymers—persistent models and new concepts, *Chem. Rev.* **110**, 4724, https://doi.org/10.1021/cr900226k.

[12] Shirakawa H., Louis E.J., MacDiarmid A.G., Chiang C.K., Heeger A.J. (1977) Synthesis of electrically conducting organic polymers: halogen derivatives of polyacetylene, (CH), *J. Chem. Soc. Chem. Commun.* 578, https://doi.org/10.1039/C39770000578.

[13] Mike J.F., Lutkenhaus J.L. (2013) Recent advances in conjugated polymer energy storage *J. Polym. Sci. Part B Polym. Phys.* **51**, 468, https://doi.org/10.1002/polb.23256.

[14] Nakahara K., Iwasa S., Satoh M., Morioka Y., Iriyama J., Suguro M., Hasegawa E. (2002) Rechargeable batteries with organic radical cathodes, *Chem. Phys. Lett.* **359**, 351, https://dx.doi.org/10.1016/S0009-2614(02)00705-4.

[15] Nishide H., Iwasa S.b, Pu Y.-J., Suga T., Nakahara K., Satoh M. (2004) Organic radical battery: nitroxide polymers as a cathode-active material, *Electrochimica Acta* **50**, 827, https://doi.org/10.1016/j.electacta.2004.02.052.

[16] Lee J.Y., Ong L.H., Chuah G.K. (1992) Rechargeable thin film batteries of polypyrrole and polyaniline, *J. Appl. Electrochem.* **22**, 738, https://doi.org/10.1007/BF01027503.

[17] Naegele D., Bittihn R. (1988) Electrically conductive polymers as rechargeable battery electrodes, *Solid State Ion* **28–30** Part 2, 983, https://dx.doi.org/10.1016/0167-2738(88)90316-5.

[18] Qu J., Katsumata T., Satoh M., Wada J., Igarashi J., Mizoguchi K., Masuda T. (2007) Synthesis and charge/discharge properties of polyacetylenes carrying 2,2,6,6-Tetramethyl-1-piperidinoxy radicals, *Chem. – Eur. J.* **13**, 7965, https://doi.org/10.1002/chem.200700698.

[19] Armand M.B. (1985) Utilization of conductive polymers in rechargeable batteries, *Solid State Batteries* (C.A.C. Sequeira, A. Hooper, Eds). Springer Netherlands, Dordrecht, pp. 363–375, https://doi.org/10.1007/978-94-009-5167-9_24.

[20] Mohammadi A., Inganäs O., Lundström I. (1986) Properties of polypyrrole-electrolyte-polypyrrole cells, *J. Electrochem. Soc.* **133**, 947, https://doi.org/10.1149/1.2108770.

[21] Panero S., Spila E., Scrosati B. (1996) A new type of a rocking-chair battery family based on a graphite anode and a polymer cathode, *J. Electrochem. Soc.* **143**, L29, https://doi.org/10.1149/1.1836446.

[22] Tourillon G., Garnier F. (1982) New electrochemically generated organic conducting polymers, *J. Electroanal. Chem. Interfacial Electrochem.* **135**, 173, https://doi.org/10.1016/0022-0728(82)90015-8.

[23] Kaufman J.H., Chung T.-C., Heeger A.J., Wudl F. (1984) Poly(Thiophene): a stable polymer cathode material, *J. Electrochem. Soc.* **131**, 2092, https://doi.org/10.1149/1.2116025.

[24] Zhan L., Song Z., Zhang J., Tang J., Zhan H., Zhou Y., Zhan C. (2008) PEDOT: cathode active material with high specific capacity in novel electrolyte system, *Electrochimica Acta* **53**, 8319, https://doi.org/10.1016/j.electacta.2008.06.053.

[25] Macdiarmid A.G., Chiang J.C., Richter A.F., Epstein A.J. (1987) Polyaniline: a new concept in conducting polymers, *Proc. Int. Conf. Sci. Technol. Synth. Met.* **18**, 285, https://doi.org/10.1016/0379-6779(87)90893-9.

[26] Ray A., Asturias G.E., Kershner D.L., Richter A.F., MacDiarmid A.G., Epstein A.J. (1989) Polyaniline: doping, structure and derivatives, *Proc. Int. Conf. Sci. Technol. Synth. Met.* **29**, 141, https://doi.org/10.1016/0379-6779(89)90289-0.

[27] Bernard M.-C., Hugot-Le Goff A., Zeng W. (1997) Characterization and stability tests of an all solid state electrochromic cell using polyaniline, *Synth. Met.* **85**, 1347, https://doi.org/10.1016/S0379-6779(97)80265-2.

[28] Taguchi S., Tanaka T. (1987) Fibrous polyaniline as positive active material in lithium secondary batteries, *J. Power Sources* **20**, 249, https://doi.org/10.1016/0378-7753(87)80119-2.

[29] Pablo J., Eric L., Olivier A., Dominique G., Bernard L., Joël G. (2017) Lithium n-doped polyaniline as a high-performance electroactive material for rechargeable batteries, *Angew. Chem. Int. Ed.* **56**, 1553, https://doi.org/10.1002/anie.201607820.

[30] Liang Y., Tao Z., Chen J. (2012) Organic electrodes: organic electrode materials for rechargeable lithium batteries, *Adv. Energy Mater.* **2**, 702, https://doi.org/10.1002/aenm.201290037.

[31] Häupler B., Wild A., Schubert U.S. (2015) Carbonyls: powerful organic materials for secondary batteries, *Adv. Energy Mater.* **5**, 1402034, https://doi.org/10.1002/aenm.201402034.

[32] Lee M., Hong J., Seo D.-H., Nam D.H., Nam K.T., Kang K., Park C.B. (2013) Redox cofactor from biological energy transduction as molecularly tunable energy-storage compound, *Angew. Chem. Int. Ed.* **52**, 8322, https://doi.org/10.1002/anie.201301850.

[33] Zhao Q., Guo C., Lu Y., Liu L., Liang J., Chen J. (2016) Rechargeable lithium batteries with electrodes of small organic carbonyl salts and advanced electrolytes, *Ind. Eng. Chem. Res.* **55**, 5795, https://doi.org/10.1021/acs.iecr.6b01462.

[34] Renault S., Gottis S., Barrès A.-L., Courty M., Chauvet O., Dolhem F., Poizot P. (2013) A green Li-organic battery working as a fuel cell in case of emergency, *Energy Environ. Sci.* **6**, 2124, https://doi.org/10.1039/c3ee40878g.

[35] Wang S., Wang L., Zhu Z., Hu Z., Zhao Q., Chen J. (2014) All organic sodium-ion batteries with $Na_4C_8H_2O_6$, *Angew. Chem. Int. Ed.* **53**, 5892, https://doi.org/10.1002/anie.201400032.

[36] Hanyu Y., Honma I. (2012) Rechargeable quasi-solid state lithium battery with organic crystalline cathode, *Sci. Rep.* **2**, 453.

[37] Wang S., Wang L., Zhang K., Zhu Z., Tao Z., Chen J. (2013) Organic $Li_4C_8H_2O_6$ nanosheets for lithium-ion batteries, *Nano Lett.* **13**, 4404, https://doi.org/10.1021/nl402239p.

[38] Walker W., Grugeon S., Vezin H., Laruelle S., Armand M., Wudl F., Tarascon J.-M. (2011) Electrochemical characterization of lithium 4,4[prime or minute]-tolane-dicarboxylate for use as a negative electrode in Li-ion batteries, *J. Mater. Chem.* **21**, 1615, https://doi.org/10.1039/C0JM03458D.

[39] Deunf É., Moreau P., Quarez É., Guyomard D., Dolhem F., Poizot P. (2016) Reversible anion intercalation in a layered aromatic amine: a high-voltage host structure for organic batteries, *J. Mater. Chem. A.* **4**, 6131, https://doi.org/10.1039/c6ta02356h.

[40] Deunf É., Dupré N., Quarez É., Soudan P., Guyomar D., Dolhem F., Poizot P. (2016) Solvation, exchange and electrochemical intercalation properties of disodium 2,5-(dianilino)terephthalate, *Cryst. Eng. Comm.* **18**, 6076, https://doi.org/10.1039/c6ce01112h.

[41] Deunf É., Jiménez P., Guyomard D., Dolhem F., Poizot P. (2016) A dual–ion battery using diamino–rubicene as anion–inserting positive electrode material, *Electrochem. Commun.* **72**, 64, https://dx.doi.org/10.1016/j.elecom.2016.09.002.

[42] Doeff M.M., Visco S.J., De Jonghe L.C. (1992) Thin film rechargeable room temperature batteries using solid redox polymerization electrodes, *J. Electrochem. Soc.* **139**, 1808, https://doi.org/10.1149/1.2069502.

[43] Doeff M.M., Lerner M.M., Visco S.J., De Jonghe L.C. (1992) The use of polydisulfides and copolymeric disulfides in the Li/PEO/SRPE battery system, *J. Electrochem. Soc.* **139**, 2077, https://doi.org/10.1149/1.2221181.

[44] Tsutsumi H., Okada K., Fujita K., Oishi T. (1997) New type polyamides containing disulfide bonds for positive active material of lithium secondary batteries, *Proc. Eighth Int. Meet. Lithium Batter.* **68**, 735, https://doi.org/10.1016/S0378-7753(96)02578-5.

[45] Iwasa S., Yasui M., Nishi T., Nakano K. (2012) Developement of organic radical battery.

[46] Audebert P., Bidan G., Lapkowski M. (1986) Reduction by two successive one-electron transfers of anthraquinone units bonded to electrodeposited poly(pyrrole) films, *J. Chem. Soc. Chem. Commun.*, 887, https://doi.org/10.1039/C39860000887.

[47] Aydın M., Esat B., Kılıç Ç., Köse M.E., Ata A., Yılmaz F. (2011) A polythiophene derivative bearing TEMPO as a cathode material for rechargeable batteries *Eur. Polym. J.* **47**, 2283, https://doi.org/10.1016/j.eurpolymj.2011.09.002.

[48] Casado N., Hernández G., Veloso A., Devaraj S., Mecerreyes D., Armand M. (2016) PEDOT radical polymer with synergetic redox and electrical properties, *ACS Macro Lett.* **5**, 59, https://doi.org/10.1021/acsmacrolett.5b00811.

[49] Amaike M., Iihama T. (2006) Chemical polymerization of pyrrole with disulfide structure and the application to lithium secondary batteries, *Synth. Met.* **156**, 239, https://doi.org/10.1016/j.synthmet.2005.11.007.

[50] Abruña H.D., Matsumoto F., Cohen J.L., Jin J., Roychowdhury C., Prochaska M., van Dover R.B., DiSalvo F.J., Kiya Y., Henderson J.C., Hutchison G.R. (2006) Electrochemical energy generation and storage. fuel cells and lithium-ion batteries. *Bull. Chem. Soc. Jpn.* **80**, 1843 (2007), https://doi.org/10.1246/bcsj.80.1843.

[51] Naoi K., Kawase K., Mori M., Komiyama M. (1997) Electrochemistry of Poly(2,2′-dithiodianiline): a new class of high energy conducting polymer interconnected with S S Bonds, *J. Electrochem. Soc.* **144**, L173, https://doi.org/10.1149/1.1837715.

[52] Wang G., Yang X., Sun Y., Bao H., Li X. (2009) Aniline-based disulfide/aniline copolymers as a high energy-storage material, *Macromol. Chem. Phys.* **210**, 2118, https://doi.org/10.1002/macp.200900324.

[53] Häringer D., Novák P., Haas O., Piro B., Pham M. (1999), Poly(5-amino-1,4-naphthoquinone), a novel lithium-inserting electroactive polymer with high specific charge, *J. Electrochem. Soc.* **146**, 2393, https://doi.org/10.1149/1.1391947.

[54] Holze R., Wu Y.P. (2014) Intrinsically conducting polymers in electrochemical energy technology: trends and progress, *Electrochem. Electroact. Mater.* **122**, 93, https://doi.org/10.1016/j.electacta.2013.08.100.

[55] Barrès A.-L., Geng J., Bonnard G., Renault S., Gottis S., Mentré O., Frayret C., Dolhem F., Poizot P. (2012) High-potential reversible Li deintercalation in a substituted tetrahydroxy-p-benzoquinone dilithium salt: an experimental and theoretical study, *Chem. - Eur. J.* **18**, 8800, https://doi.org/10.1002/chem.201103820.

[56] Liang Y., Zhang P., Chen J. (2013) Function-oriented design of conjugated carbonyl compound electrodes for high energy lithium batteries, *Chem Sci.* **4**, 1330, https://doi.org/10.1039/C3SC22093A.

[57] Zhu Z., Chen J. (2015) Review—advanced carbon-supported organic electrode materials for lithium (sodium)-ion batteries, *J. Electrochem. Soc.* **162**, A2393, https://doi.org/10.1149/2.0031514jes.

[58] Kwon M.-S., Choi A., Park Y., Cheon J.Y., Kang H., Jo Y.N., Kim Y.-J., Hong S.Y., Joo S.H., Yang C., Lee K.T. (2014) Synthesis of ordered mesoporous phenanthrenequinone-carbon via π-π interaction-dependent vapor pressure for rechargeable batteries, *Sci. Rep.* **4**, 7404.

[59] Lee M., Hong J., Kim H., Lim H.-D., Cho S.B., Kang K., Park C.B. (2014) Organic nanohybrids for fast and sustainable energy storage, *Adv. Mater.* **26**, 2558, https://doi.org/10.1002/adma.201305005.

[60] Ravet N., Michot C., Armand M. (1997) Novel cathode materials based on organic couples for lithium batteries, *MRS Proc.* **496**, https://doi.org/10.1557/PROC-496-263.

Chapter 5

Electrolytes and Separators

Jean-Frédéric Martin, Djamel Mourzagh, Thibaut Gutel
and Hélène Rouault

To determine the importance of the electrolyte and the separator in the Li-ion system, their cost contribution can be examined. According to recent cost analyses [1] that consider all materials, processes and services involved in the production of Li-ion cells, the electrolyte accounts for 6% of the total cost of the unit cell, while the separator accounts for 5%.

If these figures appear much larger than the mass ratios of these components within the cell, it is, above all, because these two components guarantee the key performance of the accumulator. Indeed, the service life, temperature stability, safety and performance of the final battery are largely conditioned by the choice and quality of the electrolyte and the separator. In fact, although these products are mass-produced by large firms, usually in Asia, sometimes in USA, these products require expensive manufacturing or purification processes.

Research is therefore active in these areas, especially since electrolyte composition optimization is often the last hurdle to overcome before a decision is made to commercialize a technology. This is the case, for example, for electrolytes compatible with "high potential" positive electrodes or those operating at extreme temperatures. It is also a major topic of interest for more innovative "post-lithium ion" chemistries (all solid state batteries, lithium metallic anode, sodium-ion, lithium sulfur, organic active materials, etc.).

Today, various solutions are investigated: liquid electrolytes with separators, gelled membranes, polymer electrolytes and ceramic or composite electrolytes. This chapter deals with liquid electrolytes. Other types of electrolytes are described in chapter 8: "all-solid" batteries. Moreover, this chapter is limited to the case of electrolytes for Li-ion batteries – those used for other technologies are described in the corresponding chapters.

DOI: 10.1051/978-2-7598-2555-4.c005
© Science Press, EDP Sciences, 2021

5.1 Liquid Electrolytes

An electrolyte is defined as an ionic conductive but electrically insulating medium. A lithium salt dissolved in a mixture of solvents remains the simplest solution. Such an option is, by far, the most widespread in commercial batteries: it meets the specifications of use at ambient temperature for power levels usually required by most applications. This subchapter gives an overview of the general properties of electrolytes and questions their limitations. It also discusses ionic liquids, which, although not yet widely represented industrially, may have the potential to replace organic solvents in order to achieve cleaner and safer batteries.

5.1.1 *Lithium Salts and Organic Solvents*

Kang Xu's famous literature review [2] is an excellent starting point for the study of liquid electrolytes. It describes in detail the individual components at stake. Readers who wish to learn more in this domain should refer to it.

5.1.1.1 *Basic Properties and General Observation*

Two fundamental properties are sought for an electrolyte in a battery:

(1) The first is its electrochemical stability. Water, for example, cannot be used in Li-ion systems which present a high voltage difference between the positive and negative electrodes, primarily because the electrochemical stability window of water is extremely narrow, and at any pH value it is not possible to use graphite as a negative electrode [3]. Fortunately, there are many organic solvents that offer much better intrinsic stability, which is further enhanced by spontaneous or induced passivation phenomena.

   For example, most electrolytes today contain ethylene carbonate (EC) as a solvent, and lithium hexafluorophosphate $LiPF_6$ as a salt. The presence of such components in the electrolyte leads to passivation mechanisms that happen to be very useful in the Li-ion battery: EC passivates the graphite used as the negative electrode material by creating a protective and ion-conducting layer (SEI = solid electrolyte interphase [4]) on the surface of the material through decomposition (figure 5.1). The $LiPF_6$ salt, on the other hand, creates an AlF_3 layer on the aluminium current collector [5] of the positive electrode, thus preventing its anodic decomposition during cycling.

(2) The other essential property of the electrolyte is its kinetic behaviour, which can be broken down into different characteristics, including conductivity and transport number. These quantities are mainly derived from the permittivity and viscosity of solvents, as well as their interaction with the salt ions. This has led to the emergence of some optimized solvent mixtures [7].

Here again, extrinsic effects, which condition the kinetic performance of the solution, have to be taken into account, such as the affinity with the separator and the electrodes, including the charge transfer resistance or the SEI resistance [8].

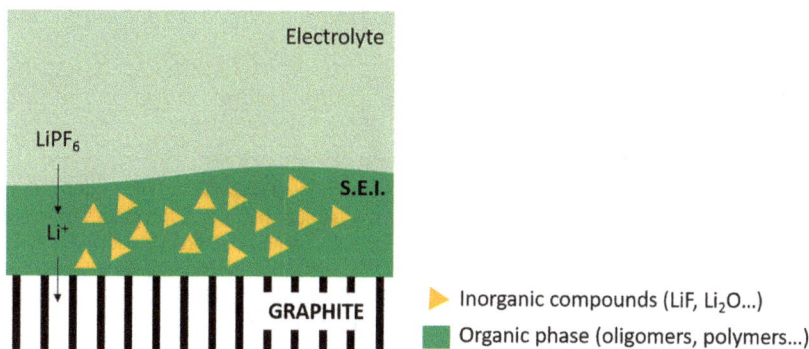

FIG. 5.1 – Illustration of the passivating interphase on the graphite surface (SEI). Once formed by reduction of the electrolyte on the graphite surface, this composite layer, which is ionically conductive, still allows lithium ions to access to the graphitic planes, but prevents direct contact of the electrolyte with the reducing surface. Some articles propose particular organization or architecture [6] (not shown here) of this layer, generally studied by XPS or FTIR. In addition, many studies focus on its optimization *via* polymerizing electrolyte additives or specific formation steps during the first cycles of the battery life, in order to improve its stability and conductivity.

To these already complex characteristics, it is important to add a time dimension in order to better assess the performance of a selected solution. The electrolyte is usually the component that is most sensitive to ageing [9] and is in contact with sometimes extremely large active surfaces (conductive carbon additive, nanoscale materials), which increases the rate of interference reactions. The chemical, electrochemical and thermal degradations of solvents and salt generate internal resistances and irreversibly decrease the cell capacity [10]. These failures are summarized in figure 5.2. Their study is essential, although delicate, because of the time required to observe their symptoms. However, such developments can be accelerated, for example, by irradiating the systems [11]. Identifying solutions to remedy each of these degradations remains an important source of innovation.

The entanglement of the different phenomena at play in a Li-ion battery is extremely strong and the electrolyte is the hub of balances and imbalances that make the system either work or die at the same time. The conclusions of *ex-situ* studies or overly restricted modelling can therefore be quickly thwarted. Thus, concerning the resolution of "real-life" problems, the pragmatic and systemic approach (benchmarking/high throughput/post mortem analyses) is obviously to be favoured [12]. It is also important to quickly transpose a pre-initial study in coin-type test cells to systems more representative of the final object in order to avoid the often delicate extrapolations of scale (ideally flexible cells of several tens of mA.h are a minimum [13]). Finally, although research generally refuses to be driven by chance, serendipity has sometimes brought important results, for example, LiBOB (lithium bis-oxalato borate) that was originally tested as a salt capable of

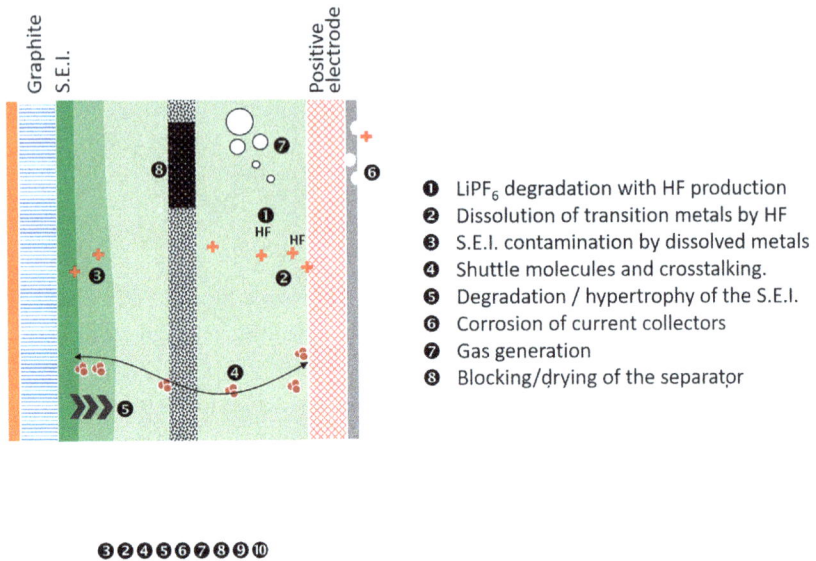

Fig. 5.2 – Diagram illustrating degradation in the electrolyte. The LiPF$_6$ salt, sensitive to the presence of moisture traces and temperature, forms HF ❶, which leads to the dissolution of the active materials and current collectors (❷ and ❻) with the possibility of poisoning the SEI layer ❸. In addition, solvent molecules or impurities may react on an electrode resulting in the formation of soluble or insoluble degradation products. If the products are soluble, they may form "shuttle" mechanisms or may react on the opposite electrode ("cross-talking") ❹ and lead to a loss of system efficiency. If they are insoluble, they can clog the porosities of the separator ❽ and destabilize the SEI ❺. Finally, a consequence of these degradations can be the formation of gaseous species ❼ that will cause the cell to swell.

replacing LiPF$_6$ and that ultimately proved to be an excellent additive for the formation of SEI [14].

5.1.1.2 State of the Art, Its Limitations and Research to Overcome Them

Today, the vast majority of commercially available Li-ion cells are activated with a mixture of linear and cyclic alkyl carbonates (figure 5.3) in which LiPF$_6$ has been added as a salt (for a TESLA cell, for example, see [15]) as well as additives for the protection of the negative electrode. Such a solution is satisfactory in terms of lifetime under normal operating conditions. The improvements made over the last 20 years now enable lithium batteries operating in appropriate conditions to reach several thousand charge/discharge cycles, at the end of which the capacity is still around 80% of the initial capacity. However, they have at least two weak points: the flammability of solvents [16] and the relatively bad high-temperature resilience of the lithium salt [17].

Concerning the issue related to the solvent flammability, some additives can delay ignition, and the use of solvents with a high flash point, for example,

Ethylene carbonate	Propylene carbonate	Dimethyl carbonate	Diethyl carbonate
EC	PC	DMC	DEC
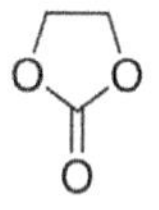	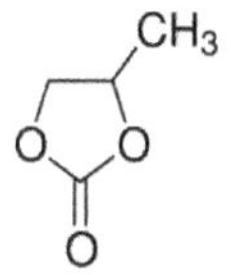	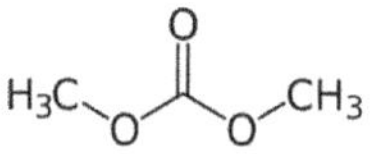	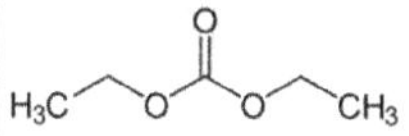

FIG. 5.3 – Main organic solvents used for Li-ion battery electrolytes.

containing heteroatoms (mainly phosphorus or fluorine), also seems beneficial [18]. Other types of solutions are inestigated using ionic liquids or even solid electrolytes, which are discussed chapter 8 of this book.

Concerning the thermal stability of the salt, even in an extremely clean and dry electrolyte, $LiPF_6$ degrades significantly above 60 °C due to autocatalytic reactions [19]. The research for more stable lithium salts than $LiPF_6$ has already led to the emergence of a number of alternatives [20] such as LiTFSI [21] (lithium bis(tri-fluoromethane) sulfonimide), LiFSI [22] (lithium bis(fluorosulfonyl)imide) or LiTDI [23] (lithium 4,5-dicyano-2-(trifluoromethyl)imidazole) to name only the most successful ones. However, the first two salts lead to serious corrosion problems with relatively low-potential on aluminum [24]. The third seems to be usable up to 4.7 V *vs.* Li^+/Li [25], but has still to be industrially tested.

Some more exotic functional specifications require the development of specific electrolytes:

- For very low temperatures: the use of carbonate mixtures or solvents with low solidification temperatures (*e.g.* esters [26]) is recommended by the Jet Propulsion Laboratory.
- For very high temperatures: up to 80 °C, a salt change can avoid too rapid degradation (also using a high potential anode of the $Li_4Ti_5O_{12}$ type to avoid too reductive potentials [27]); for higher temperatures, solid or liquid ionic electrolytes are the most studied.

The electrolyte is also interacting with a number of new materials that impose new rules, such as:

- Materials operating at high voltages (typically 4.5 V or higher), which require extended electrochemical stability windows (fluorinated solvents [28]) or protective interphases (positive electrode additives [29]).
- Silicon, which requires the formation of a specific SEI due to its high volume expansion. Fluoroethylene carbonate [30] is generally used as an additive or co-solvent, but the evolution of the interphase remains problematic and generally leads to "smothering" of the electrode by filling the porosities with the SEI.

Similarly, lithium metal is depleted during cycling due to its inevitably changing surface [31].

- Titanium-based materials, such as $Li_4Ti_5O_{12}$, which, despite having a potential compatible with the stability window of conventional electrolytes, appear to catalyze the dehydrogenation of solvent molecules and the significant formation of gases [32].

The rest of the studies concern safety aspects (polymerizing agents or redox shuttles avoiding overload problems [33]) or manufacturing processes (by acting on the wettability of the separator with surfactants for example [34]).

To conclude, waiting for possible new "all-solid" electrolytes which would inevitably compete with the solutions described above, particularly in terms of safety, liquid and/or gelled electrolytes will probably remain, for at least the ten years to come, the preferred ionic conduction medium in terms of cost and processability for lithium-ion batteries. For the same reasons, they will probably remain unavoidable for certain applications, even after ceramic or polymeric electrolytes may be put on the market.

Furthermore, between a liquid electrolyte impregnating a separator and an "all-solid" electrolyte, which still requires development efforts, industrials develop and use intermediate solutions known as "hybrid liquid-polymer systems", which are gelled electrolytes that are less volatile than their liquid counterparts. Such a product has been developed by CEA Tech's Liten institute from fluorinated polymers (PVDF) specially developed by Solvay [35, 36].

In such a system, PVDF is first solubilized, then mixed with the electrolyte before undergoing cross-linking at the time of coating. The electrolyte is thus trapped in a network of polymer chains. Much safer than purely liquid systems, this technology works in a range of comparable temperatures, and is used on conventional roll-to-roll equipment in an anhydrous environment. The electrochemical tests results performed on 500 mAh cells show performances equivalent to conventional, liquid electrolytes lithium-ion batteries at moderate discharge rates. This option could also significantlty reduce the electrode fabrication cost and simplify the process by avoiding the solvent recovery step.

5.1.2 Lithium Salts and Ionic Liquids

Ionic Liquids (IL) are salts, which, by definition, are liquid at temperatures below 100 °C. They consist of an organic cation (figure 5.4) associated with an organic or

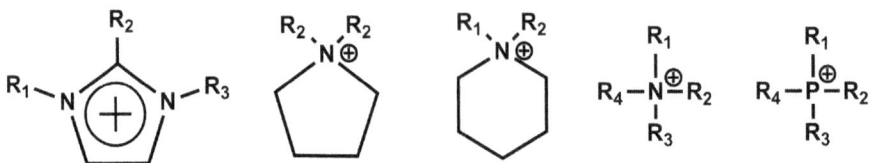

FIG. 5.4 – Most frequently encountered cations (R representing alkyl chains).

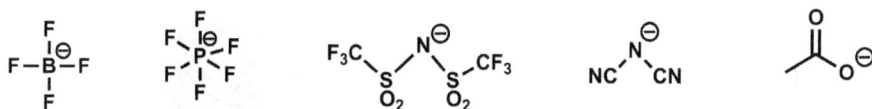

FIG. 5.5 – Most frequently encountered anions.

inorganic anion (figure 5.5), and this wide variety of combinations (>106) allows their physicochemical and electro-chemical properties to be adjusted to the intended application.

First described by Walden *et al.*, in 1914 [37], ionic liquids did not really develop until the 1990s with the discovery of moisture-stable anions. Apart from their outstanding thermal stability and low vapor pressure [38], these compounds have very large electrochemical stability windows and high ionic conductivities. They have therefore been extensively investigated as solvents for battery electrolytes. Solutions of lithium salts in ionic liquids improve the safety of batteries under abusive conditions [39].

It should be noted that it is preferable to use a lithium salt with the same anion as that of the ionic liquid in order to avoid spontaneous ion exchange phenomena (according to the theory of hard/soft acids and bases) within the solution. However, the cost and viscosity at room temperature still hamper the use of these ILs in commercial electrolytes.

5.2 Separators

Liquid electrolytes require a intermediate layer that effectively physically separates the electrodes to avoid short cicuit. This role is assigned to the separator. In the context of this document, separators are not presented exhaustively, the aim being rather to describe their role and to position them on the market.

5.2.1 Properties of Separators

The separator for Li-ion batteries is an essential component for the adequate functioning of the battery. It consists of a microporous film (figure 5.6) placed between the positive and negative electrodes. It provides a physical barrier while maintaining a good ionic conductivity. It must have a good mechanical strength and must be chemically and electrochemically inert towards the electrolyte and the electrodes. Battery separators are generally produced by two different routes: a "dry way" by extrusion and stretching; and a "wet way" using solvent extraction methods for plasticizers. The majority of commercial separators are made of polyolefin [40]. The main advantages are good mechanical properties, good chemical stability and acceptable cost. The separators are made of monolayers of polyethylene (PE) or polypropylene (PP), bilayers or triple layers alternating the two components (*e.g.* PP/PE/PP).

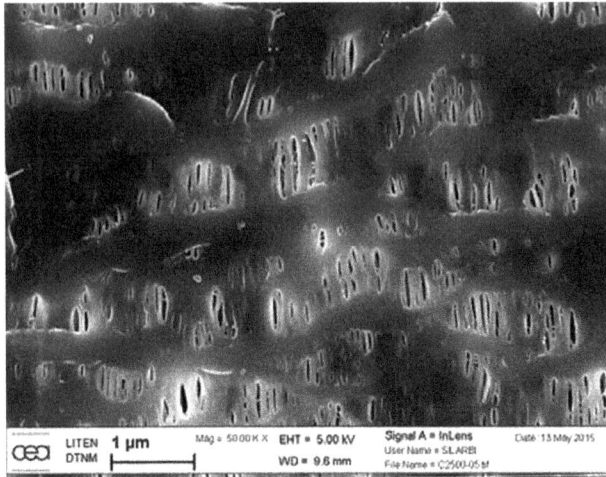

FIG. 5.6 – Scanning electron microscope view of a Celgard © 2500 separator showing microporosity.

Three-layer separators are widely used in Li-ion batteries because they provide a good level of safety. If the battery heats up, the polyethylene melts (melting temperature around 130 °C) and blocks the pores of the polypropylene, which has a higher melting temperature (around 160 °C). As a result, the separator isolates the two electrodes (shutdown), which limits possible thermal runaway. However, this type of separator is usually not sufficient to fully prevent a runaway because temperatures above 160 °C are often reached.

5.2.2   The Separator Market

The market for separators for Li-ion batteries (figure 5.7) is mainly dominated by ASAHI (17% market share), TORAY (14% market share), SUMITOMO (14% market share), and Celgard (9% market share).

In the last five years, new grades of separators for Li-ion batteries have appeared on the market, such as:

- Ceramic-coated separators developed by companies such as Celgard, TORAY, Evonik, Porous Power technologies, etc.... To limit or even prevent thermal runaway, manufacturers coat their separators with a thin film containing high melting temperature compounds. Ceramics such as alumina Al_2O_3 (melting temperature 2050 °C), silica SiO_2 (melting temperature 1600 °C) are most often used.
- Aramid-coated separators (developed by TEIJIN and Sumitomo Chemical) are designed to improve the safety of batteries, especially in the event of an internal short circuit. Aramid melting point is higher than 1000 °C.

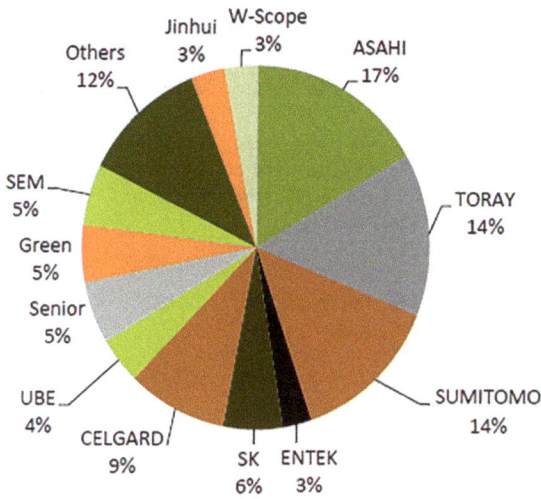

FIG. 5.7 – World market for lithium-ion battery separators in 2021 (source: Christophe PILLOT, AVICENNE Energy 2021).

- The nanofiber and microfiber separators developed by DreamWeaver offer the same quality/safety as conventional separators at a lower cost.

5.2.3 *Cost and Security*

The price of the separator has been falling steadily in recent years, from 20% of the price of the unit cell to almost 5% in 2021, thanks in particular to the increase in production volumes. Separators commercialy available, based on ceramic, aramid and even cellulose nanofibers show good behavior during abusive tests (especially during the nail test), thus considerably limiting the risk of thermal runaway.

Conclusion

This brief overview of liquid electrolyte technologies for Li-ion batteries does not aim at describing the proliferation of scientific initiatives in this field. However, although a large number of new electrolyte molecules have been developed since the 1990s, it is interesting to note that "historic" mixtures are still of interest today and that a real breakthrough, mainly in terms of safety, has yet to be achieved. From the point of view of separators, progress has been significant: there are now finer separators (<15 μm), less expensive, and offering a much higher level of safety than a few years ago.

Bibliography

[1]   Avicenne Energy – The Rechargeable Battery Market: Value Chain and Main Trends 2019-2030-IBS – 11 March, 2021.
[2]   *Chem. Rev.* (2004) **104**, 4303; *Chem. Rev.* (2014) **114**, 11503.
[3]   *Mater. Chem. A* (2014) **2**, 9025. On the other hand, strategies exist for using water in some suitable Li-ion systems. See for example *Electrochem. Comm.* (2017) **82**, 71.
[4]   *J. Electrochem. Soc.* (1979) **126** 2047.
[5]   *Electrochim. Acta* (2009) **55**, 288.
[6]   *Electrochim. Acta* (2010) **55**, 6332.
[7]   *J. Electrochem. Soc.* (1999) **146**, 486.
[8]   *J. Electrochem. Soc.* (2018) **165**, A361.
[9]   *J. Power Sources* (2015) **273**, 83.
[10] *J. Power Sources* (2005) **147**, 269.
[11] *Nat. Comm.* (2015) **6**, 6950.
[12] *J. Power Sources* (2014) **251**, 311.
[13] *J. Power Sources* (2014) **270**, 68.
[14] *Electrochem. Solid-State Lett.* (2005) **8**, A365.
[15] *J. Electrochem. Soc.* (2017) **164**, A3503.
[16] *Electrochem. Soc. Interface* (2012) **21**, 45.
[17] *J. Electrochem. Soc.* (2005) **152**, A2327.
[18] Abe K. (2014) *"Nonaqueous electrolytes and advances in additives" in modern aspects of electrochemistry.* vol. 58, Springer, New York, NY.
[19] *J. Electrochem. Soc.* (2016) **163**, A1095.
[20] *Chem. Eur. J.* (2011) **17**, 14326.
[21] *J. Power Sources* (2011) **196**, 9743.
[22] *J. Power Sources* (2011) **196**, 3623.
[23] *J. Power Sources* (2011) **196**, 8696.
[24] *J. Electrochem. Soc.* (2000) **147**, 4399.
[25] *J. Power Sources* (2015) **294**, 507.
[26] *J. Electrochem. Soc.* (2002) **149**, A361.
[27] *J. Power Sources* (2012) **216**, 192.
[28] *Energy Environ. Sci.* (2013) **6**, 1806.
[29] *J. Power Sources* (2013) **237**, 229.
[30] *Sci. Rep.* (2017) **7**, 6326.
[31] *Adv. Sci.* (2016) **3**, 1500213.
[32] *Sci. Rep.* (2012) **2**, 913.
[33] *Energy Environ. Sci.* (2016) **9**, 1955.
[34] *J. Membr Sci.* (2016) **503**, 25.
[35] IBA (2016), Nantes, March 20–25, R. Pieri *et al.*
[36] MACRO (2018) Cairns, July 1–5, H. Rouault *et al.*
[37] *Bull. Acad. Imper. Sci.* (1914) 405.
[38] *Ionic Liq. Synth.* 2nd Ed., (2008) Wiley-VCH, Weinheim.
[39] *J. Power Sources* (2009) **194**, 601; *Chem. Soc. Rev.* (2013) **42**, 5963; *Energy Environ. Sci.* (2014) **7**, 232; *MRS Bull.* (2013) **38**, 548; *J. Power Sources* (2010) **195**, 2419; *Nat. Mater.* (2009) **8**, 621; *Electrochim. Acta* (2006) **51**, 5567.
[40] *Chem. Rev.* (2004) **104**, 4419.

Chapter 6

Na-ion Batteries: Should/Can Lithium be Replaced?

Loïc Simonin, Virginie Simone, Laure Monconduit and Sébastien Martinet

Li-ion batteries are currently the most suitable technology for storing electricity for portable electronics and e-mobility. Nevertheless, the very sharp demand growth is increasingly affecting not only raw material resources such as lithium but also (and above all) transition metals (Co, Ni, Cu) and fluorine. The availability of cobalt, nickel and lithium in particular does not seem unlimited and the resources of these materials are relatively localized. From this perspective, the question of replacing Li-ion technology with less critical materials-intensive technology arises. We will try to suggest in this chapter that Na-ion technology could meet these requirements under certain conditions.

6.1 General Aspects

6.1.1 Should Lithium be Replaced?

6.1.1.1 Lithium Resources

Lithium is the 33rd most abundant element of the periodic table, and represents about 20 ppm by mass of the earth crust [1]. It is naturally in the state of salt (mainly chlorides) in mines and partially or completely dried salt lakes [2]. One can also find it in low-concentration in sea water [1]. Apart from sea water, the distribution of its resources is quite inhomogeneous and is concentrated in a few countries of the world, notably in South America, Asia and North America. Table 6.1 presents the list of lithium-bearing countries. The table shows that only 7 countries share the major part of lithium production in the world and that Australia and Chile alone

DOI: 10.1051/978-2-7598-2555-4.c006

TAB. 6.1 – Major Lithium-holding countries by Annual Quantity produced (tons) [2].

	Resources	Reserves	Production 2016	Production 2017
Australia	5 000 000	2 700 000	14 000	18 700
Chile	8 400 000	7 500 000	14 300	14 100
Argentina	9 800 000	2 000 000	5800	5500
China	7 000 000	3 200 000	2300	2300
Zimbabwe	500 000	Nc	1000	1000
Portugal	100 000	Nc	400	400
Brazil	180 000	48 000	200	200
USA	6 800 000	35 000	–	–
Bolivia	9 000 000	Nc	–	–
Canada	1 900 000	Nc	–	–
DR. Congo	1 000 000	Nc	–	–
Russia	1 000 000	Nc	–	–
Serbia	1 000 000	Nc	–	–
A. Czech	840 000	Nc	–	–
Spain	400 000	Nc	–	–
Mali	200 000	Nc	–	–
Mexico	180 000	Nc	–	–
Austria	50 000	Nc	–	–
Total	**53 000 000**	**16 000 000**	**38 000**	**43 000**

account for 75% of the world production in 2017. One should also note that some countries, such as Bolivia and the United States, have very large untapped (or undertapped) resources. As a result, developments in this market need to be closely monitored in the years ahead.

As with most metals, the relative concentration of lithium production sites and the relative lithium scarcity has led to some volatility in raw material prices. For example, the price of lithium carbonate has increased from $2000. ton^{-1} in the early 2000s to $17 000. ton^{-1} in 2018. Between January, and March, 2021, its cost grew by 88%, from 6700 up to $ 12600.ton^{-1} [2]. This has generated controversy over its availability for more than a decade. This debate is still not settled. Indeed, the literature often suggests that the economic risk of depletion of lithium resources is moderate or zero [3], while some publications suggest that a real economic issue may arise in the coming years [4]. The reality seems to be halfway through. Indeed a technology such as "Na-ion" could become very attractive as it is free of critical metals like lithium but can also avoid cobalt and nickel or more generally the use of other sensitive materials.

6.1.1.2 Lithium Cost

In the context of large-scale deployment of energy storage for applications such as electric mobility, renewable energy storage and support for the electricity grid, the cost of stored energy expressed in $. kWh^{-1} is a central issue. As a result of the digital revolution of the 2000s, and especially because of the growth of the electric (EV) and hybrid (HEV) vehicles market, the demand for Li-ion batteries is growing very strongly. The growth in battery production has two notable consequences: The first is the very rapid and continuous decrease in the cost of stored energy at pack level: about $2600. kWh^{-1} in 2000 compared to about $100–120. kWh^{-1} in 2021 [5]. The second consequence is the fluctuations observed of the cost of raw materials, including lithium carbonate, because of the possible difficulties of the offer to follow the materials demand rapid increase. All in all, production costs are falling sharply thanks to process improvements, large manufacturing scale effects, R&D cost amortization, etc., which largely offsets the upward trend in the cost of raw materials.

From this point of view, the share of the lithium carbonate cost, which was totally negligible in the early 2000s represents in 2021 8%–9% of the cost of a $ 100 kWh^{-1} EV battery pack (figure 6.1).

On the other hand, the market price per ton of sodium carbonate is around $150 ($ 0.15.kg^{-1}), and it is unlikely to fluctuate, due to the abundance of sodium on earth – because it is the seventh most abundant element in the earth's crust the fourth most abundant element in seawater – and it is homogeneous distributed. Its share in the cost of a Na-ion battery pack would therefore remain negligible and a gain on the total cost of the battery pack would be possible.

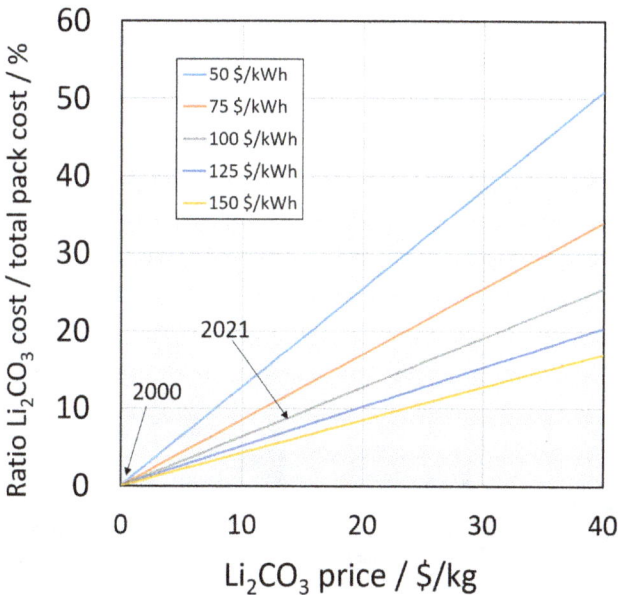

FIG. 6.1 – Share of the Lithium carbonate raw material cost in the total cost of a Li-Ion battery pack for an electric vehicle, at various battery pack cost levels. Assumptions : chemistry : NMC 111/Graphite, embedded energy 25 kWh. Such a learning curve is generally followed by the cost of materials used in a rapidly growing production, until the material's share represent the major part of the cost of the final product.

More generally, the same assessment has to be made carefully for other sensitive or even critical materials used in Lithium-Ion batteries, such as Cobalt, Nickel, Copper, etc.), taking into account all parameter *e.g.* the change with time of the chemical composition of the cathode active material : to increase energy density and reduce cell cost, compositions change with higher nickel and lower cobalt content are used, which modifies continuously the respective share of the materials in the cell cost [5].

6.1.2   Can Lithium be Replaced? Towards a 100% Abundant Element-Based Battery

The question is therefore legitimate: is a battery using only abundant elements on earth possible? All raw materials used in the industry can be considered to be extracted from the following four natural environments: earth crust (metals, minerals, oil, gas, etc.), biomass (organic materials, carbon, etc.), seawater (chlorine, sodium, hydrogen, etc.) and atmosphere (Dioxygen, dinitrogen, noble gases, etc.). From this perspective, the abundance of elements in these four media is reported on figure 6.2. One can note that only 17 elements are present in a proportion greater than 1000 ppm (0.1%) by mass in at least one of these media, we will consider them to be "abundant" among them; 5 are present at more than 100 000 ppm (10%), we will consider them to be "extremely abundant".

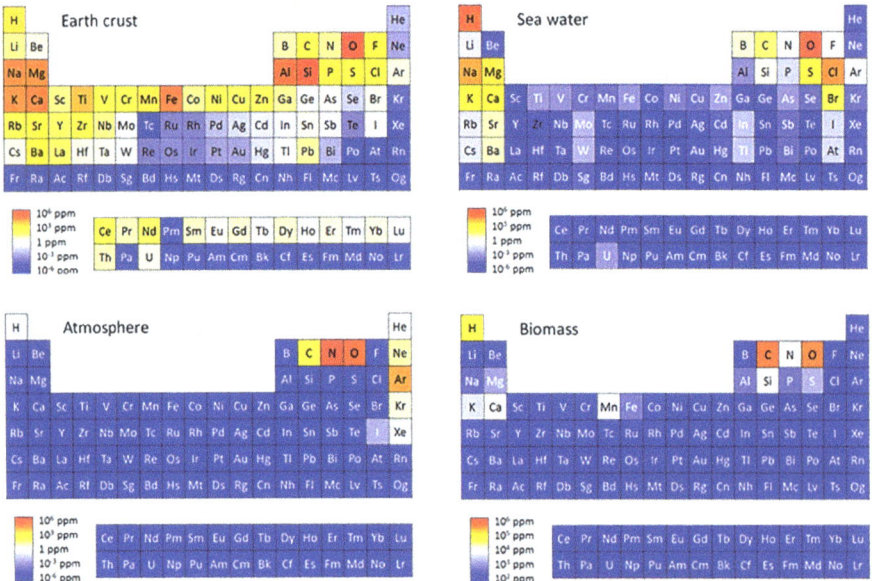

FIG. 6.2 – Abundance of elements on earth, in the earth crust (top left) [1], in biomass (top right) [6], in seawater (bottom left) [1] and in the atmosphere (bottom right) [7].

The 17 abundant or extremely abundant elements are as follows: oxygen, silicon, carbon, hydrogen, nitrogen, sodium, potassium, titanium, manganese, iron, aluminium, phosphorus, magnesium, chlorine, calcium, sulfur and argon. Let us consider whether it is possible to make a battery using only items from this list. Today, as Larcher and Tarascon point out [8], the elements necessary for making a conventional Li-ion cell are Li (charge carrier), Ti, Mn, Fe, Co, Ni (electroactive elements), Cu, Al (current collectors, packaging), C, H, O, F, P (negative electrode, conductive additives, electrode binders, electrolyte, separator, packaging, anionic framework of active materials).

The non-abundant elements of this list are lithium, cobalt, nickel, copper and fluorine. If lithium appears to be replaced by sodium (with a lower energy density for the time being) and copper current collectors replaced by aluminium (see Part 2), it seems much more complicated to handle the case without electroactive components such as cobalt and nickel. Among the three abundant transition elements, iron and titanium have very low oxidation potentials and manganese suffers instability during battery cycling. Finally, fluorine, which is present in many of the key components of the battery such as electrolyte and electrode binders, remains one of the main obstacles to making a "100% abundant" battery.

6.2 The Na-ion Technology

6.2.1 Brief History

In the late 1970s, the first lithium metal rechargeable batteries (using a metallic lithium anode) were introduced, but many issues with the development of negative electrode dendrites were hindering their development [9]. In the 1980s, the concept of "rocking chair" batteries involving Lithium ion intercalation materials at both electrodes was introduced. At the same time, many research groups conducted studies involving other alkaline ions [10–14]. Nevertheless, SONY's commercialization of the first Li-ion batteries in 1991 led scientists to focus on this technology, leaving aside the Na-ion technology. It was only after the digital revolution of the 2000s and the development of electromobility, involving fears about lithium resources availability that the scientific community put back efforts on Na-ion batteries' research [15].

6.2.2 Operating Principle

The operation of a sodium-ion battery is very similar to that of a lithium-ion battery (figure 6.3). Oxidation occurs at the negative electrode during discharge: an electron flows through the external circuit, compensated by the release of a Na^+ ion. This ion migrates through the electrolyte to the positive electrode where the reduction takes place. As with lithium-ion, the electrolyte can be liquid, polymer or ceramic based. During the recharge of the accumulator, the reverse reaction takes place: a reduction

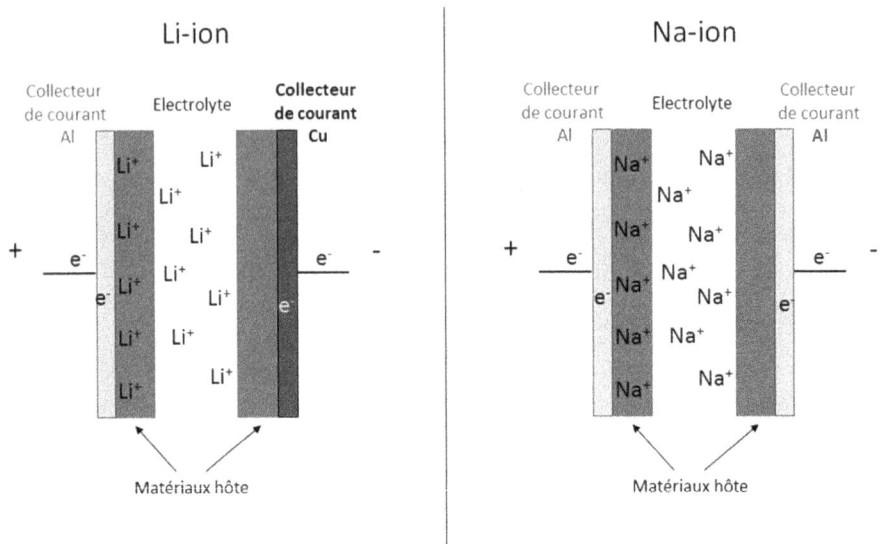

FIG. 6.3 – Operation's principle of a sodium-ion cell (comparison with a Li-ion cell).

occurs at the negative electrode and oxidation occurs at the positive electrode [16]. An electrolyte-soaked separator is interposed between the two electrodes to avoid short circuits. Finally, as already mentioned, the use of aluminum as a current collector for the negative electrode is possible, even at low potential.

Electroactive materials allow the insertion of ions in a reversible manner. This property is due to the presence of vacancies in the crystallographic structure of the insertion material, which can accommodate ions. The three known reactions (intercalation, alloy and conversion) for lithium-ion are also found for sodium-ion.

The main weakness of this technology compared to Li-Ion is its lower energy density. The redox potential of the Na^+/Na couple is 0.3 V above that of the Li^+/Li couple, which reduces the battery's voltage, and thus the energy density. Besides, the negative and positive electrode materials appear to pose a supply issue. At the negative electrode, Sodium ions do not easily intercalate in graphite, so that other materials have to be found; and at the positive electrode, layered oxides so far used in Sodium-Ion batteries have high specific capacities but lower mean discharge voltage than materials used in lithium-ion batteries. The search for materials that are competitive in terms of energy density thus remains one of the major challenges of this technology.

In conclusion, Na-ion technology remains an emerging technology with several issues to overcome. However, proximity with Li-ion technology and the decade-long knowledge of Li-ion batteries suggest that the development of sodium-ion batteries may be faster [17].

6.3 State of the Art

6.3.1 Negative Electrode Materials

6.3.1.1 Graphite

It is well admitted that graphite, the most common negative electrode for Li-ion batteries, cannot be used as an electrode in Na-ion batteries. The insertion of sodium ions is very modest in graphite (capacity of about 20 mAh.g^{-1}) [18]. The reason often given is the size of Na$^+$ ion, which is too large compared to Li$^+$ ion. However, K$^+$ ion, which has even greater radius, is known to form a very stable intercalation compound with graphite [19]. In fact, as Nobuhara *et al.*, revealed, [20] the problem comes from thermodynamics. Indeed, through simulations on the formation of an AC_x compound (where A is an alkaline element), this group has shown the very high instability of NaC_x compounds (for $x < 12$), in particular due to the excessive stretching of C–C bonds. However, according to recent publications, it would appear that, with an adequate spacing of graphene sheets, insertion is possible [21, 22].

6.3.1.2 Hard Carbon

To overcome the impossibility of inserting sodium ions in graphite, the best alternative seems to be the use of the hard carbon family [23, 24]. Hard carbons are generally defined as disordered non-graphitizable carbons as opposed to soft carbons that can be graphitized at high temperatures (>2000 °C). In fact, in a hard carbon, the planes of graphene are unaligned, which makes it difficult to stack them regularly, whereas in a soft carbon, these planes already have some alignment [25]. It should be noted that the ability of a disordered carbon to graphitize depends on the disorder present in the precursor used for its synthesis. Carbon-rich precursors such as hydrocarbons and petroleum residues thus yield soft carbon while the more complex structure of organic macromolecules and heteroatom-rich polymers yields hard carbon [26].

Several research groups have studied the insertion of alkaline ions in this type of carbon. In the case of lithium, Mochida *et al.*, propose a mechanism involving 5 types of sites. A first site (site I) corresponds to the transfer of load to the surface, a second site (site II) corresponds to the insertion of lithium between the planes of graphene, and the other three sites (III, IV and V) correspond to the voids and micro-porosities left between the graphene planes clusters [25]. It should be noted that insertions I and II occur at higher potentials, while insertions III, IV, and V occur at lower potential. These mechanisms were verified by Hori *et al.* by photo-electron spectroscopy using hard X-ray [27].

In the case of sodium, the insertion mechanisms are still discussed in the literature [28, 29]. Nevertheless, the majority of studies seem to be moving towards consensus on the attribution of electrochemical phenomena to both parts of the electrochemical curve. Figure 6.4 shows a typical curve for inserting sodium into hard carbon. The first sloping part of the electrochemical curve (region 1) would be attributed to insertion between the graphene planes, while the flat to low potential

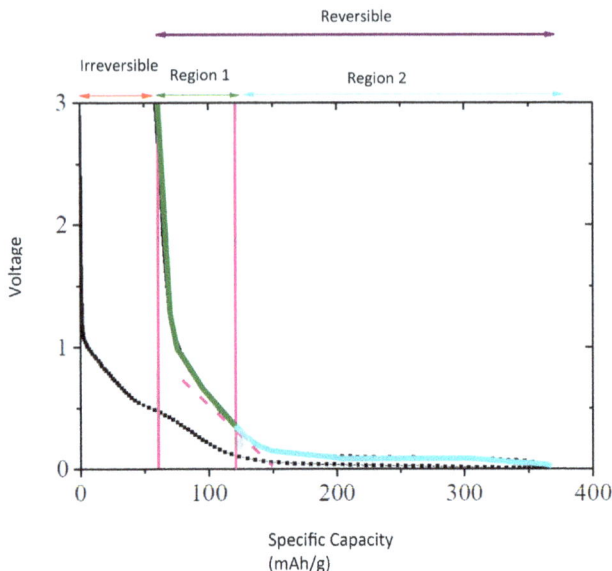

FIG. 6.4 – Typical electrochemical curve of a hard carbon cut into three distinct parts – irreversible capacity, sloping capacity (Region 1) and flat region (Region 2) [32] Permission Grenoble University Alps (Thesis V. Simone).

curve (region 2) might correspond to insertion in the microporosity. Several studies tend to confirm this mechanism [30, 31].[1]

6.3.2   Non-Carbon Materials

In addition to the carbon materials, two other families of compounds can be used as negative electrodes for sodium-ion accumulators. The use of materials forming alloys or sodium conversion compounds is envisaged to improve energy density performance, while titanium-based oxides (titanates) are used to improve power performance. However, the use of the latter leads to a reduction in the overall voltage at the terminals of the complete cell due to the high operating potential of titanates, as is already the case for Li-ion batteries.

Titanium-based compounds were studied in the 1980s: hard carbons with TiS_2 [33] and NASICON $NaTi_2(PO_4)_3$ [34]. At the time, however, they were considered to be positive electrode materials, and used with sodium metal as negative electrode. More recently, titanium oxides were the subject of numerous studies, including TiO_2 [35] and $Na_2Ti_3O_7$ [36]. Very good performances (figure 6.5) were achieved at

---

[1] *Author's note*: Since the publication of the original French version of this book, the mechanism tends to be more controversial. Many publications tend to propose insertion in between the graphene planes at lower potential.

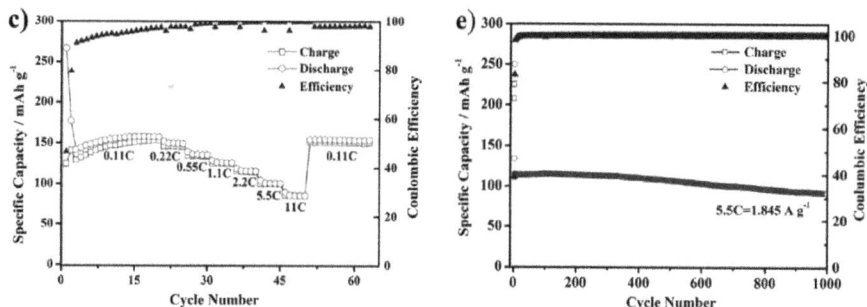

FIG. 6.5 – Cycling performance of the TiO_2 anatase in Na-ion configuration. Left: Capacity according to the regime. Right: Capacity stability over 1000 cycles. From [35] authorization Elsevier (*J. Power Sources*).

high discharge regime, with nearly 60% capacity available at 11C for TiO_2 [35], but the limitations in the selection of positive electrodes did not allow for a sodium ion cell based on these materials to be competitive with other technologies available on the market.

For alloys, the elements Sn, Ge, P and Sb can form sodium-rich phases [37]. It is important to note that silicon is not included in this category because of a diffusion barrier that is too high for sodium (lithium seems to be more promising from this point of view, see chapter 3). NaSi alloy is described in the literature; however, it appears difficult to synthesize electrochemically in batteries [38, 39]. The advantage of these compounds is that they offer high theoretical capacities, higher than hard carbons, which can exceed 500 mAh.g^{-1}, while offering sufficiently low operating potentials. In the case of tin, the $Na_{15}Sn_4$ composition may be reached, *i.e.* 490 mAh.g^{-1}, and in the case of phosphorus, it is possible to reach 2597 mAh.g^{-1} with Na_3P formation. However, as shown in figure 6.6, these very high capacities are associated with very large volume variations (more than 420% in the case of tin [40]) which appear to make their practical use as Na-ion active materials in the state of knowledge impossible and require optimization strategies (material structuring, electrode optimization...).

However, there are exceptions, such as antimony (Sb). As shown in figure 6.7, excellent stability has been demonstrated for a capacity (580 mAh g^{-1}) close to theoretical capacity, even at high discharge regime. These performances are higher than those of the same electrode tested in Li-ion battery [41–43]. Various *operando* characterization techniques (XRD, PDF, XAS) or *ex situ* (XPS) [44–47] reveal complex electrochemical mechanisms, involving the formation of mainly amorphous intermediate phases, which seem to minimize the consequences of volume changes. However, the proven criticality of Sb resources appears to be a barrier to its development.

In terms of conversion materials, there are many similarities between the Li-ion and Na-ion batteries with a high capacity that can be achieved (in the order of 1000–2000 mAh g^{-1}) and maintained over a few dozen cycles when a suitable electrode formulation is used. However, the large potential hysteresis (electrode

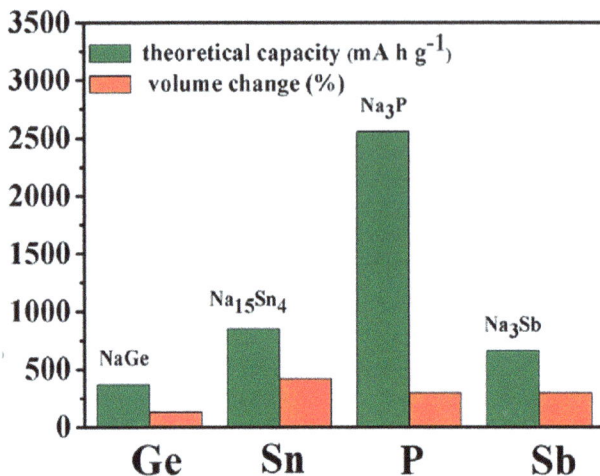

FIG. 6.6 – Specific capabilities and expected volume expansions for some p-block elements used as electrode in Na-ion batteries [37] J. of Materials Chemistry A RSC authorization).

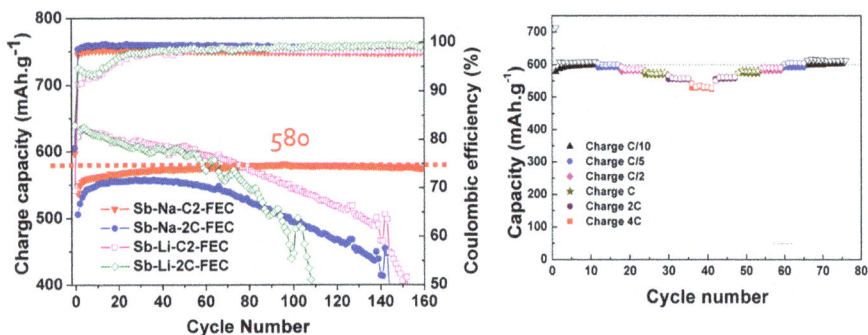

FIG. 6.7 – (Left) Specific capacities of Sb/Li and Sb/Na batteries at two different cycling rates, 2C and C/2. (Right) Sb/Na capacity at different C/10 – 4C rates. [41] ACS authorization (JACS).

polarization between charge and discharge), identified in Li-ion systems as being related to different reduction and oxidation reaction paths, is also present in Na-ion systems, and penalizes cycling performance.

6.3.3   *Positive Electrode Materials*

6.3.3.1   *Layered Oxides*

As with lithium-ion, A_xMO_2 general-formula layered oxides, where A is an alkaline ion and M is one or several transition metals, have many advantages for Na-ion [48].

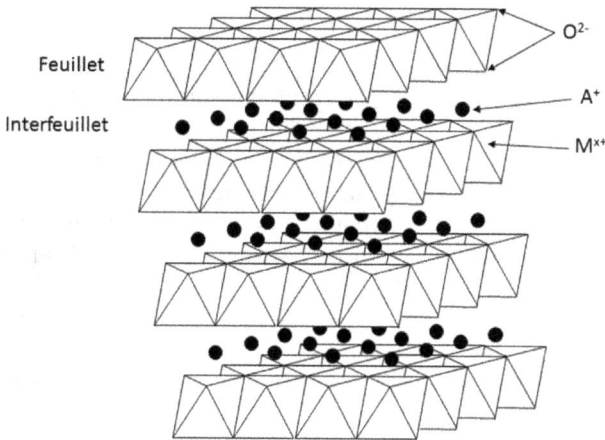

FIG. 6.8 – Typical structure of an A_xMO_2 lamellar oxide where A is an alkaline ion and M is a transition element.

Their structure can be seen as the alternative stacking of transition metal "layers" and alkali-occupied "interlayers" (figure 6.8). They are widely used as cathode materials in Li-ion batteries, such as $LiCoO_2$, $LiNiO_2$, or mixed materials such as $LiNi_{1/2-y}Mn_{1/2-y}Co_{2y}O_2$ and $LiNi_{1-y-z}Co_yAl_zO_2$ generalized "NMC".

As with lithium batteries, equivalent sodium ion lamellar oxides are good candidates for Na-ion battery electrodes. However, unlike in the case of lamellar materials containing lithium, there is a significant variation in the structure of these materials in relation to the composition of transition metals and sodium. To differentiate them, a nomenclature has been established by Delmas *et al.*, [49]. This nomenclature consists of a letter representing the nature of the site of insertion of the alkali ion (T for tetrahedral, O for octahedral and P for prismatics) and a number corresponding to the sequence of stacking of the transition metal layers. Thus, if $LiCoO_2$ and the majority of lithiated lamellar oxides have an O3 structure, sodiated compounds may generally adopt P2, P3 or O3 structures [50]. Prismatic structures are generally more stable for low sodium levels, whereas octahedral structures are stable for compositions close to $NaMO_2$ stoichiometry and lose stability during great levels of deintercallation of the alkali ion. This is due to the easy migration of the transition metals from the layer to the interlayer. Therefore, in practice, lithiated compounds are never completely disinserted. In the case of sodiated oxides, phase O3 achieves higher specific capacities [48]; however, it is more unstable and has poorer cycling behavior than phase P2. In all cases, the average discharge voltage is lower than for lithiated oxides, resulting in relatively low energy densities. Recently, however, an O3-type oxide with light lithium doping has shown encouraging performance in terms of energy density and cyclability [51].

6.3.3.2 Polyanionic Materials

These compounds are generally considered to have remarkable structural stability and have the advantage of offering electrode potential adjustable to the inductive effect of the anionic group. On the other hand, they are often less electronic conductors than oxides, and then may require the addition of electronic conductors (carbon in most of the cases) by surface modification techniques.

For this second family of positive electrode materials, the interested reader will be able to consult the very detailed review carried out by C. Masquelier and L. Croguennec [52].

The first example of sodium intercalation in a 3D polyanionic material was done by Delmas *et al.*, in 1987 [34] in the case of $NaTi_2(PO_4)_3$. Later, with the high interest in polyanionic materials generated by the discovery in 1997 by Goodenough *et al.*, of $LiFePO_4$ for Li-ion batteries [53], other compounds were studied for sodium, mainly phosphates $(PO_4)^{3-}$, pyrophosphates $(P_2O_7)^{4-}$, fluorophosphates [54] or sulfates $(SO_4)^{2-}$.

Sodium analog of the olivine structure compound $LiFePO_4$ was rapidly investigated. It exists in 3 different forms, olivine, maricite and amorphous. The maricite-type structure is not suitable for the diffusion of sodium ions with the presence of PO_4 polyhedrons that completely block sodium ions, thus not allowing a practical compound to be used. In contrast, olivine and amorphous compounds are electrochemically active with interesting practical capacities, respectively, 120 and 150 mAh.g^{-1} [55, 56]. Despite a higher capacity, the amorphous compound suffers from a lower discharge voltage of 2.4 V compared to 2.8 V for olivine. In both cases, the energy density remains insufficient for practical application.

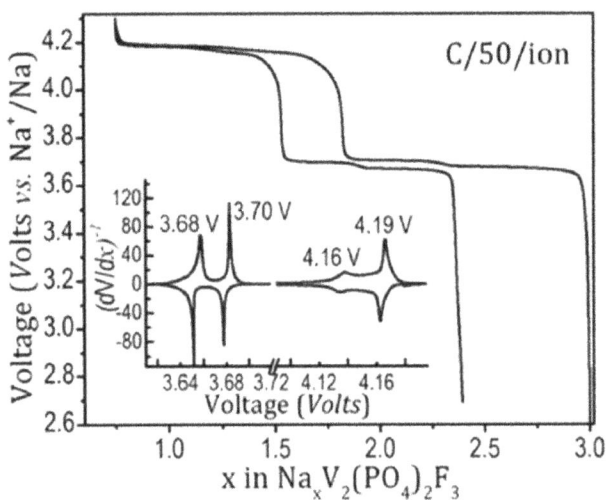

FIG. 6.9 – Cycling of a $Na_3V_2(PO_4)_2F_3$//Na cell at C/50. Insert: Reverse Derivative Curve (dV/dx) − 1 illustrating the presence of several electrochemical processes [57] ACS authorization (Chem. Mater.).

Another promising compound, $Na_3V_2(PO_4)_2F_3$ (NVPF), was first studied by Barker *et al.*, [54]. This compound has a high operating voltage with 2 plateaus at 3.8 and 4.3 V, respectively, for a specific capacity of 120 mAh.g^{-1} (see figure 6.9), leading to a high specific energy that makes NVPF the reference material as an alternative to lamellar oxides in the current state of research. It should be noted, however, that vanadium may cause problems of abundance and toxicity for the deployment of this electrode material on a large scale.

6.3.4   Electrolytes and Interfaces

As with Li-ion batteries, many electrolytes exist, and they can react with electrodes, knowing that the SEI ("Solid Electrolyte Interphase") created at the interface directly impacts performance. Prior knowledge of Li-ion systems is valuable in determining the appropriate solvent for each type of Na-ion battery. Electrolytes require a high conductivity of the order of 1 mS.cm^{-1}. Like electrode materials, electrolyte must preferably have good thermal stability over a wide range of temperatures, be based on sustainable chemistry and abundant elements. As with Li-ion systems, most electrolytes correspond to one salt (or more, $NaClO_4$, $NaBF_4$, $NaPF_6$, NaTFSI...) dissolved in a mixture of organic solvents (EC, PC, DMC, DEC...). Na-ion based electrolytes appear to show the same trends as their Li counterparts. $NaPF_6$ dissolved in a mixture of alkyl carbonates is the most commonly encountered electrolyte. However, the role of additives has yet to be clarified. Not being able to use graphite in Na-ion cells can simplify the choice of electrolyte. PC appears to be a key solvent, allowing for the establishment of a stable SEI. The use of FEC as an additive appears necessary although its mechanism is not yet fully clarified [58, 59].

Ionic liquids (Ils) could eventually become viable alternatives for Na-ion operation, which are of interest in terms of safety, although their cost and purity problems remain a limit (see chapter 5). Hybrid electrolytes combining organic/Il hybrids make a compromise between efficiency and safety, and could lower material costs. For example, using NaTFSI in EC: PC: Pyr13TFSI, for a hard carbon electrode (HC), a better SEI (as compared to a single IL), more mechanically and electrochemically stable, obtained for example a capacity of 180 mAh g^{-1} to C/10 over 40 cycles [60].

It is necessary to perfectly characterize the SEI, which determines the performance of the Na-ion batteries as well as that of the Li-ion batteries (*cf.* chapter 5). Analyses for hard carbons show a great similarity of electrode/electrolyte-formed species, but with a larger amount of inorganic species than for lithium. In addition, this layer is rough and not uniform in the case of sodium and smoother and slightly thicker for lithium. In the case of alloying and conversion materials, a thick SEI consisting of carbonates (Na_2CO_3 and $NaCO_3R$) covering the electrode at the end of the discharge has been shown; which is partially dissolved during oxidation. The mechanisms of formation and evolution of this layer are only partially clarified [61].

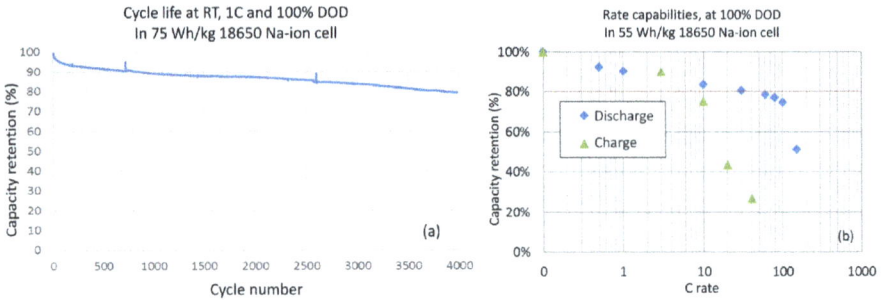

FIG. 6.10 – 18 650 Cell Performance: (a) 1C cyclability for cells of 75 Wh.kg^{-1} and (b) power ability for cells of 55 Wh.kg^{-1} [66]. Willey permission (small methods).

6.4 Full System Performance

There are currently few examples of full system tests of this battery technology. The majority of studies are on layered oxide/hard carbon. It should be noted that it is difficult to compare the performances of the different systems because capacities can be reported to the mass of cathode, anode, or even total active materials.

In 2011, the group of Komaba prepared coin-cells for the system $NaNi_{0,5}Mn_{0,5}O_2$/hard carbon and demonstrated capacities of about 200 mAh.g^{-1} of hard carbon [62]. More recently, Johnson's group showed capacities in the order of 100 mAh.g^{-1} of positive electrode for the $NaNi_{1/3}Mn_{1/3}Co_{1/3}O_2$/Hard carbon system [63]. In terms of applications, this system has been industrialized by FAR-ADION based in the United Kingdom. The cells developed are pouch-type cells of type 10 Ah; the energy density is predicted to be approximately 125 Wh.kg^{-1} for a lifetime of about 100 to a few thousand cycles depending on the application concerned [64].

Polyanionic materials technologies, while less studied, remain highly promising in terms of durability. From this point of view, they are the subject of particular attention by several French research teams, including the CEA and RS2E laboratories. In 2015, the $Na_3V_3(PO_4)_2F_3$ system was tested for the first time in a 18 650 cell [65] More recently, we showed a life span of 4000 cycles and a power density of more than 3.8 kW.kg^{-1} [66]. These results are presented in figure 6.10. This study led to the creation of the French spin-off company TIAMAT.

6.5 Outlook

6.5.1 *Low Cost Approach*

One of the main advantages of sodium is its abundance and homogeneous geolocation. Indeed, it accounts for 2% by mass of the entire earth's crust. The other interest is the replacement of copper as a current collector with a negative electrode with aluminum. From this point of view, a low-cost approach is possible for this system.

For positive electrode materials, the main barrier to the development of low-cost materials is the presence of non-abundant and/or expensive transition metals (V, Co, Ni, etc.). The relatively low energy density linked to the low capacity and/or average discharge potential is also a barrier to achieving competitive costs reduced to kWh. In the future, it will be necessary to work on all three aspects, eliminate/limit the presence of critical transition metals by content with abundant metals such as manganese, iron and titanium while preserving high discharge capacities and potentials.

In the case of negative electrode materials, hard carbon remains the most relevant candidate, but its cost is much higher than graphite: the main reason is the use of expensive precursors in its synthesis. In this respect, alternative precursors must be used. The use of bioresources is one of the approaches envisaged, especially from unrecovered agricultural waste.

Electrolyte remains a very important part of the cost of batteries: in this respect, alternative electrolytes must be assessed against Na-ion.

6.5.2 High Power Approach

The NVPF/hard carbon system shows remarkable performance in terms of load and discharge power. The conductive properties of active materials make it a very good candidate for "power" applications, such as fast charging, applications with high power calls at room temperature, or even at low temperature. In order to assess these systems under good conditions, electrolytes adapted to low temperatures must be identified. In addition, modeling work must be carried out in order to predict and achieve the best possible energy/power balance.

In conclusion, Na-ion battery technology, which has been booming in terms of scientific research in recent years, seems to be at a crossroads. Although it appears to be mature thanks to the proximity to the Li-ion technology and presents very good power and cyclability performances, its commercial deployment seems to depend on future developments in the lithium and transition metal resource markets.

References

[1] Haynes W.M. (2016) Abundance of elements in the Earth's crust and in the sea, *CRC Handb. Chem. Phys.* **97**, 2652.

[2] https://www.agenceecofin.com/metaux/2203-86395-les-prix-du-lithium-s-envolent-en-chine-pour-le-premier-trimestre-de-l-annee-2021.

[3] The Long Term Pain of New Supply (2018) Morgan Stanley.

[4]   What if I Told You... Lithium is the new gasoline (2015) Goldman Sachs Global Investment Research.

[5]   Avicenne Energy – The Rechargeable Battery Market: Value Chain and Main Trends 2019-2030-IBS – 11 March, 2021.

[6]   Vassilev S.V., Baxter D., Andersen L.K., Vassileva C.G. (2010) An overview of the chemical composition of biomass, *Fuel* **89**, 913.

[7]   NASA earth factsheet.

[8]   Larcher D., Tarascon J.-M. (2014) Towards greener and more sustainable batteries for electrical energy storage, *Nat. Chem.* **7**, 19.

[9]   Yoshio M., Brodd R.J, Kozawa A. (2010) *Lithium-ion batteries: science and technologies.* Springer Science & Business Media.

[10] Tarascon J.-M. (1986) Electrochemical insertion of lithium and sodium in the two crystallographic forms of a new molybdenum chalcogenide phase Mo15Se19, *Solid State Ionics* **18–19**, 768.

[11] Delmas C., Braconier J.-J., Fouassier C., Hagenmuller P. (1981) Electrochemical intercalation of sodium in Na_xCoO_2 bronzes, *Solid State Ionics* **4**, 165.

[12] Molenda, Delmas C., Hagenmuller P. (1983) Electronic and electrochemical properties of Na_xCo_2 cathode, *Solid State Ionics* **10**, 431.

[13] Shacklette L.W., Jow T.R., Townsend L. (1985) Rechargeable electrodes from sodium cobalt bronzes, *J. Electrochem. Soc.* **135**, 2669.

[14] Tarascon J.-M., Hull G.W. (1986) Sodium intercalation into the layered oxides Na_xMoO_4, *Solid State Ionics* **22**, 85.

[15] Ellis B.L., Nazar L.F. (2012) Sodium and sodium-ion energy storage batteries, *Curr. Opin. Solid State Mater. Sci.* **16**, 168.

[16] Doublet M.L. (2009) Batteries Li-ion: conception theorique, *Techniques de l'Ingénieur.*

[17] Ponrouch A., Monti D., Boschin A., Steen B., Johansson P., Palacin M.R. (2015) Non-aqueous electrolytes for sodium-ion batteries, *J. Mater. Chem. A* 3, 22.

[18] Hong S.Y., Kim Y., Park Y., Choi A., Choi N.S., Lee K.T. (2013) Charge carriers in rechargeable batteries: Na ions vs. Li ions, *Energy Environ. Sci.*, **6**, 2067.

[19] Eftekhari A., Jian Z., Ji X. (2017) Potassium secondary batteries, *ACS Appl. Mater. interfaces* **9**, 4404.

[20] Nobuhara K., Nakayama H., Nose M., Nakanishi S., Iba H. (2013) First-principles study of alkali metal-graphite intercalation compounds, *J. Power Sources* **243**, 585.

[21] Cao Y., Xiao L., Sushko M.L., Wang W., Schwenzer B., Xiao J., Nie Z., Saraf L.V., Yang Z., Liu J. (2012) Sodium ion insertion in hollow carbon nanowires for battery applications, *Nano Lett.* **12**, 3783.

[22] Wen Y., He K., Zhu Y., Han F., Xu Y., Matsuda I., Ishii Y., Cumings J., Wang C. (2014) Expanded graphite as superior anode for sodium-ion batteries, *Nat. Commun.* **5**, 1.

[23] Irisarri E., Ponrouch A., Palacin M.R. (2015) Review - hard carbon negative electrode materials for sodium-ion batteries, *J. Electrochem. Soc.* **162**, A2476.

[24] Balogun M.S., Luo Y., Qiu W., Liu P., Tong Y. (2016) A review of carbon materials and their composites with alloy metals for sodium ion battery anodes, *Carbon* **98**, 162.

[25] Mochida I., Ku C.H., Korai Y. (2001) Anodic performance and insertion mechanism of hard carbons prepared from synthetic isotropic pitches, *Carbon* **39**, 399.

[26] Dahbi M., Yabuuchi N., Kubota K., Tokiwa K., Komaba S. (2014) Negative electrodes for Na-ion batteries, *PCCP* **16**, 15007.

[27] Hori H., Shikano M., Kobayashi H., Koike S., Sakaebe H., Saito Y., Tatsumi K., Yoshikawa H., Ikenaga E. (2013) Analysis of hard carbon for lithium-ion batteries by hard X-ray photoelectron spectroscopy, *J. Power Sources* **242**, 844.

[28] Bommier C., Surta T.W., Dolgos M., Ji X. (2015) New mechanistic insights on Na-ion storage in nongraphitizable carbon, *Nano Lett.* **15**, 5888.

[29] Tsai P.C., Chung S.C., Lin S.K., Yamada A. (2015) Ab initio study of sodium intercalation into disordered carbon, *J. Mater. Chem. A* **3**, 9763.

[30] Simone V., Boulineau A., de Geyer A., Rouchon D., Simonin L., Martinet S. (2016) Hard carbon derived from cellulose as anode for sodium ion batteries: dependence of electrochemical properties on structure, *J. Energy Chem.* **25**, 761.

[31] Stevens D.A., Dahn J.R. (2000) In situ small-angle X-ray scattering study of sodium insertion into a nanoporous carbon anode material within an operating electrochemical cell, *J. Electrochem. Soc.* **147**, 4428.

[32] Simone V. (2016) Développement d'accumulateurs sodium-ion, Thèse de Doctorat, *Université Grenoble Alpes*.

[33] Newman G.H., Klemann L.P. (1980) Ambient Temperature Cycling of an Na-TiS Cell, *J. Electrochem. Soc.* **127**, 10.

[34] Delmas C., Cherkaoui F., Nadiri A., Hagenmuller P. (1987) A nasicon-type phase as intercalation electrode: NaTi2(PO4)3, *Mater. Res. Bull.* **22**, 5.

[35] Wu L., Buchholz D., Bresser D., Gomes Chagas L., Passerini S. (2014) Anatase TiO_2 nanoparticles for high power sodium-ion anodes, *J. Power Sources* **251**, 379.

[36] Senguttuvan P., Rousse G., Seznec V., Tarascon J.-M., Palacín M.R. (2011) $Na_2Ti_3O_7$: lowest voltage ever reported oxide insertion electrode for sodium ion batteries, *Chem. Mater.* **23**, 4109.

[37] Liu Y., Cao K., Zhao Y., Jiao L., Wang Y., Yuan H. (2015) Update on anode materials for Na-ion batteries, *J. Mater. Chem. A* **3**, 17899.

[38] Shimizu M., Usui H., Fujiwara K., Yamane K., Sakaguchi H. (2015) Electrochemical behavior of SiO as an anode material for Na-ion battery, *J. Alloys Compd.* **640**, 440 (Aug. 2015).

[39] Xu Y., Swaans E., Basak S., Zandbergen H.W., Borsa D.M., Mulder F.M. (2016) Reversible Na-ion uptake in Si nanoparticles, *Adv. Energy Mater.* **6**, 1501436.

[40] Liu Y., Zhang N., Jiao L., Tao Z., Chen J. (2015) Ultrasmall Sn nanoparticles embedded in carbon as high-performance anode for sodium-ion batteries, *Adv. Funct. Mater.* **25**, 214.

[41] Darwiche A., Marino C., Sougrati M.T., Fraisse B., Stievano L., Monconduit L. (2012) Better cycling performances of bulk Sb in Na-ion batteries compared to Li-ion systems: an unexpected electrochemical mechanism, *J. Am. Chem. Soc.* **13**, 20805.

[42] Jow T.R., Shacklette L.W., Maxfield M., Vernick D. (1987) The role of conductive polymers in alkali-metal secondary electrodes, *J. Electrochem. Soc.* **134**, 1730.

[43] Darwiche A., Dugas R., Fraisse B., Monconduit L. (2016) Reinstating lead for high-loaded efficient negative electrode for rechargeable sodium-ion battery, *J. Power Sources* **304**, 1.

[44] Chevrier L., Ceder G. (2011) Challenges for Na-ion negative electrodes, *J. Electrochem. Soc.* **158**, A101.

[45] Ellis L.D., Hatchard T.D., Obrovac M.N. (2012) Reversible insertion of sodium in tin, *J. Electrochem. Soc.* **159**, A1801.

[46] Ellis L.D., Wilkes B.N., Hatchard T.D., Obrovac M.N. (2014) In situ XRD study of silicon, lead and bismuth negative electrodes in nonaqueous sodium cells, *J. Electrochem. Soc.* **161**, A416.

[47] Allan P.K., Griffin J.M., Darwiche A., Borkiewicz O.J., Wiaderek K.M., Chapman K.W., Morris A.J., Chupas P.J., Monconduit L., Grey C.P. (2016) Tracking sodium-antimonide phase transformations in sodium-ion anodes: insights from operando pair distribution function analysis and solid-state NMR spectroscopy, *J. Am. Chem. Soc.* **138**, 2352.

[48] Han M.H., Gonzalo E., Singh G., Rojo T. (2015) A comprehensive review of sodium layered oxides: powerful cathodes for Na-ion batteries, *Energy Environ. Sci.* **8**, 81.

[49] Delmas C., Fouassier C., Hagenmuller P. (1980) Structural classification and properties of the layered oxides, *Physica B + C* **99**, 81.

[50] Clément R.J., Bruce P.G., Grey C.P. (2015) Review-manganese-based P2-type transition metal oxides as sodium-ion battery cathode materials, *J. Electrochem. Soc.* **162**, A2589.

[51] Oh S.M., Myung S.T., Hwang J.Y, Scrosati B., Amine K., Sun Y.K. (2014) High capacity O3-type Na[Li0.05(Ni0.25Fe0.25Mn0.5)0.95]O2 cathode for sodium ion batteries, *Chem. Mater.* **26**, 6165.

[52] Masquelier C., Croguennec L. (2013) Polyanionic (Phosphates, silicates, sulfates) frameworks as electrode materials for rechargeable Li (or Na) batteries, *Chem. Rev.* **113**, 6552.

[53] Pahdi A.K., Nanjundaswamy K.S., Goodenough J.B. (1997) Phospho-olivines as positive-electrode materials for rechargeable lithium batteries, *J. Electrochem. Soc.* **144**, 1188.

[54] Gover R.K.B., Bryan A., Burns P., Barker J. (2006) The electrochemical insertion properties of sodium vanadium fluorophosphate, $Na_3V_2(PO_4)_2F_3$, *Solid State Ionics* **177**, 1495.

[55] Moreau P., Guyomard D., Gaubicher J., Boucher F. (2010) Structure and stability of sodium intercalated phases in olivine FePO4, *Chem. Mater.* **22**, 4126.

[56] Li C., Miao X., Chu W., Wu F., Tong D.G. (2015) Hollow amorphous $NaFePO_4$ nanospheres as a high-capacity and high-rate cathode for sodium-ion batteries, *J. Mater. Chem.* A **3**, 8265.

[57] Bianchini M., Fauth F., Brisset N., Weill F., Suard E., Masquelier C., Croguennec L. (2015) Comprehensive investigation of the $Na_3V_2(PO_4)_2F_3$–$NaV_2(PO_4)_2F_3$ system by operando high resolution synchrotron X-ray diffraction, *Chem. Mater.* **27**, 3009.

[58] Simone V., Lecarme L., Simonin L., Martinet S. (2017) Identification and quantification of the main electrolyte decomposition by-product in Na-ion batteries through FEC: towards an improvement of safety and lifetime, *J. Electrochem. Soc.* **164**, A145.

[59] Dugas R., Ponrouch A., Gachot G., David R., Palacin M.R., Tarascon J.M. (2016) Na reactivity toward carbonate-based electrolytes: the effect of FEC as additive, *J. Electrochem. Soc.* **163**, A2333.

[60] Monti D., Ponrouch A., Palacin M.R., Johansson P. (2016) Towards safer sodium-ion batteries via organic solvent/ionic liquid based hybrid electrolytes, *J. Power Sources* **324**, 712.

[61] Bodenes L., Darwiche A., Monconduit L., Martinez H. (2015) The solid electrolyte interphase a key parameter of the high performance of Sb in sodium-ion batteries: comparative X-ray photoelectron spectroscopy study of Sb/Na-ion and Sb/Li-ion batteries, *J. Power Sources* **273**, 14.

[62] Komaba S., Murata W., Ishikawa T., Yabuuchi N., Ozeki T., Nakayama T., Ogata A., Gotoh K., Fujiwara K. (2011) Electrochemical Na insertion and solid electrolyte interphase for hard-carbon electrodes and application to Na-ion batteries, *Adv. Funct. Mater.* **21**, 3859.

[63] Kim D., Lee E., Slater M., Lu W., Rood S., Johnson C.S. (2012) Layered Na [Ni1/3Fe1/3Mn1/3]O2 cathodes for na-ion battery application, *Electrochem. Commun.* **18**, 66.

[64] https://www.faradion.co.uk/technology-benefits/strong-performance/.

[65] Cailloce L. (2015) Batterie sodium-ion: une révolution en marche, *J. du CNRS*.

[66] Broux T., Fauth F., Hall N., Chatillon Y., Bianchini M., Bamine T., Leriche J.-B., Suard E., Carlier D., Reynier Y., Simonin L., Masquelier C., Croguennec L. (2018) High rate performance for carbon-coated $Na_3V_2(PO_4)_2F_3$ in Na-ion, *Small Methods* 1800215.

Chapter 7

Metal-Sulfur Batteries

Céline Barchasz, Frédéric Le Cras, Fabien Perdu and Rémi Dedryvère

In order to increase the gravimetric energy density of batteries while reducing their cost and environmental impact, research is turning to the use of elemental sulfur (or counterparts) at the positive electrode. This active material, although very promising, suffers from several limitations related to the unconventional discharge mechanism that is very different from lithium-ion systems. Moreover, and contrarily to Li-ion, the system usually requires the use of a metallic anode such as lithium, sodium or magnesium, to compensate the absence of lithium in the positive electrode.

7.1 The Metal-Sulfur Cell

7.1.1 Advantages and Comparison with Other Technologies

The metal-sulfur battery consists of a composite positive electrode based on elemental sulfur (S_8) and a metal negative electrode (lithium, sodium or magnesium, figure 7.1) [1, 2]. Sulfur allows the exchange of 16 electrons per S_8 molecule, resulting in a high theoretical specific capacity (1675 mAh.g_{Sulfur}^{-1}) [3], well above the values achievable in Li-ion systems (maximum 250 mAh.$g_{cathode\ material}^{-1}$). As a result, the theoretical gravimetric energy density of the various metal-sulfur systems is particularly high (table 7.1). In addition, unlike the materials used in conventional Li-ion systems, sulfur has the advantage of being inexpensive (100 €.ton^{-1}), abundant, non-critical and non-toxic – which explains the growing interest in this new battery technology [4, 5].

DOI: 10.1051/978-2-7598-2555-4.c007

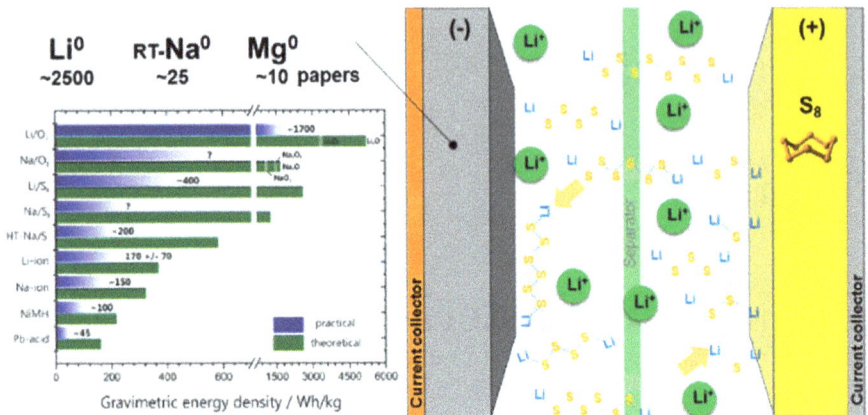

FIG. 7.1 – Schematic representation of a metal-sulfur cell, with dissolution of active material as polysulfides in the electrolyte while cycling.

7.1.2   *Working Mechanism of the Metal-Sulfur Cell*

The principle of the lithium–sulfur battery was proposed by Herbert and Ulam in 1957 [11]. The first papers were published in the 1970s [12–14], in particular thanks to the pioneering work of Peled *et al.* in the 1980s [15, 16]. Work on Na–S and Mg–S systems is much more recent, mainly illustrated by the work of Muldoon *et al.* and Fichtner *et al.* [7, 17].

The discharge mechanism of metal-S_8 batteries differs from that of Li-ion systems, since they do not involve insertion, conversion, or alloying reactions with the metal. In metal-sulfur systems, the charging and discharging reactions relate on multi-stage electrochemical processes, in which the active material successively changes from the solid state (S_8 or M_xS_2) to the soluble state (formation of different M_xS_n polysulfides intermediate products), depending on the state of charge of the battery [18, 19]. As a result, the electrolytes used for metal-sulfur batteries also differ from those used in Li-ion systems (formulations are usually based on $LiPF_6$ salt and carbonates, which are reactive with polysulfides [20]). They are generally composed of ether solvents and LiTFSI salt, more occasionally of sulfones or ionic liquids [1].

The operation of lithium–sulfur "Li–S" technologies is very similar to that of sodium–sulfur "Na–S" batteries (illustrated in section 2.3): dissolution of the active material during the discharge in the presence of liquid electrolyte, formation of polysulfide intermediate species, etc. The two systems have a number of common limitations: low practical capacity, limited cyclability, appearance of an electrochemical shuttle mechanism in charge, relatively high self-discharge, low electronic conductivity of sulfur species, negative electrode reversibility.

However, there are significant differences between Li–S and Na–S technologies: the Na–S battery operating at room temperature offers a lower cell voltage than Li–S counterparts, which reduces the energy density of the battery and limits its practical performances. In addition, the formation of solid polysulfide intermediate

TAB. 7.1 – Comparison of the performances, advantages and limitations of Li–S, "room temperature" (RT) Na–S and Mg–S technologies [6–10]. TBD: "To Be Determined", no proof of concept and no full cell demonstrated yet.

Technology	Theoretical energy density		Expected performances			Advantages	Limitations
	$Wh/kg_{active\ material}$	$Wh/L_{active\ material}$	Wh/kg_{cell}	$mAh/g_{active\ material}$	Cell voltage		
Li-ion	560 ($LiCoO_2$)	2800 ($LiCoO_2$)	≤220	150 ($LiCoO_2$)	3.9 V	Commercial	Presence of Co Limited energy density
Li-ion perspectives	900 (NMC)	2000–3500 (NMC)	≤300	250 (NMC)	3.6 V	Conventional Li-ion	Presence of Ni, Co Cost of cathode raw materials
Li-ion perspectives	654 (LNMO)	3000 (LNMO)	≤250	135 (LNMO)	4.85 V	Conventional Li-ion	"High voltage" electrolytes required
Li–S	2500 (Li_2S)	2900 (Li_2S)	300–600	1675 (S_8)	2.2 V	Low potential Li (performances)	Formation of Li dendrites (safety, cyclability)
(RT) Na–S[1]	1275 (Na_2S)	1580 (Na_2S)	TBD		1.85 V	Na abundance Suppression of Cu current collector (↘cost)	Reactivity and processing of Na Cell voltage (performances)
Mg–S	1700 MgS	>4000 MgS	TBD		1.7 V?	No formation of Mg dendrites (safety) Volumetric energy density	Cell voltage Cyclability of Mg electrode (performances)

[1]"High temperature" sodium–sulfur technology, working above 300 °C, is not considered in this chapter.

species during the discharge, as well as the volume expansion associated with sodium polysulfides, leads to further limitations in terms of cycling performances [18, 19]. Finally, the high reactivity of sodium metal, even in an anhydrous environment, poses real problems of implementation and integration in an industrial process. This explains why most of the work is focused on Li–S technology, as shown by the important number of publications and patents on the subject, particularly since 2010 [21].

The magnesium–sulfur "Mg–S" technology differs from the two previous technologies due to the use of a magnesium negative electrode, which requires the use of very specific families of electrolytes allowing the reversible deposition of metallic magnesium (organomagnesium or organohaloaluminates) [8].

7.1.3 The (Li,Na)-ion Sulfur Cell

There is also a version of (Li,Na)-ion sulfur [22, 23] cell which, like conventional Li-ion batteries, uses a positive electrode in its reduced and lithiated state (here Li_2S or Na_2S) with a negative electrode composed of silicon or carbon. In principle, these systems prevent the problems of cyclability and safety related to the use of a metal negative electrode. However, the use of a negative electrode, which then operates at a higher potential than metal foils, leads to a decrease in the operating voltage compared to metal-sulfur systems (1.8 V for the Li_2S–Si system, for example, compared to 2.1 V for Li–S). Consequently, the expected energy densities for (Li,Na)-ion sulfur systems are lower than for metal-sulfur targets (*e.g.* 300–400 $Wh.kg^{-1}$ cell for the Li_2S–Si system compared to 300–600 $Wh.kg^{-1}$ cell for Li–S) [24].

7.2 Technology State of the Art and Performances

7.2.1 Main Actors

The maps of the main European and World players in the R&D on metal-sulfur technologies are given in figure 7.2.

7.2.2 Understanding the Complex Mechanism

This chapter describes in more detail the working mechanism of lithium–sulfur batteries using a liquid electrolyte. It should be noted that solid-state metal-sulfur batteries also exist, in which the sulfur active material is not solublilized during cycling. Their particular operation mechanism is also presented below.

The Li–S system is based on different operating mechanisms from that of conventional lithium batteries, since most of the species that are electrochemically active (polysulfides) are soluble in the liquid electrolytes. Thus, unlike in conventional Li-ion batteries, most redox phenomena involve a liquid phase reacting at the electrode surface (usually composed of carbon) and within its porous network. The operation of metal-sulfur systems is therefore highly dependent on the composition

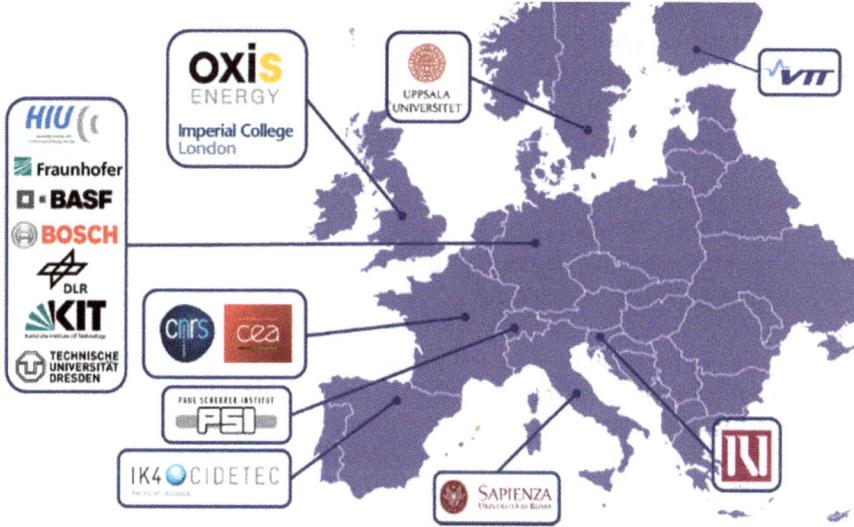

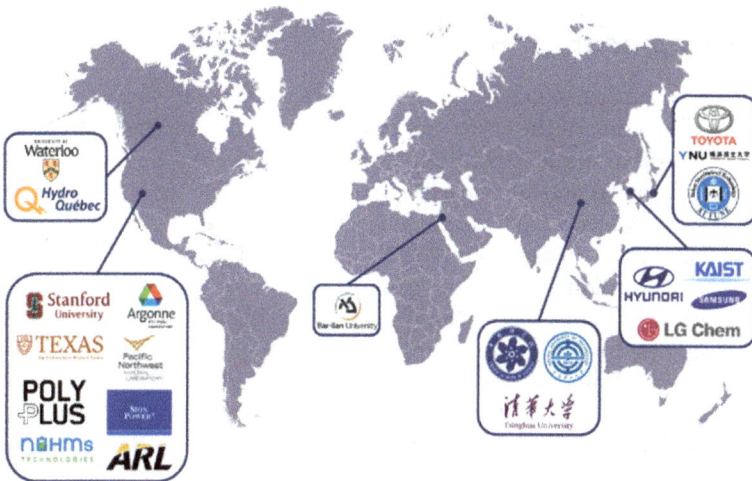

FIG. 7.2 – Maps of the main European and World players in Li–S, (RT) Na–S and Mg–S technologies.

of the liquid electrolyte. In addition, the usual carbonate solvents are generally prohibited because of their reactivity with lithium polysulfides (formation of organic thio-carbonates) [20].

A consequence of the electrochemically active species solubility is related to their possible diffusion and migration through the cell, which has an impact on the other battery components. In turn, researchers have therefore developed characterization methods to identify and locate sulfur species while cycling, in particular using *in situ* or *operando* methods.

Thanks to these characterization methodologies, considerable progress has been made in the understanding of redox processes during charging and discharging, and identification of the sulfur-reduction products.

It has been shown experimentally, thanks to the use of different characterization methods, such as X-ray absorption spectroscopies, UV–visible, Raman, NMR, XPS, EPR, etc. that sulfur reduction is a multistep process. This multistep process starts with the opening of the S_8 ring and the formation of long-chain polysulfides S_n^{2-} (figure 7.3), followed by a decrease in the average length of the n-chains, finally leading to the formation of the insoluble final discharge product Li_2S [26–33]. The formation of the $S_3^{\cdot-}$ anion radical was also demonstrated [29, 31].

The development of new techniques, now available at large scale facilities (synchrotrons), such as X-ray phase contrast tomography (figure 7.4), has allowed a better understanding of the migration of sulfur species in the different parts of the electrode and the mechanisms leading to irreversible loss of active material during cycling [34].

7.2.3   Development Strategies

Since the operation of other metal-sulfur technologies is relatively similar to the one of Li–S system (dissolution of the active material during the discharge, formation of polysulfides intermediate species), these technologies also share a certain number of similar limitations: low practical capacity, limited cyclability, appearance of an electrochemical shuttle mechanism during charging, relatively high self-discharge, low electronic conductivity of the sulfur species, limited reversibility of the negative electrode (illustrated in figure 7.5). Consequently, the teams working on Na–S and Mg–S technologies are usually also involved in the development of the Li–S system, such as the Fraunhofer Institute, the Helmholtz Institute at Ulm, the University of Texas at Austin or the Bar Ilan University.

Similarly, many electrode optimization strategies, first validated for the Li–S technology (figure 7.6) [35–37], are often relevant and adopted for Na–S and Mg–S systems [38], such as:

– Confinement of the sulfur active species within a microporous carbon matrix [39].
– Use of carbon nanotube papers in combination with a catholyte, with the formula M_xS_n, to produce a cathode with a high specific surface area [40].
– Use of compatible electrolyte formulations based on ethers [17].

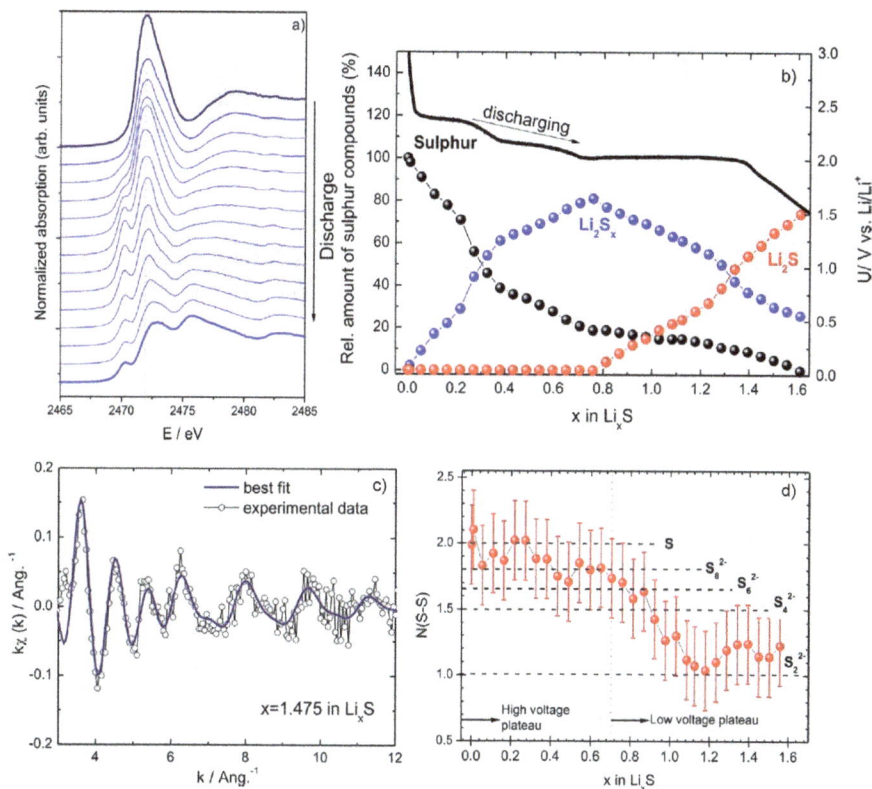

FIG. 7.3 – (a) Operando X-ray Absorption Spectroscopy (XANES) measurements of a Li–S cell during the discharge, et (b) percentages of the different electroactive species during the discharge determined by XANES. (c) Exploitation of EXAFS signal and (d), evaluation of the polysulfide chain length Li_2S_n during the discharge determined by EXAFS [25].

- Use of "barrier" membranes in the electrolyte, based on perfluorinated polymers or carbons, aimed at preventing or limiting the contact between polysulfides and the reducing negative electrode [41].
- The use of electrolyte additives or protective layers to protect the lithium negative electrode [42].

Table 7.2 summarizes and compares the major results that have been reported for different metal-sulfur technologies. It is important to note that the different results are sometimes difficult to compare, since the test conditions and the cell design (button cells, prototypes) are often different, while the output quantities are not always identical too (capacity in $mAh.g^{-1}$ of sulfur, capacity expressed in Ah or energy densities in $Wh.kg^{-1}$, see figure 7.7). Nevertheless, the most representative results of the different actors are listed in this table, in $Wh.kg^{-1}$ of cell ($Wh.kg_{cell}^{-1}$) if available, or in $mAh.g^{-1}$ of sulfur ($mAh.g_{sulfur}^{-1}$) if not available.

FIG. 7.4 – Phase contrast X-ray tomography of a carbon–sulfur electrode. Visualization of the sulfur electrode morphology changes during cycling, leading to the loss of electrical contacts as well as electroactive sulfur species [34].

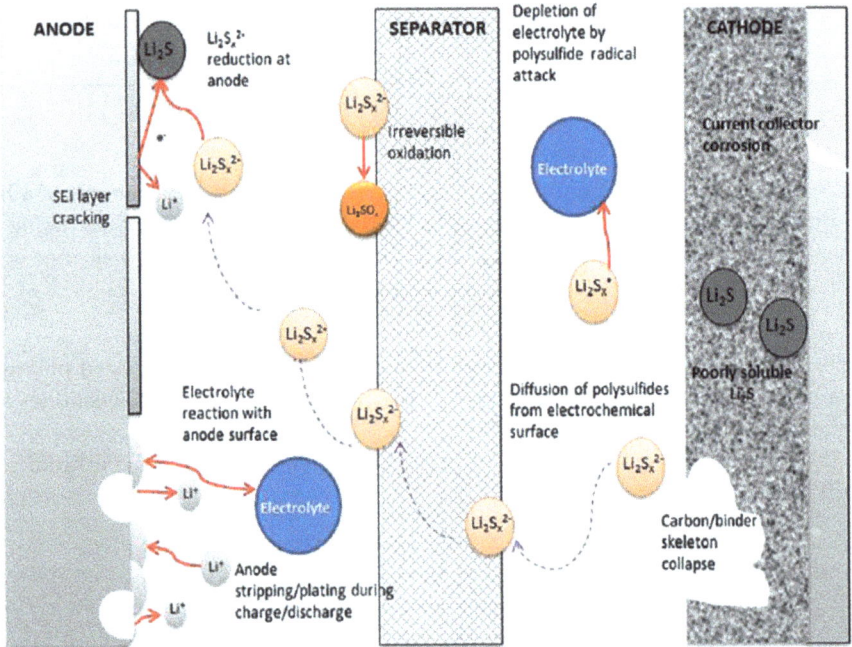

FIG. 7.5 – Schematic representation of a lithium–sulfur cell [18].

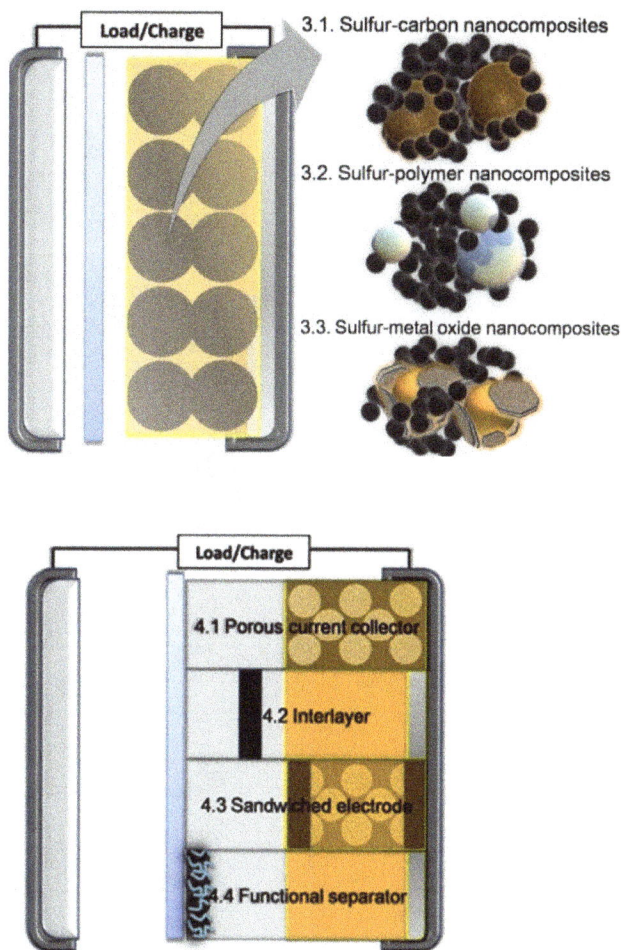

FIG. 7.6 – Examples of the strategies that have been proposed in the literature to address the Li–S cell limitations: at the positive electrode level (sulfur composites, left) and at the cell level (battery component, right) [37].

Figure 7.7 illustrates the importance of the positive sulfur electrode composition (the synthesis has been made on a selection of 84 most representative papers), and the great variability and dispersion of the results reported in the literature. While the electrode loadings vary between 0.1 and 10 $mg_{sulfur}.cm^{-2}$, considering only the capacity in relation to the quantity of active material (sulfur) is not sufficient to reflect the electrode performance, which can be significantly reduced if the carbon content in the electrode is too high ($mAh.g_{cathode}^{-1}$) or the current collector too heavy ($mAh.g_{cathode+collector}^{-1}$). For a complete comparison, these last two parameters are the most relevant to be compared from an applicative point of view ($Wh.kg^{-1}$).

TAB. 7.2 – Comparison of major results reported in 2016 for Li–S, (RT) Na–S and Mg–S technologies.

Technology	Approach(es)	Performances			Maturity	Actor(s)	[Ref]
		Energy density (Wh.kg^{-1}) Or practical capacity (mAh.g$_{Sulfur}^{-1}$)	Cyclability	Other			
Li-S	UltralightCell	300 Wh.kg^{-1}	>80	2C 2.1 V	TRL6 Prototypes (12 Ah)	OXIS Energy	[43]
Li-S	LonglifeCell	150 Wh.kg^{-1}	>1400	2C 2.1 V	TRL6 Prototypes (10 Ah)	OXIS Energy	[44]
Li-S	Protected lithium metal (Licerion®)	500 Wh.kg^{-1}	400 cycles	AD	TRL6 Prototypes (20 Ah)	SION Power	[45]
Li-S	Protected lithium metal, aqueous electrolyte	400 Wh.kg^{-1}	AD	AD	AD	POLYPLUS	[46]
Li-S	Organic/inorganic Gelled electrolyte	1400 (1st)	>250	3C	TRL3	NOHMS	[47]
Li-S	Use of fluorinated ethers	1200 (50th) 1600 (1st) 1200 (50th)	>50	2.1 V C/10 1.9 V	Small cell TRL3 Small cell	Argonne	[48]
Li-S	Porous carbon layer with high specific surface	1500 (1st)	>200	2C	TRL3	Univ. Texas	[49]
Li-S	3D carbon-based current collector	1000 (200th) 800 (1st)	>1000	2.1 V 0,2C	Small cell TRL3	Univ. Tsinghua	[50]
Li-S	Polar host structure at the cathode	700 (500th) 800 (1st)	>1500	2.1 V C	Small cell TRL3	Univ. Waterloo	[51]
Li-S	Transition metal electrode additive as redox mediator	800 (200th) 1400 (1st)	>300	2.1 V C/2	Small cell TRL3	BASF	[52]
Li-S	Multimodal porous carbon	1000 (300th) 1200 (1st) 1000 (100th)	>100	2 V 2C 2.1 V	Prototypes TRL3 Prototypes	Fraunhofer	[53]

Tab. 7.2 – (continued).

Li–S	Vertically aligned carbon nanotubes and catholyte	1100 (1st)	>100	C/2	TRL3	CEA	[54]
Li–S	Fluorinated reduced graphene interlayer	800 (100th) 1200 (1st)	>100	2.1 V C/10	Prototypes TRL3	NIC	[55]
Li–S	Carbon foam and catholyte	700 (100th) 1600 (1st)	>50	2 V C/10	Small cell TRL3	LRCS	[56]
Li–S	Nanostructured carbon/sulfur composite	1000 (50th) 1200 (1st)	>40	2 V C/100	Small cell TRL3	Chinese Academic of Science	[57]
Li–S	Solvent-in-Salt electrolyte	900 (40th) 1200 (1st)	>50	2.1 V C/12	Prototypes (primary) TRL3	Yokohama National University	[58]
Li–S	Thio-LISICON solid electrolyte	700 (50th) 2000 (1st)	20	2.1 V C/100?	Small cell TRL3	Tokyo Institute of Technology	[59]
Li–S	POE-LiTf polymer electrolyte	500 (20th) 1000 (1st)	>30	1.5 V 2 V	Small cell TRL3	La Sapienza Univ	[60]
Li–S	Oxide type additive in the cathode	750 (30th) 1200 (1st)	>300	C/10	Small cell TRL3	Stanford	[61]
Li–S	Polydopamine dendrimer binder	1000 (300th) 1200 (1st) 900 (100th)	>100	2.1 V C/5 2.1 V	Small cell TRL3 Small cell	PNNL	[62]
Li–S	Li_3PS_4 solid electrolyte	1300 (1st) 1200 (50th)	>50	C/60 1.5 V	TRL3 Small cell	Konan Univ	[63]

TAB. 7.2 – (continued).

Technology	Approach(es)	Performances			Maturity	Actor(s)	[Ref]
		Energy density (Wh.kg$_{coll}^{-1}$) Or practical capacity (mAh.g$_{Sulfur}^{-1}$)	Cyclability	Other			
Li–S	LiBH$_4$–LiCl electrolyte	1500 (1st) 600 (5th)	>5	C/33 2 V	TRL3 Small cell	Tohoku Univ	[64]
Na–S	Use of glyme electrolyte	1000 (1st)	>40	0.5C	TRL3	Gyeongsang Univ	[65]
Na–S	Hard carbon negative electrode	800 (5th) 1000 (1st) 400 (500th)	>1000	1.8 V	Small cell TRL3	Fraunhofer	[53]
Na–S	Carbon fibers + Na$_2$S, Nafion® based protective membrane	600 (1st)	>100	0.3C	TRL3	Texas Univ	[66]
Na–S	Confinement du soufre dans un carbone microporeux	500 (100th) 1600 (1st)	>200	1.8 V 1.4 V	Small cell TRL3	Chinese Academy of Sciences	[39]
Mg–S	Use of glyme electrolyte, understanding the discharge mechanism	1000 (20th) 800 (1st)	<5	1.65 V	Small cell TRL3	HIU	[17]
Mg–S	Lithium salt mediator, understanding the discharge mechanism	200 (5th) 1000 (stab)	>30	1.7 V/1 V	Small cell TRL3	US Army	[9]
Mg–S	Graphene/sulfur composite	1000 (1st) 200 (5th)	<5	1 V	Small cell TRL3 Small cell	HIU	[67]

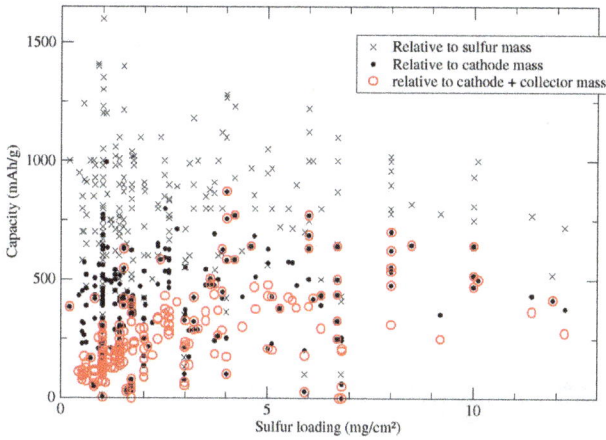

FIG. 7.7 – Cartography of practical capacity values, in mAh.g$_{sulfur}^{-1}$ (cross), in mAh.g$_{cathode}^{-1}$ (point) and in mAh.g$_{cathode+collecteur}^{-1}$ (circle), in the literature (selection of 84 representative papers), with courtesy of ESA (LiSSA project, ESA Contract n° 4 000 117 343/16/NL/HK).

7.2.4 All-Solid-State Metal-Sulfur Batteries

The electrochemical operation of all-solid-state metal sulfur batteries differs, as there should not be any formation of solubilized polysulfide intermediates in the electrolyte [68]. The direct conversion of sulfur to Li$_2$S has been reported, resulting in a single discharge plateau around 2 V. This approach is advantageous, as it avoids dissolution of the active material during cycling and the occurrence of shuttle mechanism. It should also allow to limit the formation of dendrites during charging and to improve the battery safety [69]. However, it is necessary to accommodate the volume expansion of the positive electrode during discharge and to compensate for the low electronic conductivity of the sulfur-containing materials [70, 71]. The properties required for these solid electrolytes are generally the same as for liquid electrolytes, including high ionic conductivity ($>10^{-3}$ S.cm^{-1} at 25 °C), electrochemical stability towards electrode materials, and the possibility of forming thin (<100 μm) electrolyte membranes at a moderate price (<10 \$.m^{-2}) [72]. The main solid electrolytes currently being studied for Li–S applications are Li$_2$S–P$_2$S$_5$, Li$_2$S–SiS$_2$, Li$_{3.25}$Ge$_{0.25}$P$_{0.75}$S$_4$ thio-LiSICon, Li$_6$PS$_5$Br argyrodite or LiBH$_4$ [73].

7.2.5 Industrial Actors

Until then, industrial players have been mainly focusing on the development of Li–S technology, which is more mature than Na–S and Mg–S systems. Several companies are already in the pre-industrialization phase of the Li–S technology (OXIS Energy, SION Power, PolyPlus, figure 7.8), thanks to the financial support of major

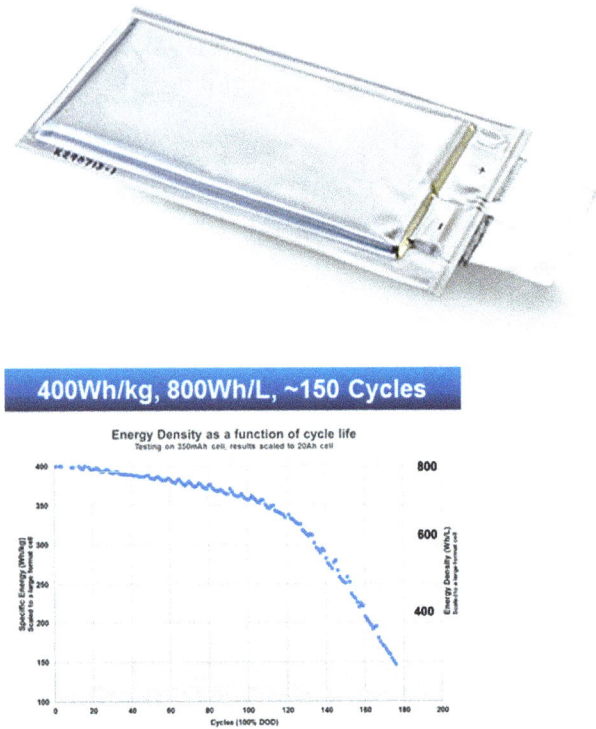

FIG. 7.8 – (Left) Picture of a 20 Ah cell developed by OXIS Energy. (Right) Performance of a 10 cm × 10 cm × 1 cm cell developed by SION Power [43–45].

industrial partners. No industrial player seems to invest on Na–S and Mg–S technologies yet. Moreover, no performance at the prototype level has been reported in the literature so far.

OXIS Energy company is particularly active and collaborates with many research centers such as Imperial College London, the Fraunhofer Institute and Airbus [74, 75]. Outside Europe, BASF has invested in SION Power American company in 2012 [76], and is developing Li–S technology through several collaborations, notably with the Canadian University of Waterloo and the Israeli University of Bar Ilan. SION Power has also received support from LG Chem for its protected lithium metal technology [77].

Concerning solid electrolyte Li–S batteries, NagaseChemtex Corp, which is particularly active in this field, is seeking to develop a low-cost solution with a material having high ionic conductivity. In particular, it has been shown that P_2S_5, which activates during the first charge, $Li_{1.5}PS_{3.3}$ ($\sigma_{Li+} \sim 2.10^{-4}$ S.cm^{-1}) and Li_2S. P_2S_5.LiI ($\sigma_{Li+} \sim 3.10^{-4}$ S.cm^{-1}) can be viable solutions for the production of all-solid Li–S batteries [78–80]. All-solid Li-ion sulfur systems have also been studied, but the resulting low cell voltage (1.5 V) appears to be incompatible with the target of high energy density [80, 81].

7.3 Perspectives and Applications

High expectations are placed on the Li–S system, which is expected to be one of the most mature "post Li-ion" systems, as evidenced by industrial interests and proofs of concept. In addition to the high theoretical specific energy density of the positive electrode, this system has also the advantage of using lithium metal, which allows higher cell voltage compared to Na–S and Mg–S analog systems.

The Na–S system is similar to Li–S (operation, limitations). Despite the advantages of low cost and abundance of sodium, the expected energy density is lower (lower voltage), and the interest of this technology has not been clearly demonstrated yet. For the Mg–S system, it presents the problems of both sulfur (ageing, dissolution) and magnesium (reversibility of the anode, complexity of the electrolytes). In addition, the cell voltage is even lower than that of sodium batteries. Nevertheless, the use of this divalent metal could still make it possible to limit the formation of dendrites and even modify the properties of the polysulfides formed. In any case, work on these two other metal-sulfur systems remains essentially limited to the field of fundamental research.

Because of the high gravimetric energy density expected for Li–S technology, the first targeted applications are in the field of space and aeronautics, as well as military applications, as shown by the first proof of realization (*e.g.* Zephyr HAPS) in particular by SION Power company [82]. At the same time, as no gain in volumetric energy density is expected for Li–S and Na–S technologies compared to best Li-ion systems, it is unlikely that these technologies will be relevant for portable and embedded electronic applications. Only Mg–S technology could theoretically compete with Li-ion technology in this respect, provided that it becomes a reality. Moreover, if significant progress in terms of cycling life are made, metal-sulfur technologies could eventually become a relevant technology for stationary storage (*e.g.* storage of renewable energies) as a significant decrease in the cost in €.kWh^{-1} is compared to Li-ion, or even conventional Pb/acid or alkaline systems. A further reduction in cost and an enhanced cyclability performance could also open their use to applications such as electric vehicles.

Lithium-sulfur technology has benefited from major advances in recent years, leading to the development of high-performance demonstrators and increasing maturity. Prospects are promising in terms of gravimetric energy density, while progresses are expected in terms of cycle life, mainly related to the metal anode cyclability issues. Sodium-sulfur and magnesium-sulfur systems remain less mature so far, although they may constitute advantageous alternatives to the use of lithium.

Improving the lifetime of these systems could enable to compete with high energy density lithium-ion technologies in the future, for applications such as electric vehicles or stationary electricity storage, thanks to their expected low cost and the absence of critical materials.

Bibliography

[1] Manthiram A. *et al.* (2014) *Chem. Rev.* **11**, 11751.
[2] Pope M.A. *et al.* (2015) *Adv. Energy Mater.* **5**, 1500124.
[3] Nazar L. *et al.* (2010) *J. Mater. Chem.* **20**, 9821.
[4] Report on Critical raw materials for the EU (2014).
[5] https://www.sunsirs.com/fr/prodetail-427.html.
[6] Janek J. *et al.* (2015) *Beilstein J. Nanotechnol.* **6**, 1016.
[7] Muldoon J. *et al.* (2011) *Nat. Commun.* **2**, 427.
[8] Muldoon J. *et al.* (2014) *Chem. Rev.* **114**, 11683.
[9] Gao T. *et al.* (2015) *J. Am. Chem. Soc.* **137**, 12388.
[10] Amine K. *et al.* (2014) *MRS Bull.* **39**, 395.
[11] US 3 043 896, Herbert D., Ulam J. (1962).
[12] BE 695 984, Rao M.L.B., Mallory Battery Ltd. (1966).
[13] US 3 532 543, Nole D.A., Ross V., Aerojet-General Corp. (1970).
[14] Rauh R.D., Abraham K.M., Pearson G.F., Surprenant J.K., Brummer S.B. (1979) *J. Electrochem. Soc.* **126**, 523.
[15] Peled E., Gorenshtein A., Segal M., Sternberg Y. (1989) *J. Power Sources* **26**, 269.
[16] Yamin H., Penciner J., Gorenshtein A., Elam M., Peled E. (1985) *J. Power Sources* **14**, 129.
[17] Fichtner M. *et al.* (2014) *Adv. Energy Mater.* 1401155.
[18] Wild M. *et al.* (2015) *Energy Environ. Sci.* **8**, 3477.
[19] Janek J. *et al.* (2013) *J. Power Sources* **243**, 758.
[20] Gao J. *et al.* (2011) *J. Phys. Chem. C* **115**, 25132.
[21] Hagen M. *et al.* (2015) *Adv. Energy Mater.* 1401986.
[22] Cho J. *et al.* (2015) *Adv. Energy Mater.* **5**, 1500110.
[23] Manthiram A. *et al.* (2015) *Chem. Eur. J.* **21**, 4233.
[24] Novak P. *et al.* (2015) *J. Electrochem. Soc.* **162**, A2468.
[25] Dominko R. *et al.* (2015) *J. Phys. Chem. C* **119**, 19001.
[26] Pascal T. *et al.* (2014) *J. Phys. Chem. Lett.* **5**, 1547.
[27] Patel M.U.M. *et al.* (2013) *ChemSusChem* **6**, 1177.
[28] Barchasz C. *et al.* (2012) *Anal. Chem.* **84**, 3973.
[29] Wu H.-L. *et al.* (2015) *ACS Appl. Mater. Interfaces* **7**, 1709.
[30] Hannauer J. *et al.* (2015) *ChemPhysChem* **16**, 2755.
[31] Huff L.A. *et al.* (2015) *Surf. Sci.* **631**, 295.
[32] Xiao J. *et al.* (2015) *Nano Lett.* **15**, 3309.
[33] Vizintin A. *et al.* (2015) *Chem. Mater.* **27**, 7070.
[34] Zielke L. *et al.* (2015) *Sci. Rep.* **5**, 10921.
[35] Chen R. *et al.* (2015) *Chem. Commun.* **51**, 18.
[36] Liang J. *et al.* (2016) *Energy Storage Mater.* **2**, 76.
[37] Manthiram A. *et al.* (2015) *Arch. Mater. Sci.Eng.* **75**, 70.
[38] Manthiram A. *et al.* (2015) *Small* **11**, 2108.
[39] Xin S. *et al.* (2014) *Adv. Mater.* **26**, 1261.
[40] Manthiram A. *et al.* (2014) *J. Phys. Chem. C* **118**, 22952.
[41] Kaskel S. *et al.* (2014) *Chem. Commun.* **50**, 3208.
[42] Kaskel S. *et al.* (2016) *Adv. Energy Mater.* 1502185.
[43] https://oxisenergy.com/wp-content/uploads/2016/10/OXIS-Li-S-Ultra-Light-Cell-v4.02.pdf.
[44] https://oxisenergy.com/wp-content/uploads/2016/10/OXIS-Li-S-Long-Life-Cell-v4.01.pdf.
[45] https://www.sionpower.com/technology-licerion.php.
[46] https://ppbcadmin.webfactional.com/page/technology#_lithium-sulfur.
[47] Archer L. *et al.* (2016) *J. Mater. Chem. A* **4**, 14709.
[48] Azimi N. *et al.* (2015) *ACS Appl. Mater. Interfaces* **7**, 9169.
[49] Manthiram A. *et al.* (2015) *Adv. Mater.* **27**, 1694.
[50] Peng H.-J. *et al.* (2016) *Adv. Funct. Mater.* **26**, 6351.
[51] Nazar L. *et al.* (2016) *ACS Nano* **10**, 4111.

[52]   Janek J. *et al.* (2016) *Adv. Energy Mater.* **6**, 1501636.
[53]   Kaskel S. *et al.* (2016) *Carbon* **107**, 705.
[54]   Liatard S. *et al.* (2016) *Electrochim. Acta* **187**, 670.
[55]   Vizitin A. *et al.* (2015) *Chem. Mater.* **27**, 7070.
[56]   Morcrette M. *et al.* (2016) *Solid State Sci.* **55**, 112.
[57]   Ma Y. *et al.* (2015) *Sci. Rep.* **5**, 14949.
[58]   Watanabe M. *et al.* (2016) *ACS Appl. Mater. Interfaces* **8**, 27803.
[59]   Nagao M. *et al.* (2016) *J. Power Sources* **330**, 120.
[60]   Hassoun J. *et al.* (2016) *Ionics* **22**, 2341.
[61]   Cui Y. *et al.* (2016) *Nature Commun.* **7**, 11203.
[62]   Xiao J. *et al.* (2016) *Nano Energy* **19**, 176.
[63]   Kinoshita S. *et al.* (2014) *Solid State Ionics* **256**, 97.
[64]   Unemoto A. *et al.* (2015) *Nanotechnology* **26**, 254001.
[65]   Ryu H. *et al.* (2011) *J. Power Sources* **196**, 5186.
[66]   Manthiram A. *et al.* (2016) *Chem. Mater.* **28**, 896.
[67]   Fichtner M. *et al.* (2016) *Nanoscale* **8**, 3296.
[68]   Wu B. *et al.* (2016) *J. Mater. Chem. A* **4**, 15266.
[69]   Urbonaite S. *et al.* (2015) *Adv. Energy Mater.* **5**, 1500118.
[70]   Lin Z. *et al.* (2015) *J. Mater. Chem. A* **3**, 936.
[71]   Lin Z. *et al.* (2013) *Angew. Chem. Int. Ed.* **52**, 7460.
[72]   McCloskey B.D. (2015) *J. Phys. Chem. Lett.* **6**, 4581.
[73]   Zhang S. *et al.* (2015) *Adv. Energy Mater.* **5**, 1500117.
[74]   https://oxisenergy.com/category/press-releases/.
[75]   https://www.aliseproject.com/.
[76]   https://www.sionpower.com/media-center.php?code=sion-power-announces-50-million-equity-investment-.
[77]   https://www.sionpower.com/media-center.php?code=lg-chem-acquires-rightsto-sion-power-technology.
[78]   Nagata H. *et al.* (2014) *Energy Technol.* **2**, 753.
[79]   Nagata H. *et al.* (2016) *J. Power Sources* **329**, 268.
[80]   Nagata H. *et al.* (2014) *J. Power Sources* **264**, 206.
[81]   Nagao M. *et al.* (2015) *J. Power Sources* **274**, 471.
[82]   https://www.sionpower.com/pdf/articles/Sion_Power_Zephyr_Flight_Press_Release_2014.pdf.

Chapter 8

All Solid-State Batteries

Frédéric Le Cras, Vasily Tarnopolskiy, Céline Barchasz,
Renaud Bouchet and Didier Devaux

8.1 Introduction and Overview

The main motivation for all-solid-state batteries (ASSB) is the improvement of safety of large-format Li-ion batteries used in electric transport and stationary applications [1]. Here, it is worth mentioning, that even non-flammable electrolyte cannot guarantee total safety at abuse conditions. As shown in [2], a charged ASSB can still have small heat generation in abuse conditions caused by the exothermic reaction between the anode materials and the oxygen released by the positive electrode.

Along with safety considerations, the use of an immobile and less reactive solid electrolyte should, in principle, allow the use of advanced electrode materials, such as metallic Li anode or high-voltage cathodes, which could improve energy density performance. Additionally, an ASSB (All Solid State Battery) is expected to have a wider operating temperature range, as a solid electrolyte does not freeze or boil, and a longer lifetime, as there is no dissolution of interfaces [3] or shuttling of decomposition products. Finally, using different, specifically designed, solid electrolytes for the anode and the cathode should offer a better electrochemical stability at each electrode.

For the last 50 years (figure 8.1), the main research directions are investigating solid electrolytes with high ionic conductivity (Li^+, Na^+); production processes of polymeric or ceramic materials; and are focusing on the behaviour of solid interfaces between electrolytes and electrode materials, including their modifications for faster ionic transport [3–5].

In the ideal case, the monolithic electrochemical cell is made of a metallic lithium anode; a dense separator made of a polymeric and/or a ceramic solid electrolyte; and a composite cathode containing the active material; a solid electrolyte, and an electronically conductive additive (figure 8.2). The cell design is of course conditioned by

DOI: 10.1051/978-2-7598-2555-4.c008

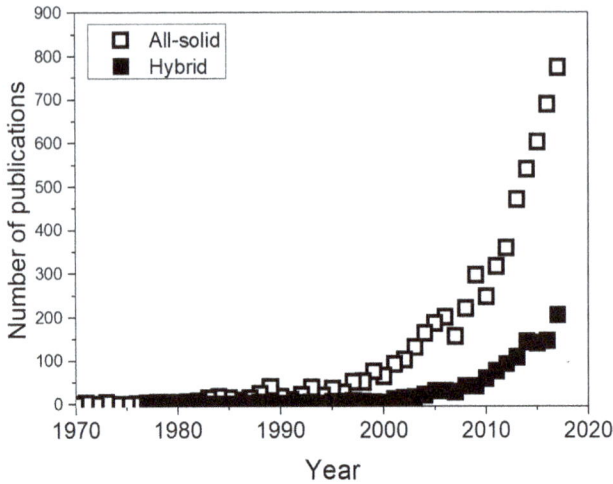

FIG. 8.1 – Evolution of the number of publications on solid electrolytes including hybrid materials (composed of polymers and ceramics conductors), Scopus.

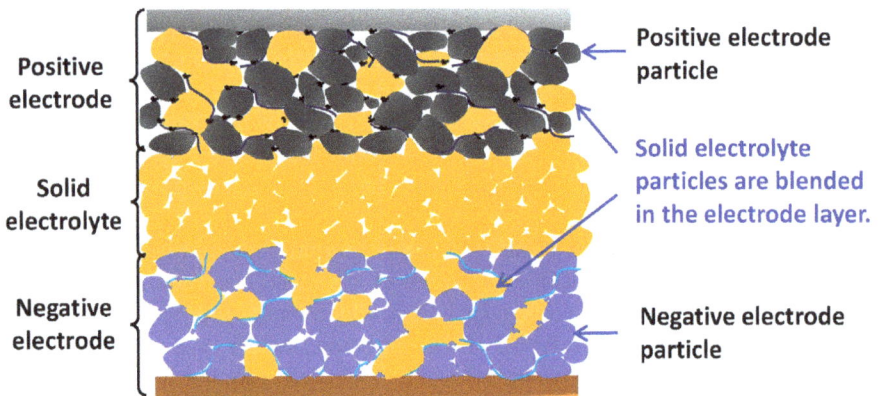

FIG. 8.2 – A principle schematic of all-solid-state Li-ion [6].

the materials used. For example, when a rigid ceramic material is selected, the cell format is preferably of the multilayer, prismatic type.

The successful development of a solid electrolyte, requires to fulfil the following specifications:

– An ionic conductivity similar to that of liquid electrolytes ($\sigma \sim 10^{-2}$ S.cm^{-1} at room temperature), including a high Li$^+$ transference number. The electronic conductivity must be very low for solid electrolytes used as separators, but may advantageously be high for SE's used in composite electrodes as pore fillers.

- A very good electrochemical stability at low and high voltages in contact with metallic anodes and high-voltage cathodes.
- A relatively inexpensive synthesis and an easy handling and cell implementation: solid electrolytes must be densified to play the role of a separator, and be easily integrated in a composite electrode.
- Preferably, a low density in order to achieve the highest possible gravimetric energy density of the cell.

Another key element to achieve the high performance of the cell is to insure the fast ionic transport through the solid interfaces (such as low grain boundary resistance between particles of solid electrolyte and electrode materials). The quality of the ionic transport through these interfaces depends on the chemistry of electrode/electrolyte boundaries, effective contact surface area between different components, and the geometry of the conduction pathways in each phase. It is also important to optimize mechanical properties of each component including the volume changes of electrode active materials during cycling. The effects of charge double layers must be taken into account especially for the cathode, as they influence the distribution of charge carriers in the vicinity of the electrode, and consequently the overall cell resistance.

8.2 Main Families of Solid Ionic Conductors

In theory, various ionic conductors can be used in the same cell to meet the requirements described above [7]. Special attention must be paid to interfaces between different solid electrolytes, to avoid the formation of non-conductive phases.

A number of families of solid ionic conductors have been identified: polymers, ceramics, glasses, hybrid materials.

8.2.1 Polymeric Solid Electrolytes

Polymer electrolytes have many interesting properties. Firstly, they can be shaped as needed thanks to various commercially available automatized industrial processes such as extrusion or stamping. They are also soft and lightweight materials, and their thicknesses (of a dozen micrometers) can be precisely controlled [8]. In addition, they do not present any saturation vapor pressure and their degradation temperature is usually higher than 250 °C. Concerning their mechanical properties, unlike solid-state inorganic electrolytes, one of the main interests of polymers lies in their flexibility and capability to wet interfaces, so that they can easily accommodate the large volume change of the active materials [9]. Among various macromolecular backbones, polyethers, and in particular poly(ethylene oxide) (PEO), are the most studied materials [103]. The chemical formula of PEO is given in figure 8.3a.

The polarized ether groups are Lewis basic and can complex Li^+ cations, which facilitates the dissolution of Li salt such as Lithium bis(trifluoromethanesulfonyl)

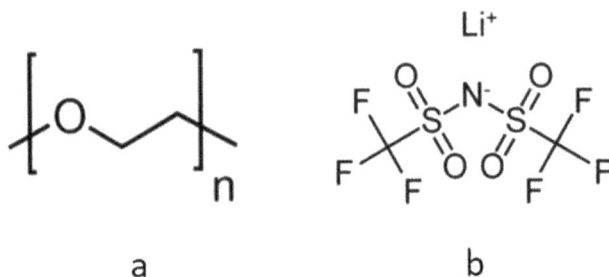

FIG. 8.3 – A typical polymer electrolyte. (a) Poly(ethylene oxide) (PEO). (b) Lithium bis (trifluoromethanesulfonyl)imide (LiTFSI).

imide (LiTFSI) (*cf.* figure 8.3b). Furthermore, ethers tend to form stable passive layers on the Li metal, permitting an adequate functioning of the Li/electrolyte interface [10, 11]. The use of a polymer electrolyte, therefore, permits to envision batteries using a negative electrode made of metallic lithium (or Li metal). Li metal is an interesting material due its high theoretical gravimetric capacity (3861 Ah.kg^{-1}) and its low reduction potential (−3.04 V *vs.* Standard Hydrogen Electrode). For comparison, Lithium-sulfur batteries (Li metal negative electrode and sulfur based positive electrode) should deliver a theoretical gravimetric and volumetric energy density in the order of 2.5 kWh.kg^{-1} (6 times higher than Li-ion) and 2.2 kWh.l^{-1} (2 times higher than Li-ion), respectively [12]. Today, Bollore Technology (Blue Solutions co.) has industrially developed a battery based on a Li metal polymer (LMP) comprising initially of a vanadium oxide (LiV_3O_8) then an iron phosphate ($LiFePO_4$) based positive electrode for an electric vehicle (Bluecar) application.

Nevertheless, PEO is a semi-crystalline polymer and below its melting temperature, its ionic conductivity is very low. Its use becomes only possible at higher temperatures, typically above 60 °C. Conductivities higher than 1 mS.cm^{-1} are achieved at temperatures above 80 °C, corresponding to the working temperature of the LMP batteries. However, at such temperatures, complexes of PEO-Li salt (mainly LiTFSI) are viscous liquids and their mechanical properties are not good enough to hinder Li dendrite growth at the negative Li metal electrode. These limitations highlight the difficulty to reach a good compromise between ionic conductivity and mechanical properties. Indeed, in polymer electrolytes, ionic transport is linked to the segmental dynamic and the free volume redistribution that increases with temperature [13]. Thus to get a high ionic mobility, the polymer matrix must be flexible, inducing low mechanical properties. Despite a great stability at low potential (at the Li metal negative electrode) the high potential stability is typically limited around 3.8 V *vs.* Li$^+$/Li. In this research area, the main objectives are to design polymer electrolytes with enhanced conductivity at low temperatures, high mechanical resistance, and stable to operate up to 4.5 V *vs.* Li$^+$/Li.

The main strategies for developing high performance, multifunctional polymers are listed below:

The first issue is related to the ionic conductivity, and notably that of Li^+ ions. One important property is the Li^+ transference number (t^+). Usually t^+ has a low value, below 0.3, because most of the polymer hosts act as Lewis bases. They interact strongly with Li^+ cations whereas the anions are weakly solvated. One approach consists in developing new anions that can facilitate salt dissociation due to a large charge delocalization all over their structure and with a large volume to hinder their mobility, such as the super $TFSI^-$ anion developed by Prof. Zhou [14]. Matrices with borate based additives act also as Lewis acids to increase salt dissociation, and, by limiting the anion mobility, cause an increase in t^+. For example, the combination of these two approaches with macro-anion base of polyoctahedral-silsesquioxane, POSS-phenyl7(BF_3L)$_3$, leads to a t^+ value of 0.6 [15].

The second issue is the lowering of the PEO melting temperature. One simple approach, widely explored, consists of using PEO oligomers grafted onto a synthetic macromolecular backbone. Thus, grafted PEO cannot crystallize anymore and the conductivity at low temperature is increased up to 5–8 10^{-5} S.cm^{-1} at 25 °C. Unfortunately, in such conditions, these materials are viscous liquids [16]. Another strategy is to create structural defects onto the macromolecular PEO chain to break its stereo-regularity and thus prevent its crystallization [17]. Other strategies use nano-particles to achieve a good compromise between mechanical properties and conductivity. Inorganic nano-particles (Al_2O_3, TiO_2, $BaTiO_3$, etc.) can be used. In particular, interesting results were reported by Hu *et al.* on nano-particles using mesoporous $LiAlO_2$ nano-sheets [15]. The ionic conductivity of the PEO comprising 15 weight % of nano-particles reaches 2.24 10^{-5} S.cm^{-1} at 25 °C, a value two orders of magnitude higher than the PEO in the absence of nano-particles. These materials also show a large enhancement regarding the electrochemical stability up to 5.0 V *vs.* Li^+/Li.

Organic materials such as cellulose nano-fibres are also reported [16]. They can reduce the degree of crystallinity which favors conductivity while strongly increasing Young's modulus of PEO matrices. The main difficulty of these solutions is to obtain a homogeneous distribution of the nano-particles. Therefore, numerous works have recently proposed to functionalize these nano-particles using PEO oligomers in order to facilitate their dispersion [17].

An elegant solution to overcome the paradigm between ionic conductivity/mechanical properties consists in using block copolymers, with one block providing mechanical properties such as polystyrene (PS) and another block made of PEO doped with a salt ensuring ionic conduction. These materials present a nano-phase separation due to the immiscibility between the blocks, leading to mesostructured materials. Thus, the volume ratio between the blocks and the chemical nature of the blocks permits the optimization of the properties of these electrolytes [18]. The group of Balsara published a series of studies on PS-PEO diblocks. For diblocks with 50% PEO, they reported that both the conductivity and mechanical properties increase with the total copolymer molar mass (M_n), in opposition to the behavior of homopolymers for which the conductivity decreases with M_n [19]. They also show a higher resistance to dendrite growth, with high M_n due to the increase in Young's modulus [20]. In addition, Bouchet *et al.,* proposed to directly graft the anion, TFSI⁻, onto the PS backbone to produce triblock, PSTFSILi-PEO-PSTFSILi, with single-ion conduction ($t^+ = 1$) displaying 1.2 10⁻⁵ S.cm⁻¹ at 60 °C, and with outstanding power performances at 60 °C in Li metal battery [21].

To conclude this panorama on polymer electrolytes, other macromolecular backbones based on carbonates or acrylonitrile [22] have been recently reported in *"polymer in salt"* configuration, *i.e.* with a large amount of salt, typically corresponding to one salt molecule for two monomers. Materials based on nitriles are quite interesting in terms of conductivity but are weakly stable at low potential. Conversely, polymers based on carbonates present a very good electrochemical stability, notably at high potential, permitting to envision their use with active materials operating at more than 4 V *vs.* Li⁺/Li [23]. However, with such high quantities of salt, these materials have poor mechanical properties and must be associated with porous separators, or with organic or inorganic additives to ensure their role as a separator.

8.2.2 *Inorganic Solid Electrolytes*

Compared to dry polymer electrolytes, inorganic superionic conductors can exhibit higher ionic conductivities at room temperature, and even challenge conventional liquid electrolytes with values reaching 10⁻² S.cm⁻¹ (figure 8.4).

They have also many theoretical assets since they can possibly:

(i)   increase significantly the safety of the cell as they are not flammable;
(ii) allow the use of lithium and sodium metallic negative electrodes: some compounds are electrochemically (meta)stable towards the metallic electrode;

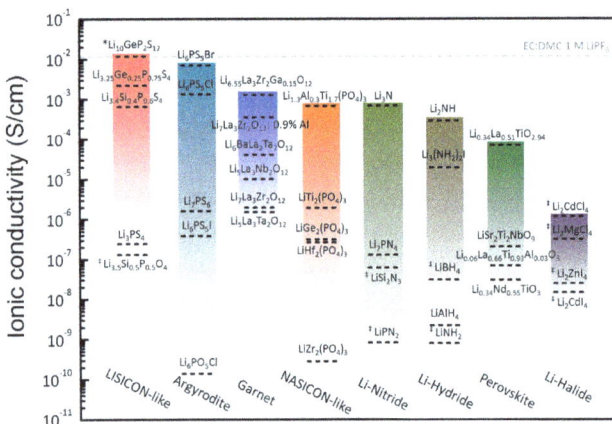

FIG. 8.4 – Ionic conductivities for the main structural groups of ceramic Li⁺ conductors⁺ Reprinted with permission from Bachmann *et al.* [4]. Copyright (2016) American Chemical Society [4].

(iii) allow the use of a positive electrode operating at higher voltage: better anodic stability can be achieved;

(iv) consequently, contribute to enhance the energy density of the cell, and also;

(v) extend the cycle life and the calendar life of the cell, due to a reduced reactivity towards electrode materials [24].

Nevertheless, the use of pure ceramic electrolytes is facing major issues. In particular, those related to (i) the densification of the components; (ii) the assembly of full all-solid-state cells and (iii) the damages of mechanical origin (fractures, stress) induced during cycling by the volume changes of the active material. Additional concerns have grown recently when it has been demonstrated that lithium dendrites can actually propagate inside these stiff electrolytes [25–27], thus jeopardizing the possibility to reintroduce the lithium metal negative electrode.

8.2.2.1 Oxides

Various families of oxide materials have been thoroughly studied in recent years. Among them, NASICON-type materials ($Li_{1.3}Al_{0.3}Ti_{1.7}(PO_4)_3$ (LATP), $LiGe_2(PO_4)_3$ (LAGP)), perovskite-type compounds ($Li_{0.34}La_{0.5}TiO_{2.94}$ (LLTO)) and garnet-type materials (Garnet, $Li_7La_3Zr_2O_{12}$ (LLZO)) [28] have drawn much attention. The electrochemical instability of Ti^{4+} below 1.5 V/Li^+/Li, and the cost of germanium, finally lead to focus most of the recent research on LLZO and related compounds, especially as they are also exhibiting the highest bulk ionic conductivities among oxides [29, 30]. The high temperature cubic form of this material, synthesized at 1200 °C, has an ionic conductivity of 10^{-4} S.cm^{-1}, but it is not stable at room temperature (RT). Nevertheless, the addition of few % of Al, Ga or Ge was found to be effective in stabilizing the cubic form and to allow to reach ionic conductivities as high as $1.3 \cdot 10^{-3}$ S.cm^{-1} at RT. Insight into the structural evolutions caused by these substitutions, and their consequences on the ionic transport inside these material, is the aim of most of the current studies [31, 32]. The main limitation for the use of this class of fast ionic conductors is the difficulty to sinter (or co-sinter) and to densify them below 1200 °C (which inevitably leads to the formation of interphases between the electrode material and the LLZO, and favours the release of Li_2O). The use of sub-micron size LLZO particles [33], the addition of sintering agents (Li_3BO_3, Li_4SiO_4) [34, 35] or the optimization of the precursor mixing step [36] can possibly allow to lower the temperature required for the sintering/densification to 800–950 °C. Besides, hot pressing is also a means to obtain dense LLZO pellets [37, 38] at moderate temperatures, but possibly at the expense of a carbon contamination originating from the die material. Another issue with LLZO is its surface reactivity with ambient air, causing H^+/Li^+ exchange in the material, associated with a fast formation of a lithium oxide/hydroxide/carbonate layer on the LLZO [39]. The implementation of a whole processing of the solid electrolyte material under inert atmosphere being costly, alternative strategies aimed at reducing the reactivity of LLZO using doping have been proposed [40].

It is nevertheless possible to take advantage of the formation the Li_2CO_3 layer. It was shown in particular that the presence of Li_2CO_3, that was unavoidable on nanometre-size LLZO particles, was found to favour their sintering and the densification process during their annealing at low temperatures [41] due to the melting of the carbonate phase (the LLZO material being recovered at the same time).

8.2.2.2   *Oxyhalides with Anti-Perovskite Structure*

More recently, new lithium oxyhalides Li_3OX (X = Cl, Br) with an anti-perovskite structure displayed promising Li^+ ionic conducting properties [42]. In their crystalline state, these materials exhibit a high ionic conductivity at room temperature (1–2 10^{-3} S.cm^{-1}) together with a low activation energy (~ 0.2 eV), and negligible electronic conductivity. Additionally, their low melting temperature is expected to facilitate the manufacturing of thin films and/or their co-sintering with electrode materials. Their ionic conductivity can be enhanced by the generation of different types of defects: LiX deficiency (Schottky), Li^+ in interstitial (Frenkel), vacancies on the Li^+ site by introduction of divalent ions $Li_{3-2x}M_xOX$ (M = Ca^{2+}, Mg^{2+}, Ba^{2+}) [43] or protons ($Li_{3-x}H_xOX$ materials, such as Li_2OHCl) [44], generation of variable local environments using two types of halogens (*e.g.* $Li_3OCl_{1-y}F_y$ [45]) or by making glassy materials *via* melt-quenching. The formation of a glassy compound can be also induced by the presence of water during the synthesis [46, 47], as it is the case for $Li_{2.99}Ba_{0.005}OCl_{1-x}(OH)_x$ which exhibits a high ionic conductivity ($2.5 \cdot 10^{-2}$ S.cm^{-1}), associated with a very low activation energy (0.06 eV). Note that similar properties have been reported for sodium-based counterparts also. Depending on the material, different ionic conduction mechanisms may be involved [48–50]. Nevertheless, the one involving vacancies on the Li site seems the most likely. Finally, different studies are suggesting that these compounds are stable *vs.* the lithium metal electrode, but also at high potential (gap >4.7 eV).

8.2.2.3   *Borohydrides – Boranes*

Most of the research on these boron-based compounds is dedicated to lithium borohydride $LiBH_4$ [51–55]. The stabilization of the high temperature phase at RT by nano-confinement of $LiBH_4$ in a porous silica matrix has allowed to reach an ionic conductivity as high as $\sigma = 2 \cdot 10^{-3}$ S.cm^{-1}. Nevertheless, excellent ionic conduction properties have been also evidenced for various compounds based on anions such as $B_{10}H_{10}{}^{2-}$, $B_{12}H_{12}{}^{2-}$ and $CB_9H_{10}{}^{2-}$. These large anions, having degrees of freedom in terms of orientation in the structure, allow to generate an extended interstitial transport network for Li^+ and Na^+ ions [54–56]. Thus, high temperature forms of some compounds ($Li_2CB_9H_{10}$: T > 80 °C, $Na_2CB_9H_{10}$: T > 50 °C) can exhibit ionic conductivities as high as 10^{-1} S.cm^{-1} at 100 °C. Unfortunately, only the high temperature form of the sodium-based compound remains stable at RT ($\sigma \sim 3.10^{-2}$ S.cm^{-1}); for the lithium-based compound new means for stabilizing the HT form at RT still remain to be found.

8.2.2.4   *Sulfide Solid Electrolytes: Glasses and Ceramics*

The main advantage of sulphide solid electrolyte over ceramic oxides is their capability to be densified at room temperature (their Young modulus is between those of oxides and polymers). Also, due to higher polarizability and larger size of sulphide anion compared to oxygen, the ionic conductivity of sulphides is higher than that of oxides [36]. On the other hand, sulphides are less stable at high potentials, unstable on air forming toxic H_2S, which is a safety hazard during manufacturing or in the case of battery failure. The most studied families of sulphide solid electrolytes are presently glass–ceramic system $Li_2S–P_2S_5$ [57] and a crystalline material $Li_{10}GeP_2S_{12}$. Glasses of composition $xLi_2S\cdot(1 - x)P_2S_5$ ($x = 0.7$–0.8) are prepared by calcination/quenching or by mechano-chemical synthesis and recrystallization at 200–300 °C. The stoichiometric material $Li_7P_3S_{11}$ has a particularly high ionic conductivity at room temperature, which can be further maximized by reducing the grain boundary resistance using nucleation/crystallization [58] and controlled densification [59], reaching $\sigma_{Li+} \sim 1.7\cdot10^{-2}$ S.cm^{-1}. The processes of ionic transport in glass ceramic materials are multiple and complex [60]. As for crystalline $Li_{10}GeP_2S_{12}$, having a similar composition as thio-LISICON ($Li_{3.25}Ge_{0.25}P_{0.75}S_4$), but a different structure, it achieves the ionic conductivity of $\sim 10^{-2}$ S.cm^{-1} [61]. It has been proved by now [62] that Li-ion diffusion at room temperature is isotropic in all three dimensions which is a positive point. It has also been verified that the substitution of Ge by cheaper isovalent elements, such as Si or Sn, only moderately affects the conductivity value [63]. Unfortunately, the electrochemical stability window of these compositions is rather narrow (2.0–2.6 V/Li$^+$/Li) [64], which prevents using these solid electrolytes in direct contact with standard positive and negative electrode active materials (such as Li and $LiCoO_2$ typically exhibiting 0 V and $\sim$ 4 V *vs.* Li$^+$/Li. At low potential, the reduction of phosphorus and germanium leads to the formation of various phases, such as Li_3P, Li_2S and M/Li_xM (M = Si, Ge, Sn) [65]. At high potential, in contact with layered oxides, an oxidation of S^{2-} produces sulphites, sulphates and phosphates [66, 67]. Generally, both experimental and theoretical studies show that although more ionically conductive, sulphides are less stable than oxides at interfaces with active materials (figure 8.5) [68–71].

8.2.3 *Hybrid Solid Electrolytes*

Hybrid solid electrolytes combine highly conductive ceramics with polymers. The motivation is here to maintain the expected properties (such as conductivity, process ability, and mechanical properties) while trying to minimize the drawbacks (such as the electrochemical instability and the grain boundary resistance). Typically those electrolytes are composed of mixtures like POE: $LiClO_4$ with LAGP [72, 73] or LLZO [74, 76]. It has been confirmed that the transport of lithium ions occurs through the particles of ceramic electrolyte and not *via* interfacial ceramic/polymer zones [77].

In [78], the conductivity of PAN is increased by an order of magnitude by adding only 3% wt. of ceramic nanowires and by another order of magnitude by their alignment in the direction of ionic transport (figure 8.6).

In addition to conductivity boost, hybrid electrolytes are reported to suppress dendrites [76, 79]. As shown in figure 8.7, cycling in Li/Li symmetric cell leads to

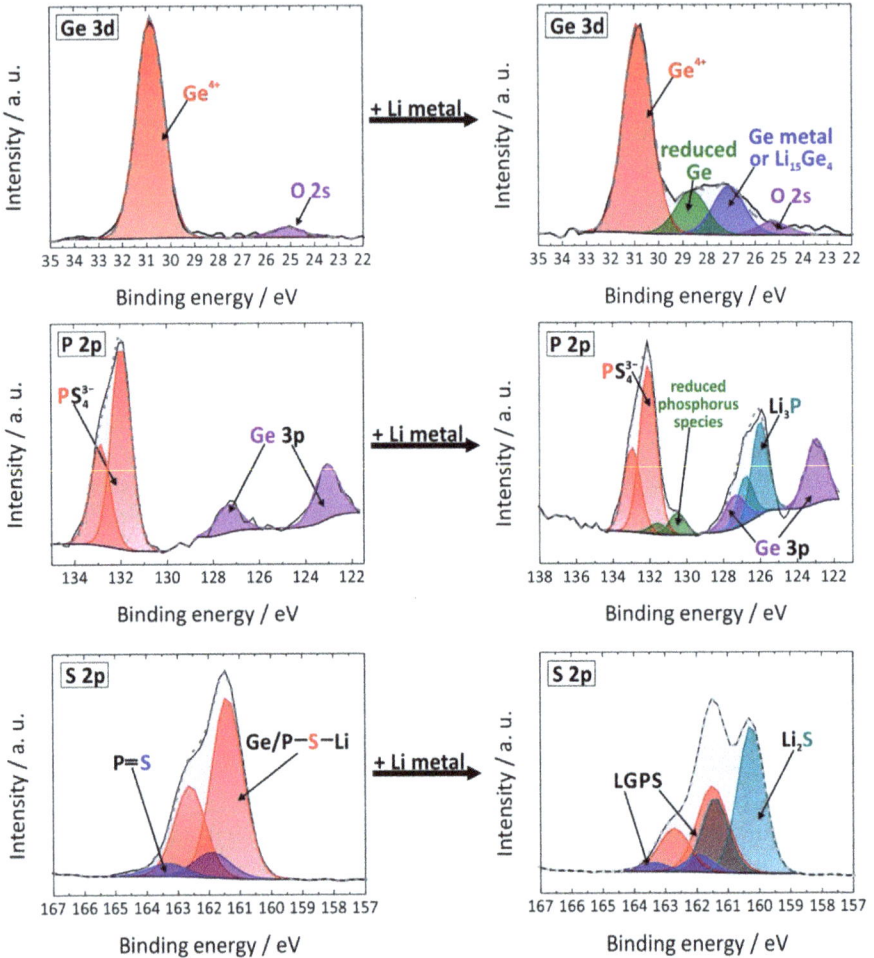

FIG. 8.5 – XPS Analysis exhibiting the reactivity of $Li_{10}GeP_2S_{12}$ solid electrolyte at the lithium interface to form Ge or Li_xGe and $Li_3P + Li_2S$ Reprinted with permission from Wenzel *et al.* [68]. Copyright (2016) American Chemical Society. [68].

dendrite propagation which is no more visible in the presence of PEO-LLZO solid electrolyte on the surface of lithium.

Finally, the presence of a ceramic skeleton prevents the loss of integrity of the hybrid membrane, which insures the cell's safety. In [76], PEO/LiTFSI – LLZO a hybrid membrane has been manufactured and tested (figure 8.8).

As shown in figure 8.8, the presence of ceramic phase does not stop the burning out of a polymeric component but the ceramic skeleton stays intact, preventing from a short circuit during the thermal runaway.

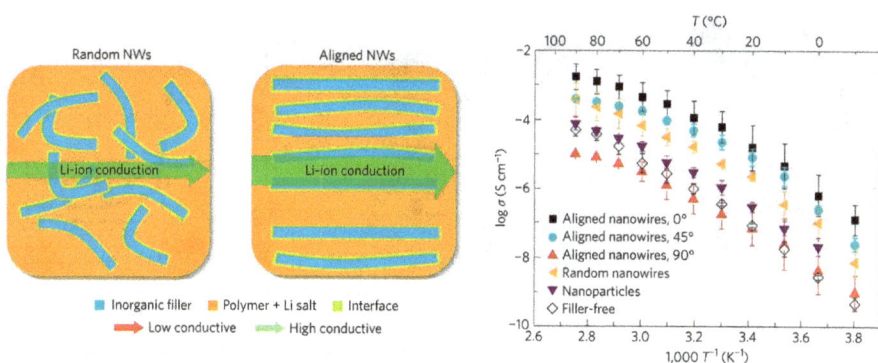

FIG. 8.6 – Random and aligned orientations of $Li_{0.33}La_{0.557}TiO_3$ ceramic nanowires in polyacrylonitrile (on the left) and the corresponding conductivity, together with the data for the composite electrolyte with nanoparticles and the filler-free electrolyte. Reprinted by permission from Liu *et al.*: Nature [78]. Copyright (2017). [78].

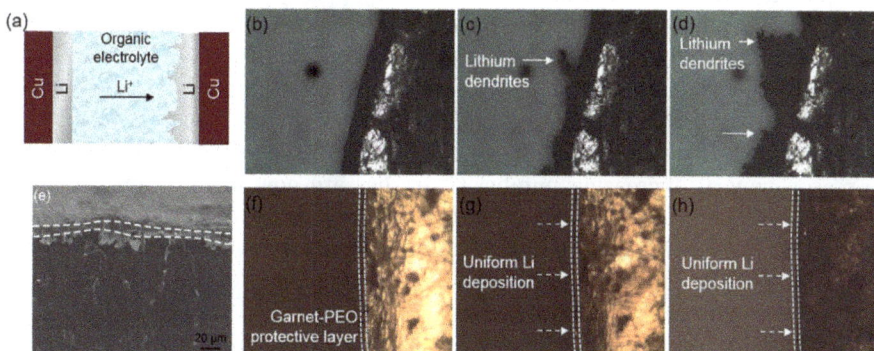

FIG. 8.7 – Dendrite observation in [79]. Reprinted by permission from Yang *et al.*: Springer. Copyright (2017). (a) General scheme of cell. (b) Optical imaging of lithium plating on bare lithium (initial state). (c) Lithium plating on bare lithium (after plating of 0.5 mAh.cm^{-2}). (d) Lithium plating on bare lithium (after plating of 1 mAh.cm^{-2}). (e) Cross-sectional SEM image illustrating the thickness of PEO-LLZO film on Li surface. (f) Optical imaging of lithium plating on PEO-LLZO-protected lithium (initial state). (g) Lithium plating on PEO-LLZO-protected lithium (after plating of 0.5 mAh.cm^{-2}). (h) Lithium plating on PEO-LLZO-protected lithium (after plating of 1 mAh.cm^{-2}).

8.3 Electrochemical Stability of Solid Electrolytes

Until recently, most of the research on all-solid-state batteries was aimed at the discovery and the development of highly conductive ceramics. Few materials that exhibit ionic conductivities close to the ones of liquid electrolyte have been identified; but assessment of their behaviour in all-solid-state electrochemical cells puts in

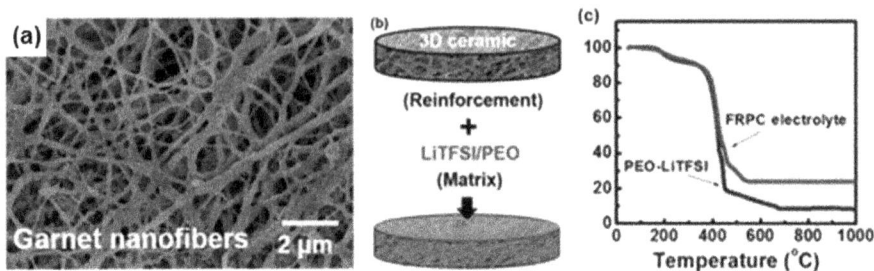

FIG. 8.8 – Fabrication and thermal properties of a flexible hybrid solid-state electrolyte composed of electrospun garnet LLZO nanofibers and LiTFSI/PEO [76]. (a) SEM image of the garnet nanofiber network. (b) Schematic procedure to manufacture the hybrid lithium-ion–conducting membrane. (c) TGA curves of LiTFSI–PEO polymer compared to hybrid electrolyte membrane.

evidence the major role played by the electrode/electrolyte interface on the limitation of the ionic transport in the cell [80, 81]. Transport limitation through the electrode/electrolyte interface can be due to the spontaneous formation of an interphase due to either a chemical reaction between the two materials, particularly during a co-sintering step, or to electrochemical reactions induced by the operation of the cell. The formation of a space-charge layer at the interface between the two ionic conductors may also hinder the ionic transport across this interface [82]. This is the reason why an increasing number of studies are now focusing on the characterization of the interface formed in different systems, on the modelling/prediction of the different phenomena expected to occur in the contact zones between the different constituents of the electrode and the solid electrolyte, and on the possible ways to control its properties.

The possible occurrence of chemical and electrochemical reaction has been assessed from a thermodynamic point of view for various electrode/electrolyte material couples, using experimental data and/or DFT (Density Functional Theory) [70, 83–87]. This brings out that, contrary to the common belief, most of the inorganic electrolytes selected due to their good ionic conductivity are not thermodynamically stable towards the lithium metal electrode nor towards an electrode operating at high voltage; and that the interposition of a ion-conductive buffer layer, acting as an artificial SEI/CEI, has to be envisaged in most cases in order to circumvent this issue (figure 8.9).

Most of these *ab initio* stability/instability forecasts have been confirmed experimentally, by XPS [88] and STEM-EELS [89, 90] in particular. Considering the reactivity of solid electrolytes with the lithium negative electrode, it has been confirmed that $Li_{10}GeP_2S_{12}$ decomposes, forming Ge, Li_3P et Li_2S [68], $Li_7P_3S_{11}$ is split in Li_3P et Li_2S components [88], and that a poorly conductive layer of the "low-temperature" form of LLZO forms between c-LLZO and lithium [91] (a similar t-LLZO interphase also forms during co-sintering of c-LLZO/LiCoO$_2$ [92]). These results are mostly qualitative; hence, they neither allow to determine the actual

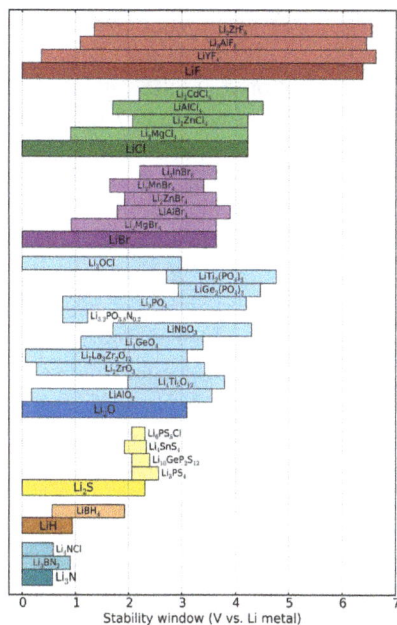

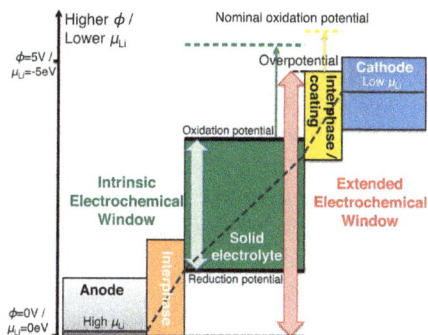

FIG. 8.9 – Theoretical electrochemical stability windows of various Li^+ ionic conductor based on DFT calculations (left) Reprinted with permission from Richard *et al.* [70]. Copyright (2015) American Chemical Society. [70]. Interposition of electrochemically stable (either on the anodic or cathodic side) ionic conductive coatings on electrodes as a means to circumvent the limitations of the electrolyte stability window (right) Reprinted with permission from Zhu *et al.* [85]. Copyright (2016) American Chemical Society. [85].

variation of the ionic conductivity induced by the formation of these interphases, or to predict the possible growth rate of the latter. As for positive electrodes, oxide ionic conductors are generally stable up to 4.5 V/Li^+/Li. Nevertheless, a space-charge layer may develop at such a high voltage that can hinder ionic transport through the solid electrolyte/electrolyte interface. Attenuation of the space-charge layer effects can be achieved by interposing a particular ionic conductor buffer layer [93] or, as it has been reported in the $Li_7La_3Zr_2O_{12}$/$LiCoO_2$ and LiPON/$LiNi_{0.5}Mn_{1.5}O_4$ systems, a thin layer of a ferroelectric material (respectively $LiNbO_3$ [94] and $BaTiO_3$ [95]).

8.4 All-Solid-State Cells

Among the large number of publications related to all-solid-state batteries, papers that describe the manufacturing and the operation of actual all-solid-state cells are relatively scarce. When reading the latter, one can generally notices that only incomplete sets of data, relative to compositions, dimensions, and performance, are provided. It is therefore quite impossible to determine the actual level of performance of a particular type of prototype cell and to make comparisons between different cells chemistries or designs. As a matter of fact, most advanced all-inorganic solid-state prototypes are reported by some companies or RTOs (Samsung R&D Japan [66, 96, 97], Toyota, Ningbo Institute of Materials Technology & Engineering, China [98, 99], KETI Korea [100], Tokyo Institute of Technology [101]).

The most frequent cell design is based on a sulphide electrolyte such as a thio-LiSiCOn or a Li_2S–P_2S_5 material, a NCA, $LiCoO_2$ or $LiNi_{0.5}Mn_{1.5}O_4$ cathode material covered by an appropriate coating (Li_2O–ZrO_2, $LiNbO_3$), and a lithium metal or Li-In alloy negative electrode. Uniaxial pressing is commonly used to prepare the components and to assembly the complete cell. Nevertheless, alternative processes that are aimed at making less rigid cells, by introducing additional binders [96] or polymer fibre reinforcement [100], are currently assessed. Other publications deal with LLZO-based batteries, all facing issues related to materials sintering [35, 99–102]. The behaviour of the cells during electrochemical cycling is generally studied over a limited number of cycles (<100). Besides, only the evolution of the specific capacity of the sole cathode material is disclosed, that again does not allow to compare the performances of different prototypes.

8.5   Academic & Industrial Players

The Toyota company has given a new momentum to the research in the field of all-solid-state batteries from 2010. Thanks to the expertise of M. Tatsumisago's team (Osaka University) in the field of solid state ionic conductors, and later with R. Kanno (Tokyo Institute of Technology), the company has assessed various cell materials and designs to improve significantly the safety and the energy density of batteries for automotive applications. So far, the Li/Li_2S–P_2S_5/NCA system seemed to be their first choice.

Japanese cell manufacturers have accompanied this impulse: Panasonic, Murata (which have taken over Sony's battery subsidiary), Hitachi Zosen... In Korea, Samsung that followed 3 years later seems to make similar choices in their Japanese research centre, and Hyundai seems to be involved in the field. In China, developments are mainly disclosed by academic research organisations (Academy of Sciences: Shanghai Institute of Ceramics, Ningbo Institute of Materials Technology & Engineering), which are by far the more active in terms of publication on this topic in the world. Besides, the National University of Singapore is particularly involved in the development of Argyrodites electrolytes (Prof. S. Adam's team). In North America, academic research is mainly led by G. Ceder's group at MIT, then at Berkeley (ab initio modelling and calculations), by N. Dudney's team at Oak Ridge National Laboratory (solid electrolytes, thin film materials), by University of Texas in connection with J.B. Goodenough (Antiperovskite electrolytes), and by S. Meng's group at the University of San Diego (interface characterization and high resolution microscopy). As for industrial players, there are some start-up companies involved in the manufacturing of all-solid-state thin film batteries or microbatteries since many years (FrontEdge Technologies, Cymbet, Prologium; one of them, Infinite Power

Solution has been absorbed by Apple). Others, like Solid Energy and Quantumscape are developing 'bulk-type' all-solid-state batteries (a joint venture between Volkswagen and the second one has been settled in 2018). SEEO, which is developing lithium metal polymer cells, has also been recently bought by Bosch. As for General Motors, it is difficult to grasp their view about the development of all-solid-state batteries, but they are actually involved in the field (ceramic and/polymer electrolyte cells).

In Europe, recent initiatives have been taken by German research centres, such as: Karlsruhe Institute of Technology (M. Fitchner, J. Janek), Helmotz Institute Ulm (ceramics processes, fluoride batteries), Technical Universities of Munich and Darmstadt (W. Jägermann, thin films and XPS characterization), Jülich research centre (Perovskite and NaSiCon ionic conductors), Giessen University (J. Janek's group, sulphide-based electrolytes), to name a few. In France, research on all-solid-state inorganic batteries is mainly carried out at the Institute for Chemistry of Condensed Matter of Bordeaux (ICMCB; microbatteries, thin film and bulk-type electrolytes), the Laboratory of Reactivity and Chemistry of Solids (LRCS, Amiens; Argyrodites & various families of inorganic ion conductors) and CEA Grenoble (All-solid state batteries and components, microbatteries). As an main industrial player in this field, Blue Solutions, a subsidiary of Bollore's group, has been manufacturing Lithium-Metal-Polymer cells for electric cars and stationary applications in Brittany since 1993.

Bibliography

[1] Xiayin Y., Huang B., Yin J., Peng G., Huang Z., Gao C., Liu D., Xu X. (2016) All-solid-state lithium batteries with inorganic solid electrolytes: review of fundamental science, *Chinese Phys.* **25**, 018802.

[2] Inoue T., Mukai K. (2017) Are all-solid-state lithium-ion batteries really safe? Verification by differential scanning calorimetry with an all-inclusive microcell, *ACS Appl. Mater. Interfaces* **9**, 1507.

[3] Jung Y.S., Oh D.Y., Nam Y.J., Park K.H. (2015) Issues and challenges for bulk-type all-solid-state rechargeable lithium batteries using sulfide solid electrolytes, *Isr. J. Chem.* **55**, 472.

[4] Bachman J.C., Muy S., Grimaud A., Chang H.-H., Pour N., Lux S.F., Paschos O., Maglia F., Lupart S., Lamp P., Giordano L., Shao-Horn Y. (2016) Inorganic solid-state electrolytes for lithium batteries: mechanisms and properties governing ion conduction, *Chem. Rev.* **116**, 140.

[5] Kim J.G., Son B., Mukherjee S., Schuppert N., Bates A., Kwon O., Choi M.J., Chung H.Y., Park S. (2015) A review of lithium and non-lithium based solid state batteries, *J. Power Sources* **282**, 299.

[6] Sakuda A. (2018) Favorable composite electrodes for all-solid-state batteries, *J. Ceramic Society of Japan* **126**, 675.

[7] Tarnopolskiy V., Azais P., Chapuis M., Daniel L. (2017) Electrochemical battery with lithium including a specific solid electrolyte, *Brevet FR3040241*.

[8] Armand M. (1994) The history of polymer electrolytes, *Solid state ionics* **69**, 309.

[9] Whittingham M.S. (2004) Lithium batteries and cathode materials, *Chem. Rev.* **104**, 4271.

[10] Xu W., Wang J., Ding F., Chen X., Nasybulin E., Zhang Y., Zhang J.-G. (2014) Lithium metal anodes for rechargeable batteries, *Energy Environ. Sci.* **7**, 513.

[11] Bouchet S. Lascaud M. Rosso (2003) An EIS study of the anode Li/PEO-LiTFSI of a Li polymer battery, *J. Electrochem. Soc.* **150**, A1385.

[12] Bruce P.G., Freunberger S.A., Hardwick L.J., Tarascon J.-M. (2012) Li–O$_2$ and Li–S batteries with high energy storage, *Nat. Mater.* **11**, 19.

[13] Devaux D., Bouchet R., Glé D., Denoyel R. (2012) Mechanism of ion transport in PEO/LiTFSI complexes: effect of temperature, molecular weight and end groups, *Solid State Ionics* **227**, 119.

[14] Ma Q., Zhang H., Zhou C., Zheng L., Cheng P., Nie J., Feng W., Hu Y.-S., Li H., Huang X., Chen L., Armand M., Zhou Z. (2016) Single lithium-ion conducting polymer electrolytes based on a super-delocalized polyanion, *Angew. Chem. Int. Ed.* **55**, 2521.

[15] Hu L., Tang Z., Zhang Z. (2007) New composite polymer electrolyte comprising mesoporous lithium aluminate nanosheets and PEO/LiClO4, *J. Power Sources* **166**, 226.

[16] Azizi Samir M.A.S., Alloin F., Gorecki W., Sanchez J.-Y., Dufresne A. (2004) Nanocomposite polymer electrolytes based on poly(oxyethylene) and cellulose nanocrystals, *J. Phys. Chem. B.* **108**, 10845.

[17] Boaretto N., Bittner A., Brinkmann C., Olsowski B.-E., Schulz J., Seyfried M., Vezzù K., Popall M., Di Noto V. (2014) Highly conducting 3D-hybrid polymer electrolytes for lithium batteries based on siloxane networks and cross-linked organic polar interphases, *Chem. Mater.* **26**, 6339.

[18] Devaux D., Glé D., Phan T.N.T., Gigmes D., Giroud E., Deschamps M., Denoyel R., Bouchet R. (2015) Optimization of block copolymer electrolytes for lithium metal batteries, *Chem. Mater.* **27**, 4682.

[19] Panday A., Mullin S., Gomez E.D., Wanakule N., Chen V.L., Hexemer A., Pople J., Balsara N.P. (2009) Effect of molecular weight and salt concentration on conductivity of block copolymer electrolytes, *Macromolecules* **42**, 4632.

[20] Schauser N.S., Harry K.J., Parkinson D.Y., Watanabe H., Balsara N.P. (2015) Lithium dendrite growth in glassy and rubbery nanostructured block copolymer electrolytes, *J. Electrochem. Soc.* **162**, A398.

[21] Bouchet R., Maria S., Meziane R., Aboulaich A., Lienafa L., Bonnet J.-P., Phan T.N.T., Bertin D., Gigmes D., Devaux D., Denoyel R., Armand M. (2013) Single-ion BAB triblock copolymers as highly efficient electrolytes for lithium-metal batteries, *Nat. Mater.* **12**, 452.

[22] Zhang Q., Liu K., Ding F., Liu X. (2017) Recent advances in solid polymer electrolytes for lithium batteries, *Nano Res.* **10**, 4139.

[23] Zhang J., Zhao J., Yue L., Wang Q., Chai J., Liu Z., Zhou X., Li H., Guo Y., Cui G., Chen L. (2015) Safety-reinforced poly(propylene carbonate)-based all-solid-state polymer electrolyte for ambient-temperature solid polymer lithium batteries, *Adv. Energy Mater.* **5**, 1501082.

[24] Abraham K.M. (2015) Prospects and limits of energy storage in batteries, *J. Phys. Chem. Lett.* **6**, 830.

[25] Aguesse F. *et al.* (2017) *ACS Appl. Mater. Interfaces* **9**, 3808.

[26] Porz L. *et al.* (2017) *Adv. Energy Mater.* 1701003.

[27] Basappa R. *et al.* (2017) *J. Power Sources* **363**, 145.

[28] Ren Y., Chen K., Chen R., Liu T., Zhang Y., Nan C.-W. (2015) Oxide electrolytes for lithium batteries, *J. Am. Ceram. Soc.* **98**, 3603.

[29] Thangadurai V., Narayanan S., Pinzaru D. (2014) Garnet-type solid-state fast Li-ion conductors for Li batteries: critical review, *Chem. Soc. Rev.* **43**, 4714.

[30] Thangadurai V., Pinzaru D., Narayanan S., Baral A.K. (2015) Fast solid-state Li ion conducting garnet-type structure metal oxides for energy storage, *J. Phys. Chem. Lett.* **6**, 292.

[31] Rettenwander D., Redhammer G., Preishuber-Pflügl F., Cheng L., Miara L., Wagner R., Welzl A., Suard E., Doeff M.M., Wilkening M., Fleig J., Amthauer G. (2016) Structural and electrochemical consequences of Al and Ga cosubstitution in Li$_7$La$_3$Zr$_2$O$_{12}$ solid electrolytes, *Chem. Mater.* **28**, 2384.

[32] Wagner R., Redhammer G.J., Rettenwander D., Senyshyn A., Schmidt W., Wilkening M., Amthauer G. (2016) Crystal structure of garnet-related Li-ion conductor Li$_{7-3x}$Ga$_x$La$_3$Zr$_2$O$_{12}$: Fast Li-ion conduction caused by a cifferent cubic modification? *Chem. Mater.* **28**, 1861.

[33] Afyon S., Hänsel C., Rupp J.L.M. (2015) A shortcut to garnet-type fast Li-ion conductors for all-solid state batteries, *J. Mater. Chem. A* **3**, 18636.

[34] Janani N., Deviannapoorani C., Dhivya L., Murugan R. (2014) Influence of sintering additives on densification and Li^+ conductivity of Al doped $Li_7La_3Zr_2O_{12}$ lithium garnet, *RSC Adv.* **4**, 51228.

[35] Ohta S., Seki J., Yagi Y., Kihira Y., Tani T., Asaoka T. (2014) Co-sinterable lithium garnet-type oxide electrolyte with cathode for all-solid-state lithium ion battery, *J. Power Sources* **265**, 40.

[36] Kumar P.J., Nishimura K., Senna M., Düvel A., Heitjans P., Kawaguchi T., Sakamoto N., Wakiya N., Suzuki H. (2016) A novel low-temperature solid-state route for nanostructured cubic garnet $Li_7La_3Zr_2O_{12}$ and its application to Li-ion battery, *RSC Adv.* **6**, 62656.

[37] Kim Y., Jo H., Allen J.L., Choe H., Wolfenstine J., Sakamoto J., Pharr G. (2016) The effect of relative density on the mechanical properties of hot-pressed cubic $Li_7La_3Zr_2O_{12}$, *J. Am. Ceram. Soc.* **99**, 1367.

[38] David I.N., Thompson T., Wolfenstine J., Allen J.L., Sakamoto J. (2015) Microstructure and li-ion conductivity of hot-pressed cubic $Li_7la_3Zr_2O_{12}$, *J. Am. Ceram. Soc.* **98**, 1209.

[39] Sharafi A., Yu S., Naguib M., Lee M., Ma C., Meyer H.M., Nanda J., Chi M., Siegel D.J., Sakamoto J. (2017) Impact of air exposure and surface chemistry on $Li–Li_7La_3Zr_2O_{12}$ interfacial resistance, *J. Mater. Chem. A* **5**, 13475.

[40] Gai J., Zhao E., Ma F., Sun D., Ma X., Jin J., Wu Q., Cui Y. (2018) Improving the Li-ion conductivity and air stability of cubic $Li_7La_3Zr_2O_{12}$ by the co-doping of Nb, Y on the Zr site, *J. Eur. Ceram. Soc.* **38**, 1673.

[41] Yi E., Wang W., Kieffer J., Laine R.M. (2017) Key parameters governing the densification of cubic-$Li_7La_3Zr_2O_{12}$ Li+ conductors, *J. Power Sources* **352**, 156.

[42] Zhao L.D.Y. (2012) Superionic conductivity in lithium-rich anti-perovskites, *J. Am. Chem. Soc.* **134**, 15042.

[43] Braga M.H., Ferreira J.A., Stockhausen V., Oliveira J.E., El-Azab A. (2014) Novel Li_3ClO based glasses with superionic properties for lithium batteries, *J. Mater. Chem.* **A2**, 5470.

[44] Hood Z.D., Wang H., Pandian A.S., Keum J.K., Liang C. (2016) Li_2OHCl crystalline electrolyte for stable metallic lithium anodes, *J. Am. Chem. Soc.* **138**, 1768.

[45] Deng Z., Radhakrishnan B., Ong S.P. (2015) Rational composition optimization of the lithium-rich $Li_3OCl_{1-x}Br_x$ anti-perovskite superionic conductors, *Chem. Mater.* **27**, 3749.

[46] Li Y., Zhou W., Xin S., Li S., Zhu J., Lü X., Cui Z., Jia Q., Zhou J., Zhao Y., Goodenough J.B. (2016) Fluorine-doped antiperovskite electrolyte for all-solid-state lithium-ion batteries, *Angew. Chem. Int. Ed.* **55**, 9965.

[47] Braga M.H., Murchison A.J., Ferreira J.A., Singh P., Goodenough J.B. (2016) Glass-amorphous alkali-ion solid electrolytes and their performance in symmetrical cells, *Energy Environ. Sci.* **9**, 948.

[48] Mouta R., Melo M.Á.B., Diniz E.M., Paschoal C.W.A. (2014) Concentration of charge carriers, migration, and stability in Li_3OCl solid electrolytes, *Chem. Mater.* **26**, 7137.

[49] Lu Z., Chen C., Baiyee Z.M., Chen X., Niu C., Ciucci F. (2015) Defect chemistry and lithium transport in Li_3OCl anti-perovskite superionic conductors, *Phys. Chem. Chem. Phys.* **17**, 32547.

[50] Mouta R., Diniz E.M., Paschoal C.W.A. (2016) Li^+ interstitials as the charge carriers in superionic lithium-rich anti-perovskites, *J. Mater. Chem.* **A4**, 1586.

[51] Unemoto A., Matsuo M., Orimo S. (2014) Complex hydrides for electrochemical energy storage, *Adv. Funct. Mater.* **24**, 2267.

[52] Blanchard D., Nale A., Sveinbjörnsson D., Eggenhuisen T.M., Verkuijlen M.H.W., Suwarno, Vegge T., Kentgens A.P.M., de Jongh P.E. (2015) Nanoconfined $LiBH_4$ as a fast lithium ion conductor, *Adv. Funct. Mater.* **25**, 184.

[53] Unemoto A., Ikeshoji T., Yasaku S., Matsuo M., Stavila V., Udovic T.J., Orimo S. (2015) Stable interface formation between TiS_2 and $LiBH_4$ in bulk-type all-solid-state lithium batteries, *Chem. Mater.* **27**, 5407.

[54] Tang W.S., Matsuo M., Wu H., Stavila V., Unemoto A., Orimo S., Udovic T.J. (2016) Stabilizing lithium and sodium fast-ion conduction in solid polyhedral-borate salts at device-relevant temperatures, *Energy Storage Mater.* **4**, 79.

[55] Tang W.S., Matsuo M., Wu H., Stavila V., Zhou W., Talin A.A., Soloninin A.V., Skoryunov R.V., Babanova O.A., Skripov A.V., Unemoto A. Orimo S., Udovic T.J. (2016) Liquid-like ionic conduction in solid lithium and sodium monocarba-closo-decaborates near or at room temperature, *Adv. Energy Mater.* **6**, 1502237.

[56] Tang W.S., Unemoto A., Zhou W., Stavila V., Matsuo M., Wu H., Orimo S., Udovic T.J. (2015) Unparalleled lithium and sodium superionic conduction in solid electrolytes with large monovalent cage-like anions, *Energy Environ. Sci.* **8**, 3637.

[57] Tatsumisago M., Hayashi A. (2014) Sulfide glass-ceramic electrolytes for all-solid-state lithium and sodium batteries, *Int. J. Appl. Glass Sci.* **5**, 226.

[58] Busche M.R., Weber D.A., Schneider Y., Dietrich C., Wenzel S., Leichtweiss T., Schröder D., Zhang W., Weigand H., Walter D., Sedlmaier S.J., Houtarde D., Nazar L.F., Janek J. (2016) In situ monitoring of fast Li-ion conductor Li7P3S11 crystallization inside a hot-press setup, *Chem. Mater.* **28**, 6152.

[59] Seino Y., Ota T., Takada K., Hayashi A., Tatsumisago M. (2014) A sulphide lithium super ion conductor is superior to liquid ion conductors for use in rechargeable batteries, *Energy Environ. Sci.* **7**, 627.

[60] Wohlmuth D., Epp V., Wilkening M. (2015) Fast Li-ion dynamics in the solid electrolyte Li7P3S11 as probed by 6, 7Li NMR spin-lattice relaxation, *Chem. Phys. Chem..* **16**, 2582.

[61] Kamaya N., Homma K., Yamakawa Y., Hiriyama M., Kanno R., Yonemura M., Kamiyama T., Kato Y., Hama S., Kawamoto K., Mitsui A. (2011) A lithium superionic conductor, *Nat. Mater.* **10**, 682.

[62] Weber D.A., Senyshyn A., Weldert K.S., Wenzel S., Zhang W., Kaiser R., Berendts S., Janek J., Zeier W.G. (2016) Structural insights and 3D diffusion pathways within the lithium superionic conductor Li10GeP2S12, *Chem. Mater.* **28**, 5905.

[63] Kato Y., Saito R., Sakano M., Mitsui A., Hirayama M., Kanno R. (2014) Synthesis, structure and lithium ionic conductivity of solid solutions of $Li_{10}(Ge_{1-x}M_x)P_2S_{12}$ (M = Si, Sn), *J. Power Sources* **271**, 60.

[64] Han F., Gao T., Zhu Y., Gaskell K.J., Wang C. (2014) A battery made from a single material, *Adv. Mater.* **27**, 3473.

[65] Tarhouchi I., Viallet V., Vinatier P., Ménétrier M. (2016) Electrochemical characterization of $Li_{10}SnP_2S_{12}$: an electrolyte or a negative electrode for solid state Li-ion batteries? *Solid State Ionics* **296**, 18.

[66] Visbal H., Aihara Y., Ito S., Watanabe T., Park Y., Doo S. (2016) The effect of diamond-like carbon coating on $LiNi_{0.8}Co_{0.15}Al_{0.05}O_2$ particles for all solid-state lithium-ion batteries based on Li2S–P2S5 glass-ceramics, *J. Power Sources* **314**, 85.

[67] Sakuda A., Hayashi A., Tatsumisago M. (2010) Interfacial observation between $LiCoO_2$ electrode and $Li_2S-P_2S_5$ solid electrolytes of all-solid-state lithium secondary batteries using transmission electron microscopy, *Chem. Mater.* **22**, 949.

[68] Wenzel S., Randau S., Leichtweiß T., Weber D.A., Sann J., Zeier W.G., Janek J. (2016) Direct observation of the interfacial instability of the fast ionic conductor $Li_{10}GeP_2S_{12}$ at the lithium metal anode, *Chem. Mater.* **28**, 2400.

[69] Sakuma M., Suzuki K., Hirayama M., Kanno R. (2016) Reactions at the electrode/electrolyte interface of all-solid-state lithium batteries incorporating Li–M (M = Sn, Si) alloy electrodes and sulfide-based solid electrolytes, *Solid State Ionics,* **285**, 101.

[70] Richards W.D., Miara L.J., Wang Y., Kim J.C., Ceder G. (2016) Interface stability in solid-state batteries, *Chem. Mater.* **28**, 266.

[71] Yokokawa H. (2016) Thermodynamic stability of sulfide electrolyte/oxide electrode interface in solid-state lithium batteries, *Solid State Ionics* **126**.

[72] Jung Y.-C., Park M.-S., Doh C.-H., Kim D.-W. (2016) Organic-inorganic hybrid solid electrolytes for solid-state lithium cells operating at room temperature, *Electrochim. Acta* **218**, 271.

[73] Zhao Y., Huang Z., Chen S., Chen B., Yang J., Zhang Q., Ding F., Chen Y., Xu X. (2016) A promising PEO/LAGP hybrid electrolyte prepared by a simple method for all-solid-state lithium batteries, *Solid State Ionics* **295**, 65.

[74] Langer F., Bardenhagen I., Glenneberg J., Kun R. (2016) Microstructure and temperature dependent lithium ion transport of ceramic-polymer composite electrolyte for solid-state lithium ion batteries based on garnet-type $Li_7La_3Zr_2O_{12}$, *Solid State Ionics* **291**, 8.

[75] Choi J.-H., Lee C.-H., Yu J.-H., Doh C.-H., Lee S.-M. (2015) Enhancement of ionic conductivity of composite membranes for all-solid-state lithium rechargeable batteries incorporating tetragonal $Li_7La_3Zr_2O_{12}$ into a polyethylene oxide matrix, *J. Power Sources* **274**, 458.

[76] Fu K., Gong Y., Dai J., Gong A., Han X., Yao Y., Wang C., Wang Y., Chen Y., Yan C., Li Y., Wachsman E.D., Hu L. (2016) Flexible, solid-state, ion-conducting membrane with 3D garnet nanofiber networks for lithium batteries, *Proc. Nat. Acad. Sci. U.S.A.* **113**, 7094.

[77] Zheng J., Tang M., Hu Y.-Y. (2016) Lithium ion pathway within $Li_7La_3Zr_2O_{12}$-Polyethylene oxide composite electrolytes, *Angew. Chem. Int. Ed.* **55**, 12538.

[78] Liu W., Lee S.W., Lin D., Shi F., Wang S., Sendek A.D., Cui Y. (2017) Enhancing ionic conductivity in composite polymer electrolytes with well-aligned ceramic nanowires, *Nat. Energy* **2**, 17035.

[79] Yang C., Liu B., Jiang F., Zhang Y., Xie H., Hitz E., Hu L. (2017) Garnet/polymer hybrid ion-conducting protective layer for stable lithium metal anode, *Nano Res.* **10**, 4256.

[80] Luntz A.C., Voss J., Reuter K. (2015) Interfacial challenges in solid-state Li-ion batteries, *J. Phys. Chem. Lett.* **6**, 4599.

[81] Wu B., Wang S., Evans W.J., Deng D.Z., Yang J., Xiao J. (2016) Interfacial behaviours between lithium ion conductors and electrode materials in various battery systems, *J. Mater. Chem. A* **4**, 15266.

[82] Chen C., Guo X. (2016) Space charge layer effect in solid state ion conductors and lithium batteries: principle and perspective, *Acta Chim. Slov.* **63**, 489.

[83] Miara L., Windmüller A., Tsai C.-L., Richards W.D., Ma Q., Uhlenbruck S., Guillon O., Ceder G. (2016) About the compatibility between high voltage spinel cathode materials and solid oxide electrolytes as a function of temperature, *ACS Appl. Mater. Interfaces* **8**, 26842.

[84] Miara L.J., Richards W.D., Wang Y.E., Ceder G. (2015) First-principles studies on cation dopants and electrolyte|cathode interphases for lithium garnets, *Chem. Mater.* **27**, 4040.

[85] Zhu Y., He X., Mo Y. (2015) Origin of outstanding stability in the lithium solid electrolyte materials: Insights from thermodynamic analyses based on first-principles calculations, *ACS Appl. Mater. Interfaces* **7**, 23685.

[86] Zhu Y., He X., Mo Y. (2016) First principles study on electrochemical and chemical stability of solid electrolyte–electrode interfaces in all-solid-state Li-ion batteries, *J. Mater. Chem. A* **4**, 3253.

[87] Sumita M., Tanaka Y., Ikeda M., Ohno T. (2016) Charged and discharged states of cathode/sulfide electrolyte interfaces in all-solid-state lithium ion batteries, *J. Phys. Chem. C* **120**, 13332.

[88] Wenzel S., Leichtweiß T., Krüger D., Sann J., Janek J. (2015) Interphase formation on lithium solid electrolytes - an in situ approach to study interfacial reactions by photoelectron spectroscopy, *Solid State Ionics* **278**, 98.

[89] Wang Z., Santhanagopalan D., Zhang W., Wang F., Xin H.L., He K., Li J., Dudney N., Meng Y.S. (2016) In situ STEM-EELS observation of nanoscale interfacial phenomena in all-solid-state batteries, *Nano Lett.* **16**, 3760.

[90] Santhanagopalan D., Qian D., McGilvray T., Wang Z., Wang F., Camino F., Graetz J., Dudney N., Meng Y.S. (2014), Interface limited lithium transport in solid-state batteries, *J. Phys. Chem. Lett.* **5**, 303.

[91] Ma C., Cheng Y., Yin K., Luo J., Sharafi A., Sakamoto J., Li J., More K.L., Dudney N.J., Chi M. (2016) Interfacial stability of Li metal-solid electrolyte elucidated *via* in situ electron microscopy, *Nano Lett.* **16**, 7030.

[92] Park K., Yu B.-C., Jung J.-W., Li Y., Zhou W., Gao H., Son S., Goodenough J.B. (2016) Electrochemical nature of the cathode interface for a solid-state lithium-ion battery: interface between $LiCoO_2$ and garnet-$Li_7La_3Zr_2O_{12}$, *Chem. Mater.* **28**, 8051.

[93] Haruyama J., Sodeyama K., Han L., Takada K., Tateyama Y. (2014) Space−charge layer effect at interface between oxide cathode and sulfide electrolyte in all-solid-state lithium-ion battery, *Chem. Mater.* **26**, 4248.

[94] Kato T., Hamanaka T., Yamamoto K., Hirayama T., Sagane F., Motoyama M., Iriyama Y. (2014) *In situ* $Li_7La_3Zr_2O_{12}/LiCoO_2$ interface modification for advanced all-solid-state battery, *J. Power Sources* **260**, 292.

[95] Yada C., Ohmori A., Ide K., Yamasaki H., Kato T., Saito T., Sagane F., Iriyama Y. (2014) Dielectric modification of 5 V-class cathodes for high-voltage all-solid-state lithium batteries, *Adv. Energy Mater.* **4**, 1301416.

[96] Ito S., Fujiki S., Yamada T., Aihara Y., Park Y., Kim T.Y., Baek S.-W., Lee J.-M., Doo S., Machida N. (2014) A rocking chair type all-solid-state lithium ion battery adopting Li_2O–ZrO_2 coated $LiNi_{0.8}Co_{0.15}Al_{0.05}O_2$ and a sulfide based electrolyte, *J. Power Sources* **248**, 943.

[97] Ulissi U., Agostini M., Ito S., Aihara Y., Hassoun J.(2016) All solid-state battery using layered oxide cathode, lithium-carbon composite anode and thio-LISICON electrolyte, *Solid State Ionics* **296**, 13.

[98] Yin J., Yao X., Peng G., Yang J., Huang Z., Liu D., Tao Y., Xu X. (2015) Influence of the Li-Ge-P-S based solid electrolytes on NCA electrochemical performances in all-solid-state lithium batteries, *Solid State Ionics* **274**, 8.

[99] Peng G., Yao X., Wan H., Huang B., Yin J., Ding F., Xu X. (2016) Insights on the fundamental lithium storage behavior of all-solid-state lithium batteries containing the $LiNi_{0.8}Co_{0.15}Al_{0.05}O_2$ cathode and sulfide electrolyte, *J. Power Sources* **307**, 724.

[100] Nam Y.J., Cho S.-J., Oh D.Y., Lim J.-M., Kim S.Y., Song J.H., Lee Y.-G., Lee S.-Y., Jung Y. S. (2016) Bendable and thin sulfide solid electrolyte film: a new electrolyte opportunity for free-standing and stackable high-energy all-solid-state lithium-ion batteries, *Nano Lett.* **15**, 3317.

[101] Oh G., Hirayama M., Kwon O., Suzuki K., Kanno R. (2016) Bulk-type all solid-state batteries with 5 V class LiNi0.5Mn1.5O4 cathode and $Li_{10}GeP_2S_{12}$ solid electrolyte, *Chem. Mater.* **28**, 2634.

[102] Ahn C.-W., Choi J.-J., Ryu J., Hahn B.-D., Kim J.-W., Yoon W.-H., Choi J.-H., Lee J.-S., Park D.-S. (2014) Electrochemical properties of $Li_7La_3Zr_2O_{12}$-based solid state battery, *J. Power Sources* **272**, 554.

[103] Xue Z., He D., Xie X. (2015) Poly(ethylene oxide)-based electrolytes for lithium-ion batteries, *J. Mater. Chem. A.* **3**, 19218.

Chapter 9

Supercapacitors: From Material to Cell

Philippe Azaïs

Introduction

The storage of electricity *via* charges stored at the metal/electrolyte interface has been studied quite extensively since the 19th century, but its practical use *via* electrochemical double layer capacitors (EDLC) really began in the 1950s with studies carried out by General Electric. The material used to maximize the contact surface with the electrolyte was a porous carbon [1]. In 1966, the Standard Oil Company, Cleveland, Ohio (SOHIO) patented an energy storage system using a "double layer" interface [2]. According to the authors of this patent, the use of this "double layer" seems to provide a "relatively large specific capacity". In 1970, SOHIO patented a component using two carbon disks impregnated with electrolyte and attempted to commercialize this component. However, in 1971, faced with the commercial failure of this component, SOHIO abandoned the development of this technology and NEC exploited the technology under license. NEC began producing these first components under the name of "supercapacitor". These low-voltage components operating in an aqueous medium have a high internal resistance, but still find applications for memory back up, thus paving the way for this technology. From that moment on, many companies became interested in this new energy storage component and began to market it. As early as 1978, Panasonic launched its "Gold Capacitor" range dedicated to back-up applications, as was the case for NEC. In 1987, Elna starts the production of its "Dynacap" range of supercapacitors. The success of this new component prompted the United States Department of Energy (D.O.E.) to launch a study aimed at the hybridization of vehicles using this new technology. From 1992 onwards, the "Ultracapacitor Development Program" was taken over by Maxwell Laboratories. For almost two decades, industrial research has been proposing numerous solutions to improve the reliability and lifetime of these components [3].

DOI: 10.1051/978-2-7598-2555-4.c009
© Science Press, EDP Sciences, 2021

As shown in figure 9.1, the number of patents relating to this technology has grown steadily since the late 1980s, demonstrating the growing interest in this technology.

The so-called "carbon/carbon" supercapacitors operating in an organic medium are the most commercially successful and are currently the most industrially developed in the world. These industrial supercapacitors consist of two similar electrodes (containing the same active material) made from activated carbon (a very porous material) separated by a porous dielectric film (separator) impregnated with electrolyte. This electrolyte contains a large quantity of ions solubilized in an organic or aqueous solvent. This assembly is then placed in a sealed housing to prevent gas and liquid leaks.

Markets potentially using supercapacitors require low-cost, high-performance components. All supercapacitor manufacturers have been working for about 15 years to reduce the cost of the component, which is currently considered "acceptable" by customers. Nevertheless, at the level of the complete energy storage system, this price remains too high, due to the high cost of assembly and the choice of materials constituting not the cell but the system (including the BMS and the converter).

To reduce this cost, several solutions are proposed:

– Increase energy density. Increasing the energy density of the component involves either increasing the voltage of the component (which does not promote long life and acceptable self-discharge) or increasing the capacity of the active material.
– Reduce the mass and volume of the passive components in the system. The reduction of passive constituents is, as with Lithium-ion batteries, the subject of numerous developments. This pragmatic approach includes the study of the component as a whole to improve its energy density, power density, and lifetime, while considerably reducing the cost of both the electrochemical cell and the

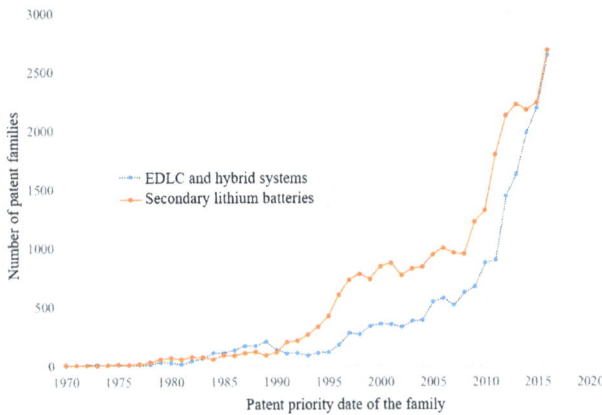

FIG. 9.1 – Number of patents published worldwide concerning supercapacitors (EDLC and hybrid systems) and comparison with rechargeable lithium systems since 1970.

energy storage system as a whole. The research associated with this theme mainly concerns the reduction of the mass of the collectors and the housing.
– Simplify the assembly process of the systems to make them lighter. This approach is generally not easy: the processes implemented must be simple to guarantee low manufacturing costs.

Finally, one of the major avenues being explored to increase energy density is to hybridize the supercapacitor by substituting one of the two carbon electrodes with an electrode of the type used in lithium batteries, in particular to meet the needs of the market for mild-hybrid vehicles.

In this chapter dedicated to supercapacitors, the core of the technology (electrode, separator, and electrolyte) is described, as well as its operating principle. Alternative solutions are also presented and compared with carbon/carbon supercapacitors. The following chapter 10 draws up an inventory of existing products currently on the market (cells and modules) and describes the performance of the components.

9.1 Operating Principle

Supercapacitors have an energy density 10–100 times higher than electrolytic and dielectric capacitors. Their operating principle is similar to that of an electrolytic capacitor: they store a large quantity of ions in a porous material. The major difference with electrolytic capacitors lies in the presence of two similar electrodes electrically separated by a separator: their operation is therefore based on the presence of two supercapacitors (one per electrode) positioned in series; the intermediate connection between the two supercapacitors is made by ion conduction between the two electrodes (figure 9.2). It is therefore important to have a conductive electrolyte to minimize the series resistance of the system. As the material used is very microporous (in the IUPAC sense, with pores smaller than 2 nm), the surface area accessible to the ions is very high. Finally, the distance between the surface and the ions (which are theoretically not supposed to react electrochemically with the active "host" material) is of the order of angström.

The mode of operation is as follows:

– As the active materials are of the same nature at both electrodes, the potential difference in the discharged state is 0 V. This means that the supercapacitors are discharged during the assembly phase.
– During the first charging phase, the ions present in the electrolyte ("free" electrolyte in the separator; between the particles of active material; and confined in the porosity of the active material) migrate under the effect of the electric field imposed between the two electrodes (this phenomenon is confirmed, for example, by *in situ* experiments/small angle neutron scattering operando [4]).
– The ions are accompanied by a procession of solvent molecules in the free environment, but they are partially desolvated in a confined environment. Unlike metal-ion batteries, the ions do not transit between the two electrodes: it is a

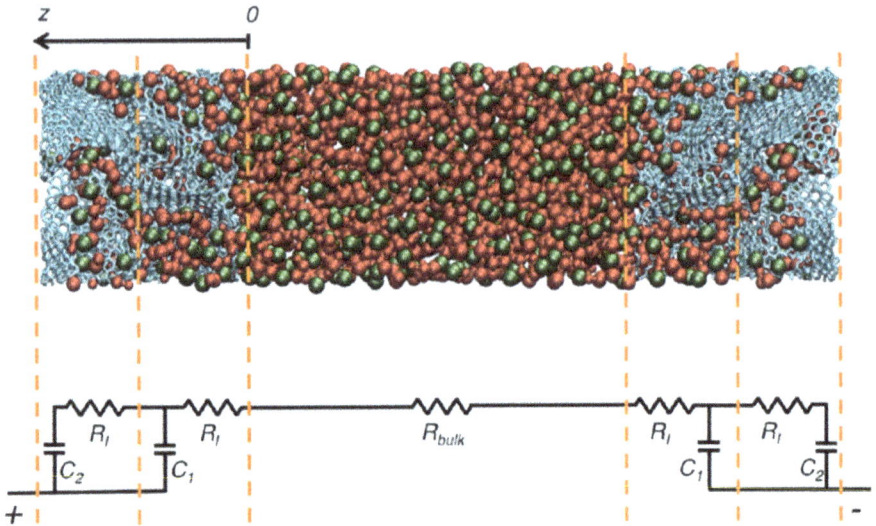

FIG. 9.2 – Modelling of a carbon/carbon supercapacitor. In this case, the BMIPF$_6$ electrolyte is an ionic liquid (therefore solvent-free). BMI$^+$ in red, PF$_6^-$ in green. In blue, the structure of the porous carbons (here, a CDC 1200) [9]. Below the figure, the equivalent electrical circuit.

local movement of the ions in the porosity. The principle of an electrochemical double layer according to the definitions specified by Helmholtz [5], Gouy [6], Chapman [7] and Stern [8] could not therefore be achieved in such a confined environment as the carbon porosity, the solvent molecules being partially absent.

- Calculation of capacitance

Standard capacitor equations give a very high capacitance per unit volume and mass; hence, the name "supercapacitor". Theoretically, one should be able to calculate the capacitance from the following equation:

$$C = \varepsilon_0 . \varepsilon_r . \frac{S}{d} \tag{9.1}$$

where ε_0 is the permittivity of the vacuum, ε_r is the relative permittivity of the dielectric material, S is the accessible surface to ions and d is the distance between the ion and the surface. Due to lack of reaction between the ions and the "host" material, energy storage is essentially driven by the mobility of the ions in the bulk (high-frequency resistance) and in the porosity (storage by confinement of the ions in the porosity).

Equation 9.1 is therefore difficult to use practically for several reasons:

- The accessible surface area S for the ions is difficult to measure.
- The distance between the ion and the surface is not measurable.

– The medium is poorly defined: the definition of ε_r is not easy.

However, if it is difficult to anticipate the capacitance value from the material constants, it is possible to measure the capacitance value (in Farad), defined according to IEC 620576 standard [10]:

$$C = I_{\text{discharge}} \cdot \frac{\Delta t}{\Delta U} \qquad (9.2)$$

where Δt is the discharge time to reach from $0.9.\,U_n$ to $0.7.\,U_n$ (*i.e.*, ΔU) at a discharge current $I_{\text{discharge}}$ defined by the application:

$$I_{\text{discharge}} = \frac{U_n}{40.R} \qquad (9.3)$$

the value of the current being that allowing an efficiency between charge and discharge of 95%.

• Calculation of resistance

The R-value of the resistance according to equation 9.3 is an underestimate of the true value: it is a high-frequency resistance that does not take into account thermal and diffusion effects in materials. Therefore, a resistance value called EDR (Equivalent Distributed Resistance) is also defined taking into account the linear constant current discharge profile (figure 9.3) in accordance with the previous equations. The current selected depends on the application, as defined in the standard [11]:

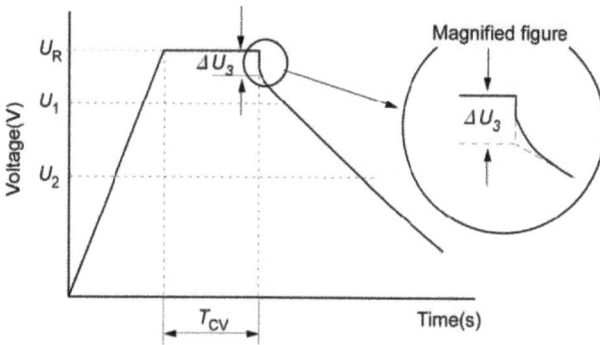

FIG. 9.3 – Discharge profile according to IEC 62391 to be taken into consideration when calculating the EDR. The T_{CV} rest time is 30 min according to the standard. To identify the ohmic drop value (ΔU_3), a linear regression by the method of least squares [12] is applied for the defined interval $U_1 = 0,8.\,U_n$ and $U_2 = 0,4.\,U_n$.

	Application	Applied current (mA)	Example of application
Class 1	Back-up	10.C	PC/small cells (1 F)
Class 2	Energy storage	$4.C.U_n$	UPS
Class 3	Power unit	$40.C.U_n$	"Mild hybrid" vehicle
Class 4	Instantaneous power unit (pulse)	$400.C.U_n$	Start-stop, cranking

EDR resistance is defined as:

$$\text{EDR} = \frac{\Delta U(\text{ohmic drop})}{I} \tag{9.4}$$

measurement step time is below 10 ms.

Other standards can be applied to measure cell resistance such as IEC 62576 (Electric double layer capacitors for use in hybrid electric vehicles – Test methods for electrical characteristics), IEC 61881-3 (Railway applications – Rolling stock equipment – Capacitors for power electronics – Part 3: Electric double-layer) and SAE J3051 (Capacitive energy storage device requirements for automotive propulsion applications capacitors). These standards have been extensively discussed by Zahng *et al.* [13] and Zhao *et al.* [14].

In the case of a complete discharge (between the maximum voltage and 0 V), the energy is calculated by:

$$E = \frac{1}{2}.C.U_{\max}^2 \tag{9.5}$$

The curve describing the evolution of the voltage *vs.* time for a discharge at constant power is in the form of a "lying" parabola and the energy restored over the same time is linear (figure 9.4).

An important value to consider for supercapacitor applications is the time constant $\tau = R.C\,(s)$. As with capacitors, this is the constant from which the cut-off frequency f_c can be calculated according to the equation:

$$f_c(\text{Hz}) = \frac{1}{2\pi.R.C} \tag{9.6}$$

At this frequency, the energy accessible in discharge is defined by:

$$E = \frac{1}{2}.C.U_{\max}^2 - \frac{1}{2}.C.(e^{-1}.U_{\max})^2 = \frac{1}{2}.C.U_{\max}^2.0,865 \tag{9.7}$$

The discharge time at constant power is calculated from the following equation:

$$t = \frac{1}{P}.C.\left(U_{\text{initial}}^2 - U_t^2\right) \tag{9.8}$$

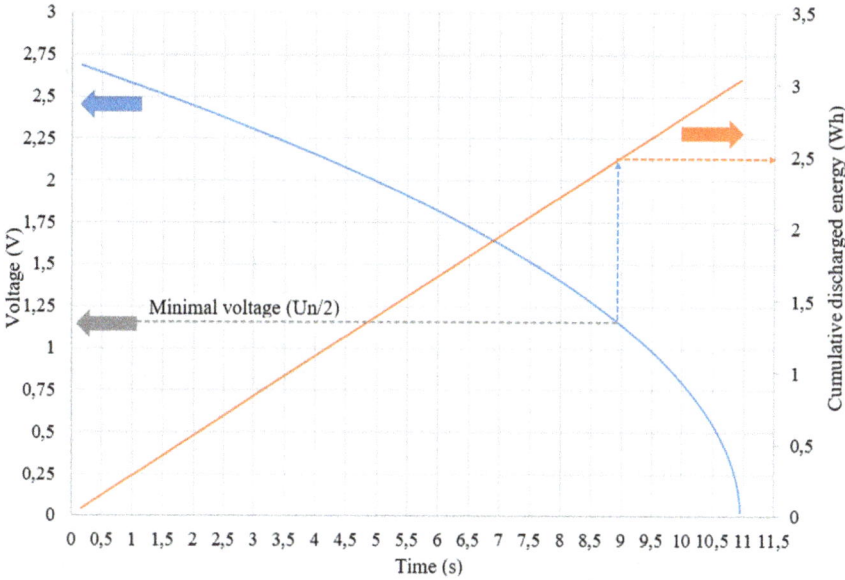

Fig. 9.4 – Discharge profile at constant power (1000 W) for a 3000 F supercapacitor cell between 2.7 and 0 V and associated cumulative recovered energy profile during discharge.

where P is the power (W), C the capacitance (F), U_{initial} is the initial discharge voltage (V), U_t is the voltage at t time (V), R is the EDR resistance (Ω) and R_{applied} is the applied resistance during the discharge (Ω).

For a constant resistance, discharge time is calculated using the following equation:

$$t = -R_{\text{applied}}.C.\ln\left(\frac{U_t}{U_{\text{initial}}}\right) \tag{9.9}$$

At a constant current, discharge time is calculated using the following equation:

$$t = \frac{C}{I}.(U_{\text{initial}} - U_t) \tag{9.10}$$

However, the usable energy is constrained by the associated converter. Therefore, the voltage range generally considered is between U_n and $U_n/2$. The energy is then ¾ of the maximum value. In addition, the energy value is to be considered at a given power value. The maximum power is defined by:

$$P_{\text{max}} = \frac{U_n^2}{4.R} \tag{9.11}$$

However, the actual accessible power is not the maximum value. The IEC 62391-2 standard defines the accessible power by:

$$P_{\text{max accessible power}} = \frac{0,12\, U_n^2}{R} \tag{9.12}$$

For more details, please refer to [15] (P. Kurtweil).

The manufacturing principle of a supercapacitor is the same as that of a lithium-ion battery. There are several steps:

(1) Realization of the electrode.
(2) Positioning of the separator (by winding or stacking).
(3) Assembly of the electrode with an interface allowing the recovery of current to the exterior.
(4) Filling the electrolyte.
(5) Closing the cell.

From an industrial point of view, two classes of components can be distinguished:

1. Cells with a capacity exceeding 350 F. These components are more particularly dedicated to urban transport, UPS (Uninterrupted Power Supply), vehicle hybridization, lifts, etc. For all these markets, the components are assembled in modules and/or energy storage systems (pack) to which an electronic balancing system between the components is added.
2. Small capacity cells (typically <50 F). They are commonly found in low-cost electronic applications such as back-up memory and consumer applications such as toys, telephony, etc. In these cases, the components are soldered directly onto electronic boards as is the case for other passive components.

The latter market is considered to be mature. The dimensions of these cells are directly inspired by electrolytic capacitors, dielectric capacitors and button cells. Improving the performance of these cells seems less crucial: cost and robustness are really the key factors for these types of cells.

It is estimated that the supercapacitor market was around US$568.2 M in 2015 and is expected to reach more than US$2 billion in 2022 (with a growth of around 20%/year) [16]. For nearly ten years, we have been witnessing a rationalization of the formats and designs of cells and modules in line with the targeted markets, which are not yet totally fixed (voltage, capacity, and dimensions) and are directly linked to the targeted application.

9.2 Carbon/Carbon Based Technology

9.2.1 *Electrode Design and Components*

Power density and lifetime are the two major advantages of supercapacitors. Therefore, it is essential to consider two crucial elements: the electrode and the cell design. Decrease in capacity, increase in resistance during ageing, and self-discharge (which improves during ageing), are the key technical factors limiting access to the organic electrolytic supercapacitor market. Typically, the electrodes consist of a current collector, an active material paste, a conductive additive and a binder. In some cases, additives are developed to increase the lifespan.

9.2.1.1   Current Collector

In order to achieve very low resistance (and therefore high power), numerous solutions have been devised. One of the most commonly used consists of depositing the active material in the form of a paste on the current collector, mainly due to the very low series resistance of the latter. Supercapacitors without a dedicated current collector can be produced (so-called "self-supporting" system). However, the series resistance is generally poor compared to that of technologies using a metallic collector. This is why the most widely industrialised technology consists of coating an ink or extruding an activated carbon-based paste onto a metal current collector using an aqueous or organic solvent: these processes are relatively economical (high rate) and are generally derived from processes already used in the field of batteries, with, however, a very large problem of adsorption of the solvent during coating and the very high shrinkage generated during the removal of the solvent. For this reason, extrusion or processes with very low solvent use (*e.g.* PTFE fibrillation) are generally the most relevant in this field.

It is very important to differentiate between a collector that can be used in an organic medium and a stable collector in an aqueous medium (acid or basic). As the vast majority of supercapacitors operate in an organic electrolytic medium (Propylene Carbonate or acetonitrile), the current collector is made of aluminium because it is inexpensive, light and its electrochemical stability is relatively high in standard electrolytes [17]. The collector most commonly used in aqueous media is stainless steel. In addition, most industrially used collectors have a specific surface finish, in order to increase the adhesion between the electrode itself (*i.e.* containing the active ingredient) and the current collector. This special surface finish is applied in various ways:

1. The current collector can be electrochemically corroded. In this case, the current collector directly forms the anode, which is used as the anode in electrolytic capacitors. This specific surface state is called etching [18, 19].
2. A very thin undercoat can be coated on a smooth aluminium current collector–etched or not. This underlayer is made from a binder [20, 21], and carbon black, carbon nanofibres [22], carbon nanotubes or graphite [23]. It must be highly conductive in relation to the electrode. The major disadvantage of this undercoat is its cost, as these collectors have been specially developed for supercapacitors and batteries.
3. A standard aluminium foil can also be used as a current collector [24]. The main advantage of this type of collector is its cost. However, it is very difficult to adhere to an electrode paste to this type of collector using conventional binders such as PVDF or PTFE while keeping the interface resistance at a very low level. However, this solution can still be used in energy type supercapacitors.

4. To improve the adhesion of the electrode to the collector, several forms of current collector have been patented:

   a. Making numerous holes through the collector to improve impregnation and reduce the distance of ion travel to prevent or reduce electrolyte localization in the supercapacitor [25].
   b. Making a current collector with an underlayer containing perforations [26].
   c. To use an aluminium grid. This type of manifold is difficult to weld with the external part of the component (cover, housing), but allows to lighten the component.

9.2.1.2   *Activated Carbons for Supercapacitors*

Activated carbon is the active material ("host" site) of supercapacitors. In 1957, the first supercapacitor was developed using a relatively non-porous carbon. Soon, many researchers realized that it was necessary to increase the surface area accessible to ions to increase the volume and mass capacity. The first commercially available porous carbons were manufactured for filtration applications and for the purification of sugars. These carbons are not of sufficient purity to be directly usable for energy storage (high surface area groups, particle size distribution unsuitable for electrode construction, pore size distribution unsuitable for the ions to be stored). Over the last fifty years, the major players in the field of supercapacitors have been trying to increase the mass and volume capacities, in particular using so-called "super-activated" carbons (such as Maxsorb activated carbon, for example). Nevertheless, these superactivated carbons have limited lifetimes because they contain many structural defects. Therefore, new carbons have been developed or modified for the supercapacitor application. Among the manufacturers involved in these developments are Kuraray Chemicals [27–29], Energ2 [30–32], Nippon Oil [33–37], GS Caltex [38], Kansai Coke [39], Calgon [40], Osaka Gas [41], Meadwestvaco [42], etc.

It is generally accepted that only 20%–30% of the porosity is actually accessible for the ions used in energy storage [43–45]. One of the most widely used carbons is Kuraray's YP17 (also called YP50F) and is widely used as a basis for comparison in the bibliography (patents [46–49] and publications [50–53]).

One of the essential characteristics of an activated carbon is its porosity. The surface of a carbon is generally measured by the B.E.T. method according to a multilayer model [54] using nitrogen adsorption at 77 K. The nitrogen molecule being smaller than the ions, the surface obtained by the BET method is thus much larger than that actually accessible to the ions (figures 9.5 and 9.6). The area or volume accessible to ions (denoted S or V respectively) at a given pore size is summarized by figure 9.6.

To increase the surface area accessible to ions, many studies have focused on developing the porosity [55–60]. However, most of the processes developed to increase this porosity also simultaneously increase the B.E.T. surface area and widen the pore size distribution. This is represented by figure 9.7. Recent works [61–65] show that there is a trade-off between the pore size distribution and the size of the ions present in the electrolyte. However, capacity optimization should not be at the

Surface comparison

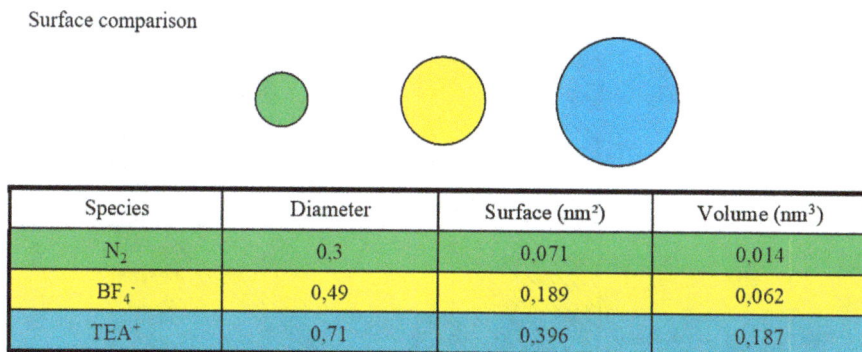

Species	Diameter	Surface (nm²)	Volume (nm³)
N_2	0,3	0,071	0,014
BF_4^-	0,49	0,189	0,062
TEA^+	0,71	0,396	0,187

FIG. 9.5 – BF_4^- and TEA^+ ion sizes and N_2 (comparative scales respected in surface area and diameter).

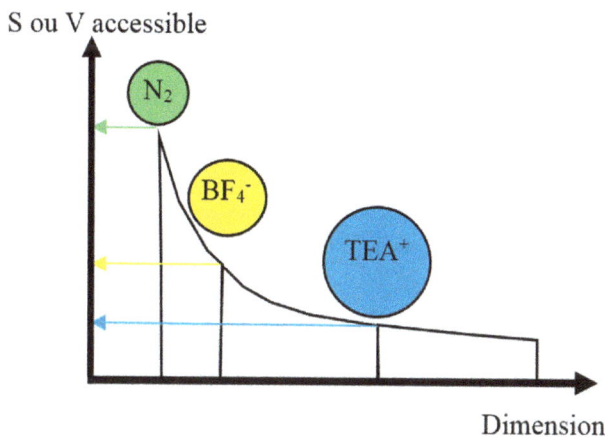

FIG. 9.6 – Representation of the volume and accessible surface area for a given species.

expense of cyclability and should not increase the EDR of the component. This leads to the conclusion that there is no correlation between the BET surface area and the specific capacity of the material, which can also be seen from the literature.

The collaborative work of the groups of P. Simon, M. Salanne, C.P. Grey and P.A. Madden in understanding how ions are stored in porous materials is crucial [66–70] on this topic [71]. These simulations confirm the first experimental results obtained by the Y. Gogotsi and P. Simon's teams on carbons derived from carbides (CDC), in terms of capacity increase with decreasing pore size. The numerical structures obtained are 3D chemically connected and cannot be reduced to a disordered stack of small aromatic sheets interacting only through scattering forces (figures 9.8 and 9.9, these carbon structures are obtained by inverse Monte-Carlo simulation starting from experimental data such as chemical composition, X-ray diffraction and X-ray scattering data). This type of modeling of the texture of

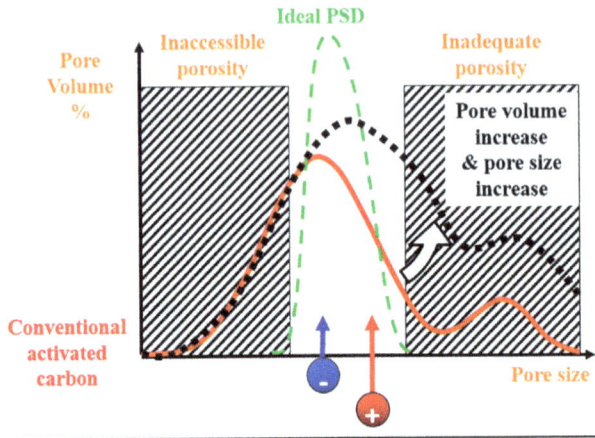

FIG. 9.7 – Pore size distribution (PSD) of a conventional activated carbon (shown in red), and the impact of the increase in pore volume (denoted by the arrow) *via* a standard activation process on the PSD. Comparison with theoretical PSD optimization. Ions are also shown (assuming the cation in red is larger than the anion in blue).

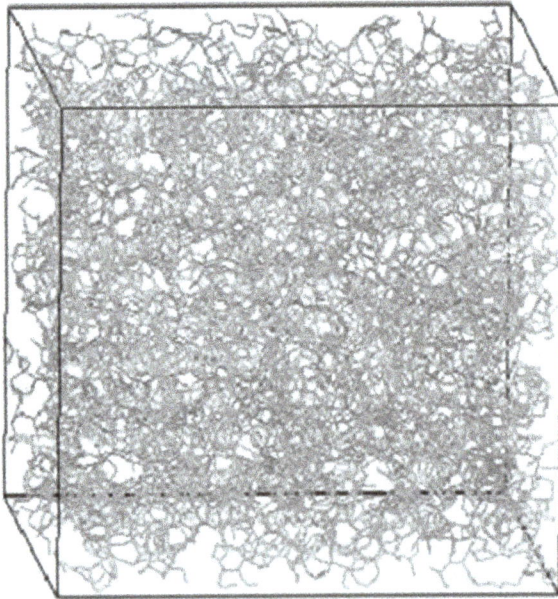

FIG. 9.8 – Atomic configuration of a sucrose coke heated to 1000 °C (CS1000), density 1.5. One line represents a C–C bond. The box size is 5 nm. The pore size is 0.4–0.6 nm.

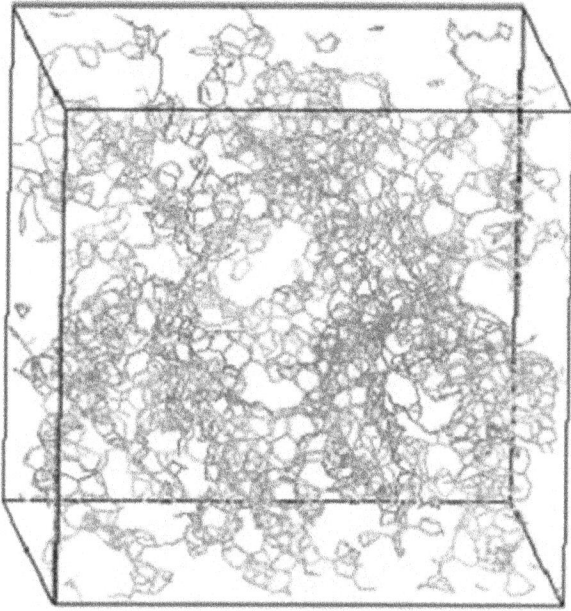

Fig. 9.9 – Atomic configuration of an activated sucrose coke, density 0.7. One line represents a C–C bond. The box size is 5 nm. The pore size is in the order of 1 nm.

porous carbons is based on a direct calculation of the energy thanks to the REBO "bond-order" potential to manage the carbon chemistry [72]. Thanks to this type of simulation, it is shown that it is not possible to retrieve the experimental data by considering the medium as an assembly of "slit" pores, infinite and unconnected, and of different sizes, thus undermining the techniques of pore size distribution calculations carried out from adsorption data [73]. Finally, in all these simulations, it is considered that the volume is fixed during the charge/discharge steps. However, this hypothesis is not necessarily verified, especially for low porous carbons, which present a swelling of structure during cycling. This swelling suggests the presence of an insertion phenomenon, notably mentioned by Takeuchi *et al.* [74] and confirmed by Hahn *et al.* [75, 76].

9.2.1.2.1 Activated Carbon Availability
The carbons can be distinguished in two categories, in particular by taking into account their activation method (physical or chemical activation) and their precursors (synthetic or natural) [77]. However, the carbons developed on a laboratory scale are generally quite different from those available on an industrial scale [78]. Nevertheless, it is important to mention the most important work published over the last two decades.

9.2.1.2.1.1 Laboratory Scale Carbonaceous Materials. Many different types of carbons have been tested as electrode material for supercapacitors [79–127]. In view of all these materials, a few comments seem necessary, especially concerning carbon nanotubes, CDCs, carbon aerogels and graphene.

- Carbon nanotubes do not have electrochemical performances superior to those of industrially available activated carbons. The price of these nanomaterials is still relatively high and is not competitive compared to industrially purified activated carbons.
- Numerous studies have been carried out with carbon aerogels [128]. The mass and gravimetric capacities of these materials are relatively low, in particular because they are generally mesoporous materials. The interest of this type of material is that it can be used directly in the form of monolith and therefore allows freeing itself from binder and from the resistance due to the particle/particle contact present in the case of powdered materials. However, this material is brittle and makes the electrode/collector contact relatively difficult to achieve.
- CDCs (carbides derived from carbides). These materials have been developed in different laboratories [129, 130]. The pore size distribution is relatively narrow and seems appropriate for supercapacitor application, making them interesting in terms of model materials. They have the advantage that they do not come from a standard activation process but are the residue of the chlorination reaction of a transition metal carbide, such as titanium. The mass yield of this chlorination to microporous carbon is low (less than 10% by mass). The final product must also be thermally reprocessed to remove traces of chlorine, which is particularly harmful to ageing. This makes this type of material expensive.
- Graphene. In theory, this material has the highest possible surface area accessible to ions for a carbonaceous material ($S_{\text{B.E.T.}} = 2675$ m^2.g^{-1} with nitrogen)[1]. However, the vast majority of articles concerning graphene for EDLCs do not actually use graphene, but rather porous disordered nanocarbons. However, a few relevant bibliographic examples can be cited: Liu *et al.* [131], Stoller *et al.* [132] and Banda *et al.* [133].

9.2.1.2.1.2 Industrially Available Active Carbonaceous Materials. Many manufacturers of activated carbons have become interested in the supercapacitor application, and are trying to improve the performance and reduce the cost of these materials.

---

[1]Considering the surface measured with nitrogen (2675 m^2.g^{-1}) and a square mesh taking into account the size of TEA$^+$ ion (limiting ion for the capacity, *i.e.* 0.71 × 0.71 nm^2), we obtain, for 1 g of graphene, $8.81.10^{-3}$ mol of ions. At 2.7 V, this represents a capacitance (F.g^{-1}) of 96 485 C × $8.81.10^{-3}$ mol/2.7 V = 314.8 F.g^{-1} (for a single electrode), *i.e.* 31.5 mF.cm^{-2}. This represents 0.16 Wh.g^{-1} or 160 Wh.kg^{-1} for the two graphene sheets (overall, only at the graphene scale, considering a charge of 1 electron/ion). By taking a ratio of 20% between the active matter/total component (case of industrial components), the maximum energy density would be about 32 Wh.kg^{-1}, *i.e.* 4 times better than the best current commercial cells.

Commercially available activated carbons are:

- Activated carbons from wood. The volume capacities of these carbons are relatively low, but these materials have the advantage of being very cheap. Their ageing performance, like that of other types of carbons, is linked to impurities, surface groups, etc. Activation of these carbons is generally carried out with water vapour [134–136].
- Activated carbons are derived from coconuts. These are the most widely used carbons in the field of supercapacitors. They offer the best cost/performance ratio (EDR, pore size distribution, purity, mass and volume capacity). Activation is carried out with steam [137].
- Activated carbons coming from oil residues (coke, pitch…) [138]. They are generally more capacitive than carbons from vegetable precursors (wood and coconut). However, they are very expensive and the surface groups are more numerous, which has the consequence of accelerating the ageing of supercapacitors [139]. Mineral impurities, on the other hand, are very few (generally less than 0.1%). Manufacturers have nevertheless found solutions to limit ageing, but these products remain expensive, mainly due to the type of activation used (often with potash) [140, 141].
- Activated carbons from sugars [142]. These carbons are still relatively uncommon but can be considered as a good compromise between carbons from coconuts and those from petroleum residues. These carbons are particularly pure but their volume capacity remains limited [143].
- Activated carbons from ex-resins (such as phenolic resins). These are by far the purest carbons. They offer attractive performance in terms of ageing, strength and volume capacity, but are extremely expensive. They are quite often overlooked by supercapacitor manufacturers because of the economic equation they generate. Two carbons manufactured by Kuraray Chemicals (RP15 and RP20) are sometimes cited in publications [144].

An important parameter to consider for ageing is the purity of the materials. According to Kuraray Chemicals, the presence of heavy metals [145] is detrimental to ageing: these metals can promote short-circuits and generate a "high" self-discharge if the quantity exceeds 50 ppm. This shows that it would not be reasonable to use activated carbon from natural precursors for supercapacitor application without specific treatment to achieve such low levels. Alkalis are important impurities for supercapacitors [146]. Sulphur [147], iron and magnesium [148] are also considered harmful impurities. Asahi Glass also recommends limiting the amounts of chromium, iron, nickel, sodium, potassium and chlorine to ensure high stability of the electrochemical system. According to the authors of this patent, the ash content should also not exceed 0.5% [149]. Other chemical species that have a particular impact on ageing are surface groups. Activated carbons contain many surface groups that can react with the electrolyte and lead to clogging of the porosity [150]. For this reason, it is essential to limit their quantity by appropriate treatments, usually thermal.

- Effect of Dissymmetrization

Given the operation of a supercapacitor, it is easily demonstrable that a key limiting factor of capacity is the size of the largest ion [151]. Indeed, the electrolyte generally used consists of an organic solvent and the ions TEA^+ and BF_4^- (tetraethylammonium tetrafluoroborate). However, these ions do not have the same size. BF_4^- has a radius of the order of 0.23 nm (between 0.22 [152] and 0.245 nm [153] without a solvating sphere). The capacitance-limiting electrode is therefore the negative electrode, *i.e.*, the one that preferentially accommodates the larger TEA^+ ions (radius between 0.348 [154] and 0.40 nm [155] without a solvating sphere). The size of the TEA^+ ion limits the storage capacity of systems using the same carbon for both electrodes. It is therefore generally considered that a carbon/carbon supercapacitor is already originally asymmetric. So it is possible to increase the energy density or improve ageing by accentuating this dissymmetry. This fact has been clearly identified using a reference electrode to determine the cathodic and anodic potentials [156–159].

This dissymmetrization has a high impact on the degradation of the electrolyte in ageing phenomena. The idea of forcing this dissymmetrization is a relatively old idea: this principle was clearly described in a Japanese patent [160] of 1986. In this patent, the aim was to adjust the potential of each electrode to minimize the ageing of a carbon/carbon supercapacitor operating in 1 M $TEABF_4$ electrolyte in propylene carbonate. The main advantage of this dissymmetrization is clearly demonstrated in table 9.1.

This principle can be used to improve strength, capacity and/or ageing. One of the best ways is therefore to adapt the pore size distribution of the activated carbon to the size of the ions [161].

Dissymmetrization is already known and demonstrated in some patent families:

– Dissymmetrization with different carbon masses at the two electrodes [162].
– Mass or volume dissymmetrization (*via* a ratio) [163, 164].
– Carbon type dissymmetrization (adaptation to ion size) [165].
– Dissymmetrization in quantity of matter [166]: this patent is exactly the same principle as the 1986 patent.
– Dissymmetrization in thickness [167–171].

An activated carbon has an electrochemical potential of the order of that of ENH (*i.e.* close to +3 V *vs.* Li^+/Li). Dissymmetrization makes it possible to modify the potential excursion of each electrode, as shown in figure 9.10.

The ageing results confirm the influence of these dissymmetries: when the positive electrode reaches high potentials, ageing is strongly accelerated, confirming the

TAB. 9.1 – Effect of dissymetrisation on ageing and performance.

	Ratio of electrode thickness +/−	Initial performance		Performance after 1000 h at 3.0 V/70 °C	
		ESR	Capacitance	ESR	Capacitance loss
Supercapacitor 1	1/0.6	6.5 Ω	1.00 F	7.4 Ω	−5%
Supercapacitor 2	1/1	5.2 Ω	1.20 F	8.9 Ω	−30%

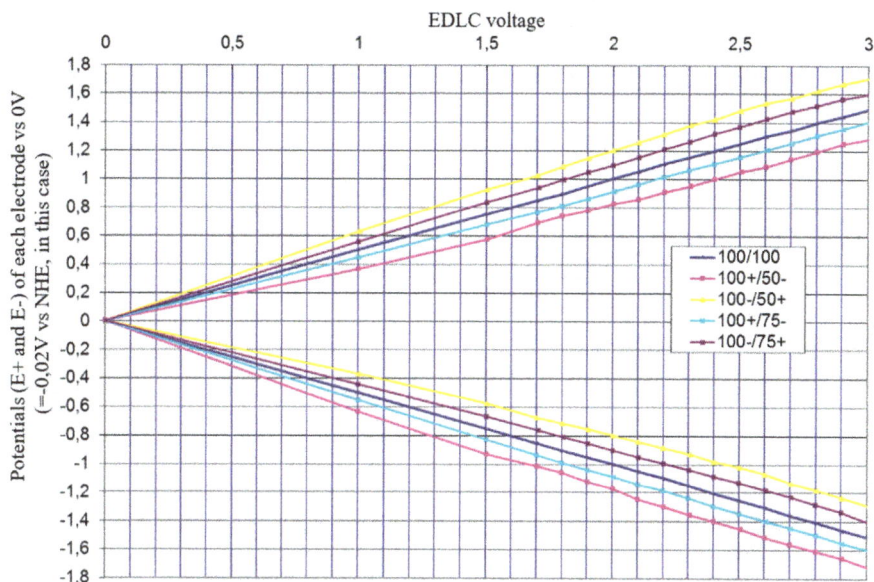

FIG. 9.10 – Distribution of the potentials of each electrode based on the same activated carbon for a given supercapacitor voltage and for different electrode ratios (100 μm, 75 μm, 50 μm). In dark blue, the 100 μm/100 μm symmetrical system. Electrolyte: TEABF$_4$ 1 M in acetonitrile.

fact that the positive electrode is responsible for most of the ageing by electrolyte degradation[2] (gas generation) and by pore plugging at the positive electrode.

Therefore, an interesting solution to optimize performance is to choose two carbons with a different pore size distribution adjusted to each ion size. Table 9.2 shows that capacitance increases and resistance decreases when the pore size distribution is smaller at the positive electrode than at the negative electrode [172]. Reversing the role of each electrode, we can see that the capacitance decreases and the resistance increases. All these facts show that there really is a compromise based on the pore size distribution to improve the performance of ultracapacitors [173].

Many other published patents claim the use of dissymmetrization (TDK [174], Maxwell [175], Nisshinbo [176, 177], CapXX [178], Ultratec [179]...). In fact, all these patents are only alternative solutions to the 1986 patent, which is free of rights.

9.2.1.2.2 *Effect of Surface Groups on Performance and Solutions Provided*

The impact of surface groups on the performance of supercapacitors is controversial: the presence of these groups seems to be desirable in an aqueous medium for reasons

---

[2]At 2.8 V and room temperature, the potential reached by the positive electrode is of the order of 4.45 V *vs.* Li$^+$/Li. By dissymmetrizing at 50 μm in the positive and 100 μm in the negative, the potential of the positive electrode reaches almost 4.65 V *vs.* Li$^+$/Li, which corresponds to the instability potential of most organic electrolytes.

TAB. 9.2 – Effect of pore size on volumetric capacitance and ESR. Electrode B has the smallest pore size distribution.

	Negative electrode	Positive electrode	Volumetric capacitance (F/cm^3)	ESR $(m\Omega)$
Supercapacitor 1	A	B	26.6	24
Supercapacitor 2	A	A	20.8	23
Supercapacitor 3	B	B	27.5	257
Supercapacitor 4	B	A	18.8	243

of potential use of pseudo-capacitance by contributing to the total capacity (redox reactions of surface groups) [180–182]. This contribution was initially foreseen by Delahay [183, 184], and then generalized for both types of electrolytic media (aqueous [185] and organic [186]) by Schultze and Koppitz in the 1970s. However, Sullivan *et al.* [187] showed that an excess of oxygen increases the resistance of materials, as confirmed both in aqueous and organic media by Momma *et al.* [188], Qiao *et al.* [189] and the Kötz group [190]. In aqueous media, presence of acid groups is harmful [191] because they are the source of gases, even at low potential (confirmed by Mayer *et al.* [192]) and they significantly reduce the service life of supercapacitors [193]. Concerning supercapacitors operating in an organic environment, Morimoto *et al.* [194] have shown that the decrease in capacity during ageing is linked to oxygen: the more the activated carbon is rich in oxygen, the higher the quantity of gas emitted (essentially hydrogen by electrolysis of water). Jow *et al.* [195] showed that the amount of surface groups present within the activated carbon has a strong impact on the usable voltage window of a supercapacitor. Therefore, in organic media, the choice must be made for a carbon low in oxygen and more specifically with few carboxyl groups in order to limit the decomposition of the electrolyte. According to Yoshida *et al.* [196], the leakage current, *i.e.*, part of the self-discharge of a supercapacitor, is related to the quantity of carboxylic acid surface groups. The self-discharge phenomenon of a supercapacitor operating in an organic medium is also related to the diffusion of ions from areas where they have accumulated during charging of the supercapacitor [197].

In summary, the surface groups of a carbon increase the capacitance but accelerate the ageing of supercapacitors. This is why many treatments have been developed to modify the surface of carbons for electrochemical applications: acid treatment [198, 199], oxidizing [200–202], electrochemical [203], thermal [204], laser [205], plasma [206, 207], polishing [208] or even solvent washing [209], hydrogenation followed by sulfonation [210], halogenation (chlorine or bromine) and dehalogenation under hydrogen [211]...

With the exception of washing treatments or simple heat treatments (nitrogen), all these proposed treatments are too expensive for the supercapacitor application. Given the high impact of surface groups, heat treatment under nitrogen seems to have become widespread among major suppliers [212–217] to offer pure materials (surface group rate generally less than 0.15 mep.g^{-1}) and inexpensive (generally less than 15 \$.kg^{-1} for high volumes).

9.2.1.3 Binders

9.2.1.3.1 Adhesion Et Cohesion: Key Parameters

The binder combines two functions: it ensures a good cohesion between the particles and it allows a good adhesion of the material on the collector, without generating a high contact resistance.

As early as 1972, the authors of one of the first patents concerning supercapacitors already mentioned that this is one of the major problems to be solved in order to guarantee a low variation of resistance during ageing and to limit the capacitance loss [218].

The amount of binder must be low for four major reasons:

1. To ensure the maximum number of particle/particle/current collector contacts.
2. The electrolyte must be able to impregnate the particles correctly: the binder must not block the intergranular spaces.
3. Most binders are insulating polymers: a high binder content can increase the resistance of the component.
4. Binders are inactive materials within the component: their presence therefore leads to a reduction in the mass and volume capacity of the component.

9.2.1.3.2 Electrode-Current Collector Interface

Coating remains the simplest technology to implement in the field of supercapacitors. In 1988, Morimoto *et al.* proposed to coat the current collector with an aqueous mixture based on PTFE and activated carbon to make a supercapacitor electrode operating in an organic medium ($TEABF_4$ in acetonitrile or sulfolane) [219]. This patent can be considered as the major patent of most of the coating patents developed to make electrodes for supercapacitors. The main advantage of these processes is to be able to control the morphology (thickness, width) of the electrode while maintaining high energy and power densities. Thus, the majority of current or former manufacturers chose this technology: Matsushita [220], Maxwell [221], SAFT [222], CEAC [223], Kureha [224]... However, this process has two major drawbacks:

(a) PTFE, FEP or PFA type binders are relatively expensive (available in aqueous or organic suspension).
(b) It is essential to modify the process quite heavily compared to the coating process that is very widely used in the field of Li-ion batteries, because the electrodes of supercapacitors contain mostly activated carbon (generally a charge rate higher than 80%). As this material is very porous compared to the crystalline materials used in batteries, it is necessary to use a particularly high quantity of solvent to obtain the same viscosity. This solvent must then be removed when drying the electrode. Therefore, an alternative process using a fibrillable binder (*e.g.* PTFE) has been developed and patented by Maxwell Tech. to solve this major problem: they claim the use of a very small amount of solvent (or even none at all) to make their current electrodes [225].

One of the most common binders in supercapacitor technology is PTFE or binders of the same family (FEP or PFA). The main reason for this use is the good electrochemical stability and its availability in aqueous medium [226]. In addition, a reduced quantity of PTFE makes it possible to produce an electrode with high mechanical strength (3%–5% is sufficient, compared to 7%–10% for other binders or binder mixtures): this results in a maximized electrode charge rate. However, this polymer is not extrudable. This type of fibrillated binder is easily detectable by scanning electron microscopy.

Water-soluble vinyl or cellulosic binders such as PVA (polyvinyl alcohol) or CMC (carboxymethylcellulose) [227] are also used for industrial coatings. Chitosan has also been tested as a binder but its cost is high [228]. However, these binders are not electrochemically stable above a voltage of 2.5 V. For this reason, they are now abandoned when the components are operated at a voltage of 2.7–2.8 V, which corresponds to a value of the order of +1.3 V/ENH. Generally, CMC is not used alone but in combination with SBR [229]. Studies have clearly shown that CMC, for example, degrades at these potentials [230]. Polyurethane has also been tested as a binder for supercapacitor electrodes [231]. This binder provides excellent adhesion to the commutator, but is difficult to use industrially to make an electrode. Polyimide is an interesting alternative solution to make a high-performance electrode from an electrochemical and mechanical point of view [232], but it must be used at high temperatures (>250 °C) [233] and is relatively expensive. However, it is potentially interesting for the realization of supercapacitors operating at high temperatures and using an electrolyte without solvents, such as ionic liquids. Recently, other water-dispersed or water-soluble binders have appeared. The interest of these binders is to demonstrate a very good electrochemical stability and a very good adhesion power on collector. One can quote in particular the binders of Zeon Corp [234–236], which have a higher electrochemical stability than homopolymer PVDFs, and JSR micro fluorinated acrylate binders [237, 238], whose adhesion is particularly high compared to standard PVDFs. It should also be noted that modified PVDFs have been launched on the market to improve adhesion by grafting acid functions [239] or maleic anhydride [240].

Finally, wettability is a parameter to be taken into account to improve the adhesion between the carbon and the binder used to manufacture the electrode [241].

An alternative process to coating is extrusion. The advantage of this high-pressure process is the use of a reduced quantity of solvent compared to coating. The polymers described above are extrudable, with the exception of PTFE and polyimide (difficult to extrude due to high temperature process).

The charge rate is generally high (at least 80%) and allows easy production of electrodes [242–244]. In conclusion, PVDF [245] (used with an organic solvent such as NMP, DMSO, THF, PC, and more recently "Equamide" as a substitute for the more and more controversial NMP...) and PTFE (in alcoholic or aqueous suspension) are efficient alternatives to CMC or PVA.

9.2.1.4 Conductive Additives

Activated carbons are not highly conductive materials. Therefore, the addition of carbon black to activated carbon in a polymer paste improves the conductivity of the supercapacitor electrode, whatever the process used [246]. A carbon black is the most widely used conductive additive industrially to improve the performance of polymers (especially in tires) and in the field of electrochemistry (Li-ion and other batteries). Generally, it is composed of more or less spherical particles (spherules) whose diameter is generally between 10 and 75 nm (primary particle). These spherules form "strings" or aggregates whose size is of the order of 50–400 nm. When this type of carbon is dispersed and mixed with other materials, the aggregates form a conductive three-dimensional network. Carbon black is generally very pure (97%–99% pure carbon) and can be considered relatively amorphous (the degree of amorphisation is very strongly driven by the synthesis and the precursor). Nevertheless, the microstructure is close to that of graphite.

Generally, the prerequisites for conductive additives are:

- Good electrical conductivity.
- High corrosion resistance.
- Very high purity.
- Low cost.
- Good thermal conductivity.
- Good dimensional and mechanical stability.
- Low density and good Processability.
- Availability in many forms.
- An easy way to make a composite.

Numerous conductive additives have been developed: carbon black [247], mesoporous carbon black (Ketjen Black), acetylene black, carbon whiskers, carbon fibers, carbon onions [248], carbon nanotubes, natural or artificial graphites, metallic fibers such as aluminum or nickel based, It is particularly advantageous to use a conductive additive with a particle size between 1 and 100 nm to allow good electronic percolation within the electrode, which is generally smaller than most particle sizes of activated carbon [249]. Ketjen Black and acetylene blacks are widely used, but do not have exactly the same performance [250]. Some carbon blacks manufactured by gasification (*e.g.*, Ketjen EC600, Ketjen EC300, or Printex XE-2) have relatively high mesoporosity [251]. A relevant review describing the taxonomy of graphites and carbons that can be used as conductive additives was published by M. Wissler [252] in 2006.

Nowadays, only conventional carbon blacks (ex-oil products and acetylene blacks) have been industrially used by supercapacitor manufacturers because they are a good compromise between cost ($<$US\$ 10.kg^{-1}) and performance [253]. However, carbon nanotubes [254–256], and nanofibers have interesting performances [257] and could be used in the future provided their price falls sharply and their purity is increased. Nevertheless, these conductive additives potentially present a problem: their morphology can generate short-circuit issues between the electrodes through the separator.

9.2.2   Electrolyte

Many electrolytes are available on the market. The ageing of a supercapacitor is strongly linked to the solvent/salt couple used. Cost, toxicity, conductivity and thermal stability are key parameters in making this choice. The two most commonly used solvents are propylene carbonate (PC) and acetonitrile. The most commonly used salt is Et_4NBF_4 (tetraethylammonium tetrafluoroborate or $TEABF_4$) and a variant also used in PC (Et_3MeNBF_4-triethylmethylammonium tetrafluoroborate or $TEMABF_4$).

9.2.2.1   Impact of Electrolyte on Performance

The energy stored in a supercapacitor is proportional to the square of the nominal voltage. This voltage is usually limited by the electrochemical stability window of the system. For solid/liquid systems, this parameter is limited by the electrochemical stability of the salt, the solvent, and the degradation of the electrode.

The electrolyte has a high impact on the performance of supercapacitors in particular because it plays a central role in the resistance of the supercapacitor and therefore has a direct influence on the power density. It also influences the capacitance value, and can also cause gas emissions during ageing [258]. Generally, the electrolyte is the limiting parameter if the electrodes are stable. In addition, for voltage values higher than 3 V, the TEA^+ ion is not sufficiently stable. This explains, among other things, the development of more electrochemically stable cations to replace the TEA^+ ion. This is the case of Honeywell ($TEMA^+$), Japan Carlit (SBP^+, or "spiro") [259–262], or PYR_{14}^+ (Merck and Solvent Innovation).

9.2.2.1.1   Conductivity

The conductivity of an electrolyte is related to the ion concentration, the ion mobility and the solvent or solvent mixture and the temperature.

9.2.2.1.1.1   Ions and Concentration Limitations.   The first criterion for the choice of salt to obtain a high-performance electrolyte is to obtain good ionic conductivity. Thus, in supercapacitors, cations are often alkylammonium type ions that have good solubility and conductivity in solvents with a high dielectric constant. These cations are preferable to alkali cations because they avoid the possible appearance of an alkali metal at the cathode in the event of an accidental overcharge, leading to a metallic deposit at the anode. Peter [263] *et al.* demonstrated that the use of a glassy carbon as a working electrode could achieve a stable cathodic potential for a tetraethylammonium ion in DMF at −3 V compared to the calomel electrode at 25 °C (−2.96 V/Ag/AgCl, −2.78 V/ENH, 0.26 V/Li/Li$^+$). Numerous studies have been carried out on the reduction of alkylammonium ions on different electrodes [264–270]. Whatever the nature of the alkyl group, the RN^{4+} cation is often reduced to alkanes, alkenes and the corresponding trialkylamines. The anions most often cited in the literature are BF_4^-, ClO_4^-, PF_6^- and $SO_3CF_3^-$. These species are characterized by relatively large sizes (except for BF_4^-), and their anodic stability evolves in the following order: $PF_6^- \geq BF_4^- > SO_3CF_3^- \geq ClO_4^-$.

Asahi Glass [271] has studied different salts and proposed the use of Et_3MeNBF_4 instead of Et_4NBF_4. TEMABF$_4$ has a higher dielectric constant than TEABF$_4$. Table 9.3 summarizes the conductivities of different salts in four different solvents (propylene carbonate, dimethylformamide, gammabutyrolactone and acetonitrile).

Due to the particularly high values of acetonitrile solutions (55–58 mS.cm^{-1} for 1 M TEABF$_4$) [272] compared to other solvents, the use of TEABF$_4$ in acetonitrile as an electrolyte quickly became widespread. Only the Japanese industry prefers to replace acetonitrile, due to its high flammability, by PC. At present, the lifetime of PC-based supercapacitors is shorter than that of those using acetonitrile at the same voltage [273].

Recently, "spiro" type salts (figure 9.11) have been developed by Japan Carlit [274] to improve the limiting concentration of TEABF$_4$ (limited to about 1.5 mol.L^{-1} at 25 °C in acetonitrile). However, this salt is relatively expensive compared to TEABF$_4$, although performance appears to be better.

TAB. 9.3 – Conductivity of organic electrolytes based on different salts (1 M, at 25 °C) in mS.cm^{-1} [275].

Electrolyte	Propylene Carbonate (PC)	g-butyrolactone (GBL)	Dimethylformamide (DMF)	Acetonitrile (ACN)
LiBF$_4$	3.4	7.5	22	18
Me$_4$NBF$_4$	2.7	2.9	7.0	10
Et$_4$NBF$_4$	**13**	18	26	**56**
Pr$_4$NBF$_4$	9.8	12	20	43
Bu$_4$NBF$_4$	7.4	9.4	14	32
LiPF$_6$	5.8	11	21	50
Me$_4$NPF$_6$	2.2	3.7	11	12
Et$_4$NPF$_6$	12	16	25	55
Pr$_4$NPF$_6$	6.4	11	19	42
Bu$_4$NPF$_6$	6.1	8.6	13	31
LiClO$_4$	5.6	11	20	32
Me$_4$NClO$_4$	2.9	3.9	7.8	7.7
Et$_4$NClO$_4$	11	16	24	50
Pr$_4$NClO$_4$	6.3	11	17	35
Bu$_4$NClO$_4$	6.0	8.1	12	27
LiCF$_3$SO$_3$	1.7	4.3	16	9.7
Me$_4$NCF$_3$SO$_3$	9.0	14	24	46
Et$_4$NCF$_3$SO$_3$	11	15	21	42
Pr$_4$NCF$_3$SO$_3$	7.8	11	15	31
Bu$_4$NCF$_3$SO$_3$	5.7	7,4	11	23
Et$_3$MeNBF$_4$	**15**	-	-	**60**

Et_4NBF_4 and Et_3MeNBF_4 are the references as salts in EDLC.

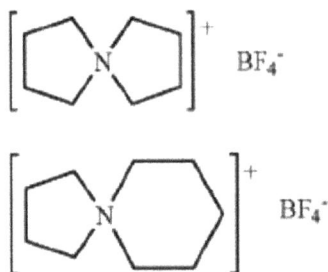

FIG. 9.11 – Examples of "spiro" type salts.

Zheng and Jow [276] showed that a minimum concentration is required to achieve maximum specific energy in a supercapacitor (data at room temperature). However, the operating temperature of supercapacitors for transport-type applications is generally between −30 and 70 °C. For some applications, a wider temperature range is required (typically −40–80 °C). Finally, for aerospace applications, an extreme operating or storage temperature of −55 °C is sometimes required. The higher the concentration, the higher the temperature at which salt crystallization and precipitation takes place. In addition, crystallization and precipitation temperatures are strongly related to the size of the pores in which the electrolyte is present. In the first approach, the electrolyte behaves in the separator and between the carbon grains as it does in the bulk. In the pores of the activated carbon, the phase change no longer behaves like the bulk: the transition takes place at a lower temperature. These confinement effects can also be identified by nuclear magnetic resonance [277, 278].

It is therefore not on this scale that the problems of precipitation and crystallization arise. The limit concentration of $TEABF_4$ in acetonitrile is relatively high (above 1.5 M) at room temperature, and salt precipitation starts below −40 °C if the concentration is 1 M [279]. This study shows that an asymmetric salt allows a wider range of temperatures of use than standard salt and the limit concentration is higher. Of course, crystallization is strongly related to the interactions between the solvent and the salt. For standard electrolytes, the difference in conductivity is not significant between 1 and 1.5 M (figure 9.12, reference combination [280, 281]).

Supercapacitors are power components. Therefore, EDR is a particularly important parameter for applications, partly related to the conductivity of the electrolyte (table 9.4). However, only part of the resistance is affected by the electrolyte conductivity for a 3000 F.

The electrolyte is one of the major components of the cost of the component: industrially, acetonitrile is a very inexpensive solvent and is a residue resulting from the synthesis of polyacrylonitrile (PAN), a polymer widely used in the production of plastic parts in China. Propylene carbonate is a solvent widely used in industry, for cleaning and stripping parts and in synthetic chemistry.

For all these reasons, most manufacturers limit the concentration of $TEABF_4$ to 1 M in acetonitrile (or in PC).

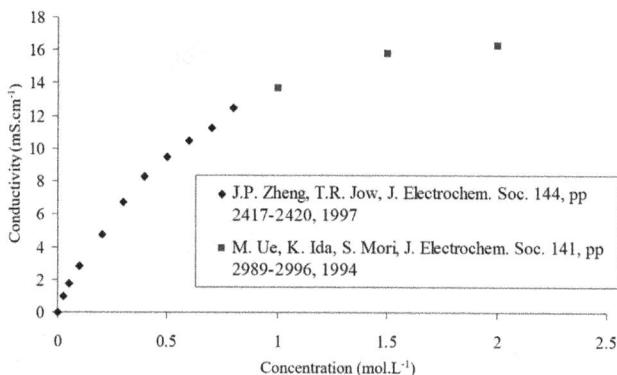

FIG. 9.12 – Conductivity of PC/TEABF$_4$ electrolyte for various concentrations at 25 °C.

TAB. 9.4 – Impact of conductivity on power and energy densities at 25 °C for a 3000 F cell. Reference 100 for 1 M TEABF$_4$ acetonitrile.

Electrolyte	Normalized capacitance	Volumetric power density	Volumetric energy density	Specific power density	Specific energy density	Conductivity
PC/TEMABF$_4$ 1 M	100	81	100	81	84	27
ACN/TEABF$_4$ 1 M	100	100	100	100	100	100

9.2.2.1.1.2 Solvents. The criteria for choosing a solvent depend on the following points:

– Its reactivity to electroactive species and/or electrode materials.
– Its dielectric constant (permittivity ε of the solvent) and its polarisability.
– Its window of electrochemical stability [282]. This parameter depends on the chosen salt and the impurities that may exist or develop in the system. In this case, a parasitic reduction phenomenon can interfere with the purely electrochemical behavior of the system, and thus limit the stability of the salt-solvent system [283].

Taking into account these different parameters, the solvents used in supercapacitors can be classified into three classes:

Aprotic polar solvents with a high dielectric constant (close to water $\varepsilon_r = 78$) such as organic carbonates, ethylene carbonate (EC, $\varepsilon_r = 89.1$) and propylene carbonate (PC, $\varepsilon_r = 69$).

Solvents with a low dielectric constant but with a strong donor character, such as ethers, dimethoxyethane (DME, $\varepsilon_r = 7.20$) and tetrahydrofuran (THF, $\varepsilon_r = 7.58$).

Aprotic solvents with an intermediate dielectric constant such as acetonitrile (ACN, $\varepsilon_r = 36.5$) or dimethylformamide (DMF, $\varepsilon_r = 37$).

9.2.2.1.2   Electrochemical Stability and Ageing
One of the most important parameters of the electrolyte is the stability of the salt
and solvent.

As demonstrated in several studies [284–286] the open circuit potential (OCV) is
approximately 3 to 3.2 V *vs.* Li/Li$^+$. Activated carbons do not have the same OCV.
This OCV depends on the surface groups and the type of carbon.

9.2.2.1.2.1   Electrochemical Stability of Ions and Solvents.   The electrochemical
stability of a solvent is strongly related to the impurities and the cathodic and
anodic potentials of this solvent [287]. For example, traces of oxygen and water have
a detrimental effect on many systems operating in organic media. If these species are
present, the electrochemical stability of the solvent-salt system is limited [288].

For a solvent to be usable, the electrochemical oxidation and reduction potentials
must be above the excursion potentials of the supercapacitor and impurities must be
controlled. The impact of water on the electrochemical stability of the electrolyte has
been demonstrated (case of the TEABF$_4$ 1 M electrolyte in acetonitrile [289, 290]).
It can easily be shown that the electrochemical stability of the electrolyte is strongly
reduced even with a small amount of water [291].

However, it is important not to jump to conclusions about the stability
(or instability) of an electrolyte on a platinum or glassy carbon electrode. Indeed,
Kötz *et al.* have shown [292] that in a real supercapacitor the amount of water does
not have much influence on the lifetime. The electrochemical stability of some
solvents [293] is shown in figure 9.13.

Jow *et al.* [295] published a comprehensive study demonstrating the importance
of electrochemical stability for many quaternary onium salts used in an organic
electrolyte with EC/DMC (1:1).

9.2.2.1.2.2   Electrolyte-Related Causes of Ageing.   If the electrolyte is pure and
its stability is high, only interactions with other constituents can degrade the electrolyte. One of the most important degradations is generated by the adsorbed water
and surface groups, and more specifically the acid groups of the activated carbon
[296]. These interactions are generated exponentially with the increase in the voltage
of the component (shown for two voltages: 2.3 V [297] and 2.5 V [298]). The consequence of these interactions is the generation of gas and the increase in resistance,
in particular *via* the blocking of the pores of the activated carbon and the separator
[299] and/or *via* the degradation of the electrode [300]. The increase in resistance is
not necessarily demonstrated on all types of components [285]. The separator may
also be the source of the problem [301]. Thus, drying the separator is a crucial step to
increase lifetime, improve capacitance and decrease resistance [302, 303].

In the electrolyte, BF^{4-} ions react with water adsorbed in the activated carbons
at the positive electrode to generate HF *via* an electrolysis process. In this way, the
protons behave as a catalyst for the decomposition reaction of propylene carbonate.
Simultaneously, the H$_3$O$^+$ ions migrate to the negative electrode and generate
hydrogen gas. In this way, electrochemical reactions promote gas generation. Ufheil
et al. [304] have demonstrated that PC can be decomposed into carbon dioxide and

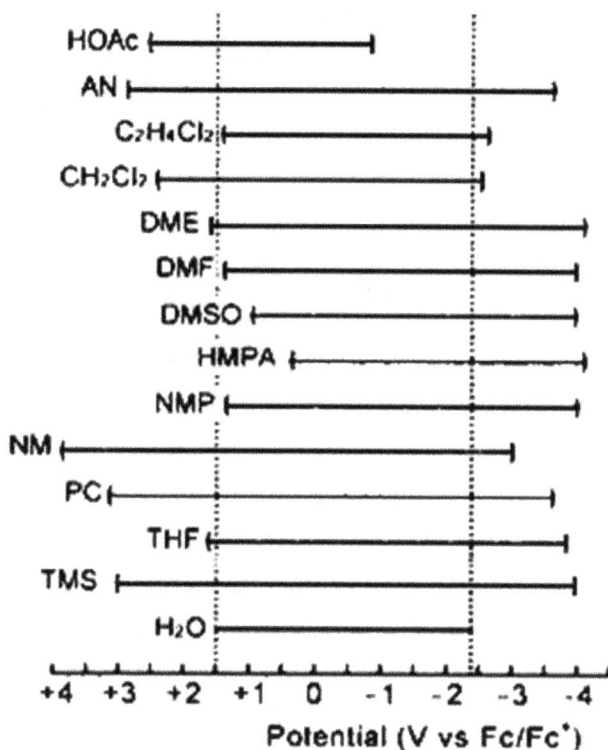

FIG. 9.13 – Potential windows of different solvents compared in a common scale (*vs.* Fc^+/Fc) obtained from cyclic voltametry on a platinum electrode at a current density of 10 $\mu A.mm^{-2}$. For redox potentials measured against other reference electrodes, refer to [294].

acetone especially when PC is subjected to a high potential. Kötz *et al.* [305] analysed the gases generated during the ageing of a supercapacitor operating as a PC electrolyte (1 M $TEABF_4$/PC). These studies demonstrate the formation of CO, CO_2 and hydrogen for operation at 2.6 V. These data had already been mentioned by Asahi Glass Co [306] as early as 2000. This work was confirmed by Naoi *et al.* [307] who separately characterized the gases and water generated at the negative and positive electrodes operating in a PC-based electrolyte. This study was carried out for an operating voltage between 2.5 and 4.0 V. In fact, the positive electrode generates a significant amount of propylene, ethylene, CO and a small amount of hydrogen, while the negative electrode generates only CO_2 and CO. The CO and CO_2 originate from the decomposition of the surface groups but also from the oxidation of the PC. Hydrogen is produced by the electrolysis of residual water. A phenomenon of exfoliation of activated carbon is also observed for a voltage of 3 V and is accompanied by the generation of hydrogen, CO and CO_2. This exfoliation phenomenon has also been studied by dilatometry by Kötz *et al.* [308].

Gases are generated during the ageing of supercapacitors. The volume of gas depends, among other things, on the salt and solvent [309]: H_2 is the predominant end-of-life gas for acetonitrile-based supercapacitors and is the result of an electrochemical reaction. It grows continuously during floating. CO is present in the initial state and also grows continuously with floating but its quantity is lower than hydrogen.

Kurzweil *et al.* [310] propose different theories to explain the degradation of the acetonitrile electrolyte:

– The authors suggest the presence of gaseous acetonitrile, water vapor and ethene during the aging of supercapacitors.
– The alkylammonium cation is degraded at high temperatures by the removal of ethene.
– The tetrafluoroborate anion is the source of fluoride, HF and boric acid derivatives.
– Hydrogen is generated by the electrolysis of water and the fluorination of carboxylic acids.

In conclusion, supercapacitors operating in an organic environment must mechanically resist a continuous increase in internal pressure and not leak. Therefore, major industrial players have patented solutions to this technological problem.

These solutions can be classified into five main categories:

– Venting the gas from the inside to the outside of the component: reversible valve [311], porous polymer membrane [312], metallic or ceramic membranes [313].
– Condensing the gases within the component using for example a getter type material [314].
– Reinforce the component by means of cover and/or housing thickeners.
– Check the opening of the component by irreversible bursting of a membrane or venting [315]. This solution has a major disadvantage: after the gases have been released to the outside, the cell can no longer store energy safely and the electrolyte can leak from the component.
– Reduce the amount of gas generated using chemical reagents. Acidic species are generated during ageing and are gas initiators: the presence of antacids therefore reduces the amount of gas generated. Honda [316, 317] has developed electrolytes containing an "antacid agent" such as an alkali silicate, carbonate or benzoate to block the parasitic reactions produced by the acids generated during ageing.

9.2.2.1.3 *Thermal Stability and Performance*

Generally, the data presented in the publications are from experiments conducted at room temperature. However, for most transport applications, the operating temperature is between −30 and 70 °C or even between −40 and 80 °C [318]. The temperature range [20 °C; 70 °C] is generally not problematic (in the initial state) because conductivity increases with temperature. However, there are important differences between electrolytes at low temperatures: PC-based components have the poorest performance and are strongly degraded at low temperatures [319]. Therefore, work has been carried out to improve the low-temperature performance of electrolytes based on PC or the dimethylcarbonate/sulfolane mixture, in particular by adding fluorobenzene [320] and using a salt of the $EMPyrBF_4$ (ethylmethylpyrroli-dinium tetrafluoroborate) type to replace $TEABF_4$ or $TEMABF_4$.

Commercial cells operating on an aqueous electrolyte basis claim to operate at -50 °C; however, these supercapacitors have very limited energy densities due to the low operating voltage. Therefore, for aerospace applications with a low operating temperature limit of -55 °C, electrolyte solutions based on a solvent mixture have been developed [321]. Supercapacitors have been characterized down to -40 °C mainly to study the impact of temperature on the operation of the electrode and to characterize the leakage current phenomena [322, 323]. These data indicate that the components still have acceptable performances at -40 °C, leaving the possibility of operating the component even at lower temperatures [324]. However, below this temperature limit, performance is degraded due to the freezing of commercial sol-vents widely used in commercial cells (propylene carbonate or acetonitrile). One of the greatest difficulties in developing electrolytes that can be used at low temper-atures is to choose solvents or solvent mixtures with particularly low melting points and at the same time maintain high ionic conductivity to minimise series resistance and not to degrade the ageing of supercapacitors. This increase in series resistance is particularly linked to the increase in viscosity of solvents at low temperatures, which leads to a decrease in conductivity. Some solvents have relevant physico-chemical performances to make electrolytes operating at low temperatures [325]: methyl formate (MF), methyl acetate (MA), ethyl acetate (EA) and dioxolane (DX).

The conductivities of the mixtures were measured at different temperatures and compared to the standard acetonitrile electrolyte. These mixtures show good initial performance at very low temperatures (-55 °C, figure 9.14).

Other solvent mixtures have also been proposed to obtain electrolytes that can be used over a wider temperature range than acetonitrile [326]. At high tempera-tures, conductivity is more favourable for high power applications. However, ageing

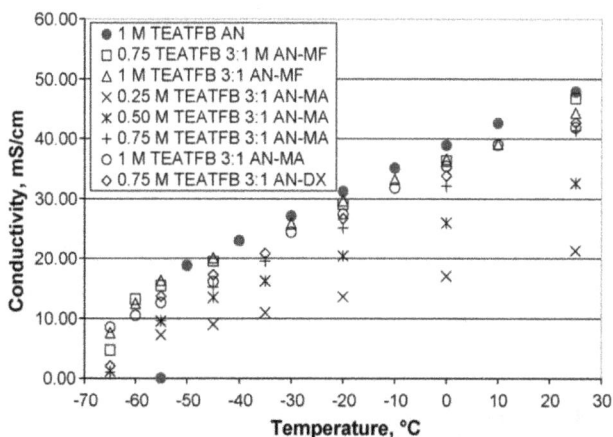

FIG. 9.14 – Conductivity *vs.* temperature for electrolytes operating at low temperatures compared to standard $TEABF_4$ 1 M acetonitrile based electrolyte.

is generally accelerated because the electrochemical degradation reactions follow an Arhenius-type law, and are therefore accelerated with temperature.

Acetonitrile can be used up to 80 °C peak in applications. Other solvents can be used as a substituent, such as sulfolane or EC, if the intended applications require operation only at high temperatures.

For wider temperature ranges, ionic liquids can be an interesting solution as electrolytes for supercapacitors [327]. These electrolytes have the advantage of being solvent-free, stable over a very wide temperature range and non-flammable. In addition, they can be used at high temperatures [328], but generally do not perform well at low temperatures (except for some specific eutectics). Unfortunately, the conductivity of ionic liquids is generally lower than propylene carbonate electrolytes [329]. Nevertheless, some ionic liquids, such as alkylimidazolium tetrafluoroborate (except $EMIBF_4$ [330]) or N-butyl-N-methylpyrrolidinium-bis-(trifluoromethane-sulfonyl) imide ($PYR_{14}TFSI$) [331] have high electrochemical stability [332–334]. However, these products are difficult to purify because they are not distillable [335–337]. Therefore, obtaining pure products generally involves successive washing, which results in a sharp increase in the price of these electrolytes and makes them difficult to use in commercial supercapacitors, even though products (cells and modules) are now available in Japan through a partnership between Nisshinbo and Japan Radio Co. [338, 339]. The electrolyte used in these components is $DEME\text{-}BF_4$ (N,N-diethyl-N-methyl-N-(2-methoxyethyl)ammonium tetrafluoroborate), which performs better than the standard propylene carbonate electrolyte [340].

In summary, ionic liquids are good models for confirming the results of numerical simulation and in particular for demonstrating that solvation of ions is not necessary for energy storage [341]. However, these electrolytes could find applications in niche markets requiring high temperature operation [342] or where the power density is relatively low but the unit voltage is high (if the purity is very high). Ionic liquids have also been tested as salts for organic solvent-based electrolytes [343, 344]. Two interesting studies have been dedicated to these mixtures and have shown encouraging results in increasing the conductivity of acetonitrile [345] and acetonitrile-free electrolytes [346].

9.2.2.1.4 *Toxicity*

Gammabutyrolactone (GBL) is an interesting solvent for supercapacitor electrolytes. However, its toxicity is questionable as it is potentially an ingested initiator of GHB, making it a regulated substance [347]. Some very commonly used solvents and their respective toxicities are presented in table 9.5.

Flammability is directly related to flash point (FP) and boiling point (BP):

– Extremely flammable liquid and vapour: FP < 23 °C and BP ≤ 35 °C.
– Highly flammable liquid and vapour: FP < 23 °C and 35 °C < BP.
– Flammable liquid and vapour: 23 °C ≤ FP ≤ 60 °C.
– Combustible liquid: 60 °C < FP ≤ 93 °C.

TAB. 9.5 – Flammability, toxicity and flash point for some solvents used in supercapacitor electrolytes. T: toxic, Xn: harmful, Xi: irritant, –: not classified.

	Flash point (°C)	Boiling point (°C)	Toxicity
Acetonitrile (ACN)	5	81.6	Xn
3-methoxypropionitrile (MPN)	66	165	Xi
Propionitrile (PN)	6	97	T
Butyronitrile (BN)	16	116	T
Propylene carbonate (PC)	123	242	Xi
Ethylene carbonate (EC)	150	248	Xi
Dimethylcarbonate (DMC)	18	90	–
Diethylcarbonate (DEC)	25	126	–
Dimethylformamide (DMF)	57	153	T
2-Pentanone (2PN)	7	102	Xi
MethylEthylketone (MEK)	−3	80	Xi
γ-Valerolactone (GVL)	96	207	Xi
γ-Butyrolactone (GBL)	104	206	Xn
MethylPropylketone (MPK)	7	102	Xi
Methylformate (MF)	−19	32	Xn
Ethylformate (EF)	−20	54	–
Ethylacetate (EA)	−4	77.1	Xi
Methylacetate (MA)	−10	56.9	Xi
Diethylsulfone (DES)	246	246	–
Dimethylsulfone (DMS)	143	238	–
Sulfolane (SL)	177	285	Xn
Dioxolane (DX)	−6	78	Xi

9.2.2.2 Issue with the Substitution of Acetonitrile

There are two problems with acetonitrile: its low flash point (5 °C in an open cup) and its toxicity (harmful). To date, acetonitrile does not have an equally effective substitute combining high conductivity [348], high thermal and electrochemical stability [349, 350], and very low cost. Acetonitrile-based mixtures have been tested and can be an interesting solution in terms of conductivity but do not circumvent the problem related to its toxicity or a potential ban [351, 352]. To get around this flammability problem, one solution is to use a flame retardant [353], but this solution is not yet used industrially.

9.2.2.3 Solid State Electrolyte

The major advantage of a solid electrolyte is that the functions of electrolyte and separator are concentrated in the same material. Bearing in mind that the electrolyte must be free of electrochemically active impurities, most solid electrolytes have been tested in systems operating in alkaline medium at low voltage, to avoid degradation in aqueous medium [354].

Supercapacitors are dedicated components for power applications. While interesting results have been found in alkaline electrolyte/PVA in aqueous medium, "all-solid" supercapacitors operating in organic medium do not show relevant results, mainly due to the high conductivity at low temperatures (as is also the case with ionic liquids) and the poor electrolyte impregnation of the electrode material, which is very porous. However, this type of component could potentially be used for electronic applications, as these applications do not require high power and operation is usually at room temperature or even high temperatures [355]. Products of this type are commercially available from AVX-Kyocera (BestCap®). For high-capacity supercapacitors, these electrolytes have not been commercially available for the reasons initially mentioned.

9.2.2.4 Electrolyte Organization in the Carbon Based Electrodes

The carbonaceous materials in the electrodes are particularly disorganized. The ions navigate in this disorganized structure, far from the concept of electrochemical double layer on a flat metal plate [356]. NMR studies carried out by Deschamps *et al.* have brought additional arguments: by demonstrating the presence of two distinct ion absorption sites confirming the effect of confinement on the mobility of ions in the porosity [357]. By performing experiments at different electrochemical potentials on supercapacitor electrodes, it is demonstrated that solvent molecules are "expelled" from the porosity when the supercapacitor is charged, in accordance with current numerical simulations.

From all these data, it can be concluded that the mechanism of energy storage in supercapacitors is finally understood:

- The electrochemical double layer does not form in a commonly accepted structure (with its diffusion layer) due to the confinement in the nanoporous material.
- Ion movements are indeed the cause of the increase in electrode charges.
- The charge to be considered is not 1 on each ion and depends on the applied potential.
- Ion/solvent interactions are also to be considered, but the desolvation of the ions is real.
- Carbon and its structure have a major effect on the storage capacity and mobility of ions. This is also confirmed by simulation studies.

9.2.3 Separators

9.2.3.1 Requirement Specifications of Separator

Requirement specifications of separator are: electrochemical stability (linked to high purity and electrochemical stability of the material(s) used), high porosity, high thermal inertia and chemical inertia with respect to the electrolyte. In addition, the separator is an inactive material in the supercapacitor. For this reason, the separator should be as thin as possible and inexpensive. However, the minimum thickness of the separator is limited by several parameters: avoid short circuit between the electrodes by means of particle contact [358] and have a sufficiently high mechanical strength to allow industrial handling (in winding or stacking).

9.2.3.2   Cellulose Based and Polymer Based Separators

In theory, supercapacitor electrodes can operate without a separator if the distance between the electrodes is maintained [359]. In practice, it is impossible to maintain the electrodes in a component without generating a short circuit. Thus, many studies have been conducted to provide high-performance separators for supercapacitors. Generally, these separators are very porous (more than 50%, even up to 80%) compared to the separators used in Li-ion batteries (generally between 30% and 50%). However, when the separator is too porous, the amount of electrolyte needed to impregnate the component is higher and the price of the component increases as well. Therefore, there is a trade-off between the price of the separator, the amount of useful electrolyte and the resistance coming from the separator. Cellulose separators, originally used in electrolytic aluminium capacitors, are currently the most widely used in supercapacitors [360]. The thickness is generally between 15 and 50 μm. The density of these separators is also low (less than 0.85). Due to the volume of the mature market for electrolytic capacitors, the price of this type of separator is relatively low and the quality is very high (very few impurities and very few manufacturing defects). However, it is necessary to dry this type of separator very properly to avoid contamination of the supercapacitor by the water adsorbed in the separator [361–363]. The drying process can be carried out thermally (and under vacuum, as conventionally done in batteries) or by washing with acetone [364].

Since the energy density and power density depend on the square of the voltage, a lot of research is currently being done to increase the operating voltage of the electrodes. Supercapacitor manufacturers (Maxwell, LS Mtron, Skeleton...) have announced industrial components operating at 3 V. However, paper separators deteriorate by oxidation at 3 V, leading to a reduction in mechanical performance of separator [365]. Also, materials substituting cellulose have been developed to operate supercapacitors at this voltage.

Numerous polymer-based separators have been developed and marketed [366]:

- Separator based on glass fibres. The dimension of the fibres is generally between 1 and 4 μm with a porosity between 70% and 90%. The thickness of this type of separator is greater than 30 μm, mainly to avoid short circuits [367]. Thinner separators have been developed but contain a binder to hold the fibres together [368]. Glass fibre separators were initially developed for AGM (Absorbed Glass Mat) lead batteries.
- Porous polypropylene films [369–371]. These separators were developed for Li-ion batteries. Most of these separators are generally not porous enough to maintain low series resistance.
- Polyethylene separators containing mineral fillers to improve thermal stability [372].

- Multi-layer separators [373]. On a polymeric separator, an ultra-thin layer of fibers prepared by electrostatic spinning is deposited. The diameter of the fibres is generally less than 1 μm. This process avoids short circuits between the electrodes.
- Separator made from thermoplastic resin and fiber pulp [374]. This type of separator has excellent mechanical strength and high permeability to ions, gases and liquids. However, the price of this separator is very high compared to cellulose separator.
- PTFE based separators. These separators are particularly inert chemically and electrochemically, but are difficult to use on an industrial scale: they are particularly plastic and therefore difficult to control in winding. In addition, their porosity is poorly adapted (anisotropic) to impregnation once the component has been wound and its price is high. For all these reasons, this type of separator is not currently available in the market.
- Polyimide separator [375] (Energain®, Du Pont de Nemours [376]). This type of separator has all the combined technical advantages of the previous types of separator: high thermal resistance (no shrinkage, even above 200 °C), very good mechanical strength, very good chemical and electrochemical inertia, and high porosity. The only major disadvantage of this separator is its particularly high price. Therefore, this type of separator has, for the time being, only found applications in niche markets, especially in very high temperature conditions in combination with ionic liquids, for example.
- Separators made of PA6,6 (DuPont de Nemours). It offers interesting performance, but has two drawbacks: firstly, it is particularly electrostatic and deformable. Secondly, it is difficult to cut.
- Separator made of polyamide charged with TiO_2, in particular to improve temperature resistance [377], and separator made from a mixture of semi-aromatic polyamide and cellulose [378]. The latter type has the advantage of being as porous and impregnable as a cellulose separator but is more electrochemically and thermally stable. In addition, the price of this type of separator is of the same order as that of cellulose.
- Aramid fiber separator. This separator is mechanically very strong and not very deformable. Its temperature resistance is also very good, but the electrical performances associated in supercapacitors are not good, which is not the case in Li-ion batteries.

Relevant work has been done on separators for supercapacitors by T.L. Wade [379]. His conclusions are as follows:

1. The separator is the most important contribution to the resistance of high power supercapacitors (at least 30% of the total resistance for four commercial separators).
2. There is a reasonably linear relationship between supercapacitor resistance and separator thickness.
3. The resistance of the supercapacitor is a function of the voltage.

9.3 Hybrid Systems

Activated carbon (AC) based supercapacitors are known to have a high power density compared to conventional batteries. Their cyclability is also high and this type of system is particularly safe compared to lithium batteries. However, these systems suffer from a major defect: their energy density remains low. The most efficient commercial systems reach 10 Wh.kg^{-1}. The energy of these systems is related to 2 parameters: the capacity and the square of the operating voltage. The technology is limited by several factors related to these parameters:

(1) The operating voltage is relatively low (generally 2.7–2.8 V) to maintain a high cycling life and a "controlled" self-discharge. This self-discharge is actually higher than that of batteries (loss of 0.2 V after 16 h at maximum voltage). Increasing the operating voltage is tantamount to increasing ageing and self-discharge. Studies have been carried out for several years to increase the operating voltage, but they are faced with a dilemma: increasing the operating voltage means changing the electrolyte and therefore increasing the cost that is the key parameter for the penetration of this type of component.

(2) The volume capacity of the material. Numerous tests have been carried out by various teams throughout the world in an attempt to increase the volume capacity of the materials. This has led to the emergence of new materials such as CDC (carbons derived from carbides). Nevertheless, these materials do not offer higher performance than the best industrially obtained activated carbons and their cost, even projected over high volumes, remains prohibitive and confines them to niche markets.

(3) Reduction of passive constituents. This approach has been undertaken by the main industrial players in the field. The performances currently achieved partly take into account these improvements through the reduction of passive components (collector, separator, casing, pouch cell technology, etc.).

In summary and taking into account the constituents of the EDLC, the optimized mass capacity should therefore not exceed 14–15 Wh.kg^{-1}, which in some power applications remains insufficient.

In addition, to increase the energy density, one of the means consists in substituting one of the two activated carbon electrodes with either a battery or pseudo-capacitance type electrode.

In this case, the discharge potential is generally different from 0 V, taking into account the difference between the two materials used.

The taxonomy of supercapacitors can be summarized as shown in figure 9.15.

The taxonomy of hybrid systems is presented in figure 9.16. Purely supercapacitor type systems (pseudo and EDLC) should be distinguished from systems combining battery and supercapacitor materials. Composite and asymmetric (supercapacitor/supercapacitor) systems have been reviewed by Choudhary *et al.* [380]. Hybrid supercapacitor/battery hybrid systems have been reviewed by Zuo *et al.* [381]. The advantages and disadvantages were discussed by Pell and Conway [382] as early as 2004.

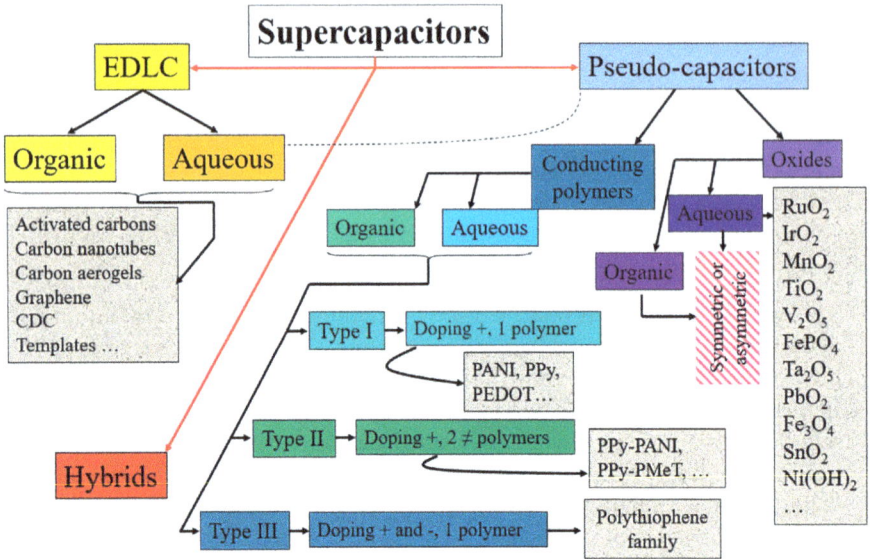

FIG. 9.15 – Taxonomy of supercapacitors.

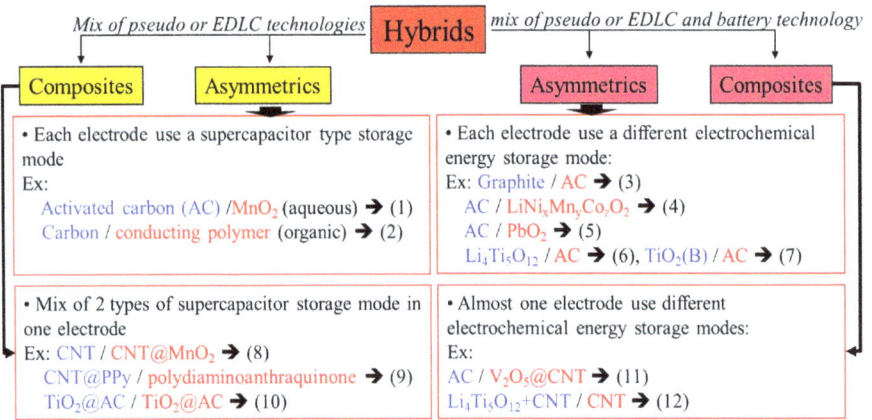

FIG. 9.16 – Taxonomy of hybrid systems. (1) [383], (2) [384], (3) [385–390], (4) [391], (5) [392–394], (6) [395], (7) [396], (8) [397], (9) [398], (10) [399], (11) [400], (12) [401].

These technologies can be distinguished as follows, taking into account the electrodes used:

1. Activated carbon/MnO_2.
2. Lead oxide PbO_2/activated carbon.
3. Nickel hydroxide NiOOH/activated carbon.
4. Graphite/activated carbon.

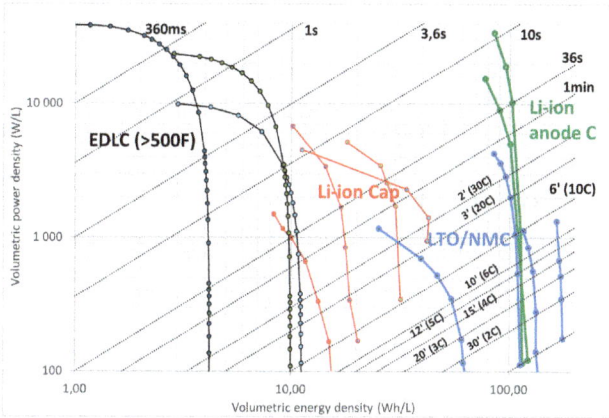

FIG. 9.17 – Volumetric diagram for the different power systems (EDLC, LIC, LTO/NMC power batteries and graphite anode based Li-ion batteries).

The literature is particularly extensive on these hybrid systems. However, the most studied (and now marketed) are those operating with a Li-ion type electrode, in particular the "Lithium-ion Capacitor (LIC)". Several companies (Yunasko, Taiyo Yuden, Maxwell Tech, JSR Micro, Samwha, Power Responder...) commercialised cells to potentially replace lead-acid batteries for vehicle starting and hybridisation applications. Figure 9.17 shows a comparison of the initial performance (Ragone diagram) of EDLCs, hybrid supercapacitors, LTO/NMC power batteries and graphite anode power batteries.

9.3.1 Activated Carbon/MnO$_2$ System

These supercapacitors operate in an aqueous medium using an electrode of activated carbon at the anode and MnO$_2$ at the cathode. The major advantage of this technology is the use of widely available (non-critical) materials and non-flammable electrolyte with a charging voltage of 2.2 V. The mass energy density remains low, even with a considerable improvement in the materials used, mainly due to the use of stainless steel, an aqueous electrolyte (low voltage, higher density than acetonitrile) and the number of additional components required to reach the application voltage. However, these modules could be relevant in stationary applications, where weight and volume are not an overwhelming problem, if the cost problems are solved and if the electrode performance is improved (80 F.cm^{-3} seems to be a value particularly difficult to achieve with current materials).

One of the major drawbacks of this technology is a difficult operation at low temperatures (below –10 °C). This is problematic for automotive applications, but advances have been made in this area. In addition, improvements have been made in terms of safety, in particular using a neutral aqueous electrolyte (neither acidic nor basic) and in terms of the electrode-collector interface.

9.3.2   Lead Oxide/Activated Carbon System

This technology also works in an aqueous medium at a voltage of around 2 V. These cells and modules have been developed and tested by CSIRO and have been marketed by Furukawa Battery. Cycling and power demonstrations have been carried out on vehicles. However, this technology does not solve the problem inherent to the use of lead and a specific design of the carbon electrode is necessary to guarantee a good lifetime. The advantage of this technology is to have the same maturity of the lead oxide electrode as that used in lead acid batteries and to be able to anticipate the failure levels very well known in this same technology.

9.3.3   NiOOH/Activated Carbon System

The proposed systems can use a nickel based electrode (NiMH battery type) and a second one of activated carbon in sulfuric acid medium [402], or a lead (PbO_2) battery electrode and a second one based on activated carbon [403] and more generally an activated carbon with one electrode and an inorganic material selected among Ru, Rh, Pd, Os, Ir, Co, Ni, Mn, Fe, Pt, and their respective alloys as well as the oxides and the alloys of oxides of the above mentioned metals [404]. The literature also includes hydrid systems operating in an aqueous medium using porous metal electrodes on the one hand and nanoporous carbons on the other hand [405]. The disadvantage of all these aqueous systems lies essentially in the fact that the operating voltage of the system remains low (generally below 2.2 V). This implies, for the realization of an energy storage module, the assembly of multiple cells of low unit voltage. The connection of these multiple cells also creates a significant additional series resistance in energy storage systems for power applications. The energy density of these complex systems is ultimately of the order of that of standard AC/AC supercapacitors. Their cost is also higher: these systems are therefore not widely marketed and only find applications in niche markets.

   This technology also works in an aqueous medium and was developed and marketed mainly in Russia many years ago by ESMA and ELIT. The energy densities of the stacked modules are of the same order as those obtained with cell-based modules operating in an organic C/C medium. However, the cyclability is lower, the cost is higher, and the maximum operating temperature is limited to 55 °C. Finally, the time constant is strongly degraded, due to the use of an oxide-based electrode (RC > 3 s).

9.3.4   Graphite/Activated Carbon System

9.3.4.1   Lithium-ion Capacitor Technology

The major interest of lithium-ion capacitors (LIC) is to effectively present an intermediate energy density between EDLCs and Li-ion batteries. The lifetime (not shown on the Ragone diagram) is about 1/10th of that of EDLCs. In addition, LICs generally have rather poor cold performance due to the use of electrolyte of the same type as that used in Li-ion (carbonate-based) batteries. The cost of LICs has decreased significantly over the last five years and is now of the same order as that of EDLCs (in $€.F^{-1}$). However, they are still facing strong competition for many applications from very long-life EDLCs and power batteries whose cost is lower ($€.Wh^{-1}$) thanks to the increase in Li-ion volumes.

These so-called "hybrid" supercapacitors consist of an electrode based on the operating principle of a battery and another electrode identical to that used in traditional activated carbon-based supercapacitors. These systems were mainly developed in Japan, mainly due to the non-use of acetonitrile, the performance of which is particularly appreciated in the manufacture of power supercapacitors. The hybrid systems developed generally have energy densities approximately twice as high as those of traditional supercapacitors. However, power is sometimes degraded and cyclability is reduced (about 200 000 cycles at room temperature, compared to 1 000 000 cycles for EDLCs).

Indeed, the negative electrode is generally made of graphite, as in a traditional Li-ion battery, and a positive electrode of activated carbon, which is therefore very porous.

The use of graphite requires, during the first charge, the creation of a so-called passivation layer (SEI for Solid Electrolyte Interphase), made from the materials present in the electrolyte, clearly described by E. Peled [406]. It has also been clearly shown that the structure, thickness and constituents of this passivation layer are strongly related to different parameters [407]. The realization of this passivation layer has the consequence of consuming electrolyte within the component and generating at least partially an irreversible capacity which can be strongly reduced if the cathode is pre-passivated by grinding.

Alternatively, the operating potential of the negative electrode can be increased relative to the reference potential Li^+/Li. This solution makes it possible to at least partially suppress the synthesis of the passivation layer, but reduces the operating voltage of the energy storage system. One of the key materials of this solution is lithium titanium oxide $Li_4Ti_5O_{12}$. The operating potential of this electrode is approximately 1.55 V *vs.* Li^+/Li. However, none of these systems can achieve high power density values unless electrode thicknesses are greatly reduced, usually to a value less than or equal to 30 μm. However, the operating temperature range is greatly reduced compared to conventional power systems such as symmetrical activated carbon/activated carbon supercapacitors operating in $TEABF_4$ medium in acetonitrile.

The use of two carbonaceous materials, at least one of which is graphite, to make a high energy density supercapacitor is not new: since 1983, Bosch [408] has been offering this operating principle with, in particular, the use of lithium perchlorate-based electrolyte. This patent predates numerous patents [409–413]: it describes the use of graphite at the negative electrode and activated carbon at the positive electrode.

One of the main problems to be solved consists in making the passivation layer at the negative electrode and simultaneously impregnating the positive electrode to implement the electrochemical double layer, in a simple and economic way. The solution generally used consists of applying a complementary lithium foil to saturate the system with lithium ions, the majority of which is consumed during the first cycles. This implementation poses safety problems and a significant additional cost. The use of a lithium foil (in alloy with another metal) to activate the passivation of the negative electrode is presented in different patents [414–419]. Another solution to deposit lithium is to evaporate it [420]. However, this process is complex and industrially expensive. Moreover, during the venting process, the deposit is easily oxidized, which is detrimental to its use in an electrochemical system.

Another solution [421] is to use a lithium foil and to implement a very precise charge/discharge profile of the first cycles.

The authors show that ageing performance is very strongly related to the presence or absence of alkali carbonate (Li_2CO_3, Na_2CO_3, K_2CO_3). The best performances are achieved in the presence of carbonates, due to the blocking of the HF produced essentially at the positive electrode (the bulk of the irreversible component coming from this electrode according to the authors). Instead of carbonates, amides or nylons can be used to improve the lifetime [422].

A proposed solution to avoid the use of a lithium foil is to saturate the surface functions of the activated carbon of the positive electrode with an electrolytic solution containing lithium. This solution has been proposed by Sanyo [423]: it consists, for example, of using LiOH in water. This step has the effect of modifying the natural potential of the material thus treated.

A solution proposed to improve the service life of a component made from graphite on the negative side and activated carbon on the positive side is to limit the potential of the negative electrode in relation to the Li^+/Li couple. The optimal potential range proposed by Sanyo [424] is 0.15–0.25 V *vs.* Li^+/Li.

Finally, work has shown that it is possible to avoid the use of a lithium foil by overconcentrating the electrolyte solution and by "stressing" the electrochemical system, in particular through successive voltage rise and fall stages [425].

Concerning the ageing of LICs, we can read with interest the thesis of El Ghossein [426] describing both the ageing mechanisms [427], the electrical modeling [428, 429], but also the tests carried out under different conditions of commercial LICs.

9.3.4.2 *Sodium-ion Capacitor Technology*

The emergence of Na-ion batteries has led players in the field of supercapacitors to work on lithium-free hydride technologies. In contrast to LIC technology, graphite is only intercalated by sodium for high stages (typically NaC_{48} *vs.* LiC_6). Cathodes based on porous or high surface area carbons do not pose any real design problem. Therefore, the main work is based on the alternative anode to graphite. The first way to increase the specific capacity of the negative electrode is to use a disordered carbon with low porosity of the hard carbon type [430]. Other more or less ordered carbonaceous materials have also been tested: MOF derived from carbon [431], carbons derived from biomass [432]... Other anode materials have been tested: $NiCo_2O_4$ [433], Ti_2C MXene [434], sodium niobiate [435], sodium titanate nanotubes [436], NaBi [437]...

However, the recurrent problem of these alternative anode materials is the very fast ageing at high current (generally less than 1000 cycles) and the cost that remains the major parameter of choice for this hybrid technology. As the capacity of the accumulator depends on the least capacitive electrode, research has also been directed towards increasing the specific capacity of the cathode, activated carbon having a specific capacity of the order of a few tens of $mAh.g^{-1}$. Examples include the work of Dall'Agnese *et al.* (use of vanadium carbide-based MXene) [438] and Wasiński *et al.* ($Na_{0,4}MnO_4$ cathode) [439].

In conclusion, this technology has the advantage of having a higher energy density than the EDLC. However, the development of this type of accumulator will be supported by the industry on condition that it has a similar energy density to LICs (therefore without using critical or strategic materials) as well as a considerable lifetime (>100 000 cycles at high current for a high temperature range, for example between –20 and +80 °C).

9.3.4.3 *Potassium-ion Capacitor Technology*

Like LICs, graphite is easily intercalated by potassium [440]. In addition, it is possible, by substituting the lithium ion-based salt with potassium ion, to obtain an innovative hybrid supercapacitor by combining a positive electrode of activated carbon with a negative electrode of a lithium-ion graphite battery. Compared to LIC [441], KIC (Potassium-ion Capacitor) has some advantages, apart from the fact that potassium is much more available and cheaper [442, 443]:

– Unlike LIC, it is possible to use acetonitrile as a solvent for the electrolyte. Also, the performances at low temperatures are intermediate between those obtained for LICs and EDLCs.
– The collector supporting the graphite can be made of aluminium. Indeed, potassium does not react with aluminum and acetonitrile does not react with aluminum.

The major disadvantage of the use of potassium is the very important exfoliation of the graphite for weak stages (stage 1 = KC_8, stage 2 = KC_{16}...). However, this exfoliation remains acceptable for a KIC, given that the intercalation stage is high (stage > 4) [444–447]. These KICs have been characterized electrochemically and spectroscopic characterizations are also in progress to elucidate the storage mechanisms involved, in particular the influence of the electrolyte and graphite morphology. The results show that at the scale of a "pouch cell", the hybrid configuration makes it possible to double the stored energy density compared to a conventional supercapacitor while maintaining a high power density (>1 $kW.kg^{-1}$). Prototypes in 18 650 format and multilayer stacked format offer superior performance to EDLCs in the same format (energy density and/or power density). In addition, these systems have excellent cyclability (up to 100 000 cycles) at high load/discharge rates ($t_{charge} = t_{discharge} = 30$ s). It is also important to note that, unlike their lithium analogues, they can be safely discharged at 0 V and do not present any risk of potassium metal plating, which makes them very promising for applications such as transport [448].

To conclude this chapter, supercapacitors offer interesting performances in terms of power density over a very wide range of temperatures and lifetime and very good robustness against abusive conditions (due to low energy density). These storage systems have been on the market for many years, but are struggling to find new markets, mainly due to the cost of the complete system (including the converter). The answer to this problem is usually to increase the energy density at the cell scale. This increase is limited for EDLCs at an acceptable cost. Therefore, current advanced solutions aim at hybridizing the system by degrading the lifetime and low temperature operation, or even by offering very high power Li-ion batteries.

Bibliography

[1] Becker H.E. (1957) US2800616, General Electric.
[2] Boos D.I. (1970) US3536963, Standard Oil Co.
[3] Miller J.M. (2007) A brief history of supercapacitors, Autum 2007, *Batteries & Energy Storage Technol.* 61.
[4] Azaïs P. (2014) Habilitation à la Direction de Recherches, INPGrenoble, July 2014.
[5] Von Helmholtz H. (1879) *Ann. Phys.* **7**, 337.
[6] Gouy G. (1910) *J. Phys.* **4**, 457.
[7] Chapman D.L. (1913) *Phil. Mag.* **6**, 475.
[8] Stern O. (1924) *Z. Elektrochem.* **30**, 508.
[9] Péan C., Merlet C., Rotenberg B., Madden P.A., Taberna P.-L., Daffos B., Salanne M., Simon P. (2014) *ACS Nano* **8**, 1576.
[10] IEC 62576:2018: Electric double-layer capacitors for use in hybrid electric vehicles-Test methods for electrical characteristics.
[11] IEC 62391 standard: Fixed electric double-layer capacitors for use in electric and electronic equipment.
[12] Deliverable D1.1, Testing methodology manual, FP7-IAPP Energy Caps EU funded project-Grant Agreement N 286210, N. Stryzhakova, Y. Maletin, S. Zelinskyi, March 2012.
[13] Zhang S., Pan N. (2015) *Adv. Energy Mater.* **5**, 1401401.
[14] Zhao J., Gao Y., Burke A. F. (2017) *J. Power Sources* **363**, 327.
[15] Kurtzweil P.(2004) AC impedance spectroscopy – a powerful tool for the characterization of materials and electrochemical power sources, the 14th Int. Sem. on D.L.C., Deerfield Beach, FL., U.S.A., December 6–8.
[16] https://www.marketsandmarkets.com/Market-Reports/supercapacitor-market-37140453.html.
[17] Hahn M., Barbieri O., Campana M., Gallay R., Kötz R. (2004) Charge-induced dimensional changes in electrochemical double layer capacitors, Florida Supercapacitor Seminar, pp. 40–48.
[18] Gwinn C.D., Stephan F.C. (1935) GB439479, The Telegraph Condenser Co.
[19] Claassen A.F.P.J., De Vries J.D. (1940) DE695770, Philips.
[20] Portet C. (2005) Ph-D Thesis, étude de supercondensateurs carbone/carbone à collecteur de courant en aluminium, November 23rd 2005, University of Toulouse.
[21] Taberna P.-L., Simon P., Sarrazin C., Fauvarque J.-F. (2002) WO200291407, Conservatoire National des Arts et Métiers.
[22] Portet C., Taberna P.-L., Simon P., Flahaut E., Laberty-Robert C. (2005) *Electrochim. Acta* **50**, 4174.
[23] Kobayashi K., Miniami K., Tachozono S. (2005) EP1672652, Hitachi Powder Metals and Japan Gore Tex.
[24] Kobayashi K. (2009) WO2009139493, Japan Gore Tex.
[25] Zhong L., Xi X. (2008) US20080089006, Maxwell Tech.
[26] Hartmut M., Scholtz T., Weber C.J. (2004) US2004264110, EPCOS.
[27] Iwasaki H., Sugo N., Hitomi M., Nishimura S., Fujino T., Oyama S., Kawabuchi Y. (2004) WO200411371, Kuraray Chemical, Kashima Oil and Honda Motor.

[28] Inoue K., Sugo N., Iwasaki H., Fujino T., Noguchi M. (2007) US20070128519, Kuraray and Honda Motor.

[29] Fujino T., Nishimura S. (2005) WO200527158, Kuraray Chemical, Kashima Oil and Honda Motor.

[30] Costantino H.R., Feaver A., Scott W.D. (2011) WO201103033, Energ2.

[31] Feaver A., Cao G. (2011) WO200861212, Energ2 and University of Washington.

[32] Costantino H.R., Feaver A., Scott W. D. (2011) WO201102536, Energ2.

[33] Tano T., Oyama T., Fujinaga I., Ono H., Fujii M. (2006) WO2006137323, Nippon Oil.

[34] Fujii M., Takeshita K., Kato H., Ikai K., Tano T., Oyama T. (2007) WO2007132936, Nippon Oil.

[35] Kitajima E., Nakano Y., Takeshita K., Ikai K., Maruyama T., Ono H., Mizuta H. (2005) WO200538836, Nippon Oil.

[36] Fujii M., Taguchi S., Ikai K., Kato H., Igarashi K., Kiuchi N., Nakamura T., Takeshita K. (2009) WO200905170, Nippon Oil.

[37] Fujii M., Taguchi S., Sanokawa Y., Ikai K. (2010) WO201032407, Nippon Oil.

[38] Lee J.H., Lee J.J., Jung K.J., Hong K.J., Oh D.H., Lee S.I., Kim B.Y. (2008) KR20080047086, GS Caltex.

[39] Shimamoto H., Yamada C., Okuyama K., Izuhara H., Hijiriyama M. (2006) WO2006109690, Kansai Coke.

[40] Yoshino Y., Akihiro I., Takeda Y., Suzuki M. (2007) WO2007145052, Calgon Mitsubishi Chemical.

[41] Nakagawa Y., Tajiri H., Mabuchi A. (2006) WO200606218, Osaka Gas.

[42] Buiel E.R. (2007) WO2007114849, Meadwestvaco.

[43] Qu D., Shi H. (1998) *J. Power Sources* **74**, 99.

[44] Pell W.G., Conway B.E., Marincic N. (2000) *J. Electroanal. Chem.* **491**, 9.

[45] Vix-Guterl C., Frackowiak E., Jurewicz K., Friebe M., Parmentier J., Béguin F. (2005) *Carbon* **43**, 1293.

[46] Hatori H., Tanaike O., Hata K. (2007) pour le NIAIT & NIAIST, WO2007020959.

[47] Gogotsi Y., Chmiola J., Yushin G., Simon P., Portet C., Taberna P.L. (2008) WO200869833, University of Drexel and University of Toulouse.

[48] Hata K., Izadi-Najafabadi A. (2011) WO2011149044, NIAIST.

[49] Aubert T., Simon P., Taberna P.L. (2006) WO2006125901, CECA.

[50] Barbieri O., Hahn M., Herzog A., Kötz R. (2005) *Carbon* **43**, 1303.

[51] Cericola D., Ruch P.W., Foelske-Schmitz A., Weingarth D., Kötz R. (2011) *Int. J. Electrochem. Sci.* **6**, 988.

[52] Chandrasekaran R., Soneda Y., Yamashita J., Kodama M., Hatori H. (2008) *J. Solid State Electrochem.* **12**, 1349.

[53] Ruch P.W., Kötz R., Wokaun A. (1999) *Electrochimica Acta* **54**, 4451.

[54] Brunauer S., Emmet P.H., Teller E. (2000) *J. Am. Chem. Soc.* **60**, 309.

[55] Okamura M. (2000) US6064562, JEOL.

[56] Murakami K., Mogi Y., Tabayashi K., Shimoyama T., Yamada K., Shinozaki Y. (2000) EP1049116, Adchemco, Asahi Glass et JFE Chemical.

[57] Morimoto T., Hiratsuka K., Sanada Y., Kurihara K. (1996) *J. of Power Sources* **60**, 239.

[58] Sonobe N., Nagai A., Aita T., Noguchi M., Iwaida M., Komazawa E. (2001) US6258337, Kureha et Honda.

[59] Nomoto S., Yoshioka K., Hirose E. (2001) EP1094478, Matsushita Electric Ind.

[60] Endo M., Noguchi M., Oki N., Oyama S. (2002) EP1113468, Honda Motor.

[61] Largeot C., Portet C., Chmiola J., Taberna P.-L., Gogotsi Y., Simon P. (2008) *J. Am.Chem. Soc.* **130**, 2730.

[62] Ania C.O., Pernak J., Stefaniak F., Raymundo-Pinero E., Béguin F. (2006) *Carbon* **44**, 3126.

[63] Chmiola J., Yushin G., Gogotsi Y., Portet C., Simon C.P., Taberna P.L. (2006) *Science* **313**, 1760.

[64] Lin R., Taberna P.-L., Chmiola J., Guay D., Gogotsi Y., Simon P. (2009) *J. Electrochem. Soc.* **156**, A7.

[65] Raymundo-Pinero E., Kierzek K., Machnikowski J., Béguin F. (2006) *Carbon* **44**, 2498.

[66] Forse A.C., Griffin J.M., Merlet C., Carretero-González J., Raji A.-R.O., Trease N. M., Grey C. P. *Nature Energy* **2**, 16216.

[67] Merlet C., Salanne M., Rotenberg B., Madden P.A. (2013) *Electrochim. Acta* **101**, 262.

[68] Merlet C., Péan C., Rotenberg B., Madden P.A., Daffos B., Taberna P.-L., Simon P., Salanne M., (2013) *Nat. Commun.* **4**, 2701.

[69] Merlet C., Forse A.C., Griffin J.M., Frenkel D., Grey C.P. (2015) *J. Chem. Phys.* **142**, 094701.

[70] Forse A.C., Merlet C., Griffin J.M., Grey C.P. (2016) *J. Am. Chem. Soc.* **138**, 5731.

[71] Merlet C., Rotenberg B., Madden P.A., Taberna P.-L., Simon P., Gogotsi Y., Salanne M. (2012) *Nature Matter.* **11**, 306.

[72] Delfour L., (2011) Ph-D Thesis, Simulation d'un supercondensateur à l'échelle atomique, June 2011, University of Marseille, https://www.cinam.univ-mrs.fr/thesis/delfour-thesis.pdf.

[73] Pikunic J., Pellenq R.J.-M., Llewellyn P., Gubbins K.E. (2005) *Langmuir* **21**, 4431.

[74] Takeuchi M., Koike K., Maruyama T., Mogami A., Okamura M. (1998) *Denki Kagaku* **66**, 1311.

[75] Ruch P.W., Hahn M., Rosciano F., Holzapfel M., Kaiser H., Scheifele W., Schmitt B., Novak P., Kötz R., Wokaun A. (2007) *Electrochim. Acta* **53**, 1074.

[76] Hahn M., Barbieri O., Campana F.P., Kötz R., Gallay R. (2006) *Appl. Phys. A* **82**, 633.

[77] Bansal R.C., Donnet J.-B., Stoeckli F. (2008) Active carbon, New York, Marcel Decker.

[78] Obreja V.V.N. (2008) *Physica E* **40**, 2596.

[79] Liu G., Kang F., Li B., Huang Z., Chuan X. (2006) *J. Phys. Chem. Solids* **67**, 1186.

[80] Tao X.Y., Zhang X.B., Zhang L., Cheng J.P., Liu F., Luo J.H., Luo Z.Q., Geise H.J. (2006) *Carbon* **44**, 1425.

[81] Gomibuchi E., Ichikawa T., Kimura K., Isobe S., Nabeta K., Fujii H. (2006) *Carbon* **44**, 983.

[82] Kim Y.J., Lee B.J., Suezaki H., Chino T., Abe Y., Yanagiura T., Park K.C., Endo M. (2006) *Carbon* **44**, 1592.

[83] Niu J., Pell W.G., Conway B.E. (2006) *J. Power Sources* **156**, 725.

[84] Kim Y.T., Mitani T. (2006) *J. Power Sources* **158**, 1517.

[85] Prabaharan S.R.S., Vimala R., Zainal Z. (2006) *J. Power Sources* **161**, 730.

[86] Rosolen J.M., Matsubara E.Y., Marchesin M.S., Lala S.M., Montoro L.A., Tronto S. (2006) *J. Power Sources* **162**, 620.

[87] Zhu Y., Hu H., Li W. C., Zhang X. (2006) *J. Power Sources* **162**, 738.

[88] Hsieh C., Lin Y. (2006) *Micropor. Mesopor. Mater.* **93**, 232.

[89] Morishita T., Soneda Y., Tsumura T., Inagaki M. (2006) *Carbon* **44**, 2360.

[90] Wei Y.-Z., Fang B., Iwasa S., Kumagai M. (2005) *J. Power Sources* **141**, 386.

[91] Wei Y.-Z., Fang B., Iwasa S., Kumagai M. (2006) *J. Power Sources* **155**, 487.

[92] Lia J., Wanga X., Huanga Q., Gamboab S., Sebastian P.J. (2006) *J. Power Sources* **158**, 784.

[93] Kaschmitter J.L., Mayer S.T., Pekala R.W. (1998) US5789338, D.O.E., Lawrence Livermore National Security LLC and University of California.

[94] Pekala R.W., Farmer J.C., Alviso C.T., Tran T.D., Mayer S.T., Miller J.M., Dunn B. (1998) *J. Non-Cryst. Solids* **225**, 74.

[95] Pröbstle H., Schmitt C., Fricke J. (2002) *J. Power Sources* **105**, 189.

[96] Firsich D.W., Ingersoll D., Delnick F.M. (1998) US5776384, Sandia.

[97] Arulepp M., Leis J., Lätt M., Miller F., Rumma K., Lust E., Burke A.F. (2006) *J. Power Sources* **162**, 1460.

[98] Fernández J.A., Arulepp M., Leis J., Stoeckli F., Centeno T.A. (2008) *Electrochim. Acta* **53**, 7111.

[99] Chmiola J., Yushin G., Gogotsi Y., Portet C., Simon P., Taberna P.-L. (2006) *Science* **313**, 1760.

[100] Lätt M., Käärik M., Permann L., Kuura H., Arulepp M., Leis J. (2010) *J. Solid State Electrochem.* **14**, 548.

[101] Wang Y., Shi Z., Huang Y., Ma Y., Wang C., Chen M., Chen Y. (2009) *J. Phys. Chem. C* **113**, 13103.

[102] Si Y., Samulski E.S. (2008) *Chem. Mater.* **20**, 6792.

[103] Vix-Guterl C., Saadallah S., Jurewicz K., Frackowiak E., Reda M., Parmentier J., Patarin J., Béguin F. (2004) *Mater. Sci. Eng. B* **108**, 148.

[104] Xing W., Qiao S.Z., Ding R.G., Li F., Lu G.Q., Yan Z.F., Cheng H.M. (2006) *Carbon* **44**, 216.
[105] Portet C., Yang Z., Korenblit Y., Gogotsi Y., Mokaya R., Yushin G. (2005) *J. Electrochem. Soc.* **156**, A1–A6.
[106] Kim C. (2005) *J. Power Sources* **142**, 382.
[107] Adhyapak P.V., Maddanimath T., Pethkar S., Chandwadkar A.J., Negi Y.S., Vijayamohanan K. (2002) *J. Power Sources* **109**, 105.
[108] Barbieri O., Hahn M., Herzog A., Kötz R. (2005) *Carbon* **43**, 1303.
[109] Mitani S., Lee S.I., Saito K., Yoon S.-H., Korai Y., Mochida I. (2005) *Carbon* **43**, 2960.
[110] Egashira M., Okada S., Korai Y., Yamaki J.-I., Mochida I. (2005) *J. Power Sources* **148**, 116.
[111] Leitner K., Lerf A., Winter M., Besenhard J.O., Viller-Rodil S., Suárez-García S., Martínez-Alonso A., Tascón J.M.D. (2006) *J. Power Sources* **153**, 419.
[112] Jiang Q., Qu M.Z., Zhou G.M., Zhang B.L., Yu Z.L. (2002) *Mater. Lett.* **57**, 988.
[113] Frackowiak E., Delpeux S., Jurewicz K., Szostak K., Cazorla-Amoros D., Béguin F. (2002) *Chem. Phys. Lett.* **361**, 35.
[114] Li W., Chen D., Li Z., Shi Y., Wan Y., Wang G., Jiang Z., Zhao D. (2007) *Carbon* **45**, 1757.
[115] Kim I., Yang S., Jeon M., Kim H., Lee Y., An K., Lee Y. (2007) *J. Power Sources* **173**, 621.
[116] Tashima D., Kurosawatsu K., Uota M., Karashima T., Otsubo M., Honda C., Sung Y.M. (2007) *Thin Solid Films* **515**, 4234.
[117] Zhang B., Liang J., Xu C.L., Wei B.Q., Ruan D.B., Wu D.H. (2001) *Matter. Lett.* **51**, 539.
[118] Niu C., Sichel E.K., Hoch R., Moy D., Tennet H. (1997) *Appl. Phys. Lett.* **70**, 1480.
[119] Frackowiak E., Jurewicz K., Delpeux S., Béguin F. (2001) *J. Power Sources* **97–98**, 822.
[120] Ma R.Z., Liang J., Wie B.Q., Zhang B., Xu C.L., Wu D.H. (1999) *J. Power Sources* **84**, 126.
[121] An K.H., Kim W.S., Park Y.S., Moon J.-M., Bae D.J., Lim S.C., Lee Y.S., Lee Y.H. (2001) *Adv. Funct. Mater.* **11**, 387.
[122] Soneda Y., Toyoda M., Hashiya K., Yamashita J., Kodama M., Hatori H., Inagaki M. (2003) *Carbon* **41**, 2680.
[123] Emmenegger C., Mauron P., Sudan P., Wenger P., Hermann V., Gallay R., Züttel A. (2003) *J. Power Sources* **124**, 321.
[124] Chen Q., Xue K., Shen W., Tao F., Yin S., Xu W. (2004) *Electrochim. Acta* **49**, 4157.
[125] Portet C., Taberna P.L., Simon P., Flahaut E. (2005) *J. Power Sources* **139**, 371.
[126] Kim Y.-J., Masutzawa Y., Ozaki S., Endo M., Dresselhaus M.S. (2004) *J. Electrochem. Soc.* **151**, E199.
[127] Tan M.X. (1999) US5993969, Sandia.
[128] Hildenbrand C. (2010) Ph-D Thesis, Nanostructured carbons from cellulose-derivative-based aerogels for electrochemical energy storage and conversion: evaluation as EDLC electrode material, University of Nice Sophia-Antipolis.
[129] Heon M., Lofland S., Applegate J., Nolte R., Cortes E., Hettinger J.D., Taberna P.-L., Simon P., Huang P., Brunet M., Gogotsi Y. (2010) *Energy Environ Sci.* **1**, 138.
[130] Chmiola J., Yushin G., Gogotsi Y., Portet C., Simon P. Taberna P.L. (2006) *Science* **313**, 1763.
[131] Liu C., Yu Z., Neff D., Zhamu A., Jang B.Z. (2010) *Nano Lett.* **10**, 4863.
[132] Stoller M.D., S. Park, Y. Zhu, J. An, R.S. (2008) Ruoff, *Nano Lett.* **8**, 3502.
[133] Banda H., Daffos B., Perié S., Chenavier Y., Dubois L., Aradilla D., Pouget S., Simon P., Crosnier O., Taberna P.-L., Duclairoir F. (2018) *Chem. Mater.* **30**, 3040.
[134] Bansal R.C., Donnet J.B., Stoeckli F. (1988) *Active carbon*, Chapter I, 1–27, Marcel DekkerEd., New-York and Basel.
[135] Alford J.A. (2007) US5926361, Meadwestvaco Corp.
[136] Lini H., Lini C. (2002) WO20024308, CECA.
[137] Ishida S., Takenaka H., Nishimura S., Egawa Y., Otsuka K. (2008) WO2008053919, Kuraray Chemical.
[138] Hijiriyama M., Yasumaru J., Ishida K. (1998) JP10199767, Kansai Coke and Chemicals.
[139] Hanioka A., Matsuda K., Yasumaru J., Asada S. (2009) JP2009184850, Kansai Coke and Chemicals.
[140] Tokuyasu A., Matsuda K., Kasu K., Asada M. (2010) JP2010105885, Kansai Coke and Chemicals.

[141] Abe H., Morohashi K. (2006) JP2006024747, Mitsubishi Gas Chemical.
[142] Buiel E.R. (2007) WO2007114849, Meadwestvaco Corp.
[143] Dietz S. (2007) Production Scale-Up of Activated Carbons for Ultracapacitors, DE-FG36-04GO1432, TDA Research.
[144] Jänes A., Kurig H., Lust E. (2007) *Carbon* **45**, 1226.
[145] Sugo N., Iwasaki H., Uehara G. (2002) EP1176617, Honda Motor et Kuraray Co.
[146] Iwasaki H., Sugo N., Hitomi M., Nishimura S., Fujino T., Oyama S., Kawabuchi Y. (2004) WO2004011371, Kuraray Chemical and Honda.
[147] Zhong L., Xi X., Mitchell P. (2008) US20080201925, Maxwell Tech.
[148] Zhong L., Xi X., Mitchell P. (2008) WO2008106533, Maxwell Tech.
[149] Morimoto T., Hiratsuka K., Sanada Y., Ariga H. (1989) JP01241811, Asahi Glass Co. and Elna.
[150] Azaïs P. (2003) Ph-D Thesis, Recherche des causes du vieillissement de supercondensateurs carbone/carbone fonctionnant en milieu électrolyte organique, 27 novembre 2003, Université of Orléans.
[151] Azaïs P. (2014) Habilitation à la Direction de Recherches, INPGrenoble, juillet.
[152] Pell W.G., Conway B.E., Marincic N. (2000) *J. Electroanal. Chem.* **491**, 9.
[153] Endo M., Kim Y. J., Ohta H., Ishii K., Inoue T., Hayashi T., Nishimura Y., Maeda T., Dresselhaus M.S. (2002) *Carbon* **40**, 2613.
[154] Conway B.E., Verall R.E., Desnoyers J.E. (1966) *Trans. Faraday Soc.* **62**, 2738.
[155] Robinson R.A., Stokes R.H. (1965) *Electrolyte Solutions*, 2ème édn. Butterworths, London.
[156] Ruch P.W., Hahn M., Cericola D., Menzel A., Kötz R., Wokaun A. (2010) *Carbon* **48**, 1880.
[157] Carl E., Landes H., Michel H., Schricker B., Schwake A., Weber C. (2004) WO2004038742, EPCOS.
[158] Nozu R., Nakamura M. (2006) EP1724797, Nisshinbo Industries.
[159] Ruch P.W., Cericola D., Foelske A., Kötz R., Wokaun A. (2010) *Electrochimica Acta* **55**, 2352.
[160] Fujiwara M., Yoneda H., Okamoto M. (1986) JP06065206, Matsushita Electric Ind.
[161] Wang L., Morishita T., Toyoda M., Inagaki M. (2007) *Electrochimica Acta* **53**, 882.
[162] Fujiwara M., Yoneda H., Okamoto M. (1986) JP06065206, Matsushita Electric Ind.
[163] Yoshida H., Nakata H. (2007) US7283349, Nisshin Spinning.
[164] Kirchner E., Landes H., Michel H., Schricker B., Schwake A., Weber C. (2008) US7440257, EPCOS.
[165] Okamura M. (1999) US6064562, JEOL.
[166] Watanabe S., Onodera H., Sakai T., Harada T., Kanno Y., Yoshimi, Takasugi S., Tamachi T. (2003) US6625008, Seiko Instrument Inc.
[167] Maletin Y., Strizhakova N., Kozachkov S., Mironova A., Podmogilny S., Sergey, Danilin V., Kolotilova J., Izotov V., Cederstrom J., Konstantinovich S., Aleksandrovna J., Vasilevitj V., Efimovitj A., Perkson A., Arulepp M., Leis J., Wallace C., Zheng J. (2002) US6697249, Foc Frankenburg Oil and Ultratec.
[168] Paul G., Pynenburg R., Mahon P., Vassallo A., Jones P., Keshishian S., Pandolfo A. (2000) EP1724796, CapXX.
[169] Mitchell P., Xi X., Zhong L., Zou B. (2008) US2008016664, Maxwell Tech.
[170] Nozu R., Nakamura M. (2006) US7426103, Nisshin Spinning.
[171] Kirchner E., Landes H., Michel H., Schriker B., Schwake A., Weber C. (2006) US20060098388, EPCOS.
[172] Okamura M. (2000) JP11067608, Advanced Capacitor Tech., JEOL and Okamura Lab.
[173] Maletin Y., Strizhakova N., Kozachkov S., Mironova A., Podmogilny S., Danilin V., Kolotilova J., Izotov V., Cederstrom J., Konstantinovich S.G., Aleksandrovna J., Vasilevitj V.S., Efimovitj A.K., Perkson A., Arulepp M., Leis J., Wallace C.L., Zheng J. (2002) US20020097549, Frankenburg Oil Company, Karbid Aozt and Ultratec.
[174] Carl E., Landes H., Michel H., Schriker B., Schwake A., Weber C. (2006) US2006098388, EPCOS/TDK.
[175] Mitchell P., Xi X., Zhong L., Zou B. (2008) US2008016664, Maxwell Tech.
[176] Nozu R., Nakamura M. (2006) EP1724797, Nisshinbo Ind.

[177] Yoshida H., Yuyama K. (2007) EP1783791, Nisshinbo Ind.
[178] Paul G., Lange, Pynenburg R. (2006) EP1724796, CapXX.
[179] Maletin Y., Strizhakova N. (2003) US2003064565, Ultratec.
[180] Sarangapani S., Tilak B.V., Chen C.P. (1996) *J. Electrochem. Soc.* **143**, 3791.
[181] Hsieh C.T., Teng H. (2002) *Carbon* **40**, 667.
[182] Shi H. (1996) *Electrochim. Acta* **41**, 1633.
[183] Delahay P. (1966) *J. Phys. Chem.* **70**, 2373.
[184] Delahay P., Holub K. (1968) *J. Electroanal. Chem.* **16**, 131.
[185] Schultze J.W., Koppitz F.D. (1976) *Electrochim. Acta* **21**, 327.
[186] Schultze J. W., Koppitz F. D. (1976) *Electrochim. Acta* **21**, 337.
[187] Sullivan M.G., Kötz R., Haas O. (2000) *J. Electrochem. Soc.* **147**, 308.
[188] Momma T., Liu X., Osaka T., Ushio Y., Sawada Y. (1996) *J. Power Sources* **60**, 249.
[189] Qiao W., Korai Y., Mochida I., Hori Y., Maeda T. (2002) *Carbon* **40**, 351.
[190] Sullivan M.G., Schnyder B., Bärtsch B., Alliata D., Barbero C., Imhof R., Kötz R. (2000) *J. Electrochem. Soc.* **147**, 2636.
[191] Nakamura M., Nakanishi M., Yamamoto K. (1996) *J. Power Sources* **60**, 225.
[192] Mayer S.T., Pekala R.W., Kaschmitter J.L. (1993) *J. Electrochem. Soc.* **140**, 446.
[193] Pillay B., Newman J. (1996) *J. Electrochem. Soc.* **143**, 1806.
[194] Morimoto T., Hiratsuka K., Sanada Y., Kurihara K. (1996) *J. Power Sources* **60**, 239.
[195] Xu K., Ding M.S., Jow T.R. (2001) *Electrochim. Acta* **46**, 1823.
[196] Yoshida A., Tanahashi I., Nishino A. (1990) *Carbon* **28**, 611.
[197] Ricketts B.W., Ton-That C. (2000) *J. Power Sources* **89**, 64.
[198] Haye G. (1965) *Carbon* **2**, 413.
[199] Matsumura Y., Hagiwara S., Takahashi H. (1976) *Carbon* **14**, 163.
[200] Hagiwara S., Tsutsumi K., Takahashi H. (1978) *Carbon* **16**, 89.
[201] Frysz C.A., Chung D.D.L. (1997) *Carbon* **35**, 1111.
[202] Noh J.S., Schwarz J.A. (1990) *Carbon* **28**, 675.
[203] Horita K., Nishibori Y., Oshima T. (1996) *Carbon* **34**, 217.
[204] Frysz C.A., Shui X., Chung D.D.L. (1994) *Carbon* **32**, 1499.
[205] Strein T.G., Ewing A.G. (1991) *Anal. Chem.* **63**, 194.
[206] Ishikawa M., Sakamoto A., Morita M., Matsuda Y., Ishida K. (1996) *J. Power Sources* **60**, 233.
[207] Takada T., Nakahara N., Kumagai H., Sanada Y. (1996) *Carbon* **34**, 1087.
[208] Wightman R.M., Deakin M.R., Kovach P.M., Kuhr W.G., Stutts K.J. (1984) *J. Electrochem. Soc.* **131**, 1578.
[209] Shui X., Frysz C.A., Chung D.D.L. (1995) *Carbon* **33**, 1681.
[210] Firsich D.W. (1999) US5993996, Inorganic Specialists.
[211] Ohsaki T., Wakaizumi A., Kigure M., Nakamura A., Marumo S., Miyagawa T., Adachi T. (1999) US5948329, Nippon Sanso Corp.
[212] Oyana S., Oki N., Noguchi M. (1999) US5891822, Honda Motor.
[213] Azaïs P. (2003) Thèse de Doctorat, Recherche des causes du vieillissement de supercondensateurs carbone/carbone fonctionnant en milieu électrolyte organique, 27 novembre 2003, University of Orléans.
[214] Watanabe F., Oshida T., Miki Y., Sugano K. (2008) JP2008195559, Mitsubishi Gas Chemical.
[215] Igai K., Ono H., Fujii M., Takeshita K., Tano T., Oyama T. (2007) JP2007302512, Nippon Oil Corp.
[216] Yamada C., Hirose E., Takamuku Y., Shimamoto H. (2007) EP1918951, Matsushita Electric Ind.
[217] Takeuchi M. (2002) EP1264797, Advanced Capacitor Tech. et JEOL.
[218] Hart B.E., Peekma R.M. (1972) US3652902, IBM.
[219] Morimoto T., Hiratsuka K., Sanada Y., Aruga H. (1988) US4725927, Asahi Glass Co. and Elna Co.
[220] Imamura K., Yamada C., Sakata M. (2006) WO2006103967, Panasonic.
[221] Farahmandi C.J., Dispennette J.M. (1997) US5621607, Maxwell Tech.

[222] Andrieu X., Josset L. (1996) WO9620504, SAFT.
[223] Bonnefoi L., Laforgue A., Simon P., Fauvarque J.F., Sarrazin C., Lailler P., Sarrau J.F. (1999) FR2793600, CEAC.
[224] Meguro K., Sato H., Tada Y. (2001) US6327136, Kureha Chemical Ind.
[225] Xi X., Mitchell P., Zhong L., Zou B. (2007) US7295423, Maxwell Tech.
[226] Ishikawa T., Suhara M., Kuroki S., Kanetoku S. (2001) US6264707, Asahi Glass Co.
[227] Imoto K., Yoshida A. (1991) US5150283, Matsushita Electric Ind.
[228] Yoshida A., Ikoma A., Imoto K. (1995) US5450279, Matsushita Electric Ind.
[229] Barusseau S., Martin F., Simon B. (1999) EP0907214, SAFT.
[230] Azaïs P. (2003) Thèse de Doctorat, Recherche des causes du vieillissement de supercondensateurs carbone/carbone fonctionnant en milieu électrolyte organique, 27 novembre 2003, University of Orléans.
[231] Hata K., Sato T. (2000) WO200056797 et WO200057439, Nisshinbo Industries.
[232] Naruse S. (2008) WO200878634, Dupont Teijin Advanced Papers.
[233] Hiratsuka K., Morimoto T., Suhara M., Kawasato T., Tsushima M. (2000) US6402792, Asahi Glass Co.
[234] Guerfi A., Kaneko M., Petitclerc M., Mori M., Zaghib K. (2007) *J. Power Sources* **163**, 1047.
[235] Terada K., Mori H. (2005) WO2005052968, Zeon Corp.
[236] Sasaki T. (2013) WO201302322, Zeon Corp.
[237] Mogi T., Honda T., Tezuka T., Yamada K. (2010) WO2010113940, JSR Corp.
[238] Itou K., Mogi T., Suzuki H., Nishikawa A. (2007) WO2007088979, JSR Corp.
[239] Abusleme J., Pieri R., Barchiesi E. (2008) WO2008129041, Solvay.
[240] Bonnet A., Werth M. (2006) FR2876712B1, Arkema.
[241] Qiao W., Korai Y., Mochida I., Hori Y., Maeda T. (2002) *Carbon* **40**, 351.
[242] Penneau J.F., Capitaine F., Le Goff P. (1998) EP0960154, Blue Solutions.
[243] Drevet H., Rey I., Le Gal G., Peillet M. (2005) WO2005116122, Blue Solutions.
[244] Drevet H., Rey I., Peillet M., Abribat F. (2006) WO2006082172, Blue Solutions.
[245] Nanjundiah C., Braun R.P., Christie R.T.E., Farahmandi J.C. (2001) WO0188934, Maxwell Tech.
[246] Zykov V.P., Panov A.A., Prudnikov P.A., Khlopin M.I., Shlyapnikov A.D. (1972) US3675087.
[247] Probst N. (1993) Chapter 8: conducting carbon black, in "carbon black: science and technology 2nd edn.", (J.-B. Donnet, R.C. Bansal, M.-J. Wang Eds). Marcel Dekker, New York, pp. 271–285.
[248] Jackel N., Weingarth D., Zeiger M., Aslan M., Grobelsek I., Presser V. (2014) *J. Power Sources* **272**, 1122.
[249] Yoshida H., Sato T., Masuda G., Kotani M., Izuka S. (2005) EP1715496, Nisshinbo Industries.
[250] Zhu M., Weber C.J., Yang Y., Konuma M., Starke U., Kern K., Bittner A.M. (2008) *Carbon* **46**, 1829.
[251] Green M.C., Taylor R., Moeser G.D., Kyrlidis A., Sawka R.M. (2009) US20090208751, Cabot Corp.
[252] Wissler M. (2006) *J. Power Sources* **156**, 142.
[253] Zhang H., Zhang W., Cheng J., Cao G., Yang Y. (2008) *Solid State Ionics* **179**, 1946.
[254] Taberna P.-L., Chevallier G., Simon P., Plée D., Aubert T. (2006) *Mat. Res. Bull.* **41**, 478.
[255] Plée D., Taberna P.-L. (2005) WO2005088657, Arkema.
[256] Taberna P.-L., Chevallier G., Simon P., Plée D., Aubert T. (2006) *Mater. Res. Bull.* **41**, 478.
[257] Portet C., Yushin G., Gogotsi Y. (2007) *Carbon* **45**, 2511.
[258] Lust E., Janes A., Arulepp M. (2004) *J. Electroanal. Chem.* **562**, 33.
[259] Ue M., Ida K., Mori S. (1994) *J. Electrochem. Soc.* **141**, 2989.
[260] Otsuki M., Endo S., Takao T. (2002) US6469888, Bridgestone.
[261] Chiba K. (2005) WO2005022571, Japan Carlit.
[262] Siggel A., Nerenz F., Palanisamy T., Poss A., Demel S. (2007) US20070049750.
[263] Dahm C.E., Peters G.D. (1996) *J. Electroanal. Chem.* **402**, 91.
[264] Simonet J., Astier Y., Dano C. (2008) *J. Electroanal. Chem.* **451**, 5.

[265] Ross S.D., Finkelstein M., Petersen R.C. (1970) *J. Am. Chem. Soc.* **92**, 6003.
[266] Ross S.D., Finkelstein M., Petersen R.C. (1960) *J. Am. Chem. Soc.* **82**, 1582.
[267] Simonet J., Lund H. (1977) *J. Electroanal. Chem.* **75**, 719.
[268] Bernard G., Simonet J. (1979) *J. Electroanal. Chem.* **96**, 249.
[269] Finkelsteinn M., Petersen R.C., Ross S.D. (1959) *J. Am. Chem. Soc.* **81**, 2361.
[270] Finkelstein M., Petersen R.C., Ross S.D. (1965) *Electrochim. Acta* **19**, 465.
[271] Kawasata T., Suhara M., Hiratsuka K., Tsushima M. (1999) US5969936, Asahi Glass.
[272] Farahmandi C.J., Dispennette J.M., Blank E., Crawford R.W. (2001) US6233135, Maxwell Tech.
[273] Azaïs P., Tertrais F., Caumont O., Depond J.-M., Lejosne J. (2009) Ageing study of advanced carbon/carbon ultracapacitor cells working in various organic electrolytes, ISEE'Cap '09, Nantes.
[274] Chiba K. (2005) WO2005022571, Japan Carlit Co.
[275] Ue M., Ida K., Mori S. (1994) *J. Electrochem. Soc.* **141**, 2989.
[276] Zheng J.P., Jow T.R. (1994) *J. Electrochem. Soc.* **144**, 2417.
[277] Deschamps M., Gilbert E., Azaïs P., Raymundo-Piñero E., Ramzi Ammar M., Simon P., Massiot D., Béguin F. (2013) *Nature Mat.* **12**, 351.
[278] Forse A.C., Griffin J.M., Wang H., Trease N.M., Presser V., Gogotsi Y., Simon P., Grey C.P. (2013) *Phys. Chem. Chem. Phys.* **20**, 7722.
[279] Degen H.-G., Ebel K., Tiefensee K., Schwake A. (2006) DE102004037601, EPCOS.
[280] Zheng J.P., Jow T.R. (1997) *J. Electrochem. Soc.* **144**, 2417.
[281] Ue M., Ida K., Mori S. (1994) *J. Electrochem. Soc.* **141**, 2989.
[282] Aurbach D., Daroux M., Faguy P., Yeager E. (1991) *J. Electroanal. Chem.* **297**, 225.
[283] Aurbach D., Youngman O., Dan P. (1990) *Electrochim. Acta* **35**, 639.
[284] Bruglachner H. (2004) Ph-D Thesis, Neue Elektrolyte für Doppelschichtkondensatoren, University of Regensburg.
[285] Xu K., Ding S. P., Jow T.R. (1999) *J. Electrochem. Soc.* **146**, 4172.
[286] Xu K., Ding M.S., Jow T.R. (2001) *J. Electrochem. Soc.* **148**, A267.
[287] Doron Aurbach (1999), Nonaqueous electrochemistry, chapter 4, CRC Press
[288] Cotton F.A., Wilkinson G. (1966) *Adv. Inorg. Chem.* 240.
[289] Farahmandi C.J., Dispennette J.M., Blank E., Kolb A.C. (1998) WO9815962, Maxwell Tech.
[290] Ricketts B.W., Ton-That C. (2000) *J. Power Sources* **89**, 64.
[291] Johann Lejosne (2009) Thèse de Doctorat, Université de Nancy.
[292] Cericola D., Ruch P.W., Foelske-Schmitz A., Weingarth D., Kötz R. (2011) *Int. J. Electrochem. Sci.* **6**, 988.
[293] Kōsuke Izutsu (2009) *Electrochemistry in nonaqueous solutions*, P. 105, Wiley Ed.
[294] Pavlishchuk V.V., Addison A.W. (2000) *Inorg. Chim. Acta* **298**, 97.
[295] Kang X., Ding M.S., Jow T.R. (2001) *J. Electrochem. Soc.* **148**, A267.
[296] Noguchi M., Oki N., Iwaida M., Aida T., Nagai A., Ichikawa Y. (2002) JP2001237149.
[297] Farahmandi C.J., Dispennette J.M., Blank E., Kolb A.C. (2000) US6094788, Maxwell Tech.
[298] Takeuchi M., Koike K., Mogami A., Maruyama T. (2002) JP2002025867, *Adv. Capacitor Tech. JEOL.*
[299] Jerabek E.C., Mansfield S.F. (2000) US6084766, General Electric.
[300] Day J.A.P., Shapiro, Jerabek E.C. (2000) US6110321, General Electric.
[301] Kimura K., Kimura F., Kobayashi T. (2001) JP3717782, Japan Vilene Co.
[302] Kimura F., Kimura K., Kobayashi T., Shimizu M. (2001) JP2001185455, Japan Vilene Co. and Power Systems Co.
[303] Azaïs P., Tamic L., Huitric A., Paulais F., Rohel X. (2009) FR2954595, Bolloré.
[304] Ufheil J., Wursig A., Schneider O.D., Novak P. (2005) *Electrochem. Commun.* **7**, 1380.
[305] Hahn M., Baertschi M., Barbieri O., Sauter J.C., Kötz R. (2004) *Elektrochem. Mater. forsch.* **29**, 120.
[306] Suhara M., Hiratsuka K. (2000) WO0016354, Asahi Glass et Honda Motor.
[307] Ishimoto S., Asakawa Y., Shinya M., Naoi K. (2009) *J. Electrochem. Soc.* **156**, A563.
[308] Hahn M., Barbieri O., Gallay R., Kötz R. (2006) *Carbon* **44**, 2523.

[309] Azaïs P., Tertrais F., Caumont O., Depond J.-M., Lejosne J. (2009) Ageing study of advanced carbon/carbon ultracapacitor cells working in various organic electrolytes, ISEE'Cap'09, Nantes.

[310] Kurzweil P., Chwistek M. (2008) *J. Power Sources* **176**, 555.

[311] Kanbe Y., Oya M. (2000) JP2000216068, NEC.

[312] Schwake A., Erhardt W., Goesmann H. (2007) DE102005033476, EPCOS.

[313] Caumont O., Depond J.M., Jourdren A., Azaïs P. (2009) WO2009112718, Blue solutions.

[314] Petersen R. O'd., Kullberg R.C., Toia L., Rondena S., Mio B.J. (2008) WO2008033560, SAES Getters.

[315] Beatty T.R. (1984) CA1209201, Union Carbide Corp.

[316] Fujino T.(2007) US7224274, Honda Motor.

[317] Fujino T. (2008) US7457101, Honda Motor.

[318] Conway B.E. (1999) Electrochemical supercapacitors. Scientific Fundamentals and Technological Applications, Kluwer-Plenum, New York.

[319] Azaïs P., Tertrais F., Caumont O., Depond J.-M. Lejosne J. (2009) Ageing study of advanced carbon/carbon ultracapacitor cells working in various organic electrolytes, ISEE'Cap'09, Nantes.

[320] Kawasato T., Ikeda K., Yoshida N., Hiratsuka K. (2006) US7173807, Asahi Glass.

[321] Brandon E.J., West W.C., Smart M.C., Whitcanack L.D., Plett G.A. (2007) *J. Power Sources* **170**, 225.

[322] Gualous H., Bouquain D., Berthon A., Kauffmann J.M. (2003) *J. Power Sources* **123**, 86.

[323] Kötz R., Hahn M., Gallay R. (2006) *J. Power Sources* **154**, 550.

[324] Brandon E.J., West W.C., Smart M.C. (2008) US20080304207, California Institute of Technology.

[325] Handbook of Chemistry and Physics (2006), 87th edn.CRC Press, Cleveland.

[326] Jänes A., Lust E. (2006) *J. Electroanal. Chem.* **588**, 285.

[327] Suna G.-H., Li K.-X., Sun C.-G (2006) *J. Power Sources* **162**, 1444.

[328] Balducci A., Dugas R., Taberna P.-L., Simon P., Plée D., Mastragostino M., Passerini S. (2007) *J. Power Sources* **165**, 922.

[329] Nanbu N., Ebina T., Uno H., Ishizawa S., Sasaki Y. (2006) *Electrochim. Acta* **52**, 1763.

[330] Morita M., Murayama I., Fukutake T., Yoshimoto N., Egashira M., Ishikawa M. (2006) 210th ECS Meeting, Abstract #136.

[331] Mastragostino M., Soavi F. (2007) *J. Power Sources* **174**, 89.

[332] Suarez P.A.Z., Selbach V.M., Dullius J.E.L., Einloft S., Piatnicki C.M.S., Azambuja D.S., De Souza R.F., Depont J. (1997) *Electrochim. Acta* **42**, 2533.

[333] Balducci A., Bardi U., Caporali S., Mastragostino M., Soavi F. (2004) *Electrochem. Comm.* **6**, 566.

[334] Frackowiak E., Lota G., Pernak J. (2005) *Appl. Phys. Lett.* **86**, 164104-1.

[335] François Y. (2006) Ph-D Thesis, Utilisation de l'électrophorèse capillaire (EC) pour la caractérisation des liquides ioniques (LI) et intérêt des LI comme nouveaux milieux de séparation en EC, November 17th 2006, University of Paris VI.

[336] Zhou Z.B., Takeda M., Ue M. (2004) *J. Fluorine Chem.* **125**, 471.

[337] Muldoon M.J., Gordon C.M., Dunkin I.R. (2002) *J. Chem. Soc, Perkin Trans.* **2**, 4339.

[338] www.nisshinbo.co.jp/r_d/capacitor/index.html.

[339] www.njrc.co.jp/product/CAPACITOR.html.

[340] Sato T., Masuda G., Takagi K. (2004) *Electrochim. Acta* **49**, 3603.

[341] Ania C.O., Pernak J., Raymundo-Pinero E., Béguin F. (2006) *Carbon* **44**, 3113.

[342] Balducci A., Dugas R., Taberna P.L., Simon P., Plée D., Mastragostino M., Passerini S. (2007) *J. Power Sources* **165**, 922.

[343] Kim Y.-J., Matsuzawa Y., Ozaki S., Park K.C., Kim C., Endo M., Yoshida H., Masuda G., Sato T., Dresselhaus M.S. (2005) *J. Electrochem. Soc.* **152**, A710.

[344] Frackowiak E. (2006) *J. Braz. Chem. Soc.* **17**, 1074.

[345] Herzig T. (2007) Ph-D Thesis, Die Synthese und Charakterisierung neuer Elektrolyte für Tieftemperatur-anwendungen in elektrochemischen Doppelschichtkondensatoren, University of Regensburg.

[346] Bruglachner H. (2004) Ph-D Thesis, Neue Elektrolyte für Doppelschichtkondensatoren, March 25th 2004, University of Regensburg.

[347] Leblanc F., Blais R., Letarte A. (2000) *Bull. Inf. Toxicol.* **16**, 5.

[348] Morimoto T., Hiratsuka K., Sanada Y., Aruga H. (1988) US4725927, Asahi Glass and Elna.

[349] Arulepp M., Permann L., Leis J., Perkson A., Rumma K., Jänes A., Lust E. (2004) *J. Power Sources* **133**, 320.

[350] Jow T.R., Xu K., Ding S.P. (1999) Work performed under contract N DE-AI07-96ID13451 by U.S. Army Research Laboratory for US Department of Energy, September 1999.

[351] Ding M.S., Xu K., Zheng J.P., Jow T. R. (2004) *J. Power Sources* **138**, 340.

[352] Brandon E. J., West W. C., Smart M.C., Whitcanack L. D., Plett G. A. (2007) *J. Power Sources* **170**, 225.

[353] Schwake A. (2004) US20040218347, EPCOS.

[354] Yang C.-C., Hsu S.-T., Chien W.-C. (2005) *J. Power Sources* **152**, 303.

[355] Despotuli A., Andreeva A. (2007) *Mod. Electron.* 7, 24. (in Russian and English).

[356] Chmiola J., Largeot C., Taberna P. L., Simon P., Gogotsi Y. (2008) *Angew. Chem. Int.* **47**, 3392.

[357] Deschamps M., Gilbert E., Azaïs P., Raymundo-Piñero E., Ammar M. R., Simon P., Massiot D., Béguin F. (2013) *Nature Mat.* **12**, 351.

[358] Richner R.P. (2001), Ph-D Thesis, Entwicklung neuartig gebundener Kohlenstoffmaterialien für elektrische Doppelschichtkondensatorelektroden, Swiss Federal Institute of Technology Zürich.

[359] Kobayashi S., Shinohara H. (1993) JP3160725, Japan Radio Co.

[360] Mizobuchi T., Yanase M., Shinsenji T. (2000) JP2000331663, NKK.

[361] Tanaka Y., Ishii N., Okuma J., Hara R. (1999) JP10125560, Honda Motor.

[362] Suhara M., Hiratsuka K., Kawasato T. (2000) JP2000040641, Asahi Glass.

[363] Wei C., Jerabek E.C., Leblanc O.H. Jr. (2001) WO0019464, General Electric.

[364] Tanaka Y., Ishii N., Okuma J., Hara R. (2001) US6190501, Honda Motor.

[365] Tsukuda T., Midorikawa M., Sato T. (2007) WO2007061108, Mitsubishi Paper Mills.

[366] Inagawa M., Inoue Y. (2000) JP11135369, NEC.

[367] Tsushima M., Morimoto T., Hiratsuka K., Kawasato T., Suhara M. (2000) US6072693, Asahi Glass.

[368] Kobayashi T., Kimura N. (2003) JP2003229328, Japan Vilene.

[369] Ito T., Tsuchiya K., Yabe K. (1989) JP2569670, Toray Industries.

[370] Fisher H.M., Wensley C.G. (1999) US6368742, Celgard.

[371] Cheon S.D., Hwang K.Y., Oh H.J., Park S.E. (2000) KR20000051312, SK Corp.

[372] Imoto H., Matsunami T., Matsunami T. (2005) JP4425595, Nippon Sheet Glass Co.

[373] Kobayashi T., Kawabe M., Kimura F., Amagasa M. (2007) US7616428, Japan Vilene.

[374] Kobayashi T., Kimura N. (2006) JP2006144158, Japan Vilene.

[375] Oya N., Asano Y., Yao S. (2003) JP2003229329, Ube Industries.

[376] Arora P., Bazzana S.F., Dennes J.T., Holowka E.P., Krishnamurthy L., Mazur S., Simmonds G.E. (2011) WO201181879, Du Pont de Nemours.

[377] Tsukuda T., Midorikawa M. (2003) JP2003309042, Mitsubishi Paper Mills.

[378] Tsukuda T. (2007) JP2007150122, Mitsubishi Paper Mills.

[379] Wade T.L. (2006) *High power carbon –based supercapacitors-chapters 3–5 and 6, 47–142*, University of Melbourne, March 2006.

[380] Choudhary N., Li C., Moore J., Nagaiah N., Zhai L., Jung Y., Thomas J. (2017) *Adv. Mater.* **29**, 1605336.

[381] Zuo W., Li R., Zhou C., Li Y., Xia J., Liu J. (2017) *Adv. Sci.* **4**, 1600539.

[382] Pell W., Conway B. (2004) *J. Power Sources* **136**, 334.

[383] Brousse T., Taberna P.-L., Crosnier O., Dugas R., Guillemet P., Scudeller Y., Zhou Y., Favier F., Bélanger D., Simon P. (2007) *J. Power Sources* **173**, 633.

[384] Energy Polymer Caapcitor, by Panasonic. Voltage range: [3.6–2.5 V]. Max energy density: 26 Wh/L//8 kW/L.

[385] Li-ion capacitor: Yoshino A., Tsubata T., Shimoyamada M., Satake H., Okano Y., Mori S., Yata S. (2004) *J. Electrochem. Soc.* **151**, A2180.

[386] Li-ion capacitor: Example of product by JSR Micro. Commercial cell working in the voltage range [3,8–2,2 V]: https://www.jsrmicro.be/sites/default/files/attachments/ultimo_lithium_ion_capacitor_brochure_0.pdf.

[387] Li-ion capacitor: Hatozaki O. (2008) Proceedings of the Advanced Capacitor World Summit 2008, San Diego, CA, USA.

[388] Li-ion capacitor: C: Naoi K., Simon P. (2008) *Interface* **17**, 34.

[389] Na-ion capacitor: Ding J., Wang H., Li Z., Cui K., Karpuzov D., Tan X., Kohandehghan A., Mitlin D. (2008) *Energy Environ. Sci.* **8**, 941.

[390] K-ion capacitor: Le Comte A., Reynier Y., Vincens C., Leys C., Azaïs P. (2017) *J. Power Sources* **363**, 34.

[391] Example of product by Samsung DI: Hi-cap, based on activated carbon/NMC, available cells: 4 Ah and 11 Ah, https://www.samsungsdi.com/automotive-battery/products/prismatic-lithium-ion-battery-cell.html.

[392] Neburchilov V., Zhang L. CNRC, https://doi.org/10.4224/23001864.

[393] Lam L.T., Louey R. (2006) *J. Power Sources* **158**, 1140.

[394] Example of product by CSIRA and Furukawa (Ultrabattery ®): https://www.csiro.au/en/Research/EF/Areas/Energy-storage/UltraBattery.

[395] DuPasquier A., Plitz I., Gural J., Menocal S., Amatucci G. (2003) *J. Power Sources* **113**, 62.

[396] Brousse T., Marchand R., Taberna P.-L., Simon P. (2006) *J. Power Sources* **158**, 571.

[397] Fenghua S., Menghe M. (2014) *Nanotechnology* **25**, 135401.

[398] Wang X., Yang C., Wang G. (2016) *J. Mater. Chem. A* **4**, 14839.

[399] Selvakumar M., Bhat D.K. (2012) *Appl. Surf. Sci.* **263**, 236.

[400] Chen Z., Augustyn V., Jia X., Xiao Q., Dunn B., Lu Y. (2012) *ACS Nano* **6**, 4319.

[401] Naoi K., Naoi W., Aoyagi S., Miyamoto J., Kamino T. (2013) *Acc. Chem. Res.* **46**, 1075.

[402] Belyakov A., Bryntsev A. (1996) RU2063085, ELIT.

[403] Belyakov A., Dashko O., Kazarov V., Kazaryan S., Litvinenko S., Kutyanin V., Shmatko P., Vasechkin V. (1999) WO9924996, ESMA.

[404] Thomas G. (1995) WO9521466, Motorola Inc.

[405] Lundblad A., Kuznetsov V., Mirzoev R. (2000) WO200002215, Alfar International Inc.

[406] Peled E. (1979) *J. Electrochem. Soc.* **126**, 2047.

[407] Verma P., Maire P., Novak P. (2010) *Electrochem. Acta* **55**, 6332.

[408] Holl W., Mandel A. (1983) DE3215126, Bosch.

[409] Matsumura T., Tsukamoto J., Kashiwara S., Saitou S. (1985) JP60182670, Japan Storage Battery and Toray Industries.

[410] Hong M. (2004) KR20040069474.

[411] Watanabe M., Tabata S., Aida T. (2006) JP2006286803, Daihatsu.

[412] Aida T., Yamada K. (2006) JP2006286841, Daihatsu.

[413] Aida T. (2006) JP2006303109, Daihatsu.

[414] Ikeda K., Hiratsuka K., Kazuhara M., Morimoto T. (1996) JP8107048, Asahi Glass.

[415] Okuyama K., Hirahara S. (1999) JP11297578, Mitsubishi Chemicals.

[416] Suhara M., Hiratsuka K., Morimoto T., Ikeda K. (1997) JP9055342, Asahi Glass.

[417] Sasaki T. (2010) WO2010024327, Zeon Corp.

[418] Natori M. (2008) JP2008130890, Hitachi Chemical.

[419] Ando N., Tasaki S., Taguchi H., Hato Y. (2003) WO03003395, Fuji Heavy et Kanebo.

[420] Fuchida H., Honjo Y., Yamakawa Y., Yamazaki H. (2008) JP2008057000, Fuchita Nano and Honshu Kinzoku.

[421] Aida T., Yagi K., Murayama I., Yamada K. (2008) JP2008103697, Daihatsu.

[422] Murayama I., Aida T., Yamada K. (2010) JP2010027736, Daihatsu.

[423] Katou K., Nakahara Y., Nonogami H. (2008) JP2008177263, Sanyo.

[424] Endo K., Nakahara Y., Nonoue H. (2008) JP2008294314, Sanyo.

[425] Anouti M., Lemordant D., Lota G., Decaux-Moueza C., Raymundo-Pinero E., Béguin F., Azaïs P. (2012) WO2012172211, CNRS and Blue Solutions.

[426] Nagham El Ghossein (2018) Ph-D Thesis, Étude et modélisation du fonctionnement et du vieillissement des Lithium-ion Capacitors, University of Lyon, December.

[427] El Ghossein N., Sari A., Venet P. (2016) 13th IEEE Vehicle Power and Propulsion Conference (VPPC2016), Hangzhou, China. P. 7791712.
[428] El Ghossein N., Sari A., Venet P. (2018) *IEEE Trans. Power Electron. IEEE* **33**, 5909.
[429] El Ghossein N., Sari A., Venet P. (2018) *IEEE Int. Conf. Ind. Technol. (ICIT)*, 1738.
[430] Kuratani K., Yao M., Senoh H., Takeichi N., Sakai T., Kiyobayashi T. (2012) *Electrochim. Acta* **76**, 320.
[431] Gu H., Kong L., Cui H., Zhou X., Xie Z., Zhou Z. (2019) *J. Energy Chem.* **28**, 79.
[432] Ding J., Wang H., Li Z., Cui K., Karpuzov D., Tan X., Kohandehghan A., Mitlin D. (2015) *Energy Environ. Sci.* **8**, 941.
[433] Ding R., Qi L., Wang H. (2013) *Electrochim. Acta* **114**, 726.
[434] Wang X., Kajiyama S., Iinuma H., Hosono E., Oro S., Moriguchi I., Okubo M., Yamada A. (2015) *Nature Comm.* **6**, 6544.
[435] Li H., Zhu Y., Dong S., Shen L., Chen Z., Zhang X., Yu G. (2016) *Chem. Mater.* **28**, 5753.
[436] Yin J., Qi L., Wang H. (2012) *ACS Appl. Mater. Interfaces* **4**, 2762.
[437] Yuan Y., Wang C., Lei K., Li H., Li F., Chen J. (2018) *ACS Cent. Sci.* **4**, 1261.
[438] Dall'Agnese Y., Taberna P.-L., Gogotsi Y., Simon P. (2015) *J. Phys. Chem. Lett.* **6**, 2305.
[439] Wasiński K., Półrolniczak P., Walkowiak M. (2018) *Electrochim. Acta* **259**, 850.
[440] Dresselhaus M.S., Dresselhaus G. (2002) *Adv. Phys.* **151**, 1.
[441] Yoshino A., Tsubata T., Shimoyamada M., Satake H., Okano Y., Mori S., Yata S. (2004) *J. Electrochem. Soc.* **151**, A2180.
[442] Azaïs P., Lejosne J., Picot M. (2014) WO2014173891, CEA.
[443] Picot M., Azaïs P. (2015) WO2015101527, CEA.
[444] Picot M., Azaïs P. (2015) GFEC, Les Karellis.
[445] Le Comte A., Picot M., Azaïs P., Perdu F. (2016) SFEC, Carqueiranne.
[446] Le Comte A., Picot M., Azaïs P., Perdu F. (2016) The World Conference on Carbon, Pennstate.
[447] Le Comte A., Reynier Y., Vincens C., Leys C., Azaïs P. (2017) AABC, Mainz, Germany.
[448] Le Comte A., Reynier Y., Vincens C., Leys C., Azaïs P. (2017) *J. Power Sources* **363**, 34.

Chapter 10

Supercapacitors: Cells and Modules

Philippe Azaïs

10.1 Cell Design

Numerous cell designs have been developed. The design is strongly linked to the target market and the target price. This section is dedicated to the description of commercial cell designs. The analysis of the components is described taking into account the targeted markets. In addition, it is very important to distinguish between small size components (button cell format, for example) and high capacity components, *i.e.* above 300–500 F, because of the targeted applications and their architectures.

Four cell manufacturing processes were used:

1. By potting then radial crimping (electrolytic capacitor process). The stack or winding is fitted with pin connectors and placed in a cylindrical or prismatic aluminium housing. The terminals are made by means of solderable wires, particularly suitable for the insertion of components on electronic boards. These components are usually closed by a butyl seal in which a "weakness" is provided. This "weakness", generally known as "venting", has the function of preventing the explosion of the component in the event of overpressure of the component through abusive overload or overheating. Typically, these supercapacitors have a high EDR and the energy density is well below 5 Wh.kg^{-1}. This technology is well suited for small supercapacitors dedicated to electronic applications.
2. Case and cover are assembled by rolling. This rolling technology, which comes from the food industry, is widely used by manufacturers of high-capacity components, such as Maxwell, LS MTron, Ioxus and Chinese players. The difficulty lies in the simultaneous management of sealing and electrical insulation *via* one or more seals. Indeed, the electrodes are connected to the housing and the cover, respectively. The performance of the components made by this process is generally much better than that of the previous process.

DOI: 10.1051/978-2-7598-2555-4.c010

3. Housing and cover are assembled by gluing. In this case, the glue replaces the gasket but has the advantage of being mechanically stronger than the flexible gaskets used in the previous design. This design has been patented by Blue Solutions [1].

4. Flexible cells. They are directly inspired by the "pouch cell" type cell designs developed in Li-ion and LiPo batteries. They are inexpensive because they largely eliminate the material associated with the cases and lids. Mass and volume energy densities are higher than previous designs, but these cells are mechanically fragile. This design was chosen by Yunasko [2] and Nisshinbo. This technology requires not to generate gas during ageing or at least to control its generation.

10.1.1 Small Cells

Small cells are almost exclusively dedicated to power electronic applications to be soldered directly onto electronic boards (examples: wireless communication for utility meters, energy storage for actuators, back-up for memory cards, power Wi-Fi for telecom stations). These components are also used in the aftermarket for audio systems (5–15 F), as a power supply for computers, as an energy storage system for toys, and as a power system for mobile phone or camera flashes (thin cells of less than 1 F). The most important players in these markets are Panasonic, Taiyo Yuden, NEC-Tokin, Elna, Seiko, Korship, Kemet, AVX-Kyocera, Cooper Buss-mann, Alumapro, CapXX, Shoei Electronics, Smart Thinker, Nichicon, Nippon Chemicon, Vina, Vishay, and Rubycon...

The growth of this technologically mature market is still important. Power and energy densities are not key parameters except for the new mobile applications, which explain in particular the development of thin cells.

These cells come in four designs:

1. "Button cell" type in standardized formats (figure 10.1).
2. Extra-fine cells, such as for example those developed by CapXX (figure 10.2). This technology is the same as that developed for flexible batteries.
3. Thin rigid prismatic cells, such as those manufactured by AVX-Kyocera (Bestcap®).
4. The wound cells. Generally, the capacity of these components is higher than that of the three previous types. The dimensions and manufacturing methods of these components are largely derived from electrolytic capacitors and dielectric capacitors (figure 10.3).

10.1.2 High-Capacity Cells

High-capacity cells do not come from standards, although efforts are currently underway to standardize dimensions and capacities: each manufacturer designs its cells taking into account its internal developments and performance optimization.

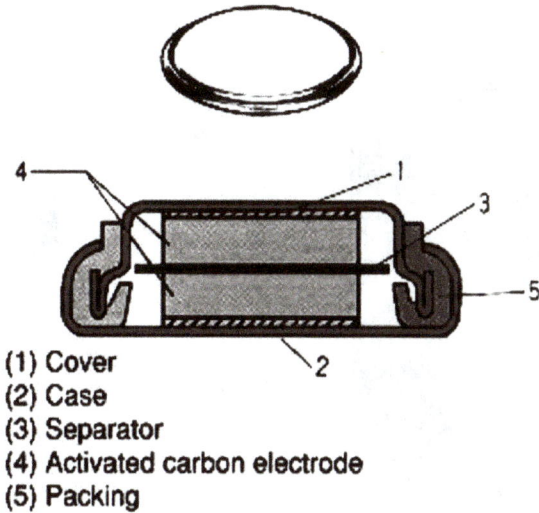

(1) Cover
(2) Case
(3) Separator
(4) Activated carbon electrode
(5) Packing

FIG. 10.1 – Cross-sectional view of a "button cell" EN/EP, Panasonic [3] ("Gold Capacitor").

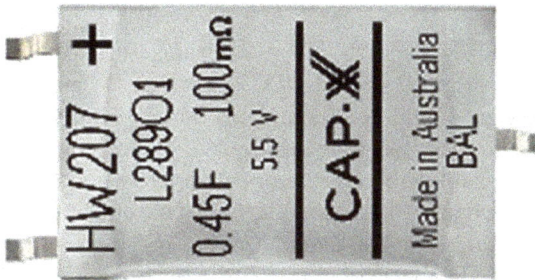

FIG. 10.2 – Example of bi-cell for mobile phone application (CapXX) (assembly of 2 cells in series in the same pouch).

Two main families of components can be distinguished:

1. High-power cells, dedicated to hybridization applications for vehicles and urban transport.
2. The so-called "energy cells", more suitable for stationary applications, such as UPS (Uninterrupted Power Supply).

10.1.2.1 High Power Cells

Numerous players manufacture high power cells such as Ioxus (USA), AVX (JP/USA), Eaton (USA), SECH SA (Switzerland/China), LS Mtron (KR), PLNano (CN), FastCap (USA), GTCap (CN), Kamcap (CN), SPSCap (CN), Skeleton (Estonia), Nippon-Chemicon (JP), Nichicon (JP), Samwha (JP/KR),

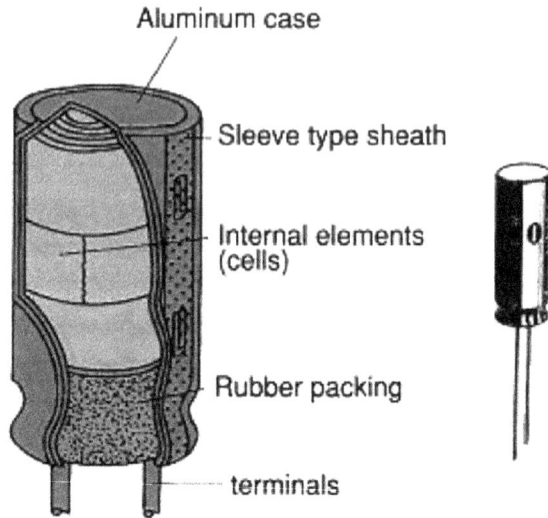

FIG. 10.3 – Small wound cells "radial type", HW/HZ series by Panasonic [1] ("Gold Capacitor").

Yunasko (Ukraine/UK)... Recently, Blue Solutions (France) and Maxwell Technologies[1] (USA) have stopped their EDLC activities. Because of the applications involved, the cell design should be as simple as possible and the series resistance as low as possible. In addition, the time constant of these components is usually less than one second. An example of a range of cells is shown in figure 10.4.

The advantage of wound cells is that they can concentrate a large electrode area in a small volume. In addition, the manufacturing process is already widely used in electrolytic batteries and capacitors. Finally, the output connections can be welded on either side of the winding, in particular by means of a laser, to cover all the turns of the winding and thus reduce the series resistance of the component. Sealing is then managed either by gluing or *via* a joint, as previously mentioned. Figure 10.5 shows, in a general way, the design used in winding. The major disadvantage associated with the assembly of supercapacitors is the need to add connecting bars between the components.

As shown in figure 10.5, intermediate pieces are added between the winding and the terminals. It is possible to remove these parts by laser welding the housing (or cover) and the winding directly: this has the effect of limiting the serial resistance of the component very strongly (Blue Solutions design). However, this removal of a relatively expensive intermediate part has the counterpart of having to perfectly control the transparency welding between a massive part and a collector of a few tens of micrometers to guarantee both the sealing of the component and a low

---

[1]Nesscap (KR) has been purchased by Maxwell Technologies in April 2017. Maxwell Technologies has been purchased by Tesla Motors in 2019.

FIG. 10.4 – Example of products (LS Mtron): 650, 1200, 1500, 2000 and 3000 F cells.

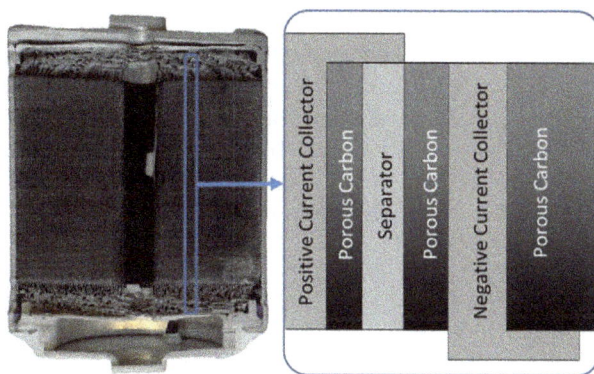

FIG. 10.5 – Cross-sectional view of a 650 F K2 serie cell (Maxwell Tech.) electrode stack diagram [5].

dispersion in the incisivity of the weld (mechanical and electrical performance). Power components generally have an electrode whose thickness outside the collector is less than or equal to 100 μm. This thickness is a compromise between energy density and power density to meet the majority of applications. The housings are made of aluminium to be able to withstand the overpressure generated during ageing and have venting (LS MTron, Maxwell, Ioxus, Blue Solutions...). Finally, these components are perfectly sealed against air and humidity to guarantee a high calendar life. In addition, reversible valves (type VRLA) are difficult to implement because the components are sensitive to oxygen and water: it is therefore necessary to avoid any back-diffusion of these species when opening the valve.

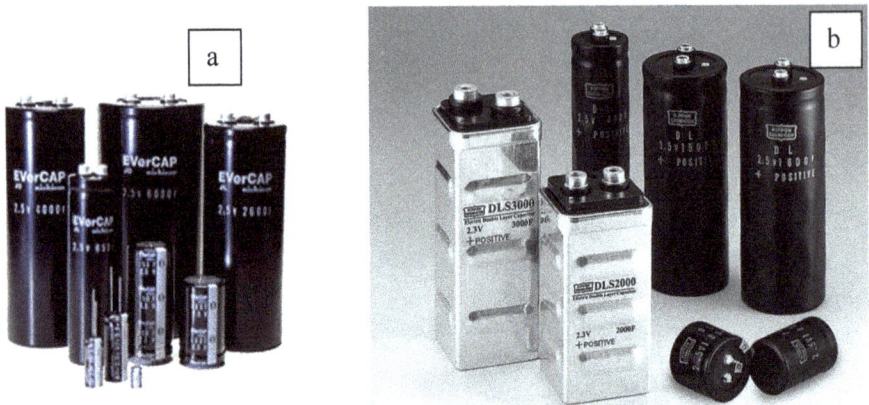

FIG. 10.6 – Examples of energy type cells: (a) Nichicon Evercap (600–4000 F), (b) Nippon Chemi-Con DLCAP™ (100–3000 F).

10.1.2.2 Energy Type Cells

The main players in the field of energy cells are Nippon Chemicon (Japan), Panasonic (Japan), Nichicon (Japan), Asahi Glass (Japan) and Meidensha (Japan). It should be noted that most supercapacitor manufacturers are also electrolytic capacitor manufacturers and have a very good knowledge of the cell industrialization. Therefore, the construction of these cells is largely based on these electrolytic components (figure 10.6). The EDR of this type of cells is higher than the "power" type for two major reasons:

1. these cells operate on a PC based electrolyte;
2. the design is not optimized for power and the thickness of these electrodes is generally greater than that used in "power" type cells.

The use of PC-based electrolyte and the targeted applications allow the manufacturers of these supercapacitors to use the same components and designs as those used in the electrolytic capacitors they also manufacture: plastic components, connectors on the same side, winding or assembly process... These cells generally have a voltage of the order of 2.3–2.5 V in order to limit ageing due to the use of PC-based electrolyte.

10.1.2.3 "Pouch" Cell Design

This type of cell is generally used for small and medium capacity cells (typically up to 1000 F). This type of design, directly inspired by the one widely used in Li-ion batteries used in portable applications and in some automotive battery packs, allows reaching a maximized energy density (gravimetric and volumetric) and is particularly used in applications requiring a flat component, such as in mobile phones

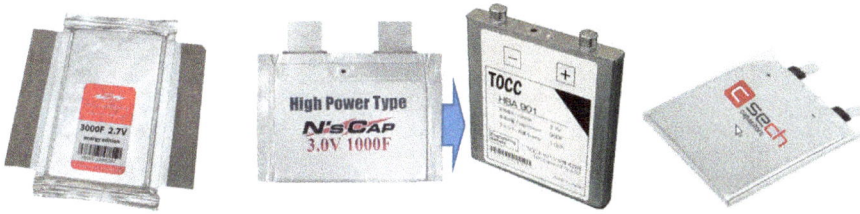

FIG. 10.7 – Examples of pouch cells (from l. to r.): Yunasko (3000 F, 2.7 V) Nisshinbo (1000 F, 3.0 V, with venting) and prismatic TOC Capacitor/TPR (2.7 V/900 F), and SECH 600 F (3.0 V).

(CapXX, figure 10.2) and electronics. Larger cells have been more developed in the past by Nisshinbo [6] (figure 10.7), and more recently by Yunasko, SECH and Ioxus.

Due to the very light packaging, these cells also have high power and energy densities. For electronic applications, the environmental conditions (extreme temperatures, resistance to crash tests...) are less severe. However, these cells suffer from major disadvantages:

- The resistance to mechanical shocks is poor.
- Heat dissipation is lower than for standard technologies.
- With gas generation, the volume of the cell is strongly modified.
- The sealing of the polymer packaging must be particularly controlled and mastered to avoid any contamination (water, oxygen).

10.1.2.4 Cells Working in Aqueous Medium

Unit cells based on aqueous electrolyte are not commercially available, mainly because of their low voltage: as a result, only modules are available on the market.

10.1.2.5 Present Performance of EDLC

One way to compare EDLC is to plot a diagram of Ragone This diagram expresses the energy density as a function of the power density (gravimetric and volumetric) at the cell level, generally at 25 °C. This diagram (figure 10.8, in gravimetric, and figure 10.9, in volumetric) does not take into account the lifetime (tests carried out when new) and is made for a complete discharge between U_n and 0 V (to be consistent with the operation of the converter, it would be more relevant to discharge between U_n and $U_n/2$, *i.e.* the ¾ stored energy).

Finally, the impact of current on capacitance and resistance is of the second order on the diagram of Ragone, as shown in figure 10.10 for 0.1 and 10 A.g^{-1} (cell) values.

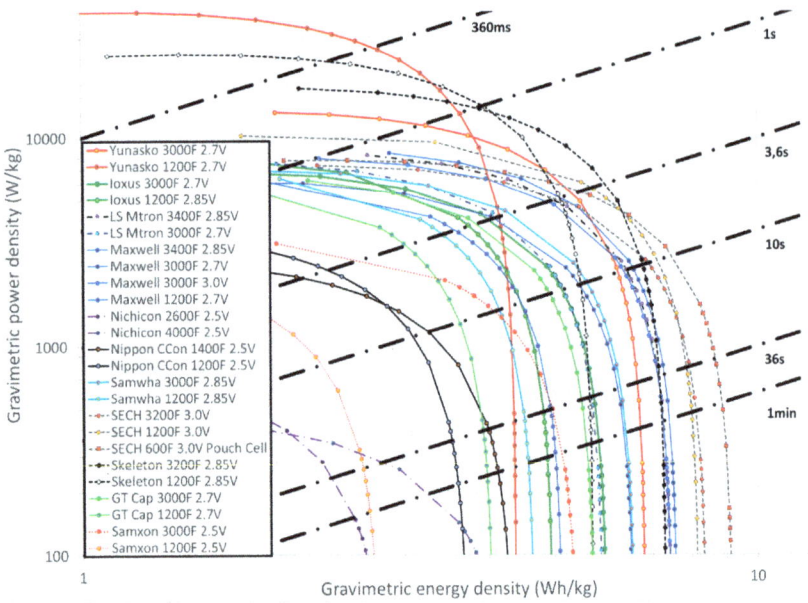

FIG. 10.8 – Gravimetric diagram of Ragone of commercial cells at room temperature for a full discharge (from U_n to 0 V). Note: Yunasko cells and SECH 600 F cell are pouch cell type. Others are cylindrical.

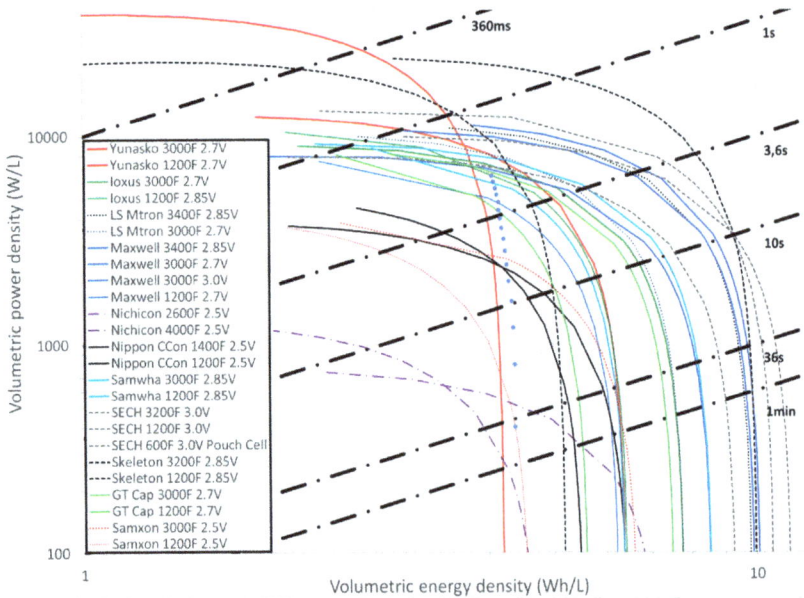

FIG. 10.9 – Volumetric diagram of Ragone of commercial cells at room temperature for a full discharge (from U_n to 0 V). Note: Yunasko cells and SECH 600 F cell are pouch cell type. The others are cylindrical.

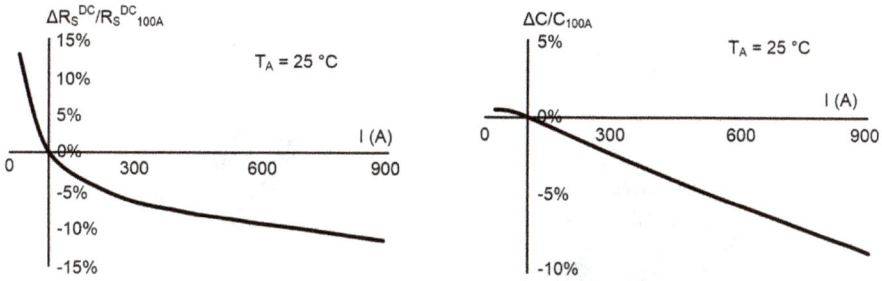

FIG. 10.10 – Variation of resistance (left) and capacitance (right) for an old generation 2600 F cell (Batscap, 2008) for different discharge current values. Note that one with a current density of 1 A.g^{-1} corresponds to 100 A on the curve.

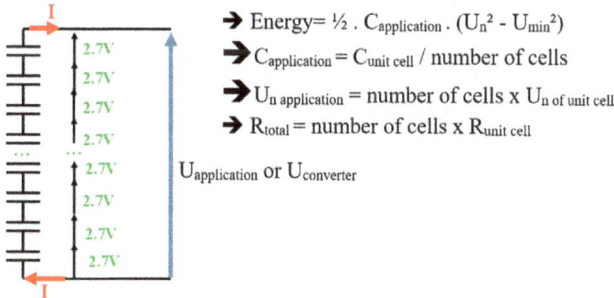

→ Energy= ½ . $C_{application}$. $(U_n^2 - U_{min}^2)$

→ $C_{application} = C_{unit\ cell}$ / number of cells

→ $U_{n\ application}$ = number of cells x $U_{n\ of\ unit\ cell}$

→ R_{total} = number of cells x $R_{unit\ cell}$

$U_{application}$ or $U_{converter}$

FIG. 10.11 – Strategy generally used to design modules from unit cells.

10.2 Design of Modules and Systems

The use of supercapacitors in high power applications can be achieved by combining the components in series, as we have just seen, in order to achieve the required voltage level. One of the benefits of increasing the voltage is the ability to increase the energy. At the same time, most applications require a DC/DC converter: a low voltage level (between 2.5 and 3.0 V) on the energy storage side and a high voltage on the application side would lead to a very important oversizing of this already expensive power converter and a lower efficiency.

The modules are dimensioned according to the targeted applications and are generally made from components with fixed capacities and whose values are now relatively fixed between manufacturers (650, 1200, 1500, 2000, 3000 F). The strategy generally adopted to design the modules is shown in figure 10.11.

The strategy generally used is to assemble cells with high capacity and low voltage. The compromise made between Umin, Umax and Umax per unit in the application makes it possible to arrive at a number of components defined by taking into account the total capacity of the application.

FIG. 10.12 – Example of a module made from supercapacitors directly soldered on an electronic board (based on 120 F LS MTron cells).

The unit capacity obtained from these equations is therefore variable for each targeted application. This is why supercapacitor manufacturers offer such a variable range of component capacities.

In addition, applications generally require a relatively high voltage and a high power level. Key parameters such as heat management, cell assembly, connection technology, balancing and electronics are discussed below.

Smaller modules are generally used in consumer applications and are soldered directly onto electronic boards, as shown, for example, in figure 10.12. Most of the applications concerned do not require high current densities (*e.g.* back up), and heat management is not a particularly critical issue.

10.2.1 Modules Based on Hard Casing Cells

These modules contain:

1. Unit cells.
2. Metallic inter-cell connections to support all the current required in the application.
3. Two terminals (input and output).
4. An electrical insulation between the components and the outer casing. This insulator can advantageously be a heat dissipator for the module (gadpad for example).
5. A balancing system.
6. Other information can be detected by sensors (temperature, pressure...).
7. A housing (plastic or aluminium).

In a module, only the cells contain active material. For this reason, the power and energy densities of the module are always lower than those of the cells included in the module. Most manufacturers have designed their modules to minimize the amount of passive material in the module, while taking into account that safety, environment and aging must not be degraded to the benefit of performance. In some cases, the geometry can be of different shapes to be easily integrated into the final application [7].

In the following paragraphs, solutions for the passive components of these modules are detailed.

10.2.1.1 *Metallic Connections between Cells*

To increase the voltage (series connections) or capacity (parallel connections), it is necessary to connect the supercapacitors to each other. These connection strips are constrained by the cell geometry and the position of the cell terminals. Given the very low series resistance of the supercapacitors and the high current that must be passed through during cycling, the control of the inter-cell connection is essential to minimize the impact on the volume while controlling the series resistance and avoiding any degradation in use. For example, if the component terminals are flat, the interconnections will also be flat [8] (figure 10.13a). Connections can be more complex if the cells are made by rolling (figure 10.13b) or if the module is tubular [9] (figure 10.13c). Figure 10.13d shows the connections that have been used in the two-component dedicated to the eHDi market of PSA made by Continental [10] (no soldering, but metal and spring washers).

The mechanical assembly between elements can be made by welding [11], by temperature difference and stripping [12], by brazing [13], by screwing [14]. Usually, the inter-element bars also play a role in the dissipation of heat to the outside of the module.

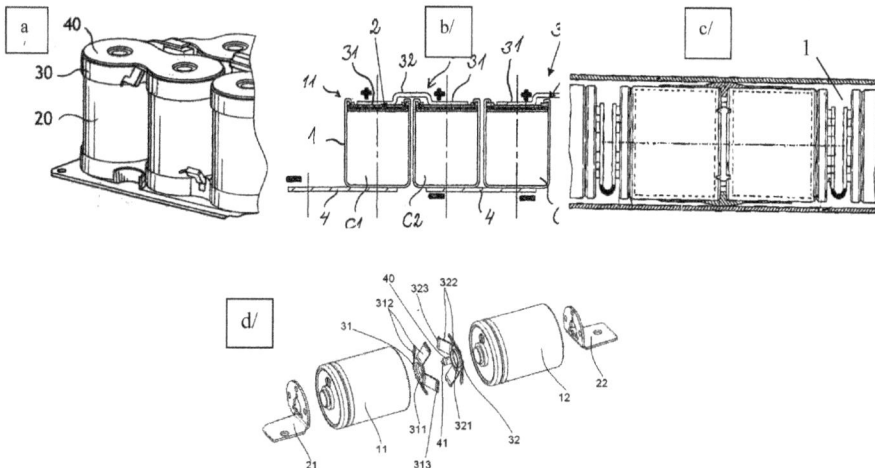

FIG. 10.13 – Different types of terminal connections: flat ((a) BlueSolutions, 40), with raised connection ((b) EPCOS, 32), U-shaped for tubular design ((c) EPCOS, 1), with spring washers (contact, (d) Continental).

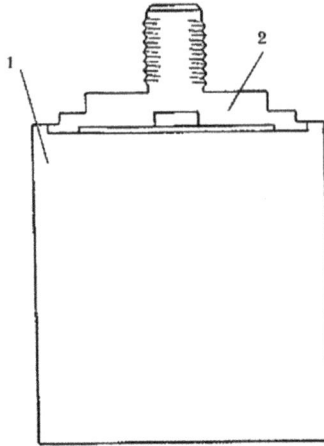

FIG. 10.14 – Terminal (1) welded on a unit cell (2).

10.2.1.2 *Module Terminals*

The terminals of the module must:

- Support the entire current of the application.
- Provide a link between the inside and outside of the module while maintaining a good seal.
- To allow the extraction of the entire current from the last and first cells while maintaining a good seal at the same time.
- Guarantee the lowest possible contact resistance.

This concept is strongly inspired by the technologies used in electrolytic capacitors [15]. One of the most commonly used means is to solder the terminals directly to the first and last cells of the module [16] (figure 10.14).

10.2.1.3 *Insulators in Module*

The lifetime of a module is linked, in particular, to the right balance of temperature and voltages between the cells within the module. In addition, it is important that the core of the module can maintain its integrity during repeated vibration and shock tests while guaranteeing a good level of safety. For this reason, additional components are added during module construction to fulfil these functions.

Many insulating materials based on polyolefins, polyesters, elastomers, etc. are available on the market.

Polyimide and PTFE are particularly suitable for high temperatures and if the thickness has to be thin, but these materials are still expensive. An electrically insulating and heat-conducting foil is usually interposed between the module core (cells + strips) and the module housing to avoid short circuits. However, this sheet must be plastically or elastically deformable under mechanical pressure to guarantee

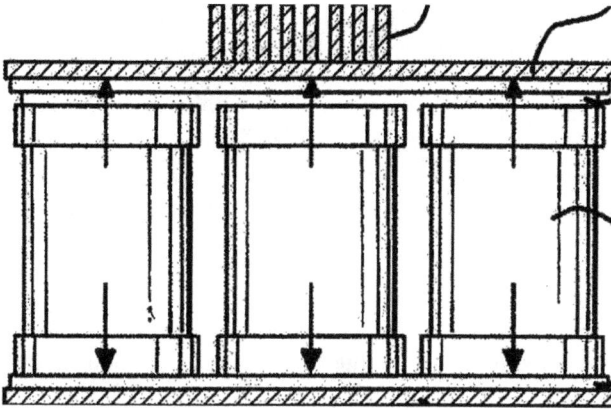

FIG. 10.15 – Insulation, heat dissipation fins (2) and cells (1). The arrows correspond to the directions of heat dissipation.

good mechanical resistance during vibration or shock tests. These insulations generally contain ceramic, silicone, wax, or a mixture of various thermally conductive substrates, and may also be in the form of multilayer coatings [17].

One of the most commonly used materials is an elastomeric insulator filled with carbon black (figure 10.15).

The elastomer layer fulfils several functions simultaneously [18]. It allows:

- An electrical insulation of the entire module core from the module housing while resisting a breakdown voltage higher than lkV.
- To absorb the geometrical dispersions of all the cells and their assembly thanks to its ability to be compressed.
- To improve heat exchange with the outside of the module.

To these polymeric materials, we also add systems already widely used in electronics: fins. They are generally positioned on the top cover to improve their efficiency.

The development of a module from a thermal point of view can be achieved by modeling [19–21].

10.2.1.4 Cell Balancing and Other Detected Information

If the thermal balancing between cells is important to guarantee a good service life, the same is true from an electrical point of view (impact of the dispersion of voltages between the cells on the ageing of the module). Therefore, balancing between the cells is necessary to even out the end-of-charge voltage of the components placed in series. The parameters of each cell (capacity, series resistance, self-discharge) are never completely identical, due to the manufacturing process and ageing variations (in particular the position of the component within the module) [22]. Moreover, if no balancing between cells is carried out, the risk of overvoltage on a cell exists and could lead to excessive ageing. The most common form of cell balancing is achieved using a current bypass when the cell reaches its rated voltage [23]. The most appropriate system to guarantee the best safety is to connect each cell to this device. This type of electronic assembly does not guarantee overload protection if the components are tested in pairs. The efficiency of this device clearly depends on the type of cycle of the application. In the case of "high current" type applications and operation at very low temperatures, it is necessary to detect other information, such as a temperature-controlled cut-off system, an electrolyte leakage detector or a hydrogen sensor. These sensors are extremely expensive and are only applicable to niche markets.

10.2.1.5 Module Casing

Figure 10.16 shows examples of modules with aluminium housings. Aluminium has many advantages for the housing: it is mechanically resistant, inexpensive, thermally interesting because it allows heat to be easily removed from the components, and relatively inert to many common chemicals. The Blue Solutions 150 F 54 V module (figure 10.16d) is dedicated to urban transport markets (tramways, metros, buses) and is made up of 3000 F unit cells. For these applications, Blue Solutions has made many efforts to ensure that this module meets the demanding specifications of this market. For example, this module is only air-cooled *via* fins positioned on only one side and complies with the "fire-smoke" standards.

These types of modules are extremely safe, but this is to the detriment of energy and power density performance (~ 3.5 Wh.L^{-1}, ~ 1 kW.kg^{-1} at 10 s): these values are relatively low (approximately halved) compared to the values obtained at the cell level (~ 9 Wh.L^{-1}, ~ 2.5 kW.kg^{-1} at 10 s).

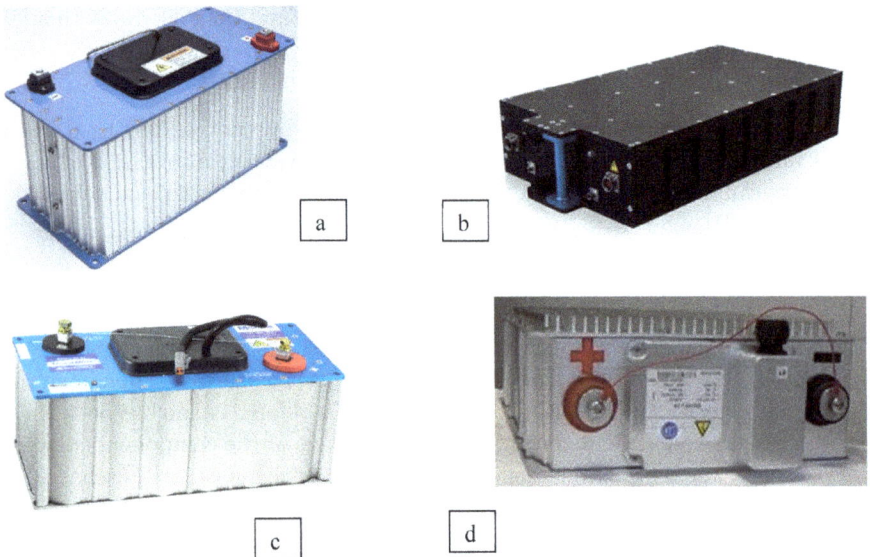

FIG. 10.16 – Examples of modules (a) LS MTron (50.4 V/166 F), (b) Skeleton Tech. (53 F/170 V), (c) Maxwell (165 F/48.6 V), (d) Blue Solutions (150 F/54 V).

10.2.2   High Capacity Modules Based on Soft Packaging Cells (Pouch Cells)

As described above, flexible cells offer optimal performance in terms of energy density and power due to the reduction of passive constituents.

In order to guarantee a high level of safety for this type of assembly, the mechanical resistance is transferred to the module housing.

Powersystems has developed three different types of modules based on pouch cells [24]. The performances of these modules are interesting for different identified applications:

- 216 V/15 F module of high power and high energy (2.3 Wh.L^{-1}, 2.9 Wh.kg^{-1}, RC = 1 s) based on 54 V/60 F unit bricks. Application: container handling crane, electric power installation, copiers, construction machinery, AGV (Automatic Guided Vehicle), stacker crane...
- Very high power 216 V/5 F module made from 4 modules of 108 V/10 F (1.2 Wh.L^{-1}, 1.5 Wh.kg^{-1}, RC = 0.2 s). Application: injection moulding machine, electric welding.
- 405 V/0.9 F modules of very high power and compact based on 2 modules of 202 V/1.8 F (1 Wh.L^{-1}, 1 Wh.kg^{-1}, RC = 0.18 s). Application: machine tools in general, robotic arm.

Other original modules have been developed by Meidensha (MeidenCap). These flat modules, realized by bipolar stacked technology (figure 10.17), have a low capacity, high voltage and interesting mass and volume energy density values [25] (table 10.1). The major disadvantage of these modules is a relatively high time constant (RC >> 1 s) mainly due to the use of PC in the electrolyte. For the time being, these modules are mainly used for the UPS market. The assembly principle of this technology means that the unit voltage of the cell is already very high

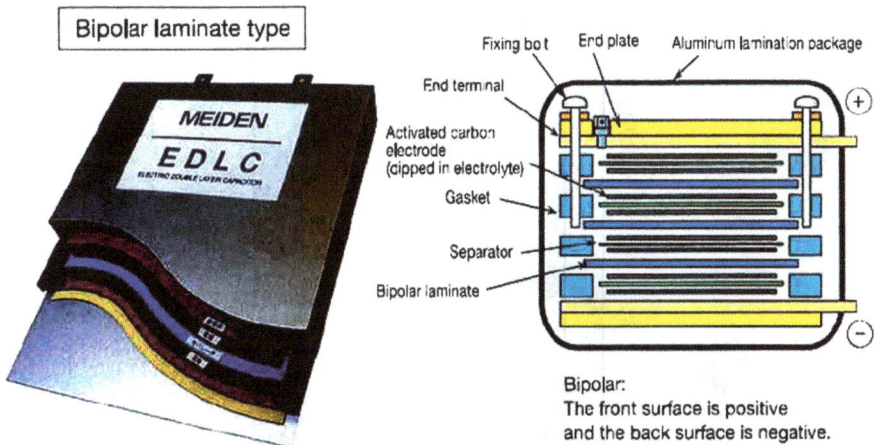

FIG. 10.17 – Structure of a bipolar Meidensha module.

TAB. 10.1 – Performances of different Meidensha modules at 25 °C.

Type	600S1-70C	600L1-70C	150S1-38C	150S2-32C
Number of stacked foils	70	70	38	32
Dimensions (W × H × L)	266 × 43 × 316 mm	266 × 43 × 316 mm	158 × 27 × 176 mm	158 × 30 × 176 mm
Weight (kg)	5.7	5.7	1.1	1.3
Nominal voltage (V)	160	160	85	72
Maximum voltage during charge (V)	175	175	95	80
Capacitance (F)	4.5	3.7	2.0	4.0
ESR (Ohm)	0.58	0.45	2.0	1.9
Gravimetric energy density (Wh.kg^{-1})	2.8	2.3	1.8	2.2
Volumetric energy density (Wh.L^{-1})	4.4	3.6	2.7	3.5
Time constant (s)	2.6	1.7	4.0	7.6

($>> 5$ V). This is tantamount to confusing the cell and the module. The electrodes are stacked and only the first and last electrodes of the module are electrically connected to the outside. The other electrodes are called "floating", *i.e.* they are self-balancing during charging or discharging. The assembly is positioned in a rigid aluminium case.

As summarized in figure 10.11, the principle generally chosen to build a super-capacitor module is to combine low-voltage, high-capacity unit cells. We have also seen that one of the major problems currently encountered is not the cost of the cell, which is already very low (around 0.005 €.F^{-1} for the automotive industry), but the assembly of the cells between them, which must withstand high currents and strong electrical insulation constraints.

A first solution consists in making a cell with several juxtaposed windings using a common electrode not connected to the outside (the "multi-track" principle) [26]. Figure 10.18 illustrates this situation.

An alternative solution to this first solution, which is more complex to implement, is to create a system containing several concentric coils [27].

To go further, it is necessary to apply a break and to recall the generally accepted postulates:

- Ageing is accelerated with the increase (combined or not) in the cell temperature and voltage.
- The self-discharge is linked to the voltage. The higher the voltage, the faster the loss of charge.

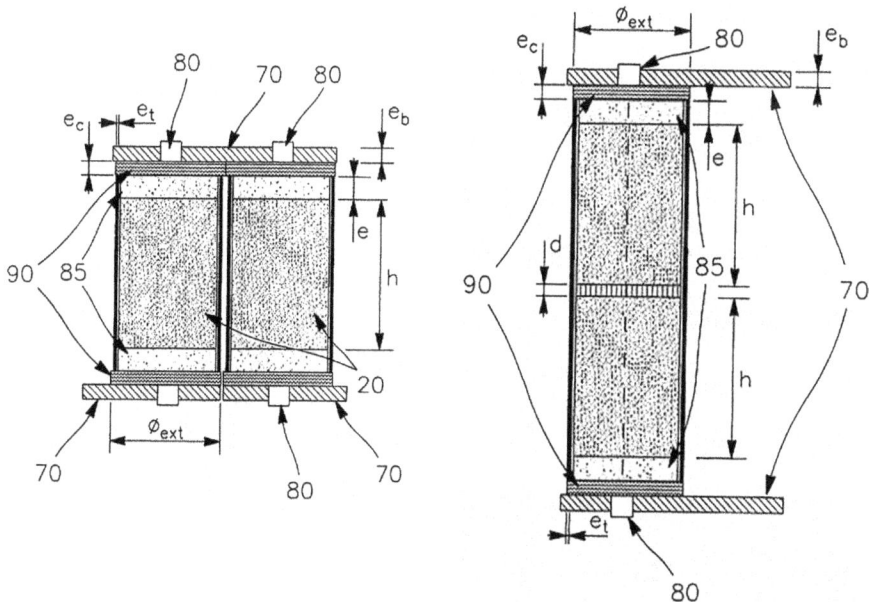

FIG. 10.18 – Illustration of the principle of a two-track component replacing two cells.

- The electrolyte is completely degraded (converted into gas) if the voltage between terminals is about 6 V (walls of the acetonitrile electrolyte).
- The storage mode of a supercapacitor consists of a displacement of ions on a very local scale.

From this last point, we can conclude that it is possible to achieve a "bipolar" type stacking without observing any major degradation, even if the voltage exceeds 6 V. This is the principle used in the two previous inventions, which confirm this hypothesis.

10.2.3 High Capacity Modules Working in Aqueous Medium

Currently, no carbon/carbon symmetrical unit cell operating with an aqueous electrolyte is industrialised, due to the very low performance of the aqueous medium. However, modules have been available and manufactured by Tavrima Canada for a few years. The design of Tavrima modules is based on a stack of circular electrodes to make cylindrical modules (figure 10.19). The energy density is of the order of 0.7 Wh.kg^{-1} and 1.1 Wh.L^{-1}: these values are particularly low compared to those obtained in an organic environment. The major advantage claimed by the manufacturer is the low series resistance, which gives the module a relatively low time constant (0.6 s) but of the same order as that obtained today for components operating in an organic medium.

The major disadvantage of this technology is the maximum operating temperature: + 55 °C is generally not sufficient for the automotive market. Moreover, the

FIG. 10.19 – ESCap 90/300 Tavrima Module (90 kJ/300 V) working in aqueous medium.

cost of the passive components (stainless steel or nickel collector, stainless steel case, etc.) is not lower than that of modules of identical capacity but operating in an organic medium at a voltage 3 times higher. In conclusion, this technology is currently underdeveloped because it is economically and technically inefficient.

Conclusions and Prospects

As explained in this chapter, it is important to differentiate between small cells and large cells, which have different target markets.

Small cells have been commercially available for many years. The degrees of freedom for improvement are low: the pressure on the cost of these components and the target markets (electronics, toys, portable...) do not expect improvements in terms of energy density or durability.

High-capacity cells have been commercially available for more than ten years. The market is growing strongly, but the future of the technology is strongly linked to improvements in cost, energy density and lifetime. Industrialization of supercapacitors is already based on low-cost components (aluminum, activated carbon from vegetable precursors, paper separator, etc.) to be competitive and meet the target markets. The cost currently reached, at the scale of the cell, is around $0.005 F^{-1} (especially for the market of cells dedicated to the automotive industry). Improvements have been achieved in 10 years: the million cycles at 20 °C is now exceeded by major manufacturers and operation under very high current is also demonstrated in various applications. For some applications, such as automotive, energy and power densities are particularly important parameters. This is even more the case when supercapacitors are used in combination or in competition with batteries. We have seen that the modules currently on offer hardly reach 5 Wh.L^{-1}, whereas high capacity cells (>3000 F) exceed 9 Wh.L^{-1}. Improvements are therefore expected at the module level, either through more compact links but allowing high currents to be passed through, or through technological breakthroughs.

The most envisaged breakthroughs are currently focused on hybrid systems, such as LICs or improved aqueous systems, and on bipolar systems, particularly applicable to carbon/carbon supercapacitors. However, regardless of the technology chosen, cost remains the key factor for market penetration. With regard to all these emerging technologies, it should not be forgotten that Li-ion technologies are advancing: energy densities are increasing almost continuously and benefit very strongly from the volume effect not yet available in large supercapacitor cells. As an indication, the current market price of the Li-ion "consumer" at the cell scale is around 150 $.kWh^{-1} with a cyclability of around 500 cycles guaranteed at 100% DoD. The cost of the cycle is therefore around $0.3 kWh^{-1}.cycle^{-1}. By way of comparison, the price currently achieved for an eHDi type two-component is around $5760 kWh^{-1} but with 1 000 000 cycles achievable, which amounts to $0.006 kWh^{-1}.cycle^{-1}.[2] This price is derisory compared to the price of batteries, but to reach this price, it is necessary to have an application requiring such a large

---

[2]Two 1200 F cells at $U_{nominal} = 2.5$ V/cell at 100% DoD, *i.e.* 7500 J = 2083 Wh. Taking into account 1 000 000 cycles and 0.005 $.F^{-1}, we can assume the above mentioned result.

number of cycles and an occupiable volume in the application. The latter is usually the stumbling block: for example, current supercapacitors would be very relevant for mild-hybrid applications, but would take up the volume of a large part of the vehicle trunk. Start-stop applications typically require around 50 000 cycles over the life of the vehicle.[3] On the other hand, urban transport applications (hybrid buses, trams, etc.) are very favourable for this type of technology, in which a life span of around 30 years (in cycles and on a calendar basis) is necessary, provided that the initial additional cost is not too high. In addition, supercapacitors have a clear advantage over their direct competitors, lead-acid batteries on the one hand and Li-ion batteries on the other: they offer almost stable performance over a wide range of temperatures. On the other hand, they require expensive power converters. Improvements are eagerly awaited on this side of the technology.

In addition, the energy density by volume must be increased. Several solutions have been or are being proposed in the case of AC/AC supercapacitors: first by adapting the ion size to the pore size. Relevant studies have been carried out over the last ten years to understand both the storage mechanism and the relevance of this approach, and second, by increasing the operating voltage. We have seen that this solution was complex because it increased self-discharge, reduced lifetime and increased dispersion in aging. A viable solution for certain niche applications, because of the additional cost, remains the use of ionic liquids with a low viscosity.

Increasing energy density can also be achieved using breakthrough technologies. LICs pose new problems: shorter lifetime than supercapacitors, lower operating temperature range, manufacturing cost of the same order as that of Li-ion batteries. They are therefore not necessarily competitive with conventional Li-ion batteries or with AC/AC supercapacitors, but these systems do have interesting safety performances: they can be totally discharged, can withstand short circuits and easily pass most safety tests to be integrated in a vehicle. Alternatives for longer life are also being strongly studied (NIC, KIC). These systems have the advantage of not potentially using critical or strategic materials. However, in order to be competitive, they must have very good cyclability, a really higher energy density than EDLCs with a comparable power density. In addition, competing technologies such as very high-power Li-ion batteries are making great progress in terms of lifetime, but have very poor power densities at low temperatures and also pose different issues in terms of safety and durability.

Bibliography

[1] Caumont O., Huibant R., Minard J.F. (2007) WO200765748, Blue Solutions.
[2] Yunasko products. Please refere to: www.yunasko.com/en/ultracapacitors.
[3] EN and EP series. This product is now discontinued since January 2020.
[4] HW and HZ series. https://industrial.panasonic.com/cdbs/www-data/pdf/RDH0000/ DMH0000COL69.pdf.

---

[3]The starter-alternator dedicated to this application is designed for 60 000 cycles.

[5]   Sarwar W., Marinescu M., Green N., Taylor N., Offer G. (2016) *J. Energy Stor.* **5**, 10.
[6]   Nisshinbo has transfered its activity to TPR/TOC Capacitor. The present design is primsatic.
[7]   Thrap G., Shelton S., Schneuwly A., Lauper P., Soliz R. (2008) US2008013253, Maxwell Technologies.
[8]   Caumont O., Depond J.-M., Juventin A.-C. (2010) EP2198472, Blue Solutions.
[9]   Goesmann H., Setz M. (2005) DE102004039231, EPCOS.
[10] Schmid T., Herdeg B., Smit A., Riepl T., Gschossmann H., Zeller R., Fenchel F., Gilch M. (2011) WO2011157629, Continental Automotive GmbH.
[11] Caumont O., Juventis-Mathes A.-C., Le Bras K., Depond J.-M. (2008) EP2145360, Blue Solutions.
[12] Thrap G., Borkenhagen J.L., Wardas M., Maheronnaghsh B. (2007) US2007054559, Maxwell Technologies.
[13] Caumont O., Paillard P., Saindrenan G. (2007) WO2007147978, Blue Solutions and Ecole Polytechnique de l'Université de Nantes.
[14] Goesmann H. (2005) DE102004030801, EPCOS.
[15] Ashino K. (1991) JP3034505, Nippon Chemicon.
[16] Setz M., Nowak S., Hoerger A. (2006) WO2006005277, EPCOS.
[17] Goesmann H., Vetter J., Mayr M., Pint S. (2009) US20090111009, BMW.
[18] Caumont O., Juventin-Mathes A.-C., Le Bras K., Depond J.-M. (2008) WO2008141845, Blue Solutions.
[19] Schiffer J., Linzen D., Sauer D.U. (2006) *J. Power Sources* **160**, 765.
[20] Lee D.H., Kim U.S., Shin C.B., Lee B.H., Kim B.W., Kim Y.-H. (2008) *J. Power Sources* **175**, 664.
[21] Guillemet P., Scudeller Y., Brousse T., Depond J.-M. (2007) *Revue internationale de génie électrique* **10**, 695.
[22] Desprez P., Barrailh G., Rochard D., Rael S., Sharif F., Davat B. (2002) EP1274105, SAFT.
[23] Venet P. Habilitation à Diriger les Recherches, "Amélioration de la sûreté de fonctionnement des dispositifs de stockage d'énergie", October 24th 2007, University Claude Bernard – Lyon1.
[24] Okamura M., Yamagishi M., Mogami A. (2000) EP1033730, Advanced Capacitor Technologies, Okamura Lab. and Powersystems.
[25] EDLC brochure de Meidensha, disponible sur internet: EDLC Catalog-E-nov1808.pdf, 2009.
[26] Azaïs P., Depond J.-M., Caumont O. (2009) FR2927728A1, Blue Solutions.
[27] Azaïs P., Depond J.-M., Caumont O. (2009) FR2927727A1, Blue Solutions.

Chapter 11

Characterization of the Electrical Performance of Li-ion Cells

Arnaud Delaille, Nicolas Guillet, Romain Tessard
and Bramy Pilipili Matadi

This chapter describes the tests commonly performed to quantify the electrical performance of individual Li-ion cells: acceptance tests, initial performance tests and ageing tests. It also describes the protocols available for measuring the resistance of batteries, which may have an impact on the measured resistance values.

11.1 Characterization of the Electrical Performance of Individual Cells

11.1.1 Acceptance Tests

The qualification of cell performance always begins with acceptance tests for all the cells in a given batch, their number being determined by the subsequent tests to be performed. A minimum of 15 cells are required to have sufficient quantity to qualify the dispersion of one batch, the maximum number depending only on the associated qualification plan, and more specifically the number of ageing and safety tests.

The purpose of the acceptance tests is to calculate the mass and volumetric energy density under nominal conditions – and thus to check whether the nominal values specified by the cell manufacturer are respected – but also to know the state of charge (SoC) at reception, to assess the dispersion of the batch, and if necessary to select the cells for subsequent tests (in case of significant dispersion among the total batch).

DOI: 10.1051/978-2-7598-2555-4.c011
© Science Press, EDP Sciences, 2021

For this purpose, these tests generally include the determination of the following quantities:

- mass;
- dimension;
- no-load voltage, also known as open circuit voltage (OCV);
- state of charge at reception, deduced from a residual discharge followed by a capacity measurement cycle;
- capacity/energy at a nominal charge/discharge rate (typically C/2–C/2 for Li-ion cells, *i.e.* charge and discharge in the order of 2 h) while respecting the end/floating voltage thresholds (keeping the voltage at the end of charging until a so-called "end-of-charge" current value, typically C/20) specified by the manufacturer, and at a nominal temperature (typically 25 °C);
- internal resistance, most often equivalent to a measurement at SoC = 50 % of the real part of the impedance at 1 kHz (see the second paragraph of this chapter on resistance measurements for more information);
- available power, calculated from the internal resistance value.

The representation of the capacity as a function of the internal resistance of the cell is generally the one that best visualizes the dispersion of the electrical performance of a given batch of cells. Figure 11.1 gives an example of the observed dispersion results on a batch of 50 cells of the same reference, from which the cells selected for the subsequent tests discussed below are selected.

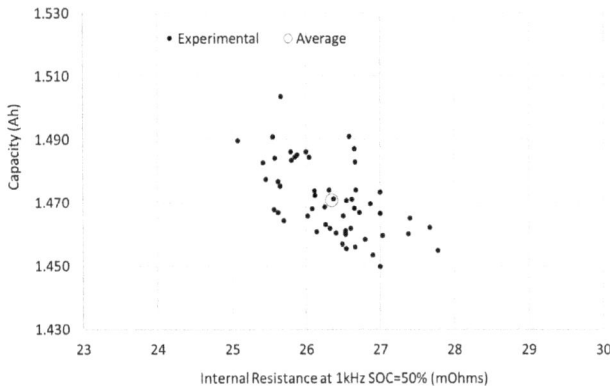

Fig. 11.1 – Illustration of the capacity values (here at C/2–C/2 and 25 °C) as a function of internal resistance (here at 1 kHz, SoC = 50 % and at 25 °C) measured on a batch of 50 Li-ion cells.

11.1.2 Beginning of Life Performance Tests

11.1.2.1 Capacity, Energy, Resistance and Power Measurements

The qualification of electrical performance at the beginning of life (BoL) of the cells consists in quantifying the following quantities:

- the capacity Q (in Ah) and the energy E (in Wh) according to various charge/discharge rates and at various temperatures;
- the resistance R (in ohms) as a function of the state of charge (SoC) and temperature, and measured for a given time, allowing the calculation of the power P (in W) available either for discharge or for charge under these specific conditions.

Since the determination of the resistance values is detailed in the second paragraph of this chapter, the following equations explain in particular how to obtain the values of capacity, energy, and power:

- Discharge capacity:

$$Q_{dish}[\text{Ah}] = \int_{\text{beg of disch}}^{\text{end of disch}} I(t) \cdot dt \qquad (11.1)$$

where I is the discharge current (A) and t is the time (s).

- Discharge energy:

$$E_{disch}[\text{Wh}] = \int_{\text{beg of disch}}^{\text{end of disch}} U(t) \cdot I(t) \cdot dt \qquad (11.2)$$

where U (Volts) is the cell voltage.

Expression that can be approximated, by considering the voltage variation in discharge at a constant average value $U_{\text{avg\_disch}}$, while keeping in mind this is an approximation:

$$E_{disch}[\text{Wh}] = U_{\text{avg\_disch}} \cdot Q \qquad (11.3)$$

The values of capacity and energy in charge are similarly gathered from the same relations, adapted to the charge.

- Discharge power at a given SoC and after a time of x seconds:

$$P_{disch}(\text{SoC}, x \text{ sec.})[\text{W}] = U_{\min} \cdot I_{\max} \qquad (11.4)$$

where:

$I_{\max}$ the smallest of the following values: $I_{\max}$ in discharge after x seconds specified by the manufacturer, *versus* $I_{\max} = \dfrac{\text{OCV}(\text{SoC}) - U_{\min\_\text{disch}}}{R_{disch}(x \text{ sec.})}$;

OCV (*Open Circuit Voltage*) the no-load voltage at the considered point of state of charge;

$U_{\text{min\_disch}}$ the voltage limit value at which the discharge is terminated;

R the internal resistance, defined by: $R_{\text{disch}}(x \text{ sec.}) = \dfrac{\text{OCV}(\text{SoC}) - U(x \text{ sec.})}{I_{\text{disch}}}$ based on a current pulse I_{disch} over a period of x seconds (refer to the second paragraph of this chapter for other possible definitions of resistance);

$$U_{\text{min}} = \text{OCV}(\text{SoC}) - R_{\text{disch}}(x \text{ sec.})$$
$$\cdot I_{\text{max}} \left(\text{equal to } U_{\text{min\_disch}} \text{ when } I_{\text{max}} = \frac{\text{OCV}(\text{SoC}) - U_{\text{min\_disch}}}{R_{\text{disch}}(x \text{ sec.})} \right).$$

- Power in charge at a given SoC and after a time of x seconds:

$$P_{\text{ch}}(\text{SoC}, x \text{ sec.})[\text{W}] = U_{\text{max}} \cdot I_{\text{max}} \qquad (11.5)$$

where:

I_{max} the smallest of the following values: I_{max} in charge after x seconds specified by the manufacturer $I_{\text{max}} = \dfrac{U_{\text{max\_ch}} - \text{OCV}(\text{SoC})}{R_{\text{ch}}(x \text{ sec.})}$;

OCV the no-load voltage at the considered point of state of charge;

$U_{\text{max\_ch}}$ the considered charging limit voltage;

R the internal resistance defined by: $R_{\text{ch}}(x \text{ sec.}) = \dfrac{U(x \text{ sec.}) - \text{OCV}(\text{SoC})}{I_{\text{ch}}}$ based on a current pulse I_{ch} over a period of x seconds;

$$U_{\text{max}} = \text{OCV}(\text{SoC}) + R_{\text{ch}}(x \text{ sec.})$$
$$\cdot I_{\text{max}} \left(\text{equal to } U_{\text{max\_ch}} \text{ when } I_{\text{max}} = \frac{U_{\text{max\_ch}} - \text{OCV}(\text{SoC})}{R_{\text{ch}}(x \text{ sec.})} \right).$$

It is important to underline here that all these quantities are highly dependent on the measurement conditions. Strictly, it is therefore necessary to always specify the conditions under which they are measured, namely the charge and discharge rates, the temperature, the considered end of charge and discharge thresholds, and the duration considered with regard to the resistance and power calculations.

During characterizations, all these parameters are generally defined according to the functional specifications of the relevant application, while respecting the limits specified by the cell manufacturer.

To illustrate this, figure 11.2 shows the available discharge capacity of a given Li-ion cell, as a function of the discharge current rate, the operating temperature, and the end-of-discharge cut-off threshold. The graphs in this figure highlight the importance and the impact of the parameters mentioned on the capacity of the specific cell.

Similarly, figure 11.3 shows the power available after 10 s, defined from the internal resistance, itself measured from a current pulse of 10 s duration. Again, the

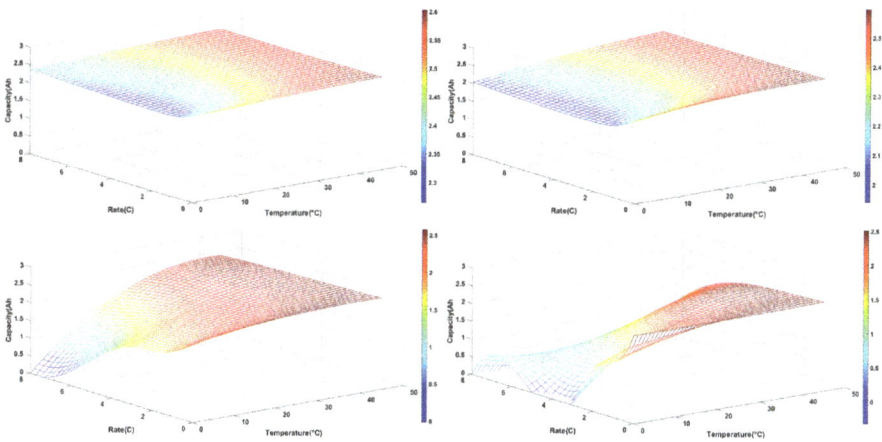

FIG. 11.2 – Illustration of the capacity values for a given cell, according to the discharge rate, temperature, and different cut-off thresholds at the end of the discharge: (a) 2.0 V; (b) 2.5 V; (c) 2.8 V; and (d) 3.0 V.

graphs in this figure highlight the influence of the SoC, of the operating temperature, and of the cut-off threshold.

11.1.2.2 *Faradic and Energy Efficiency*

The charge and discharge capacity values are used to calculate the faradic efficiency of the cells, and the charge and discharge energy values are used to calculate the energy efficiency, based on the following standard definitions:

$$\eta\_Ah = \frac{\text{Discharged Capacity}}{\text{Previously Charged Capacity}}; \quad \eta\_Wh = \frac{\text{Discharged Energy}}{\text{Previously Charged Energy}}$$

$$(11.6)$$

where η_Ah is the faradic efficiency and η_Wh the energetic efficiency.

The performance of the cell (capacity energy content...), are highly dependent on the measurement conditions, which must always be specified. The real meaning of performance values must for the same reasons be very careful controlled, to prevent *e.g.* using efficiency values greater than 1, which can be quite easily but uncorrectly reported in case of insufficient examination of the influence of measurement conditions (*e.g.* a charge made at a lower temperature than the next discharge, for example, or at a much higher rate).

Under normal use (except under abusive test conditions or ageing), the faradic efficiency of Li-ion cells can be approximated to 1, meaning the absence of any parasitic reaction. On the other hand, the energy efficiency is impacted by the higher/lower voltages in charge/discharge, at both electrodes, so that it depends mainly on the charge/discharge rate and temperature: this energy efficiency

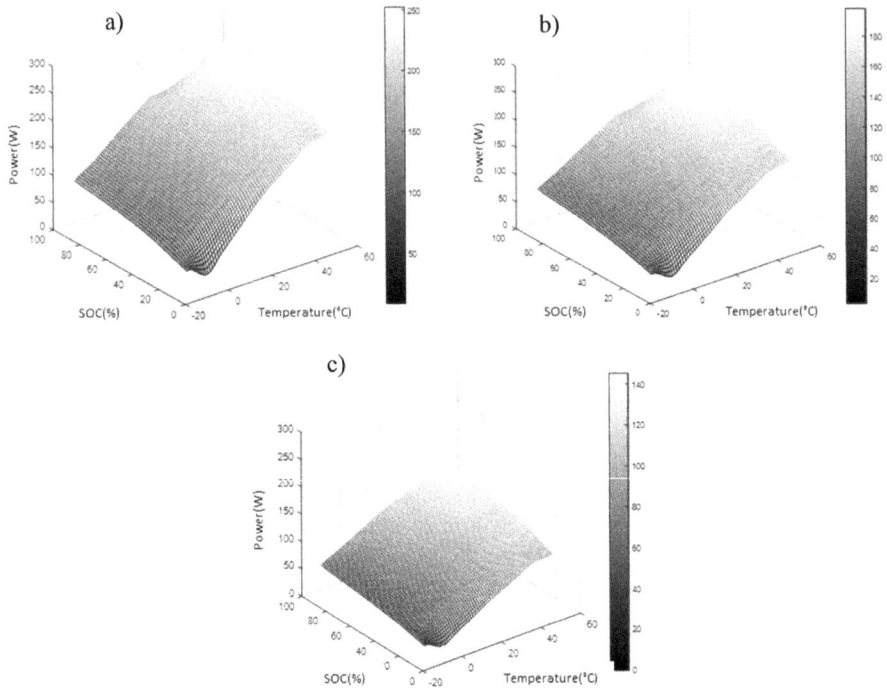

FIG. 11.3 – Illustration of the available power values in discharge of a given cell after 10 s, as a function of its state of charge (SoC), temperature, and different cut-off thresholds: (a) 2 V; (b) 2.5 V; and (c) 2.8 V.

decreases as the temperature decreases and as the charge (or discharge) rate increases.

11.1.2.3 Comparison of Beginning of Life (BoL) Performance of Li-ion Cells

Figure 11.4 illustrates the initial performance of various Li-ion cell references, based on numerous tests carried out by CEA-Liten, detailed here in terms of specific energy $(Wh.kg^{-1})$ as a function of energy density $(Wh.L^{-1})$.

The key point to have in mind is that all the values related to the electrical performance of a battery are closely related to the conditions under which they are measured. Therefore, none of these values can be used rigorously without the precise knowledge of the conditions under which they are determined (charge/discharge rates, temperature, charge/discharge limit thresholds). We will see throughout the next section that it is mandatory to discuss the state of cell health, which itself depends strongly on all the operating conditions.

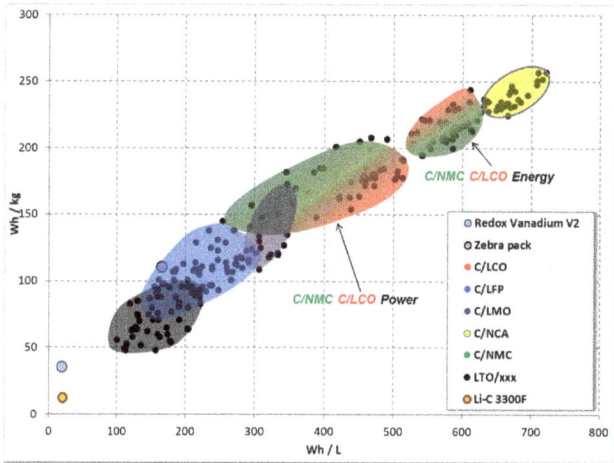

FIG. 11.4 – Gravimetric and volumetric energy densities measured on various Li-ion cells using different Li-ion technologies, at C/2–C/2 and 25 °C, from the CEA-Liten database.

11.1.3 Ageing Performance Tests

11.1.3.1 Ageing Conditions

Ageing performance test are based on exposing cells to given ageing conditions. Cell electrical performances (capacity, energy, resistance, power behaviour, energy efficiency...) are measured at given time intervals all along the ageing period.

Ageing are usually assessed under two different test conditions:

– Calendar storage: the cells are kept at rest at a given state of charge and temperature, and most often disconnected. This is also named shelf-life.
– Cycling: the cells are subjected to repeated charge/discharge cycles under specific charge/discharge current conditions, state of charge range (both the depth of discharge and the SoC range are involved) and temperature conditions. The experimental test plan depends, beyond the available resources, on the targeted application. This allows to limit the quantity of experiments to be performed.

While many tests are needed to identify dynamic endurance models to simulate cell ageing under variable conditions, it is also possible to try to set worst/best-case scenarios, in order to embed the performance of cells in shape. However, this approach quickly reaches its limits due to the cross and non-linear influences of the above-mentioned parameters, which do not allow to discriminate between one scenario and another; hence, there is a need to use more in-depth characterizations and the identification of endurance models to understand the lifetime of cells under various conditions of use.

*11.1.3.2   Ageing Follow-Up by Incremental Capacity Analysis (ICA)
            and Differential Voltage Analysis (DVA)*

Beyond monitoring performance in capacity, energy, resistance and power, it is also possible to follow the ageing of Li-ion battery performance by incremental capacity analysis (ICA).

This method consists in measuring and then graphically illustrating the capacity increments seen when applying dQ/dU voltage increments across the cell, as a function of the evolution of the charge and/or discharge, *i.e.* the voltage U, or the cumulative charge/discharge capacity, or the state of charge. A variant of this approach is the Differential Voltage Analysis (DVA), which consists in plotting, in contrast with the ICA, the voltage increments reduced to the dU/dQ capacity increments, as a function of the charge and/or discharge evolution. These two types of analysis are especially relevant as the charge/discharge current rate is low, ideally in the order of C/25. They allow approximating the cell open circuit voltage, help identifying phenomenon at stake in the electrodes, and minimize as much as possible overvoltage effects.

The main objective of these methods is to discriminate the performance losses, first with respect to the negative electrode and/or the positive electrode, and then with respect to the different ageing mechanisms (loss of active material from one and/or the other electrode, formation of passivation layers on one and/or the other electrode, etc.). Indeed, these analyses make it possible to highlight "peaks" related to the electrochemical processes of each electrode, intercalation/deintercalation in the case of Li-ion batteries. This allows following the evolution of these processes with ageing based on the modifications of these peaks, both in amplitude and in respective insertion rates.

Figure 11.5 below depicts the ICA curves obtained, respectively, on a C/LFP cell and on a C/NMC one.

In the case of a C/LFP Li-ion cell, each peak identified on the ICA curves can be attributed to the intercalation/deintercalation of the Li ions in the graphite electrode, the LFP electrode presenting no peak. For other Li-ion chemistry such as C/NMC, the assignment of peaks to each of the two electrodes requires additional measurements to study the potential of each electrode separately. This can be achieved by inserting a reference electrode inside the cell being studied, or by making half cells by collecting positive or negative electrode from used cells, and cycle them *versus* lithium metal.

To analyse ICA or DVA results efficiently, it is strongly recommended to couple them with post-mortem analyses: the opening of the cells at the end of their life and the analysis of the materials they contain make it possible to complete the understanding of the ageing modes of each of the two electrodes, to assign each of the ICA or DVA peaks to them, and to corroborate the initial assumptions concerning the ageing mechanisms.

For example, we can mention the work of Dubarry *et al.* [2] who studied the evolution of incremental capacity peaks for commercial Li-ion cells in the case of active material loss and chemical modifications. In another paper [3], the authors proposed an ageing model able to simulate different degradation mechanisms,

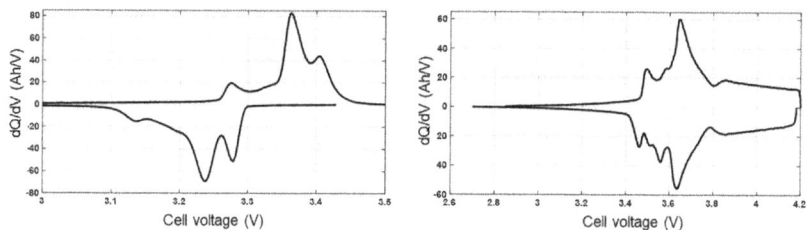

FIG. 11.5 – (a) ICA curve obtained on a C/LFP Li-ion cell, cycled at C/10 and 25 °C; (b) ICA curve obtained on a C/NMC Li-ion cell cycled at C/25 and 25 °C [1].

including the loss of cyclable lithium or the increase in ohmic resistance and their impacts on the ICA curves. These impacts and characteristics of each ageing mode have been further highlighted in the reference [4]. Finally, other authors have also used ICA measures to provide a qualitative analysis of the identification of the ageing mechanisms [1, 5–10].

11.1.3.3 Comparison of the Ageing Performance of various Li-ion Cells

Figure 11.6 illustrates the endurance tests results of various Li-ion cells made of different Li-ion technologies, assessed at the CEA-Liten. The results presented here are related to tests peformed under one given calendar condition and one given cycling condition.

11.2 Resistance Measurements of Individual Cells

11.2.1 Introduction

It is useful to pay particular attention to the resistance measurement of a cell, as it is not trivial. Depending on the method used to perform this measurement, the results may vary significantly. In addition, each method has its own technical limitations and very specific conditions to ensure the validity of the results produced. This last point is often neglected, especially when automated test campaigns are performed.

The aim of this section is to detail the different methods for measuring the internal resistance of a battery and to identify the advantages and limitations of each one. We will also carefully identify the rules to set and the possible sources of errors.

11.2.2 How to Define an Internal Resistance?

The potential difference measured across an electrochemical cell changes with the current flowing through it. This evolution is related to the overvoltages produced by

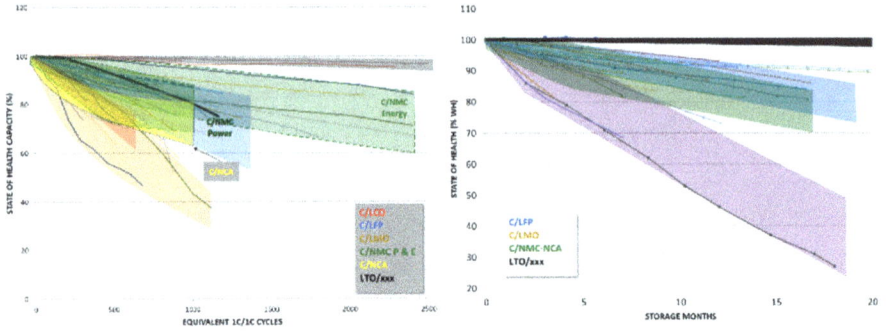

FIG. 11.6 – (Left) Evolution of the State-of-Health (SoH) status of different Li-ion cells subjected to cyclic ageing at 1C–1C, between 0 % and 100 % of SoC, and at 25 °C. (Right) SoH status of different Li-ion cells subjected to calendar ageing at SoC = 65 % and 25 °C (CEA-Liten database).

different resistive contributions within the cell. Thus, during the discharge of a battery subjected to a current i, the cell voltage (U_{bat}) is always lower than the voltage measured in open circuit under conditions close to thermodynamic equilibrium (U_{OCV}). Symmetrically, during charging, the overvoltages add to the open circuit voltage, and the voltage at the cell terminals is always higher than the open circuit voltage:

$$U_{bat} = OCV +/- \eta_{ohmic} +/- \eta_{passiv} +/- \eta_{TC} +/- \eta_{noc} \qquad (11.7)$$

where:

(+) in charge and (−) in discharge, in brackets are indicated the parameters influencing the value of the coefficient.

OCV: Open Circuit Voltage (T, SoC, SoH).

η_{ohmic}: overvoltage due to ohmic resistance (T, SoC, SoH).

η_{passiv}: overvoltage due to the presence of passivation layers on the electrodes (T, i, SoC, SoH).

η_{TC}: overvoltage related to the charge transfer of the reactions (T, i).

η_{con}: overvoltage linked to the concentration of reagents (migration, diffusion, convection, etc.) (T, i, t, SoC, SoH).

These overvoltages can be considered as linearly evolving with the current flowing through the system, *i.e.* evolving according to Ohm's law:

$$\eta = R \times i \qquad (11.8)$$

where R is the resistance value associated with each contribution.

However, this simplification requires us to consider that the various contributions are purely resistive and that they therefore have no (or very little) temporal evolution. This hypothesis may be more or less acceptable for some over voltages (ohmic resistance, passivation, charge transfer) but is difficult to consider for concentration overvoltage.

In addition, the various contributions identified may depend on several external parameters such as temperature (T), applied current (i), state of charge (SoC) and state of health (SoH).

Determining an internal resistance value of an electrochemical cell is therefore not as simple as it might seem: the measured values can vary greatly depending on the method used and the experimental conditions under which these measurements are made.

11.2.3 Different Methods of Measuring Internal Resistance

Various methods of measuring internal resistance are presented below. All these methods are applied to a single Li-ion battery cell sample, to allow the results obtained to be compared to each other.

The Li-ion battery cell used is a cylindrical format cell (18650) with a nominal capacity of 3200 mAh. Its operating voltage range is between 2.5 and 4.2 V. All the tests presented below were performed at a temperature of 25 °C (the battery is placed in a controlled climatic chamber at ±0.1 °C accuracy).

11.2.3.1 Measurements from Polarization Curves

Resistance measurement from the polarization curves $U_{bat} = f(j)$, is a method used in a classical way for electrochemical systems for which it is possible to place oneself in a stationary or quasi-stationary state (fuel cells, electrolyzers, etc.).

In the case of battery cells, a steady state is excluded since positive and negative electrode materials change when a current flows through the system. However, by taking measurements at low speed and/or for short periods, we can consider that the state of charge of the battery changes is negligible during the measurement time.

The principle of the measurement consists in carrying out a current sweep in charge and discharge, between 0 A and a maximum current which depends on the capacity of the cell. In our case, we decided to perform a current sweep at 50 mA.s^{-1}, up to 1500 mA (approximately C/2). An example of an experimental result is given in figure 11.7.

The amount of charge passing through the system during charging and discharging must be small enough so that the battery's state of charge does not change significantly during the measurement. In this example, the variation in the state of charge is 0.18 % (6.24 mAh). In addition, the imposed charge quantities must be chosen to avoid exceeding the cell voltage range authorized by the manufacturer's datasheet (2.5–4.2 V in this example). The consequence is that this method has an obvious technical limit if one wishes to measure resistance to extreme states of charge (0 % and 100 % state of charge).

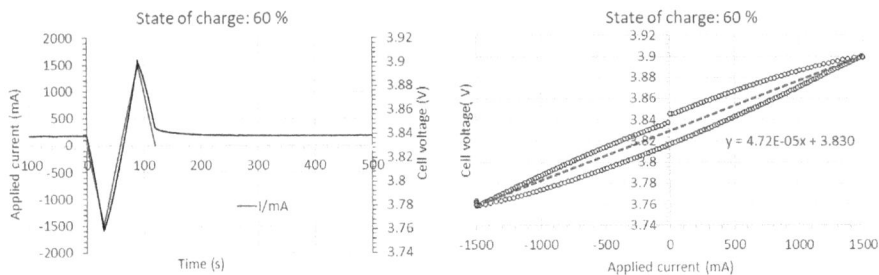

F_{IG}. 11.7 – (Left) Evolution of the voltage measured at the terminals of the cell and the current applied, over time. (Right) Evolution of the voltage measured at the terminals of the cell as a function of the applied current.

The application of symmetrical current sweep (same amount of charges applied during successive sweeps in charge and discharge) is recommended. This makes it possible to check whether the system has not been disturbed too much by the measurement, by comparing the value of the voltage at the terminals of the open circuit cell before and after the current scanning. These two values must be very close, if not identical, after a stabilization time that depends on many parameters (cell studied, temperature, amplitude and speed of current sweep...).

The measurement of the voltage at the cell terminals as a function of the applied current makes it possible to identify a linear variation and to calculate a slope, defined as the electrical resistance of the system being studied.

Figure 11.8 shows the evolution of the resistance thus calculated and the open circuit voltage of the cell studied.

It can be noted that the resistance value measured changes strongly with the state of charge, and increases significantly as one approaches extreme states of charge or discharge.

This method is simple to implement and quite accurate. Its main limitations are the waiting time required for a good stability of the system before performing the measurement (quasi-stationary state[1]) and to carry out the current sweep under pseudo-stationarity[2] conditions. Taking into account the evolution of the voltage with the applied current, this method does not theoretically make it possible to measure resistance values at 0 % and 100 % state of charge, but an adjustment of the current scanning rate and the maximum value of this current makes it possible to get close to this value.

---

[1]The rate is considered "quasi-stationary" if the variations in concentration of the reactive species and any intermediate reactions are very small, and these concentrations hardly vary over time. In practice, this results in stable operation over long periods at constant current and voltage (variations considered negligible in the time interval considered).

[2]Pseudo-stationarity conditions are considered verified if the sweep is slow enough so that there is no accumulation of reactive species or possible reactive intermediaries in the system during the measurement period.

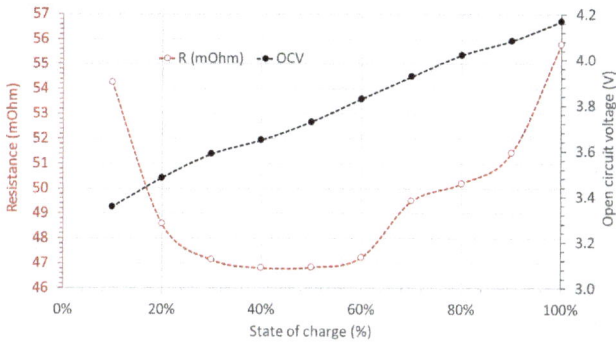

FIG. 11.8 – Evolution of the electrical resistance and open circuit voltage of a cell.

11.2.3.2 Measurement with Open Circuit Voltage Variation

This method is widely used in the field of batteries. The principle is to measure the evolution of the cell voltage as a function of the state of charge at different rates and to measure the deviation from the evolution of the open circuit voltage (no current). The voltage difference between the charge–discharge curves of the battery and the open circuit voltage value must be proportional to the current applied:

$$U_{\text{measured}} = U_{\text{OCV}} +/- Ri \tag{11.9}$$

This behavior is clearly visible in figure 11.9 showing the charge–discharge curves at different rates between C/20 and C/2.

From these measurements, it is possible to calculate a resistance value R by using equation 11.9. However, to do this, it is necessary to know accurately the evolution of the cell's open circuit voltage (OCV) as a function of the state of charge. However, the OCV is defined theoretically from the variations of two thermodynamic parameters: the enthalpy (H) and the entropy (S) of the system. It is very sensitive to temperature. However, at the highest speeds (*e.g.* C/3, C/2), although the cell is placed in a regulated thermal enclosure, the battery temperature is far from homogeneous. If the surface temperature (or skin temperature) is close to the ambient temperature, this is not the case within the cell. This can result in significant uncertainty for the value of the OCV.

The resistance calculation as a function of the state of charge for each cycling rate is proposed below (figure 11.10). A large variation in resistance value appears, especially at the lowest C/20 rate. The main reason for this error is related, on the one hand, to the difficulty of accurately defining an OCV value and, on the other hand, to the small differences between this theoretical value and the voltage measured during cycling (due to the low internal resistance of the battery).

To limit this problem, it is possible to calculate a resistance value based on the state of charge, averaged over several charging-discharging cycles at different rates. We need to calculate a linear regression of the voltage values measured at different

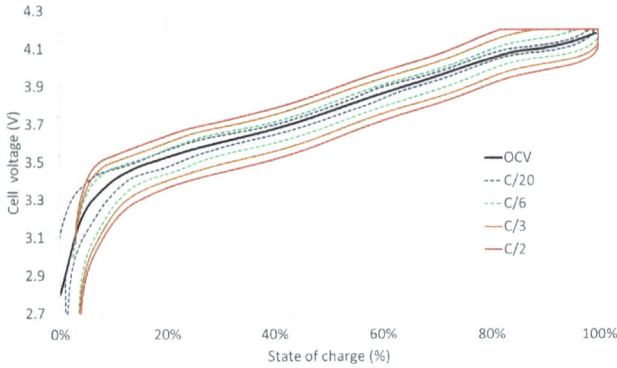

FIG. 11.9 – Comparison of the charge–discharge curves performed at different rates.

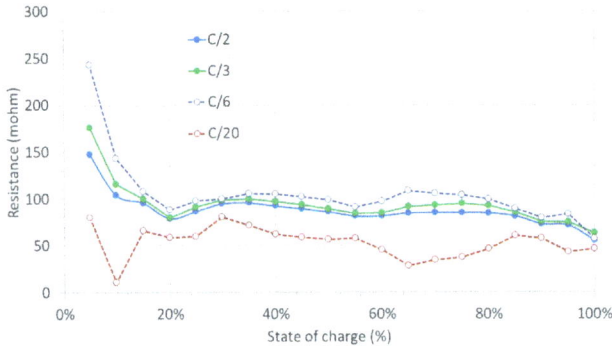

FIG. 11.10 – Calculation of state-of-charge-dependent resistance values from cycling voltage measurements at different rates (C/20–C/2).

rates for the same SoC. The slope is proportional to the electrical resistance and the intercept at the origin corresponds to a value close to the open circuit voltage.

An example is shown in figure 11.11.

It should be noted that the resistance values calculated by this method are significantly higher than those determined by the previous method (between 60 and 160 mOhms here, compared to 45–55 mOhms previously). In addition, the evolution of the resistance is very different, even if there is an increase in the calculated value at lower SoC.

This method is less accurate and takes longer to implement than the previous one. Indeed, it requires several charging and discharging cycles at different rates. On the other hand, it has the advantage of being able to be used to quantify the evolution of resistance over successive cycles applied during battery ageing tests.

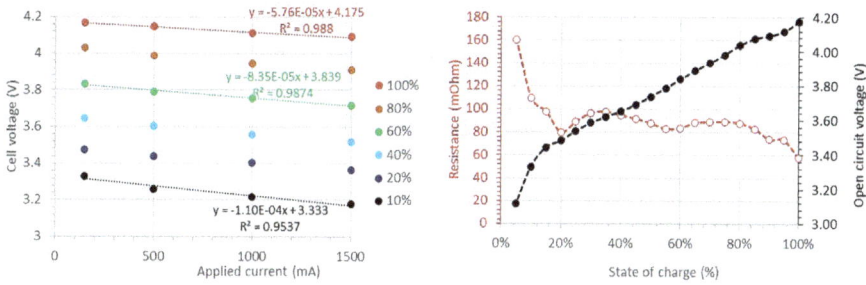

FIG. 11.11 – (Left) Evolution of the cell voltage as a function of the current applied during the various charge–discharge cycles. Values shown for different states of charge (10 %–100 %). (Right) Development of the calculated resistance value and the calculated OCV value as a function of the state of charge.

11.2.3.3 Current Pulse Measurements

The principle of this method is to place the system in a quasi-stationary state with no current, then apply a constant current for a relatively short time (typically a few tens or hundreds of seconds). During this current pulse, the voltage generally follows an evolution according to a resistive and capacitive law $(R - R/C)$ (figure 11.12). Measurements can be made at different time intervals to evaluate the evolution of resistance over time.

At any point in time, the resistance can be easily calculated from the following relation:

$$R(t) = (U(t) - U(t = 0))/i \qquad (11.10)$$

where $U(t = 0)$ is the stabilized cell's voltage before the current pulse, $U(t)$ the voltage at time t, and i the value of the applied current. The resistance $R(t)$ increases with time. An example of a result is shown in figure 11.13.

We can see that the amplitude of the current has only a limited effect on the measured resistance value. We can also check it on figure 11.14 showing the resistance values measured at different SoC.

The method is quite accurate and easily reproducible. The results are quite similar to the method based on current sweep. It also makes it possible to take into account the kinetic aspect: the evolution of the resistance value over time.

The difficulty of this method is to conform the quasi-stationarity conditions of the system before the pulse (quasi-OCV) and to limit the variation in the state of charge of the battery during the measurement (short measurement times or low currents).

Of course, it is also possible to perform the resistance measurement during the relaxation following the current pulse. The principle is the same, except that in this case only, the reference OCV will be the one measured once the system has returned to quasi-stationary mode.

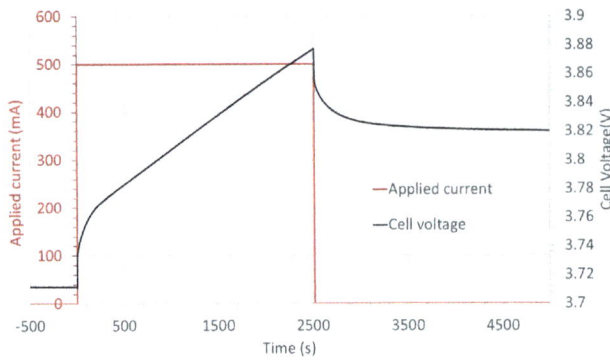

FIG. 11.12 – Evolution of the cell voltage over time during the application of a constant current pulse.

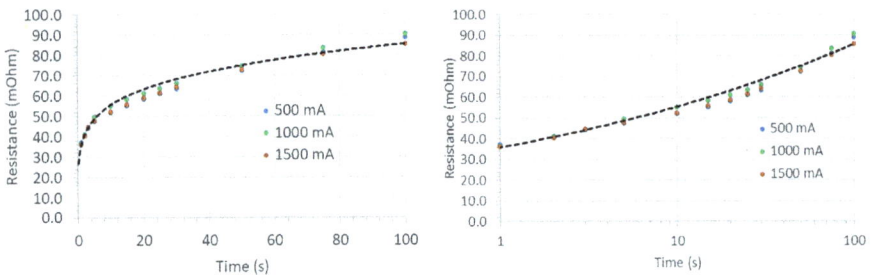

FIG. 11.13 – (Left) Evolution of the calculated resistance during pulses with different currents as a function of time. (Right) Semi-logarithmic representation of time.

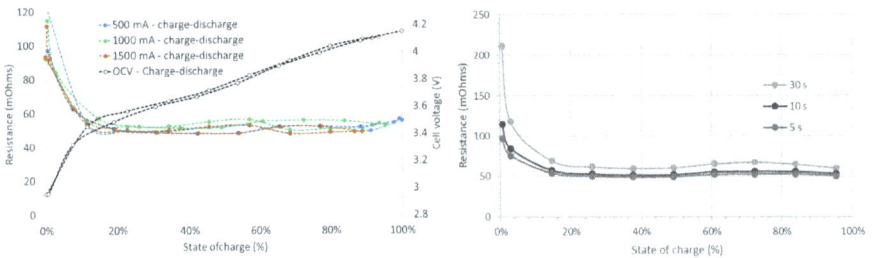

FIG. 11.14 – Evolution of the calculated resistance according to the state of charge. (Left) Measurements made with different current pulses after 10 s. (Right) Measurements performed at different time intervals.

11.2.3.4   Current Pulse Measurements and Extrapolation of Voltage Values

Based on the previous method, we can propose a solution that avoids capacitive phenomena that introduce a temporal evolution of the measured resistance. The basic assumption of this method is that the cell open circuit voltage changes linearly with the state of charge. This assumption is wrong if we consider the evolution of the OCV over the entire operating range of the cell, but it may be acceptable over a limited range of state of charge (a few % state of charge for example).

From the evolution of the voltage as a function of time during a charge or discharge, it is possible to identify a linear evolution and extrapolate this evolution to determine a voltage U_1 corresponding to the beginning of the pulse (an example is shown in figure 11.15).

The resistance thus considered is calculated from the difference between the initial cell voltage U_0 and the voltage U_1 as follows:

$$R = (U_0 - U_1)/i \tag{11.11}$$

Figure 11.16 shows the resistance values calculated as a function of the state of charge and for pulses at three different currents (500, 1000 and 1500 mA). We can note that it is difficult to determine a resistance with great accuracy. The source of the error is that the evolution of the voltage over time is not always linear. The larger the applied current and the nearer the extreme states of charge, the larger this difference in linearity is important.

This method is only applicable in very special cases, for example if batteries of the LFP/graphite or LFP/LTO type are considered, which have a linear evolution of OCV over a wide range of state of charge. It requires an optimization of the value of the current applied during the pulse and duration to confirm the starting hypothesis.

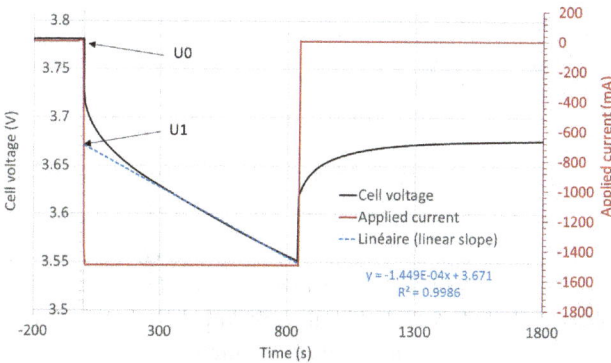

Fig. 11.15 – Evolution of the cell voltage with time during a current pulse. The extrapolation of the linear part makes it possible to determine the value U_1.

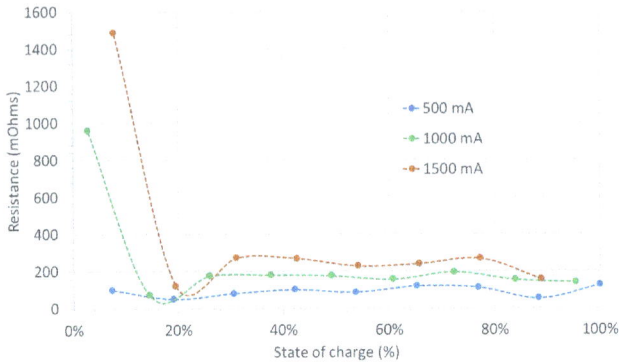

F_{IG}. 11.16 – Resistance values calculated from the linear voltage development during a current pulse at different SoC and for different pulse currents.

11.2.3.5   Electrochemical Impedance Spectroscopic Measurements

Electrochemical impedance spectroscopy (EIS) is certainly the most suitable method for measuring the electrical impedance of an electrochemical device. However, the measurements must be performed with care.

The principle of electrochemical impedance spectroscopy is to apply a small electrical signal to the system under study and to measure the electrical response to this signal. In general, this signal consists in superimposing a sinusoidal variation of voltage (PEIS: Potentiostatic Electrochemical Impedance Spectroscopy) or current (GEIS: Galvanostatic Electrochemical Impedance Spectroscopy) and measuring the evolution of the current (PEIS) or voltage (GEIS) across the cell.

- The disturbance can be generated at different frequencies, ranging from several hundreds of kHz to lower mHz. In practice, for batteries, the frequency range "f" is generally between a few kHz and mHz.
- The amplitude of this disturbance must be large enough for the system response signal to be detectable, but must also remain small enough to verify the two basic assumptions of this impedance measurement method:
  - the linearity of the response (linear evolution of the voltage with the applied current);
  - the invariance in the measurement time (the quasi-stationarity state of the system).

For each measurement performed, it is essential to confirm the validity of these two assumptions. This verification is very often (most of the time) neglected and this can lead to useless results and errors in interpretations.

Therefore, for each measurement, it is required to check the linearity of the system by performing measurements with different amplitudes of disturbing signals. The invariance of the impedance spectra with the signal amplitude is a criterion for validating the measurements.

It is also essential to check that the electric quantities of the system being studied (voltage or current) have not changed during the measurement time. This is very important, especially for measurements made at frequencies below the Hz. Indeed, long measurement times increase the risk of signal drift during the measurement time.

The impedance $Z(f)$ of the system calculated using the electrical response of the system as a function of the frequency of the generated disturbance signal is a complex number, comprising both a real component Re(Z) and an imaginary component Im(Z). Impedance values calculated from measurements made with electrical signals at different frequencies can be represented in a diagram called "Nyquist" (real part on the X-axis and imaginary part on the Y-axis) or "Bode" (frequency of the signal applied on the X-axis, module and phase on the Y-axis). An example of a result is shown in figure 11.17.

We can see that the impedance module changes strongly with the frequency of the applied signal.

An equivalent electrical circuit (EEC) has been proposed to model the electrical behavior of this cell for a frequency range from 1 kHz to 100 MHz. This equivalent electrical circuit makes it possible to report the observed inductive, capacitive and resistive phenomena. Once optimized and adjusted, it is also possible to model all impedance spectra recorded over the full SoC range (0 %–100 %, measurements made every 10 % of state of charge) with only three independent variables which are the three resistive components of the EEC (all other parameters of the inductive and capacitive components have been kept constant).

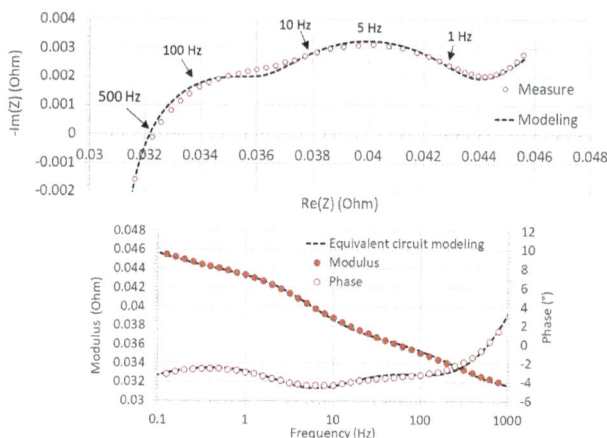

FIG. 11.17 – Example of measurement results by electrochemical impedance spectroscopy. (Top) Representation according to a Nyquist diagram. (Bottom) Representation according to a Bode diagram. The modeling of the electrical behavior by an equivalent electrical circuit is also shown in dotted line.

The first resistive component comes at high frequency (kHz range). In general, it is attributed to pure ohmic behavior (contact resistance, electronic transport in materials and ionic transport in electrolyte...). The second resistive component is easily identifiable on the Nyquist and Bode diagrams around 100 Hz. It would be attributed to charge transfer phenomenon. The third is also clearly visible at about 5 Hz. The latter could be attributed to transport phenomenon.

It is not easy to determine accurately the origin of these resistive components and even harder to identify them for sure. We will merely observe their presence and their evolution with the state of charge.

Electrical impedance spectroscopy therefore makes it possible to decompose the measured resistance value into different resistive and capacitive components that differ in their specific frequencies. Thus, in the case studied, it is easy to compare the evolution of the three identified resistive components with the state of charge of the cell and to appreciate their relative importance in the total electrical resistance of the system.

We find that the relative importance of the three resistive components is very variable (see figure 11.18). Thus, the R2 component (corresponding to a specific frequency at about 100 Hz) changes few with the state of charge. On the other hand, R3 (specific frequency at about 5 Hz) increases significantly with the state of charge. Finally, R1 (high frequency resistance) is the main resistive component of the system (75 % of the total resistance). It decreases overall with the state of charge with a minimum value at 60 % of state of charge.

In the end, the total resistive behavior takes the form of a U with higher resistance values at the extremes of state of charge (0 % and 100 %) and a minimum value between 40 % and 60 % of state of charge.

Impedance spectroscopy is an easy to implement, accurate and very powerful method for analyzing the resistive behavior of complex systems such as batteries. However, these advantages that have led to a popularization of its use, have also often led to a misuse of this method: whether through poor measurement practices (lack of consideration of basic assumptions) or data analysis (analysis by EEC that is inconsistent and/or unnecessarily complex). This can lead to serious mistakes that are often seen in the literature.

11.2.3.6 *Calorimetric Measurements*

One last method of resistance measurement with quite a different philosophy compared to the previous ones can be proposed. This method is based on measuring the heat dissipated during battery operation. The basic assumption of this method is that the current i flowing through the internal resistance of the battery called R, produces heat by the Joule effect: Ri^2. If we measure the amount of heat released during a specific period and we know the current that has been applied to the cell during this interval, it is possible to calculate an averaged internal electrical resistance value.

It is clear that this assumption cannot be verified for a complex electrochemical system such as a battery. Indeed, it only takes into account the irreversible heat produced by the system. However, chemical reactions occurring in the cell also

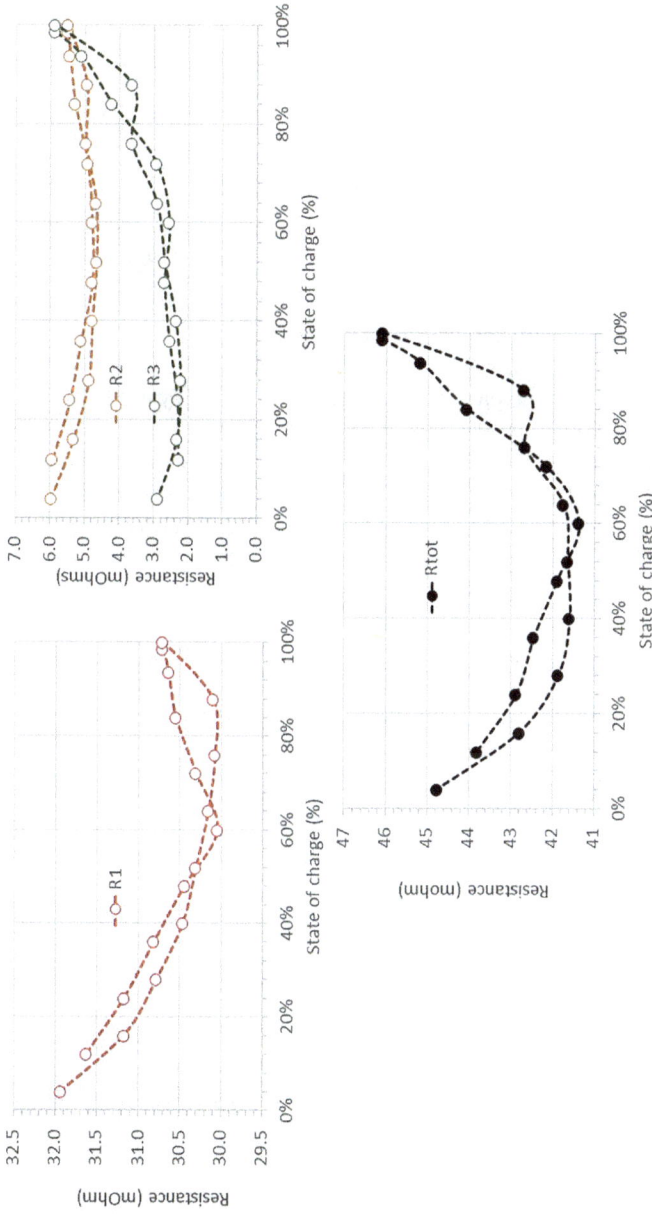

Fig. 11.18 – Evolution of the different resistive contributions identified with the state of charge of the battery. R1: resistive component of high frequency (kHz), R2: component identified around 100 Hz and R3: low frequency component identified around 5 Hz. Rtot is the sum of all the resistive components.

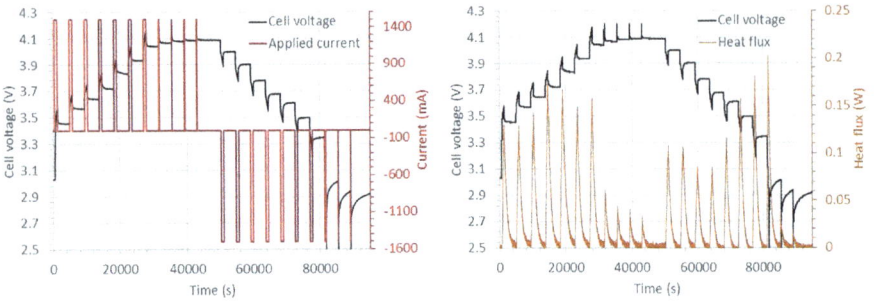

FIG. 11.19 – Example of heat flow measurements during current pulses in charge and discharge. (Left) Evolution of applied voltage and current over time. (Right) Evolution of voltage and heat flow measured as a function of time.

produce reversible heat, with exothermic reactions in one direction of operation and endothermic reactions in the other direction.

However, it is possible to limit the part of reversible phenomena by applying high current pulses over short periods.

We present here an example of results obtained with the same cell than the one used for the methods described above (figure 11.19). The heat flow measurements were carried out with a heat flow sensor wounded around the cylindrical cell.

We can see that the measured heat fluxes change with the state of charge and are not identical between charge and discharge. By integrating the measured heat fluxes, we get a quantity of heat dissipated during each pulse, thus giving the possibility to calculate an average value of electrical resistance over the duration of the pulse.

The results of these measurements are shown in figure 11.20.

We can note that the values calculated from the charge and discharge measurements differ significantly. This difference is specifically related to the reversible

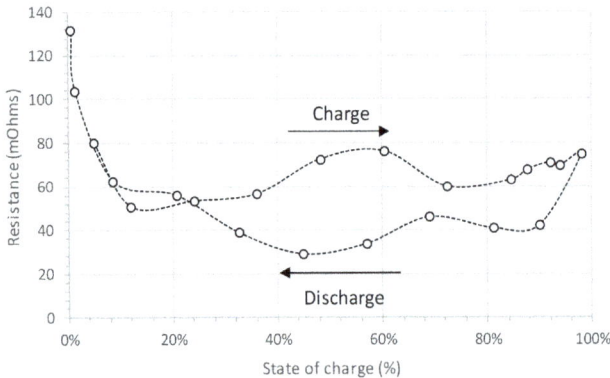

FIG. 11.20 – Evolution of the average electrical resistance calculated from heat flow measurements as a function of the state of charge of the battery.

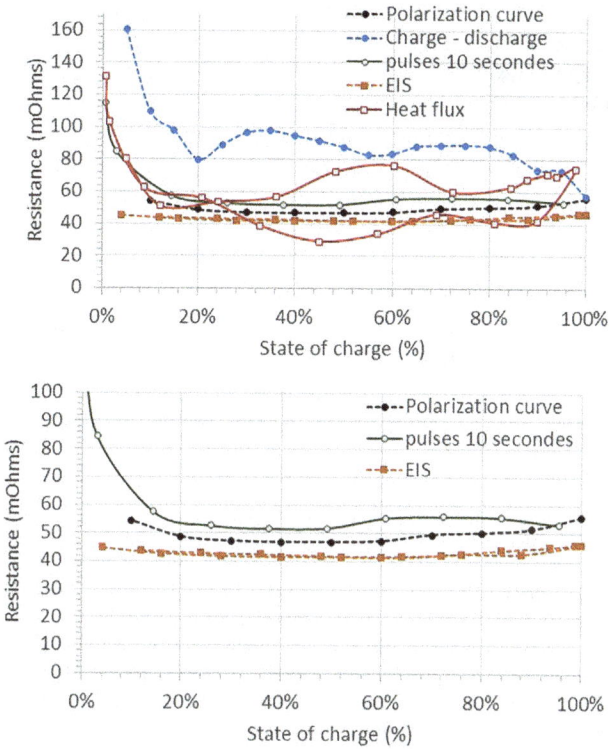

FIG. 11.21 – Evolution of the resistance values obtained by the different methods according to the state of charge of the battery. (Top) Batch of results. (Bottom) Focus on few methods.

chemical reactions that occur within the battery and which are, for this cell, generally exothermic during charging and endothermic during discharge.

However, we can note that the resistance increases strongly at low states of charge and that when we consider an average value between charge and discharge, the resistance value would be about 50 mOhms over a wide range of states of charge before increasing slightly towards the full charge.

Obviously, this measurement method, although very simple to use, does not make it possible to determine a value of the cell electrical resistance as accurately as impedance spectroscopy. However, it all depends on the aim of these measures. For example, if the purpose of these measurements is to develop an electro-thermal model of battery operation, then the calorimetric measurement method may be more relevant than the other methods presented above.

11.2.4   Conclusion

We presented several methods for measuring the internal resistance of batteries. In the light of the results presented, we can see that, for the same battery cell used with exactly the same conditions (cell placed in a climatic chamber regulated at 25 °C, operation at the same current rates), the measurement results may differ a lot, depending on the measurement methods used. Figure 11.21, on which we have represented the resistance values obtained by the different methods, illustrate this situation.

Excluding the cycling and heat flow measurement methods, the other methods give values that are quite similar. It appears that the measurement by impedance spectroscopy slightly minimizes the resistance values because we have performed measurements up to 0.1 Hz. However, as we have seen with the current pulse method, the resistance increases with the duration of the signal (which leads to an increasing resistance when the signal frequency decreases). The polarization curve and pulse methods also give very similar values. The main difference appears at low states of charge with a significant increase in calculated resistance.

These three methods can therefore be used quite easily to determine electrical resistance values. Each of them has advantages and drawbacks due to its implementation easiness.

Bibliography

[1] Thesis by Bramy Pilipili Matadi done at the CEA, "Étude des mécanismes de vieillissement des batteries Li-ion en cyclage à basse température et en stockage à haute température: compréhension des origines et modélisation du vieillissement" (completed in 12/2017).

[2] Dubarry M., Svoboda V., Hwu R., Liaw B.Y. (2006) Incremental capacity analysis and close-to-equilibrium OCV measurements to quantify capacity fade in commercial rechargeable lithium batteries, *Electrochem. Solid State Lett.* **9**, A454.DOI: https://doi.org/10.1149/1. 2221767.

[3] Dubarry M., Truchot C., Liaw B.Y. (2012) Synthesize battery degradation modes via a diagnostic and prognostic model, *J. Power Sources* **219**, 204. DOI: https://doi.org/10.1016/j. jpowsour.2012.07.016.

[4] Devie A., Dubarry M., Liaw B.Y. (2015) Overcharge study in $Li_4Ti_5O_{12}$ based lithium-ion pouch cell I. Quantitative diagnosis of degradation modes, *J. Electrochem. Soc.* **162**, A1033. DOI: https://doi.org/10.1149/2.0941506jes.

[5] Safari M. and Delacourt C. (2011) Ageing of a commercial graphite/LiFePO$_4$ cell, *J. Electrochem. Soc.* **158**, A1123.DOI: https://doi.org/10.1149/1.3614529.

[6] Han X., Ouyang M., Lu L., Li J., Zheng Y., Li Z. (2014) A comparative study of commercial lithium ion battery cycle life in electrical vehicle: Ageing mechanism identification, *J. Power Sources* **251**, 38. DOI: https://doi.org/10.1016/j.jpowsour.2013.11.029.

[7] Torai S., Nakagomi M., Yoshitake S., Yamaguchi S., Oyama N. (2016) State-of-health estimation of LiFePO$_4$/graphite batteries based on a model using differential capacity, *J. Power Sources* **306**, 62. DOI: https://doi.org/10.1016/j.jpowsour.2015.11.070.

[8] Ma Z., Jiang J., Shi W., Zhang W., Mi C.C. (2015) Investigation of path dependence in commercial lithium-ion cells for pure electric bus applications: ageing mechanism identification, *J. Power Sources* **274**, 29. DOI: https://doi.org/10.1016/j.jpowsour.2014.10. 006.

[9] Kassem M., Bernard J., Revel R., Pélissier S., Duclaud F., Delacourt C. (2012) Calendar ageing of a graphite/LiFePO$_4$ cell, *J. Power Sources* **208**, 296. DOI: https://doi.org/10.1016/j. jpowsour.2012.02.068.

[10] Sarasketa-Zabala E., Gandiaga I., Rodriguez-Martinez L.M., Villarreal I. (2014) Calendar ageing analysis of a LiFePO$_4$/graphite cell with dynamic model validations: Towards realistic lifetime predictions, *J. Power Sources* **272**, 45e57. DOI: https://doi.org/10.1016/j.jpowsour. 2014.08.051.

Chapter 12

Microstructural and Physical and Chemical Characterizations of Battery Materials

Sylvie Genies, Adrien Boulineau, Anass Benayad, Claude Chabrol, Jean-Frédéric Martin, David Brun-Buisson, Xavier Fleury, Lise Daniel, Jean-François Colin, Michel Bardet, Sandrine Lyonnard, Samul Tardif, Florence Lefebvre-Joud and Eric de Vito

The purpose of this chapter is to present some examples of characterization techniques commonly used to identify degradation mechanisms associated with performance losses of a lithium-ion cell. They provide information on the microstructure of electrode materials and the physicochemical nature of the interfaces, and allow highlighting changes before and after use.

12.1 Introduction: Characterization Methodology to Understand the Electrochemical Response of a Battery

Battery components are complex systems consisting of components of very different natures (liquid, solid, inorganic, organic, and metallic). Their characteristics at all stages of their manufacture aim at ensuring optimal performance in the fresh battery. Before as well as after use, the characterization of internal components helps to understand the mechanisms of degradation and to identify remedies. This characteristization is necessarily multi-scale, like the phenomena controlling the electrochemical response of a battery. It is performed by extracting the components of the electrochemical cell after opening and is divided into ante mortem (new cell, not yet tested) and post-mortem studies (tests are in this case performed on aged

DOI: 10.1051/978-2-7598-2555-4.c012

cells with degraded behaviour). However, in recent years, characteristization in *operando* mode have allowed monitoring the reversible and non-reversible microstructural transformations which occur inside the cells during cycling.

The performance of a Li-ion battery depends primarily on the ability of the electrodes to reversibly exchange lithium ions during the largest possible number of charge/discharge cycles. It is therefore necessary, on one hand, to know how to characterize the amount of lithium exchanged during a charge/discharge cycle, but also to identify all phenomena that might limit this exchange. This can *e.g.* be a mechanical blocking, such as contraction or expansion of electrode materials during charge/discharge cycles, which can lead to disconnection of the active material particles and thus to a loss of electrical contact, or obstruction of the separator porosity. Changes in the crystalline structure of the surface or volume electrode materials may also occur and block the insertion of lithium ions. Finally, degradation of electrolyte solvents can result in the formation of gas bubbles that locally mask the active material and can cause electrical insulation that promotes the deposit of metallic lithium. In particular, the understanding of the formation and evolution of the solid interface between the negative electrode and the electrolyte ("SEI" Solid Electrolyte Interphase, solid products generated by the electrochemical decomposition of the liquid electrolyte on the electrode/electrolyte interface) is essential in order to understand the performance and improve the life of a Li-ion cell.

During operation, the degradation mechanisms within a battery are numerous, as illustrated in figure 12.1. They occur either within the active materials of positive and negative electrodes, at the electrolyte interface (SEI), or in the electrolyte or at the separator level.

A set of methods of electrochemical, physicochemical and structural characterization of all these materials and components (positive and negative electrodes, separator, and electrolyte) must therefore be implemented at different scales to identify precisely the mechanisms that are involved and that cause degradation.

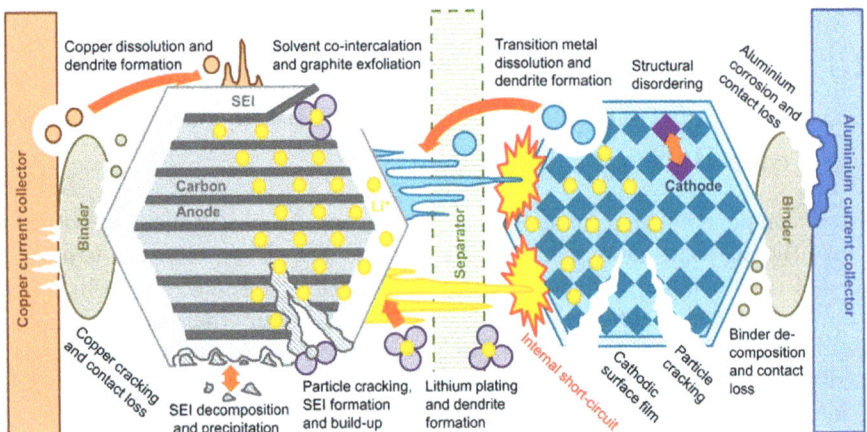

FIG. 12.1 – Degradation phenomena within a Li-ion cell [1].

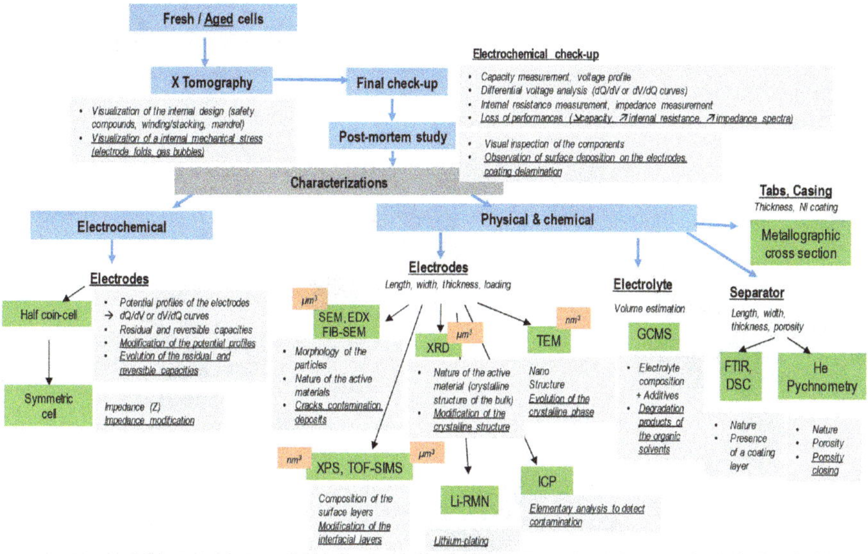

FIG. 12.2 – Diagram of the physicochemical characterization techniques associated with the internal components (see abbreviations in the glossary at the end of the chapter). The potentially expected changes after aging are highlighted [2].

Figure 12.2 illustrates the different techniques of characterization that are being studied for each of the core or surface components.

It is the complementarity of the information obtained with these techniques that helps understand the origin of the degradation and the resulting performance losses.

These techniques are generally used in post-mortem studies. The preparation of the samples requires the dismantling of a Li-ion cell in a glove box in an inert atmosphere (with argon to obtain levels of H_2O and O_2 less than 1 ppm). Thus, the internal components are not in contact with air, "freezing" the cell in the state it was during its use. The cell is always dismantled at a fully discharged state, on the one hand for safety reasons, and on the other hand, to characterize the cells at an identical state of charge between fresh and aged electrodes.

The microstructural and physicochemical characterization of a battery is most often coupled with an electrochemical characterization to control the lithiation state of each electrode and to measure their insertion capacity. Several configurations are possible, in either a complete coin cell [3], a half-coin cell [4] or a symmetrical coin cell [5]. For example, to measure residual capacity (initial capacity in lithiation for the positive electrode and in delithiation for the negative electrode) and total reversible capacity of an electrode, half-coin cells are reconstructed from the dismantled components as shown in figure 12.3. Samples are taken from the negative and positive electrodes of a cell, in the present case, Graphite/$LiNi_{0.80}Co_{0.15}Al_{0.05}O_2$ (NCA). They are cut as a pastille and then assembled in a half-coin cell facing

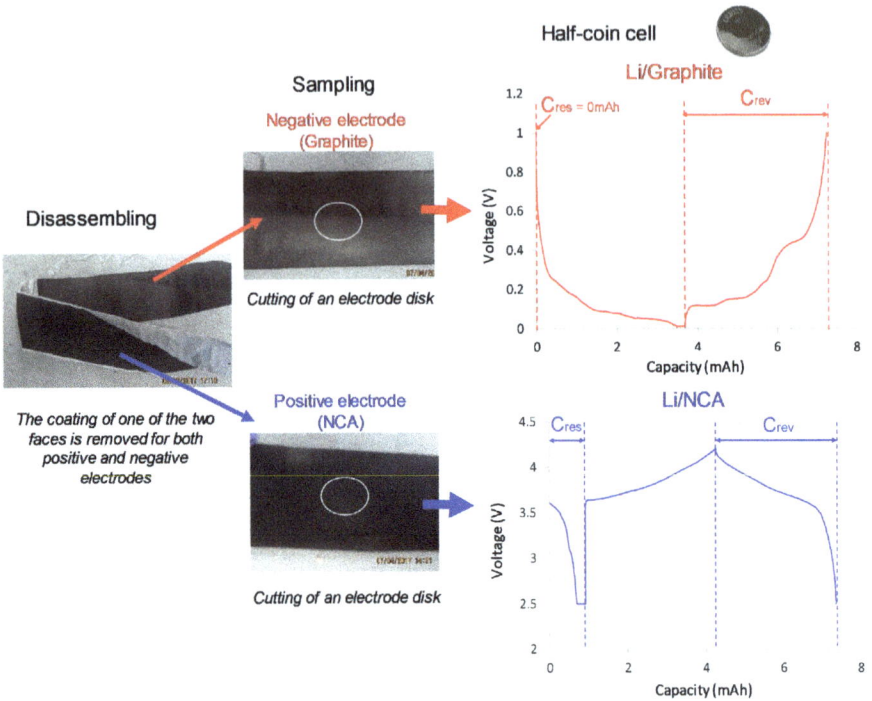

FIG. 12.3 – Electrochemical test in half-coin cell for measuring, after aging, the reversible capacity and the residual capacity of the electrodes of a Graphite/NCA cell [6].

a negative lithium metal electrode. Electrochemical tests are performed on the cells to obtain the residual and reversible capacities of each electrode. In this example, the residual capacity of the NCA/lithium button half-cell measured after ageing can be compared with the residual capacity obtained in the fresh state. The difference in capacity can be directly related to the loss of exchangeable capacity, due to irreversible consumption of lithium ions. Thus, the capacity loss observed with the complete cell can at least partially be related with *e.g.* a loss of exchangeable lithium ions consumed by parasitic reactions; the growth/reformation of the surface interfacial layer of graphite particles (SEI); the deposit of metallic lithium on the same electrode under given operating conditions; or with the loss of electronic connection of lithiated particles. Reversible capacity represents the complete electrode insertion capacity that may have been degraded after operation (*e.g.* electrical disconnection of lithiated or delithiated particles, degradation of insertion structures). Depending on the extent of these degradations, this may have had an impact on the capacity of the entire cell (in capacity or power at high current).

Once the performance loss has been quantified, the origin of the performance loss should be investigated by implementing internal component-level characterizations.

12.2 Analysis of Mechanisms Associated with Exchangeable Lithium Loss

12.2.1 SEI Formation and Li Metal Precipitation on Negative Electrode

Scanning Electron Microscopy (SEM) [7] is a topographic and morphological observation technique of a surface material. The micrographs in figure 12.4 show the constituent particles of a graphite anode [8]. The electrode in the fresh state is shown on the left image. The presence of spherical morphological graphite particles and carbon fibers is well visible. The center image corresponds to the same electrode extracted from a cell after prolonged cycling (1300 cycles at 45 °C − 184 days of test). A "coating", particularly visible on carbon fibers, appears, which can be associated with the growth of the SEI. The image on the right corresponds to the same graphite electrode of a cell after cycling, the charge process being applied according to a specific protocol. In the latter case, the presence of filaments entangled between the particles and filling the space between the particles, characteristic of the formation of a lithium metal deposit (Li-plating), is observed.

The X-ray photoelectron spectroscopy (XPS) is an extreme surface technique and is consequently one of the most direct investigation tool of the chemical and electronic structure of materials. It consists of the analysis of photoelectrons emitted

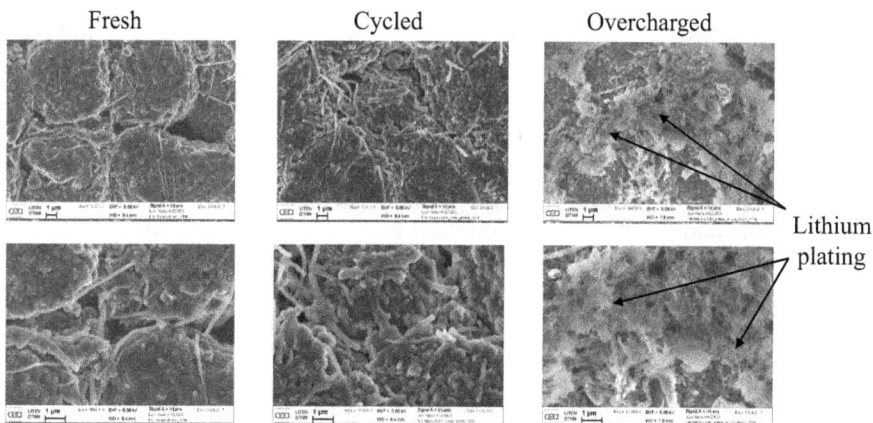

Fresh   Cycled   Overcharged

Lithium plating

FIG. 12.4 − SEM images of a fresh (left), cycled (center) and overcharged (right) graphite electrode [8]. Comparing the different images, after cycling, the growth of the SEI, and after overcharge, a Li metal deposit (images obtained using a under vacuum "transfer holder" between the glove box and the microscope chamber) is observed.

by a sample exposed to X-ray [9–11]. Given the low emission depth of the electrons analyzed (~ 50 Å), this technique is particularly well appropriate to the study of the SEI.

XPS has the advantage of being able to probe most of the core peaks from almost all the atoms that constitute the SEI, except hydrogen atoms. By precisely analyzing the energy shifts of these peaks, it is also possible to track changes in the chemical states of the elements during the cycles. Finally, the intensity of XPS peaks is directly proportional to concentration, allowing quantitative studies (with uncertainties of 10%–20%). Thus, the technique allows tracking changes in the chemical states of the SEI during cycles. Figure 12.5 illustrates the evolution of the SEI on the surface of a silicon electrode during cycling. The primary peaks of the different spectra allow determining the nature and evolution of the SEI and the SEI/Si interface (including Li_xSiO_y detection). More specifically, the evolution of the Si pattern (which is the active material of the electrode) makes it possible to better understand the SEI's thickness increase observed during cycling: after 100 cycles, silicon is no longer detected, indicating a thickness increase of the SEI compared to the early cycling stage. Associated with electrochemical characterizations (electrochemical impedance spectroscopy (EIS), voltamperometry, etc.) XPS spectroscopy finally correlates the losses of electrochemical capacity with changes in the SEI composition.

XPS analysis is therefore part of a family of surface spectroscopic analysis techniques (the depth analyzed can be between 1 and 10 nm), like Auger spectroscopy and ToF-SIMS spectroscopy. The coupling of these techniques allows obtaining morphological, chemical and structural information of the SEI [12].

The use of transmission electron microscopy (TEM) [13] allows observing the growth of the SEI on the electrode and the crystalline structure of the graphite. In figure 12.6, high-resolution MET pictures of graphite particles from fresh and cycled electrodes first illustrate the evolution of the SEI. On the surface of the fresh graphite, a thin amorphous layer of about 3 nm thickness is visible and may be associated with the SEI. The picture is characteristic of a very well crystallized graphite, consisting of large single crystalline sheets or graphene planes. After prolonged cycling, the graphite surface offers a very different appearance, with a stacking of graphite layers and SEI layers characterized by the risk of exfoliation of the edge of graphite sheets. This well-known phenomenon [14] is caused by the co-intercalation of electrolyte solvents when lithium ions are inserted and by the emission of gases when these solvents are decomposed between graphene planes. The associated schemas make it easy to read TEM images. If the presence of a thin layer of SEI is beneficial because it limits this co-intercalation of solvents when inserting lithium [14], repeated dilatations/contractions of the crystalline structure through charge/discharge cycles cause cracking in the SEI which local reconstruction consumes lithium ions. This phenomenon is associated with an irreversible loss of capacity of the electrode.

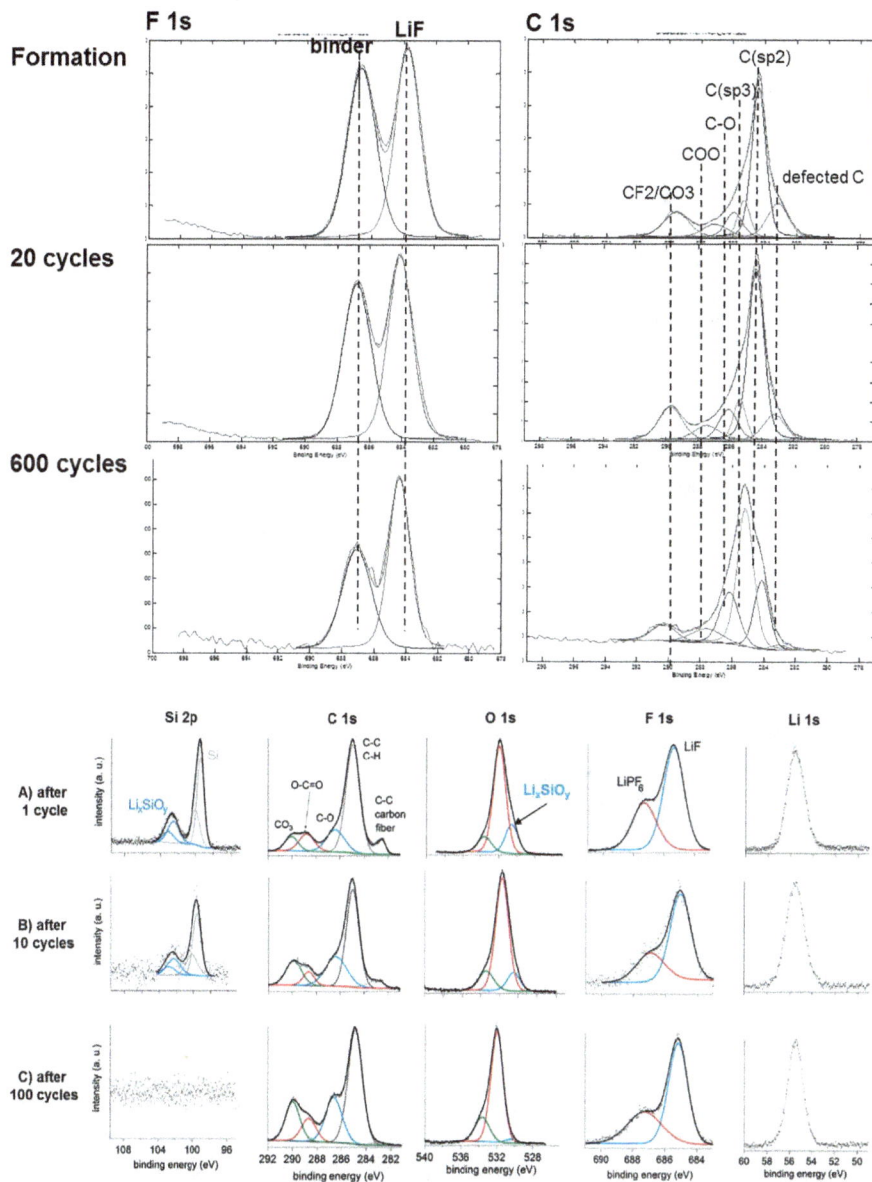

FIG. 12.5 – XPS spectra (Si 2p, C 1s, O 1s, F 1s, and Li 1s) on the Si electrode surface: (A) fresh electrode, (B) after 10 cycles, and (C) after 100 cycles. The three electrodes are analyzed at delithiated state [12].

Fresh electrode after formation   Cycled electrode

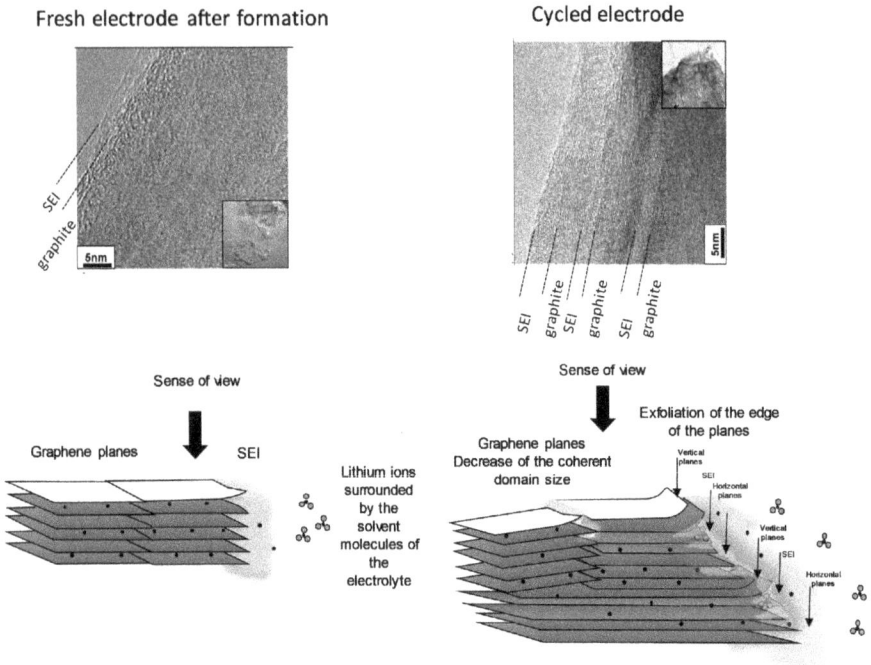

FIG. 12.6 – Images by TEM- transmission electron microscopy – high resolution of a graphite particle from a fresh electrode, immediately after formation and after cycling [15]. A succession of graphite and SEI layers characteristic of exfoliation is observed after cycling.

12.2.2 Loss of Lithium Content of Positive Electrode

Positive electrodes are often made of a lamellar material, in which lithium is inserted and disinserted into a solid solution. This is notably the case for a material such as $LiNi_xMn_yCo_zO_2$ (NMC), which has a rhombohedral structure of R-$3m$ space group and has its crystalline parameters evolving according to its lithiation state. The X-ray diffraction technique allows measuring these crystalline parameters and monitoring the electrode chemical composition during aging.

X-ray diffraction analysis (or XRD) [16] characterizes the crystalline structure of a material, the symmetries (space group) and the dimensions of its crystalline structure. With the diffractogram pattern obtained, compounds are identified from the position and relative intensity of diffraction peaks using a database of all crystalline material diffractograms. The analysis may be performed on powders or samples taken from fresh or cycled electrodes. In the latter case, the samples are prepared within a glove box and protected by a Kapton® film to avoid any reaction of the compounds with air or humidity during the analysis.

Figure 12.7 shows the diffractograms of an NMC442 electrode ($LiNi_{0.4}Mn_{0.4-}$ $Co_{0.2}O_2$) in the fresh state and after a period of calendar aging at a 100% state of charge for 15 months at 5 °C, 9 months at 25 °C and 45 °C and 4 months at 60 °C. The cell state of health (SOH) after storage period is 93% at 45 °C and 82% at 60 °C. No loss of capacity is observed after aging at 5 °C and 25 °C (SOH 100%) [17].

The *R-3m* structure is preserved after aging, but there is an evolution of the 003 diffraction line characteristic of a change in the crystalline parameters of the material. The spectra are used to quantify the variations in crystalline parameters *a* and *c*.

The coupling of X-ray diffraction *in situ* with an electrochemical test [18] allows the acquisition of diffractograms of the material at various known states of lithiation and thus a calibration curve linking the value of the crystalline parameters *a* and *c* to the state of lithiation of the material (figure 12.8).

Thus, according to the values in table 12.1, the decrease in the crystalline parameter *a* and the increase in the crystalline parameter *c* indicate a more pronounced loss of Li as the calendar aging is carried out at higher temperatures. The loss of Li is determined by the above calibration curve and can be confirmed by the electrochemical residual capacity measured in coin cell.

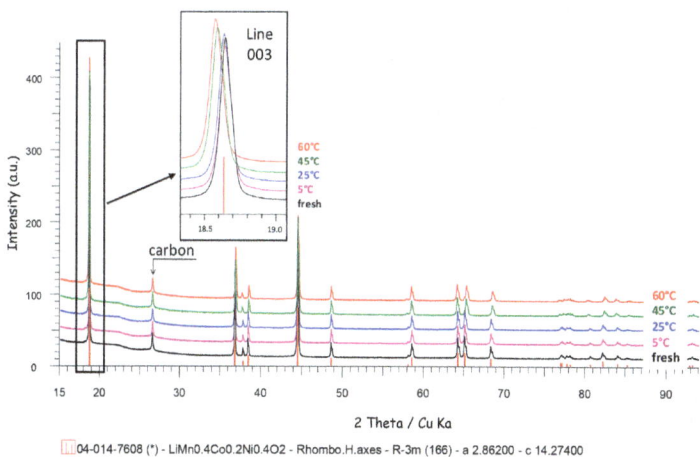

04-014-7608 (*) - LiMn0.4Co0.2Ni0.4O2 - Rhombo.H.axes - R-3m (166) - a 2.86200 - c 14.27400

FIG. 12.7 – Diffractogram of a fresh NMC442 positive electrode and after calendar aging at 100% of state of charge for 15 months at 5 °C, 9 months at 25 °C and 45 °C, and 4 months at 60 °C. A change in the 003 line, characteristic of a change in the crystalline parameters of the material, is observed.

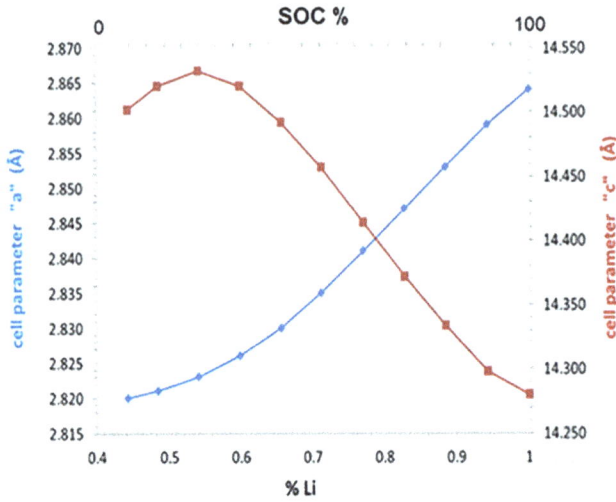

FIG. 12.8 – Calibration curve obtained by *in-situ* X-ray diffraction for a NMC442 material [19].

TAB. 12.1 – Crystalline parameters of the *R-3m* phase of the NMC442 material and associated Li content.

	a (Å)	c (Å)	Li content
Fresh state	2.862 Å	14.274 Å	0.98
5 °C	2.862 Å	14.274 Å	0.98
25 °C	2.861 Å	14.280 Å	0.96
45 °C	2.856 Å	14.314 Å	0.92
60 °C	2.854 Å	14.325 Å	0.90

12.3 Analysis of Phase Transformations that Limit Lithium Mobility

12.3.1 Microstructural Modification of a Positive Electrode

The perfect lamellar layers structure of a fresh NMC positive electrode ($LiNi_{1/3}$ $Mn_{1/3}Co_{1/3}O_2$) is presented on the TEM high-resolution image in figure 12.9. The images made on samples after cycling show that on the surface of the primary particles, the lamellar structure is disturbed, with some transition metal cations moving into the Li layer (presence of visible atoms in light gray between the alignments of white atoms as schematized).

Fresh   Cycled

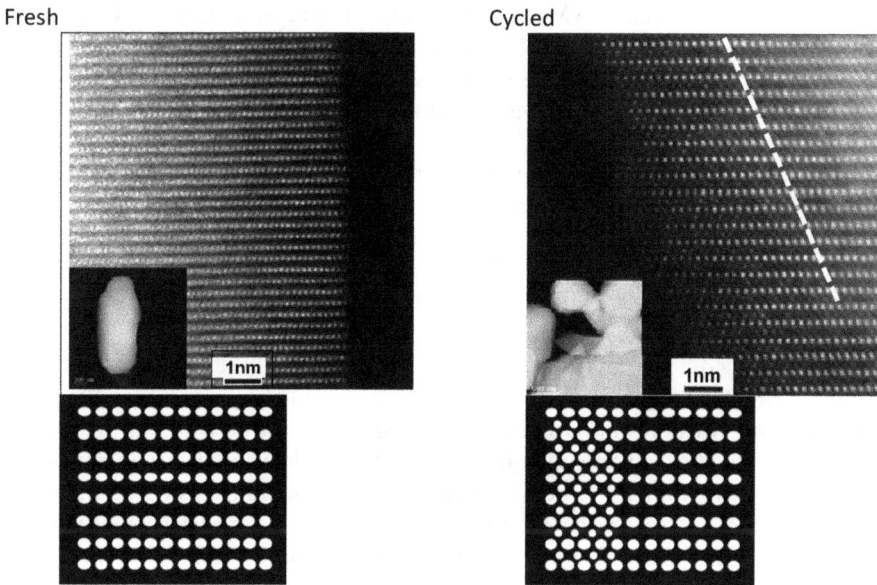

FIG. 12.9 – High resolution TEM images of a NMC particle from a fresh electrode (left figure) and after cycling (right figure) with associated schema to help understanding TEM images [20].

The presence of these cations can hinder, or limit, Li diffusion during the charge and discharge processes of the cell. This structural change thus contributes to the increased resistance of the positive electrode and to the loss of cell capacity.

12.4   Mechanical Blocking, Obstruction, Disconnection and Loss of Electrical Contact

12.4.1   Loss of Graphite Electrode Capacity in Cycling at Low Temperatures

ToF-SIMS spectrometry is also an essential tool for understanding the mechanisms of lithiation and degradation of electrodes. In the following example, it is associated with an *in situ* preparation technique (in the spectrometer chamber itself) to observe the chemical nature in the depth of the electrode itself. It is thus possible to confirm the presence of lithium ions in graphite particles, even after a complete electro-chemical delithiation, thus showing that lithiated particles are electronically disconnected and that the inserted lithium is no longer usable.

In this study, Graphite/NMC commercial cells are aged under specific conditions: electrochemical cycling at 5 °C between 0% and 100% SOC. Performance collapses after 50 cycles, while the same systems can run more than 4000 cycles at 45 °C [21]. Some cells are also cycled in a smaller state of charge range: from 10% to 90% SOC.

ToF-SIMS analyses are associated with the realization of a cross-section by FIB (Focused Ion Beam) in the spectrometer analysis chamber, allowing the distribution of different elements, in particular Li and F (SEI marker) in the electrode depth (figure 12.10). The results confirm the entrapment of lithium in graphite particles during cycling between 0% and 100% SOC, whereas an excessive presence of SEI within the electrode studied around active material particles is put in evidence.

These results differ drastically from those obtained for electrodes cycled between 10% and 90% SOC, for which lithium is not detected in the particles (in accordance with XRD and NMR results), and SEI (element F) is not detected in the particle.

These results show an excessive generation of SEI when the electrochemical cell is cycled between 0% and 100% SOC, associated with a significant degradation of the electrolyte (also demonstrated by XPS spectroscopy). This SEI promotes disconnection of graphite particles from the percolating network, trapping the intercalated lithium.

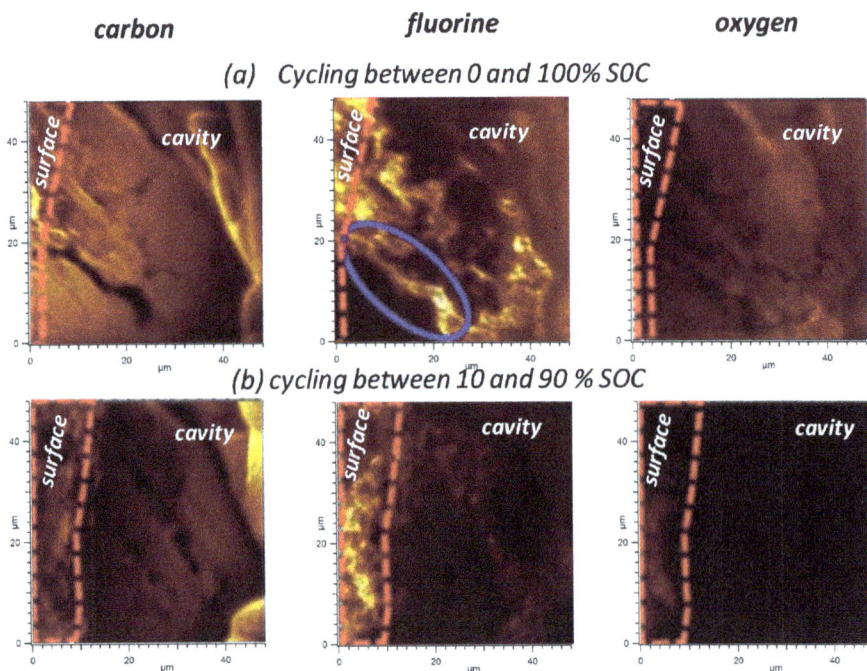

FIG. 12.10 – FIB-ToF-SIMS image of graphite electrodes. (a) Cycling between 0% and 100% SOC; and (b) cycling between 10% and 90% SOC. In the latter case, the detection of lithium associated with graphite particles, as well as that of SEI (fluorine mapping) in interstices between particles [22].

12.4.2 Exogenous Deposits

The X-microanalysis or EDX (Energy Dispersive X-ray spectrometry) [23] is particularly useful for detecting the presence of exogenous deposits. An X-ray detector associated with a scanning electron microscope provides access to the elementary microanalysis of the sample. An X-ray spectrum is obtained for the identification of chemical elements. A quantification of these elements allows determining the chemical composition of the compounds that are present.

The volume and thickness analyzed, or the ionization pear, depends on the energy of the beam and the density of the material. This volume is in the order of a few μm^3 for electrode materials.

Figure 12.11 shows the SEM image of a negative electrode with significantly degraded performance. It is composed of graphite particles of parallelepiped shape (1) with other smaller particles of different shapes (2) visible at higher magnification on the right image. Elemental EDX analysis of these particles (spot 2) shows that these particles contain copper.

By EDX mapping, not presented here, it is possible to assess the distribution of copper deposits. The copper deposit is found to be high in oxygen, which tends to show that the cell has been overdischarged. The potential of the negative electrode over-discharge causes oxidation of the copper collector and the release of copper ions into the electrolyte. During recharge, copper ions reduce to the surface of graphite particles, forming deposits that block the insertion of Li [25].

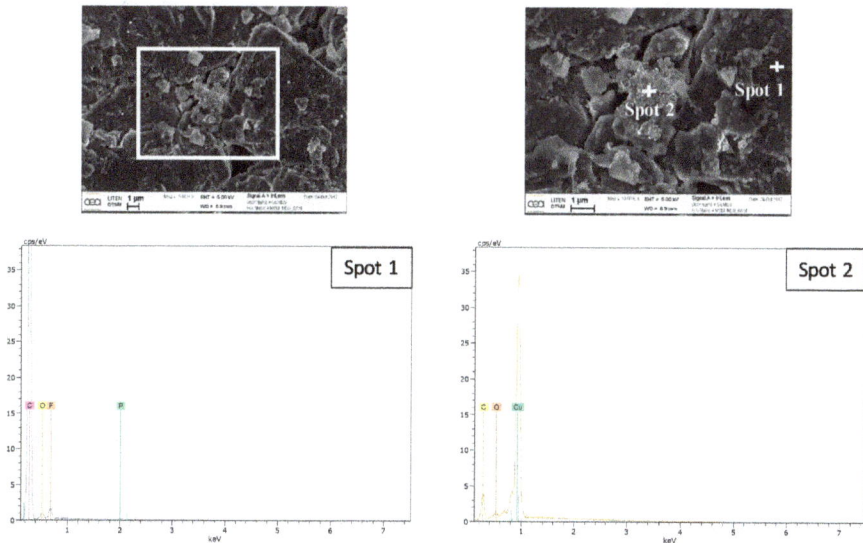

FIG. 12.11 – SEM images of a negative graphite electrode and local elementary analyzes by EDX [24].

12.5   Electrolyte Degradation

The use of gas chromatography coupled with mass spectrometry (GC–MS) [26] is particularly well suited to the study of the solvents that constitute the electrolyte. It is a technique for analyzing and quantifying volatile and semi-volatile organic compounds. It allows separating the solvents from a solution by means of a temperature controlled capillary column. Small molecules, which have low boiling temperatures, descend along the column faster than large molecules, which have higher boiling temperatures. Mass spectrometry (MS) allows the identification of the various components after separation through their mass spectrum. Their identification is possible by an interpretation of the mass spectrum or by its comparison with databases.

The evolution of the electrolyte of a Li-ion cell in the fresh state and at different stages of calendar aging at 100% SOC and 25 °C or 45 °C for 9 months, and 60 °C for 4 months (figure 12.12) is thus possible.

The electrolyte consists of two main solvents: Ethylene carbonate (EC) and methyl ethyl carbonate (EMC) and four additives: Biphenyl (BP), vinyl carbonate (VC), fluoroethylene carbonate (FEC) and 1,3-propane sultone (1,3-PS).

After a calendar aging at 25 °C, no change in the composition of the electrolyte is detected. However, at a temperature of 45 °C, dimethyl carbonate (DMC) and diethyl carbonate (DEC) are observed in relation to the decomposition of EMC. Compared with EC, biphenyl is decreased. At 60 °C, these same decomposition compounds are present, in addition to EC, producing three additional compounds.

When disassembling the two cells stored at 45 °C and 60 °C, visual inspection shows the presence of gas bubbles between electrodes, associated with the decomposition of biphenyl. This compound is indeed unstable under the storage conditions applied (100% SOC and temperature of 45 °C and above).

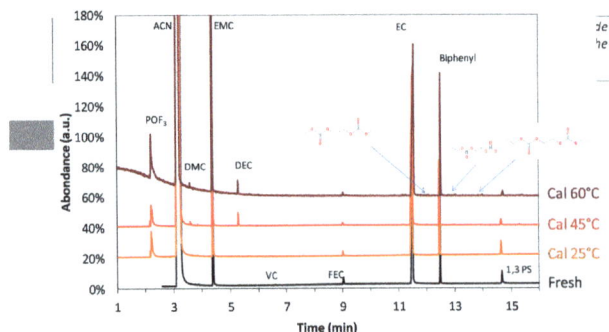

FIG. 12.12 – GC–MS electrolyte analyzes for a fresh and aged cell in 100% SOC calendar mode according to three temperatures: 25 °C, 45 °C and 60 °C [27]. The *x*-axis corresponds to the retention time of the solvents in the column.

12.6 Perspectives

Despite the development of *in-lab ex situ* analytical techniques, the *in situ/operando* characterization is used more and more often because it allows probing the dynamic chemical, structural and morphological changes in the battery during cycling. Its development remains a challenge, but represents a growing interest. The *operando* characterization for battery technology is strongly linked to cell development that allows physico-chemical characterization under an electrochemical stimulus. Significant progress has been made on electrochemical cells deve loped for optical microscope [28–30], *in situ* Raman [31, 32], *in situ* FTIR [33, 34], *operando* XRD [35–37]. In an attempt to study the dendrite growth, electrochemical cells using a dedicated sample holder have been developed recently to perform *in situ* scanning electron microscopy (SEM) and transmission electron microscope (TEM) [38–48]. Recent developments were focused on the electrical and mechanical properties of SEI evolution assessement in dynamic mode, such as scanning probe based techniques, atomic force microscopy (AFM), scanning electrochemical strain microscopy (CESM), and scanning ion conductance microscopy (SICM). These techniques open a new way to study the local properties of the SEI and their temporal evolution [49–52]. In the case of X-ray photoelectron spectroscopy (XPS), the progress of *operando* analyses faced complex technological issues such as depth analysis, and ultra-high vacuum condition. Some other improvements targeted "all solid state battery" or low vapor pressure electrolytes such as ionic liquids [53–55]. *Operando* NMR has contributed to investigate lithium metal deposition on graphites, paving the way for new perspectives regarding the understanding of the mechanisms of lithium diffusion inside the electrolyte and the active electrode [56].

Synchrotron radiation facilities have been used recently, through X-ray tomographic microscopy to observe the expansion of the silicon particles for example [57]. This approach allows the use of commercial battery which is a significant advantage to perform operando analyses.

Operando techniques represent a new perspective to probe the phenomena at the interface between electrolyte and electrodes; however, the realistic electrochemical response of the cell designed for operando purpose remains a drawback for accurate *operando* characterization.

Bibliography

[1] Birkl C.R., Roberts M.R., McTurk E., Bruce P.G., Howey D.A. (2017) Degradation diagnostics for lithium ion cells, *J. Power Sources* **341**, 373.
[2] Source CEA.
[3] Full coin-cell: system associating a negative electrode and a positive electrode as in the real configuration of a commercial cell. The amount of exchangeable lithium is contained in one or the other of the electrodes and conditions the capacity of the cell.
[4] Half coin cell: system associating an electrode facing a metallic lithium electrode. This infinite lithium source makes it possible to characterize the electrode over its entire operating range.

[5] Symmetrical coin cell: pairing of two same electrodes (either positive or negative) in the same lithiation state in order to characterize by electrochemical impedance spectroscopy. For the two other assemblies indicated, the impedance spectrum will include the contribution of the two positive and negative electrodes and would therefore complicate its analysis.

[6] Source CEA.

[7] "Microscopie électronique à balayage. Principe et équipement", par J. RUSTE, Techniques de l'ingénieur, Réf.: P865 V3.

[8] Fleury X., Noh M.H., Geniès S., Thivel P.X., Lefrou C., Bultel Y. (2018) Fast-charging of lithium iron phosphate battery with ohmic-drop compen- sation method: ageing study, *J. Energy Storage* **16**, 21.

[9] "Spectroscopies de photoélectrons: XPS ou ESCA et UPS", par G. HOLLINGER, Techniques de l'ingénieur, Réf.: P2625 V1.

[10] Madey T.E., Wagner C.D., Joshi A. (1977) Surface characterization of catalysts using electron spectroscopies: results of a round-robin sponsored by ASTM committee D-32 on catalysts, *J. Electron. Spectrosc.* **10**, 359.

[11] (a) Briggs D., Seah M.P., Ed. (1983) *Practical surface analysis*, Chap. 8, Wiley, New York; (b) Barr T.L. (1983) An XPS study of Si as it occurs in adsorbents, catalysts, and thin film, *Appl. Surf. Sci.* **15**, 1.

[12] Radvanyi E., Porcher W., De Vito E., Montani A., Franger S., Jouanneau Si Larbi S. (2014) Failure mechanisms of nano-silicon anodes upon cycling: an electrode porosity evolution model, *Phys. Chem. Chem. Phys.* **16**, 17142.

[13] "Étude des métaux par microscopie électronique en transmission (MET) - Microscope, échantillons et diffraction" par M. KARLÍK et B. JOUFFREY, Techniques de l'Ingénieur, Réf.: M4134 V1.

[14] Winter L., Besenhard J.O., Spahr M.E., Novak P. (1998) Insertion electrode materials for rechargeable lithium batteries, *Adv. Mater.* **10**, 725.

[15] Source CEA.

[16] "Caractérisation de solides cristallisés par diffraction X" par N. BROLL, Techniques de l'Ingénieur, Réf.: P1080 V2.

[17] Iturrondobeitia A., Aguesse F., Genies S., Waldmann T., Kasper M., Ghanbari N., Wohlfahrt-Mehrens M., Bekaert E. (2017) Post-mortem analysis of calendar aged 16 Ah NMC/graphite pouch cells for EV application, *J. Phys. Chem. C* **121**, 21865.

[18] Morcrette M., Chabre Y., Vaughan G., Amatucci G., Leriche J.-B., Patoux S., Masquelier C., Tarascon J.-M. (2002) In situ X-ray diffraction techniques as a powerful tool to study battery electrode materials, *Electrochemica Acta* **47**, 3137.

[19] Source CEA.

[20] Source CEA.

[21] Pilipili Matadi B., Geniès S., Delaille A., Chabrol C., de Vito E., Bardet M., Martin J.-F., Daniel L., Bultel Y. (2017) Irreversible capacity loss of Li-ion batteries cycled at low temperature due to an untypical layer hindering Li diffusion into graphite electrode, *J. Electrochem. Soc.* **164**, A2374.

[22] Source CEA.

[23] "Microscopie électronique à balayage. Principe et équipement" par J. RUSTE, Techniques de l'ingénieur, Réf.: P865 V3.

[24] Source CEA.

[25] He H., Liu Y., Liu Q., Li Z., Xu F., Dun C., Ren Y., Wang M.-X., Xie J. (2013) Failure investigation of $LiFePO_4$ cells in over-discharge conditions, *J. Electrochem. Soc.* **160**, A793.

[26] "Couplages chromatographiques avec la spectrométrie de masse. I et II" par P. ARPINO, Technique de l'Ingénieur, Réf.: P1490 V1 et P1491 V1.

[27] Source CEA.

[28] (a) Brissot C., Rosso M., Chazalviel J.N., Lascaud S. (1999) Dendritic growth mechanisms in lithium/polymer cells, *J. Power Sources* **81–82**, 925; (b) Howlett P.C., MacFarlane D.R., Hollenkamp A.F. (2003) A sealed optical cell for the study of lithium-electrode|electrolyte interfaces, *J. Power Sources* **114**, 277.

[29]   Sano H., Sakaebe H., Matsumoto H. (2011) Observation of electrodeposited lithium by optical microscope in room temperature ionic liquid-based electrolyte, *J. Power Sources* **196**, 6663.

[30]   Nishida T., Nishikawa K., Rosso M., Fukunaka Y. (2013) Optical observation of Li dendrite growth in ionic liquid, *Electrochim. Acta* **100**, 333.

[31]   Stancovski V., Badilescu S (2014) In situ Raman spectroscopic-electrochemical studies of lithium-ion battery materials: a historical overview, *J. Appl. Electrochem.* **44**, 23.

[32]   Cabo-Fernandez L., Mueller F., Passerini S., Hardwick L.J. (2016) In situ Raman spectroscopy of carbon-coated $ZnFe_2O_4$ anode material in Li-ion batteries - investigation of SEI growth, *Chem. Commun.* **52**, 3970.

[33]   Sharabi R., Markevich E., Borgel V., Salitra G., Aurbach D., Semrau G., Schmidt M.A. (2010) In situ FTIR spectroscopy study of $Li/LiNi_{0.8}Co_{0.15}Al_{0.05}O_2$ Cells with ionic liquid-based electrolytes in overcharge condition, *Electrochem. Solid State Lett.* **13**, A32.

[34]   Akita Y., Segawa M., Munakata H., Kanamura K. (2013) In-situ fourier transform infrared spectroscopic analysis on dynamic behavior of electrolyte solution on $LiFePO_4$ Cathode, *J. Power Sources* **239**, 175.

[35]   Wilson B.E., Smyrl W.H., Stein A. (2014) Design of a low-cost electrochemical cell for in situ XRD analysis of electrode materials, *J. Electrochem. Soc.* **161**, 700.

[36]   Brant W.R., Schmid S., Du G., Gu Q., Sharma N.A (2013) Simple electrochemical cell for in-situ fundamental structural analysis using synchrotron X-Ray powder diffraction, *J. Power Sources* **244**, 109.

[37]   Finegan D.P., Quinn A., Wragg D.S., Colclasure A.M., Lu X., Tan C., Heenan T.M.M., Jervis R., Brett D.J.L., Das S., Gao T., Cogswell D.A., Bazant M.Z., Di Michiel M., Checchia S., Shearing P.R., Smith K. (2020) Spatial dynamics of lithiation and lithium plating during high-rate operation of graphite electrodes, *Energy Environ. Sci.* **13**, 2570.

[38]   Orsini F., Du Pasquier A., Beaudouin B., Tarascon, J.M., Trentin M., Langenhuizen N., De Beer E., Notten P. (1999) In situ sem study of the interfaces in plastic lithium cells, *J. Power Sources* **81–82**, 918.

[39]   Gu M., Parent L.R., Mehdi B.L., Unocic R.R., McDowell M.T., Sacci R.L., Xu W., Connell J. G., Xu P., Abellan P. *et al.* (2013) Demonstration of an electrochemical liquid cell for operando transmission electron microscopy observation of the lithiation/delithiation behavior of Si nanowire battery anodes, *Nano Lett.* **13,** 6106.

[40]   Li Z., Tan X., Li P., Kalisvaart P., Janish M.T., Mook W.M., Luber E.J., Jungjohann K.L., Carter C.B., Mitlin D. (2015) Coupling in situ TEM and ex situ analysis to understand heterogeneous sodiation of antimony, *Nano Lett.* **15**, 6339.

[41]   Unocic R.R., Sun X.G., Sacci R.L., Adamczyk L.A., Alsem D.H., Dai S., Dudney N.J., More K.L. (2014) Direct visualization of solid electrolyte interphase formation in lithium-ion batteries with in situ electrochemical transmission electron microscopy, *Microsc. Microanal.* **20**, 1029.

[42]   Cheong J.Y., Chang J.H., Seo H.K., Yuk J.M., Shin J.W., Lee J.Y., Kim I.D. (2016) Growth dynamics of solid electrolyte interphase layer on SnO_2 nanotubes realized by graphene liquid cell electron microscopy, *Nano Energy* **25**, 154.

[43]   Nakashima K., Kao K.C. (1977) Study of heat treated and electron beam bombarded amorphous semi-conductor surfaces by scanning electron microscopy, *Thin Solid Films* **41**, 29.

[44]   Gale B., Hale K.F. (1961) Heating of metal/ceramic foils in an electron microscope, *Br. J. Appl. Phys.* **3**, 115.

[45]   Wang X., Zhang M., Alvarado J., Wang S., Sina M., Lu B., Bouwer J., Xu W., Xiao J., Zhang J.G. *et al.* (2017) New insights on the structure of electrochemically deposited lithium metal and its solid electrolyte interphases via cryogenic TEM, *Nano Lett.* **17**, 7606.

[46]   Zachman M.J., Tu Z., Choudhury S., Archer L.A., Kourkoutis L.F. (2018) Cryo-STEM mapping of solid–liquid interfaces and dendrites in lithium-metal batteries, *Nature* **560**, 345.

[47]   Morigaki K.I., Ohta A. (1998) Analysis of the surface of lithium in organic electrolyte by atomic force microscopy, fourier transform infrared spectroscopy and scanning auger electron microscopy, *J. Power Sources* **76**, 159.

[48] Demirocak D.E., Bhushan B. (2014) In situ atomic force microscopy analysis of morphology and particle size changes in lithium iron phosphate cathode during discharge, *J. Colloid Interface Sci.* **423**, 151.

[49] Balke N., Jesse S., Morozovska A.N., Eliseev E., Chung D.W., Kim Y., Adamczyk L., García R.E., Dudney N., Kalinin S.V. (2010) Nanoscale mapping of ion diffusion in a lithium-ion battery cathode, *Nat. Nanotechnol.* **5**, 749.

[50] Morozovska A.N., Eliseev E.A., Kalinin S.V. (2010) Electromechanical probing of ionic currents in energy storage materials, *Appl. Phys. Lett.* **96**, 22.

[51] Lipson A.L., Ginder R.S., Hersam M.C. (2011) Nanoscale in situ characterization of Li-ion battery electrochemistry via scanning ion conductance microscopy, *Adv. Mater.* **23**, 5613.

[52] Lipson A.L., Puntambekar K., Comstock D.J., Meng X., Geier M.L., Elam J.W., Hersam M.C. (2014) Nanoscale investigation of solid electrolyte interphase inhibition on Li-ion battery MnO electrodes via atomic layer deposition of Al_2O_3, *Chem. Mater.* **26**, 935.

[53] (a) Nandasiri M.I., Camacho-Forero L.E., Schwarz A.M., Shutthanandan V., Thevuthasan S., Balbuena P.B., Mueller K.T., Murugesan V. (2017) In situ chemical imaging of solid-electrolyte interphase layer evolution in Li-S batteries, *Chem. Mater.* **29**, 4728; (b) Wu X., Villevieille C., Novák P., El Kazzi M. (2018) Monitoring the chemical and electronic properties of electrolyte-electrode interfaces in all-solid-state batteries using: operando X-Ray photoelectron spectroscopy, *Phys. Chem. Chem. Phys.* **20**, 11123.

[54] Wood K.N., Steirer K.X., Hafner S.E., Ban C., Santhanagopalan S., Lee S.H., Teeter G. (2018) Operando X-Ray photoelectron spectroscopy of solid electrolyte interphase formation and evolution in $Li_2S-P_2S_5$ solid-state electrolytes, *Nat. Commun.* **9**, 2490.

[55] Morales-Ugarte J.E., Bolimowska E., Rouault H., Santos-Peña, Santini C.C., Benayad A. (2018) EIS and XPS investigation on SEI layer formation during first discharge on graphite electrode with a vinylene carbonate doped imidazolium based ionic liquid electrolyte, *J. Phys. Chem. C* **122**, 18223.

[56] Marker K., Xu C., Grey C.P., Operando NMR of NMC811/graphite lithium-ion batteries: structure, dynamics, and lithium metal deposition, *J. Am. Chem. Soc.*, Article ASAP.

[57] Pietsch P., Hess M., Ludwig W., Eller J., Wood V. (2016) Combining operando synchrotron X-ray tomographic microscopy and scanning X-ray diffraction to study lithium ion batteries, *Scientific Reports* **6**, Article number: 27994.

Chapter 13

Cell and Electrode Manufacturing Process

Guillaume Claude, Nicolas Mariage, Willy Porcher, Yvan Reynier, Dane Sotta and Florence Rouillon

This chapter describes the manufacturing steps of Li-ion and post-Li-ion battery electrodes and cells. We will discuss these processes from the point of view of critical parameters and performance.

13.1 General Principles

The manufacturing process of Li-ion batteries, and post Li-ion technologies to some extent, largely depends on the conduction properties of the electrolyte used.

Indeed, the so-called aqueous technologies using water as an electrolyte solvent (used for *e.g.* Ni–MH and lead acid batteries) are limited in voltage to about 1.5 V (2 V for lead batteries, due to the exceptional oxidation over potential of water on this metal). Above this voltage, water is oxidized on the anodic side (oxygen formation) and reduced on the cathodic side (hydrogen formation). To obtain batteries that operate between 3 and 4 V, so-called organic solvents are used, which have a much wider electrochemical stability window, but very low lithium (or sodium) salt dissolving capability, in the order of 1 mol.L^{-1}.

The results are electrolytes with a conductivity of about 10 mS.cm^{-1}, whereas it can exceed 100 mS.cm^{-1} with aqueous systems [1]. This low conductivity imposes to reduce the thickness of the electrodes down to 30-200 μm, in order to preserve the power performances capability of the battery. In contrast, aqueous systems can use electrodes several millimeters thick, manufactured by much more rustic pasting processes.

In the case of Li-ion batteries and other systems operating at more than 3 V, it is thus necessary, to keep a compact battery, to stack multiple layers of electrically

DOI: 10.1051/978-2-7598-2555-4.c013

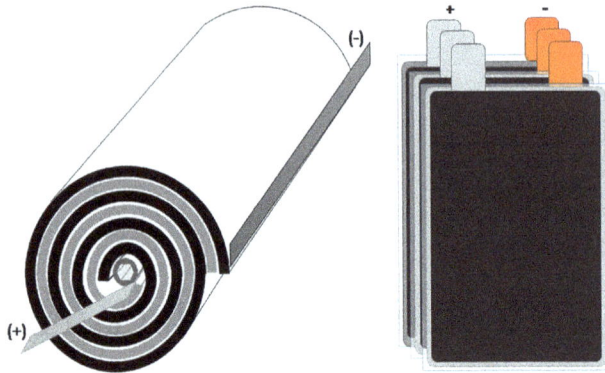

FIG. 13.1 – Wound (left) and stacked (right) electrodes.

paralleled electrodes (stacking process) or to wrap long electrodes around a cylindrical or prismatic core (winding process) [2], as shown in figure 13.1.

The consequences on the processes are multiple: the need for micron level control of the coated layer thickness, the alignment of the electrodes with a better accuracy than one mm, current collectors and separator thickness not exceeding twenty microns in most cases to preserve the active/inactive material ratio in the battery, which conditions the final energy density... On the other hand, lithium compounds (or, in the case of sodium-ion batteries, sodium as the alkaline metal) have a high reactivity with water. Therefore, the removal of any trace of water during the battery manufacturing process is essential. All components of the battery are dried and their assembly is usually carried out in a dry room. This results in a process that is inherently more expensive than for aqueous systems.

13.2   Cell Design

The standard designation of a Li-ion cylindrical battery is a five-digit number, the first four digits of which define the size of the battery and the last one its shape. In the case of an 18650 format, the final "0" indicates a cylindrical battery. The first two digits give the diameter: 18 mm. The next two digits give the length: 65 mm. For a prismatic design, the format is given as thickness/width/length in mm (for instance 16/90/140).

With regard to battery size, extended electrode lengths (between 50 cm and 1 m are required in a standard 18650 Li-ion battery (figure 13.2)) are also a limiting factor.

Given the thickness of the current collectors, the ohmic drop becomes significant beyond 50 cm length which invites, for example, to position a connector tab at each end of the electrode for a power design, against only one for an energy design [3] (figure 13.3).

FIG. 13.2 – 18650 battery made at CEA (18 mm diameter and 65 mm length).

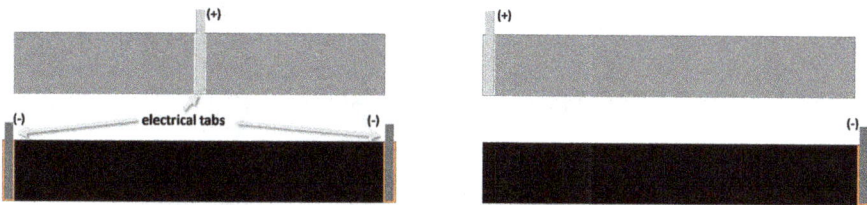

FIG. 13.3 – Two tabs negative electrode design for power 18650 cells (left) and one tab design (right) for energy 18650 cells.

The current collection becomes a serious issue for the manufacturing of batteries of larger capacity (electrode length >1 m) : it is necessary to multiply the internal tabs, which complicates the implementation, because multiple skip coating area on the electrode becomes necessary to allow their welding.

A stacked design (figure 13.4) is therefore more flexible, but this is done at the expense of a higher assembly time (a "layer by layer" process is used, instead of continuous winding).

Other factors determine the final choice of format, such as thermal and safety considerations. The larger the battery, the higher its internal temperature during a high current pulse ($P_{dissipated} = RI^2$). It is well known that batteries ageing is usually faster at higher temperatures. On the other hand, depending on the format, an uneven temperature distribution inside the cell may lead to uneven ageing, which may in turn lead to an even faster end of life. Thus, a prismatic format may be better adapted for a power battery (with a flexible or rigid packaging), rather than a cylindrical design, in order to reduce local over heating in the cell center. In figure 13.5, two CEA formats with the same capacity (16.5 Ah in LFP-G) are presented: the "50125" cylindrical and the prismatic format "25/100/125". While exactly the same electrode length is used in these two designs, the cell thickness is halved for the prismatic, greatly reducing temperature gradients inside the cell during discharge at high power.

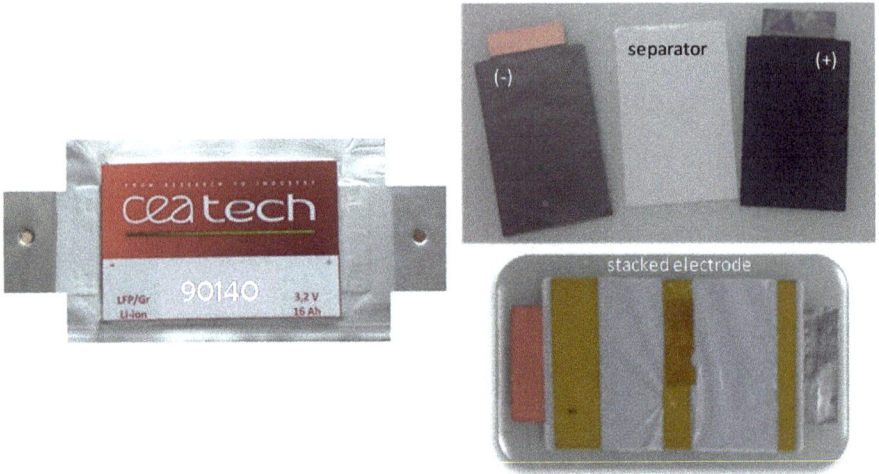

FIG. 13.4 – CEA 16 Ah "LFP-Graphiter" stacked pouch cell (16 x 90 x 140 mm).

FIG. 13.5 – CEA prismatic (25 x 100 x 125 mm) and cylindrical 50125 (50 ⌀ x 125 mm) cells.

Finally, above a given temperature (about 110 °C for Li-ion), the passivation layers present on the electrodes become unstable. This can lead to a thermal runaway phenomenon, during which the heat generated by the exothermic reactions in the battery can no longer dissipate quickly enough, and causes a temperature increase that can go beyond the melting point of aluminum (660 °C) [4]. For all these reasons, a parallel connection of small batteries is sometimes preferable to the use of a single large, more unstable battery.

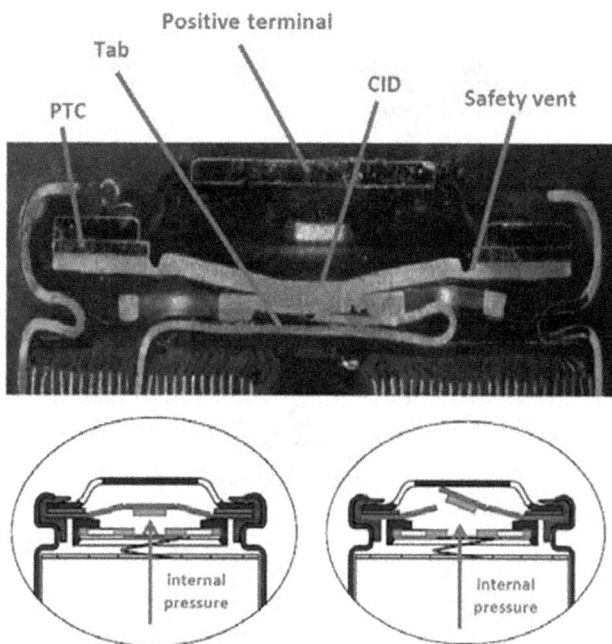

FIG. 13.6 – Li-ion cell safety devices in the top cap (top); activation of CID and safety vent (bottom).

Several safety devices can be integrated to limit these risks (figure 13.6) [5, 6]. First, most hard casing cells contain a safety vent, which, in the event of gas release, opens at a given pressure (<20 bar) to prevent a cell explosion. A current interrupt device (CID) can also be added. It is activated by the pressure increase and allows stopping an uncontrolled overcharge or external short circuit. The 18650 batteries also contain a thermally activated current limiter (PTC washer): made of a conductive charged polymer, its resistance increases significantly when heated above its glass transition temperature (generally about 125 °C). Other elements contribute to battery safety, apart from the nature of the active materials. We can mention, without being exhaustive, the separator (with "shutdown" effect or covered with a thermally stable ceramic coating), additives in the electrolyte (*e.g.* with rapid gas release in the event of overcharge to activate mechanical safety devices)…

Finally, from the manufacturer's point of view, the loss of a large battery during assembly (several tens of Ah) results in a higher cost than the loss of a cell of smaller size. The manufacture of large batteries therefore requires a better control of the implementation processes.

Li-ion battery manufacturing is divided into five main steps: electrode production, electrode calendering, electrodes and separator assembly, electrolyte filling and electrochemical formation. To ensure optimal performances, all these steps must be carried out in a controlled environment and with controlled equipment.

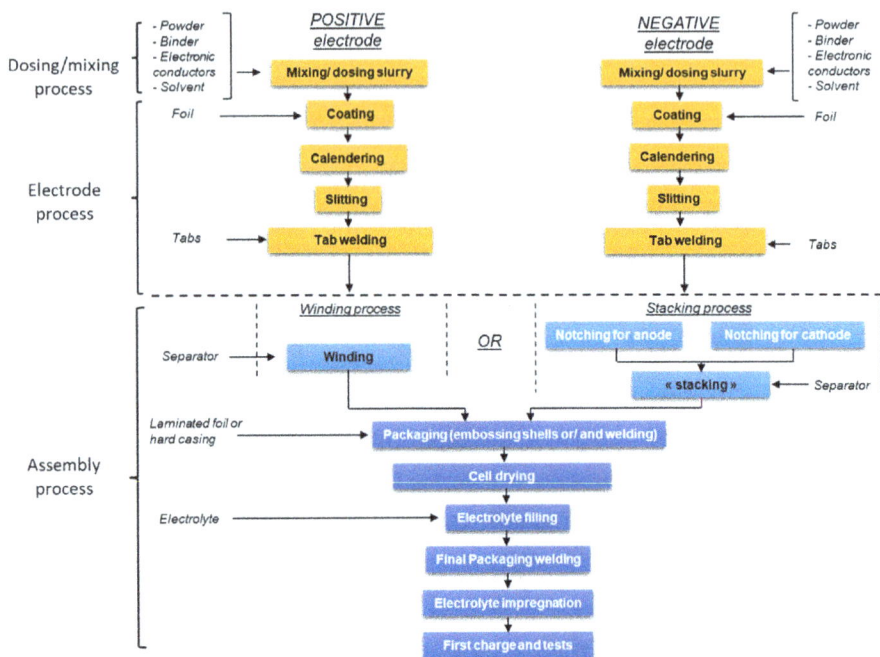

FIG. 13.7 – Main steps of Li-ion cells manufacturing.

The cell is mainly composed of a negative electrode (anode), a positive electrode (cathode) and a separator, enclosed in a housing. Figure 13.7 describes the main manufacturing steps.

13.3   Electrode Manufacturing Process

13.3.1   Electrode Formulation

An electrode is a composite mixture made of several components. The active material is its main constituent. It is electrochemically active and determines the energy density of the final system. The aim is therefore to maximize the mass ratio of active material in the electrode. A ratio greater than 90% by mass is generally targeted in electrodes produced on an industrial scale. The other components are said to be electrochemically "inactive", but provide additional properties to the electrode. One or more electronic conductors are added to improve electrical conductivity in the electrode thickness. An optimum, called the percolation threshold, is sought in order to electrically connect all active material particles between each other and with the current collectors. These conductors are generally carbonaceous materials (carbon blacks, graphites) present as fine particles or fibers. Polymeric binders are also added to provide mechanical properties to the composite electrode.

They allow the materials particles to be held together, as well as a sufficient adhesion of the electrode to its current collector. Additional flexibility properties are sought if the electrodes need to be wound during the battery assembly phase. Finally, these polymers play an important role in the slurry by contributing to the dispersion of components. Candidate polymers meeting these criteria, and being stable with respect to the electrolyte in the relevant electrochemical window, are relatively few in number. Two main approaches have been identified in the industry. The first approach is used to manufacture positive electrodes. It consists in using fluorinated polymers, such as polyvinylidene fluoride (PVDF), because of their excellent chemical and electrochemical stability at high potential. However, a PVDF slurry requires the use of organic solvents such as N-methylpyrrolidone (NMP), which raises questions in terms of cost (solvent and recycling) and health (NMP is reprotoxic). Another approach, used at industrial scale to manufacture negative graphite-based electrodes, is to disperse the ink constituents in water using water-soluble polymers. In this case, the polymer binders are, for example, carboxymethylcellulose (CMC), for its excellent dispersion properties, as well as synthetic latex polymers, which provide flexibility.

13.3.2 *Slurry Preparation*

The first step in the production of an electrode is to mix homogeneously different components. The most commonly used solution is a liquid mixture: the active material, electronic conductors and polymers are dosed and successively dispersed in a solvent to form a thick suspension called slurry or ink (figure 13.8). The solvents used on an industrial scale, *i.e.* NMP, or water, must readily dissolve polymers. The ratio between the mass of dry matter actually present in the ink and the total mass

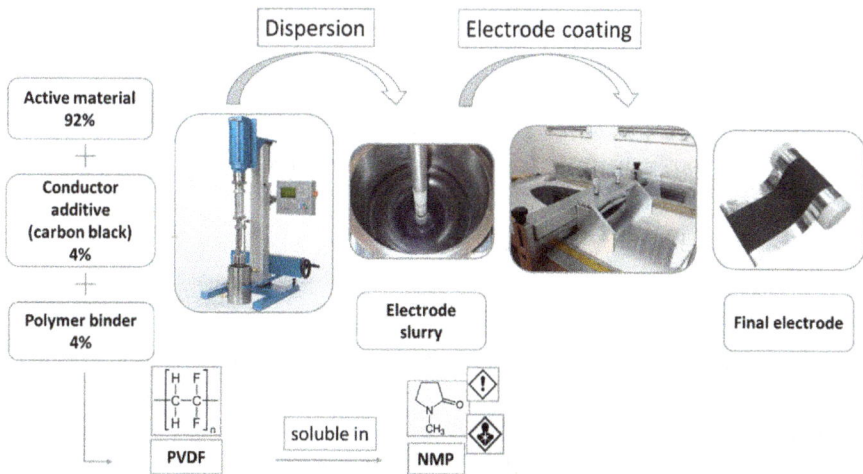

FIG. 13.8 – Principle of Electrode Slurry Making.

of the ink is called the dry content. In conventional industrial-scale coating processes, the aim is to maximize the slurry dry content (*i.e.* limit their solvent content)
in order to simplify the subsequent electrode drying steps.

13.3.2.1 *Mixing Process*

As the active materials and conductors are in powder form (with a high specific
surface area for the latter), it is necessary to apply a vigorous dispersion step to
wet all the powders and break up the agglomerates [7]. This step is achieved in batch
mixers consisting of a tank and one or more motorized axes immersed in the slurry.
There is a wide variety of possible axes, the most common being the axes in the
center of the tank with a dispersion tool of the deflocutioner type (dispersers/
homogenizers), or double helix axes sweeping the entire volume of the tank (plan-
etary mixers). Double-walled tanks with coolant are used to limit heating during ink
dispersion. Sealed tanks are sometimes used for mixtures needing controlled atmosphere or to avoid the incorporation of gases into the volume.

13.3.2.2 *Slurry Control*

Several critical parameters such as ink grain size and rheological properties are
monitored during production and determine its use during the coating phase.
A simple means of control is the thinness gauge, which makes it possible to estimate
the maximum size of the agglomerates of particles contained in the ink after dispersion. Sizes smaller than 50 μm are generally recommended so that the ink can be
used during the process. If the thinness control is not satisfactory, it is necessary to
continue dispersing the ink in the mixer or to use "refining" tools that break up the
remaining agglomerates.

The rheological properties of the ink determine its behavior when it is coated. It
is generally necessary to use inks with a rheothinning character, *i.e.* the viscosity
decreases as the shear rate increases. A low viscosity with a high shear rate (*e.g.* in
the range of 2 Pa.s to 100 s^{-1}) ensures a laminar flow of ink through the coating
tools (slot dies in particular). On the other hand, the higher viscosity at rest allows
the ink "freezing" on its collector after coating, which avoids running-type defects
and loading inhomogeneity. In practice, viscosity measurements are performed with
equipment such as viscometers or rheometers.

13.4 Electrodes

Electrode manufacturing consists in coating a mixture of active material, electronic
conductor, polymer binder and solvent, on both sides of a current collector.

For the positive electrode, a collector made of aluminum is used: aluminum is a
good electronic conductor material (which is the main property sought), it is relatively inexpensive, and stable in the potential window of the positive electrode
[3–4.5 V]. However, this stability is not intrinsic, but due to passivation by ALF$_3$

FIG. 13.9 – Electrode manufacturing steps starting from material processing.

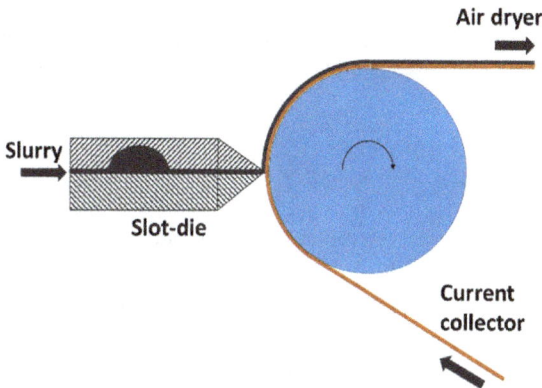

FIG. 13.10 – Slot die coating process sketch.

produced by oxidation of the lithium salt of the electrolyte ($LiPF_6$) during the first charge of the battery [8, 9]. In addition, its low density (2.7 g.cm^{-3}) limits the ratio of inactive mass in the cell. The collector thickness is currently between 20 and 12 μm. This value must be minimized to increase the energy density, but mechanical resistance limits are encountered, especially during calendering (possible breakage of the collector).

A 7–15 μm thick, copper current collector is used for the negative electrode. It has very good electrical conductivity, but is more expensive and more dense (8.9 g.cm^{-3}) than aluminum. If the use of copper is most often necessary instead of aluminum, it is because aluminum forms an alloy with lithium between 0.3 and 0 V *vs.* Li [10], which has the effect of rapidly degrading the integrity of the current collector. This is the case, for example, when the negative electrode used is graphite-based, with a potential lower than 0.3 V *vs.* Li/Li$^+$. On the other hand, aluminum can be used when active compounds such as titanates (LTO: 1.5 V *vs.* Li) are used at the anode (figure 13.9).

Coating is usually carried out with a slot-die (figure 13.10), or roll-to-roll transfer to achieve a homogeneous deposit with constant thickness.

The control of the amount of active material coated and of its mechanical characteristics are crucial, as these parameters exert a considerable influence on the battery performance.

The drying process is carried out by kilning the electrodes in an oven. This must be made gradually to eliminate any trace of solvent and avoid excessive stresses during evaporation (cracked earth effect...).

Historically, one of the most widely used polymers for ink production has been PVDF, known for its good rheological properties during implementation, its stability in the range of battery operating potentials and its lack of interaction with other battery components (electrolyte). Unfortunately, this polymer is only soluble in NMP (N-methylpyrrolidone), which is a CMR classified solvent. Therefore, it is necessary to set up a system for recovering and reprocessing this solvent during drying. For economic and environmental reasons, other alternatives to this type of solvent are being considered, in particular other types of water-based polymer binders. Dry processes are also investigated at R&D level, in order to get rid of the solvent recovery step.

13.4.1   Calendering

Since coating is achieved in the presence of solvents, its removal during drying generates porosity within the electrode. This porosity facilitates the exchange between the active materials and the lithium ions contained in the electrolyte that fills this porosity, thus allowing optimal operation of the battery. The purpose of calendering is to densify the electrode to increase the total quantity of active material in a given volume, while still allowing lithium ions to diffuse through the residual open porosity.

The calendering step is carried out by laminating the electrode between two metal cylinders (roll press) to apply a compressive stress on the coating of several hundred MPa.

In addition to adjusting the porosity, calendering generally improves the mechanical characteristics of the electrodes, that are required for the next assembly process step namely good adhesion properties of the coating on the collector and flexibility of the coating to resist mechanical winding stresses. A common method of measuring adhesion is to peel an adhesive tape applied to the surface of the coat. The tearing force, measured in $N.m^{-1}$, provides an indication of the mechanical properties: a value of several $N.m^{-1}$ is essential to ensure adequate implementation during assembly. Flexibility can be assessed by bending the electrode on mandrels of different diameters (in the order of a few millimeters).

13.5   Cell Fabrication Process

13.5.1   Slitting

Depending on the format and the design of the battery, the two electrodes (cathode and anode) are cut to the desired width using a device called a "slitter" by a knife system and a counter knife. This method is the one industrially used for all spiral batteries. Other cutting methods also exist, such as laser cutting, for laboratory applications, or punching, which is widely used for stacking technology.

13.5.2 Cell Assembly

Stacking is an assembly method consisting of successively inserting a separator in the middle of a cathode sheet and an anode sheet. The positive and negative electrode sheets are stored in two separate containers. As shown in figure 13.11, they are alternately placed on a central station with a separator coil covering each sheet. Once the stacking is complete, a station allows wrapping the separator around the electrochemical core to ensure a good mechanical hold. The assembly thus formed is sent to the next step *via* a conveyor (figure 13.12).

One of the most common techniques for cell assembly is the winding process. A positive electrode, negative electrode and two separators are wound together around a mandrel as shown in figure 13.13. The resulting electrochemical building block is called a "jelly roll".

The jelly roll thus assembled is placed in a flexible (pouch cell) or rigid (metal can) packaging depending on the application.

The flexible packaging consists of an aluminum foil (to ensure moisture tightness) laminated with different polymers on each side: usually polypropylene on one side and polyamide on the other, the aim being that the melting temperature of one side is much lower than that of the other to allow heat sealing. This packaging is, if necessary, embossed to fit with the cell format. The jelly roll is then placed into the packaging, the positive and negative tabs extending outside of the pouch and equipped with special polymer patches to be sealed with the pouch. The two sheets are then heat-sealed together to form a pouch, leaving only an opening for the electrolyte to be later introduced.

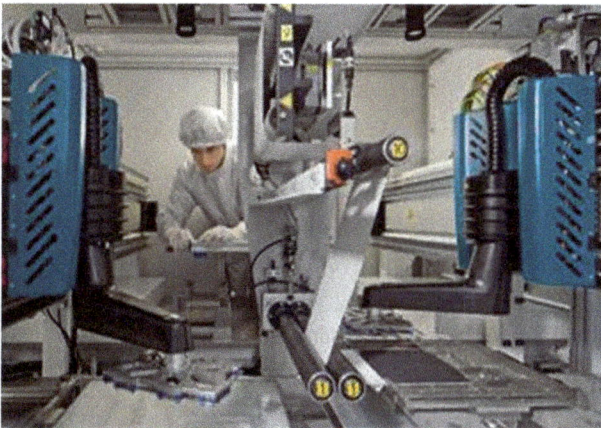

FIG. 13.11 – Stacking machine (courtesy Heddergott/TUM).

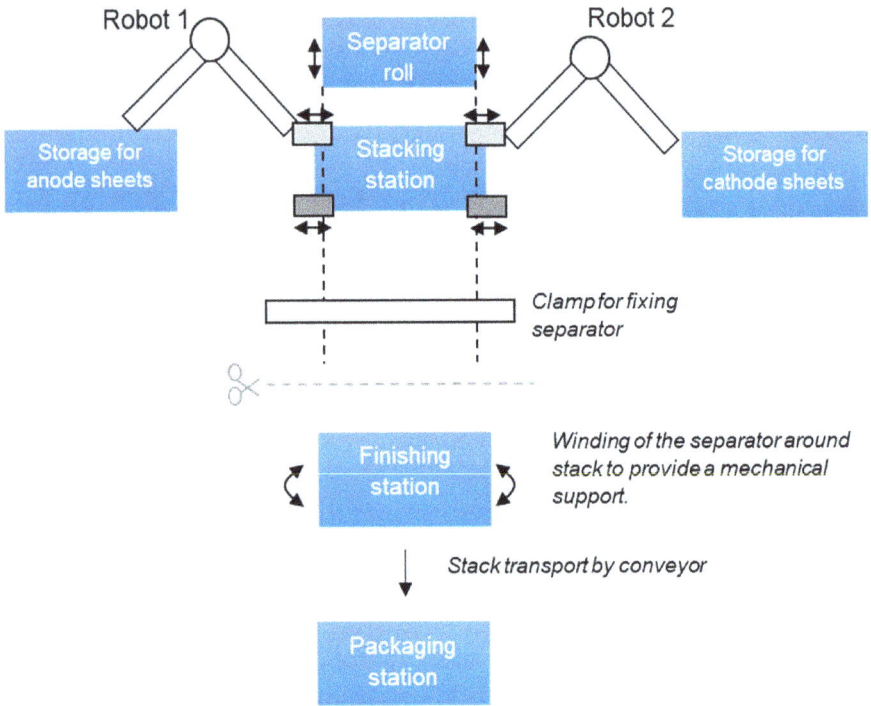

FIG. 13.12 – Stack assembly process sketch [11].

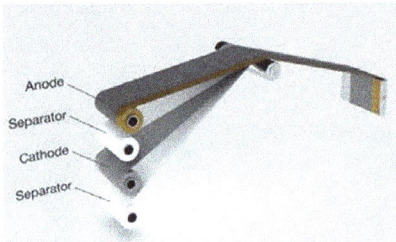

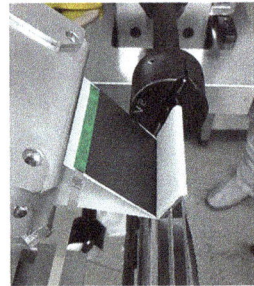

FIG. 13.13 – Jelly roll assembly using winding process.

Hard casing is commonly made of aluminum or nickel-plated iron can. The choice is based on multiple criteria: chemical resistance to electrolyte, galvanic compatibility with the cell current collectors, the mass of the material, its weldability, its cost, its mechanical resistance, etc. Unlike pouch cells, hard casing cells often contain safety devices, in particular a vent that opens in the event of abuse to prevent the battery from exploding, and a circuit breaker that stops current flow in the event of internal pressure buildup.

13.5.3 Electrolyte Filling

The last step in the manufacture of a lithium-ion battery, before the cell is electrically formed, is filling. It consists of adding the right dose of electrolyte to the cell before sealing it tightly, by means of heat sealing, riveting or welding. To facilitate the penetration of this viscous liquid, this step is often carried out under vacuum.

The electrolyte is the liquid that allows lithium ions moving between the cathode and the anode through the separator. It is usually composed of a lithium salt $LiPF_6$ dissolved at about 1 mol.L^{-1} in a mixture of solvent carbonates (EC, PC, DMC...) and additives (VC, FEC, LiTFSI...).

13.5.4 Electrical Formation

Once the assembly is completed and the cell is hermetically sealed, the formation is aimed in particular at controlling the first electrical charge. During this step, a passivation layer is formed on the surface of the negative electrode, its potential being lower than the stability range of the electrolyte. The quality of this passivation layer largely determines the life expectancy of the battery. To optimize it, additives are present in the electrolyte, the most common being vinylene carbonate (VC) for graphite-based Li-ion batteries [12]. Each manufacturer adjusts the formation parameters (current density, temperature, tensile strength, rest time) according to the battery and technology [13]. For example, gas (CO_2, H_2...) generated during the formation of the passivation layer, must be, in the case of a cell in flexible packaging, evacuated after a first formation step, and the battery resealed. On the contrary, for a hard casing cylindrical cell, the gas usually remains inside the bucket and generates an increase in pressure that must be taken into account (presence of an appropriate dead volume).

At the end of the first cycle, electrical characterization (internal resistance, capacity at nominal C-rate) is generally performed. It is followed by a self-discharge characterization (measurement of the loss of capacity in an open circuit) which is all the longer as the application in question is critical. Several weeks may be necessary to discriminate between a compliant cell and a scrap. Some manufacturers, based on these results, sort the cells into batches according to their quality.

13.6 Cells Bill of Materials and Cost Aspects

Within a battery pack, the materials that make up the cells represent about two thirds of the total mass of the pack (figure 13.14) [14]. At the cell level, the predominant materials regarding weight are, on the one hand, the active cathode material, and on the other hand, copper and aluminum used as collectors and housings.

Driven by the take-off of the EV market, the cost of a Li-ion cell decreased by a factor of 10 since 2010. In 2021 is getting closer to €100–150 per kWh [15]. Cost variations are primarily associated with the cell format, with a lower cost for

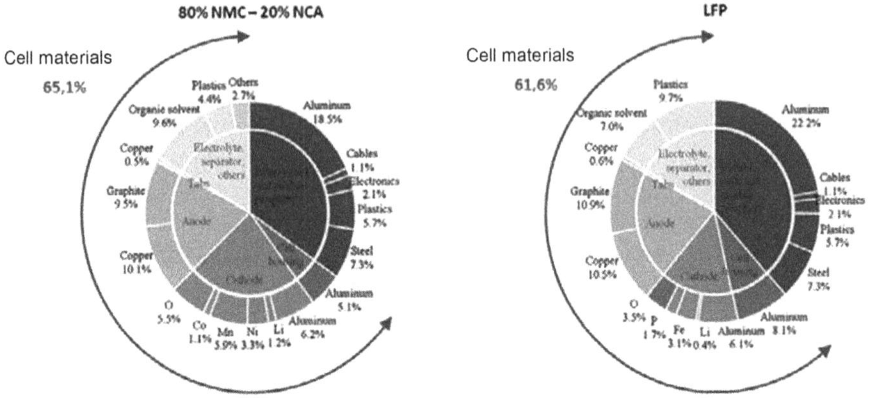

FIG. 13.14 – Generic composition for traction batteries: (left) 80% NMC – 20% NCA battery, (right) LFP battery.

standard elements such as 18650 and 21700 cylindrical cells. Cells designed for high-power applications such as hybrid vehicles are more expensive because their energy content is lower per unit mass (the active/inactive material ratio is less favorable).

The cost of a Li-ion cell is divided between materials and production, from the ink to the formation of the cell. For a Li-ion cell dedicated to electric vehicles, with an "energy" design, the proportion of materials represents more than 60% of its cost depending on production capacity, and even up to 75% for standard cylindrical elements produced by several tens of millions per month [15].

The cathode active material is the main cost contributor for cells using nickel and cobalt-based cathodes; with a price between \$25 and \$30 kg^{-1} for NMC, LCO and NCA materials. The recent performance/cost ratio improvement of batteries using iron-based material such as LFP may open new opportunities to low-cost systems for vehicle electrification and stationary applications. Regarding process, the coating and drying steps are the most expensive (30%), in terms of both depreciation and energy.

13.7   New Processes/Perspectives

For Li-ion technologies, two needs drive implementation developments: cost reduction to supply emerging mass markets such as electric vehicles, and the environmental impact of production integrating the reduction of critical materials such as cobalt, nickel or natural graphite. The trend is therefore towards increasing the production rate of machines, for example, by widening the width of acceptable electrodes (nowadays produced on strip up to 1200 mm wide). One also tries to minimize waste by taking advantage of the maximum proportion of electrodes produced. The aqueous solvent application processes are preferred over the "standard" NMP solvent used with PVDF binders. Dry processes are also explored, in order to avoid costly evaporation and recovery steps.

Most of the new emerging battery technologies require the development of new manufacturing processes. For example, for Li-air or Li-sulphur batteries, and most of the time for all-solid state electrolyte technologies, a lithium metal foil must be integrated as a negative electrode. Its thickness will not exceed 100 μm (A 2 mAh/cm^2 capacity exchange between the electrodes during cycling corresponds to the reversible consumption/deposit of a 10 μm-thick layer of metallic lithium), making electrode handling delicate. In addition, the soldering techniques must be adapted.

In the case of "all-solid state" technologies, multiple process hurdles are identified to date, and must be overcome to allow the development of these new concepts. In addition to the problems of interface between the constituents, we can mention, for example, the sensitivity of certain electrolytes to air, the need to compact the electrodes to remove all porosity, or the difficulty of creating a homogeneous and thin separator membrane (a few tens of microns at most) over large surfaces [16].

13.8 Conclusion

Organic electrolyte battery manufacturing requires a high degree of process control. This valuable expertise is usually not disclosed by manufacturers. The products' quality control is even more essential as it directly conditions operational safety. The new high energy density technologies under development, most of which use lithium metal, will only accentuate this trend.

Bibliography

[1] Xu K. (2004) *Chem. Rev.* **104**, 4303.
[2] Schröder R. *et al.* (2017) *Proc. Manufacturing* **8**, 104.
[3] Reimers J.N. (2006) *J. Power Sources* **158**, 663.
[4] Wang Q. *et al.* (2012) *J. Power Sources* **208**, 210.
[5] Mähliß J. (2017) *Lithium-ion battery technology.* Batteries Event, Nice.
[6] Balakrishnan P.G. *et al.* (2006) *J. Power Sources* **155**, 401.
[7] Wenzel V. *et al.* (2015) Challenges in lithium-ion-battery slurry preparation and potential of modifying electrode structures by different mixing processes, *Energy Tech.* **3**, 692 doi: https://doi.org/10.1002/ente.201402218.
[8] Mun J. (2010) *Electrochem. Solid State Lett.* **13**, A109.
[9] Zhang S.S., Jow T.R. (2002) *J. Power Sources* **109**, 458.
[10] Oltean G. *et al.* (2014) *J. Power Sources* **269**, 266.
[11] Shoji N., Shuji Y., Mitsunori S. (2014) HITACHI POWER SOLUTIONS, JP2014165055.
[12] Petibon R. *et al.* (2014) J. *Electrochem. Soc.* **161**, A1618.
[13] Huang C. *et al.* (2011) *J. Solid State Electrochem.* **15**, 1987.
[14] Elwert T. *et al.* (2018) *Behaviour of lithium-ion batteries in electric vehicles.* Springer International Publishing, pp. 289–321.
[15] Avicenne Energy - The Rechargeable Battery Market : Value Chain and Main Trends 2019-2030-IBS -11 March, 2021.
[16] Schnell J. *et al.* (2018) *J. Power Sources* **382**, 160.

Chapter 14

Battery System and Battery Management System (BMS)

Laurent Garnier, Julien Dauchy, Daniel Chatroux, Dimitri Gevet and Ghislain Despesse

A battery system is not just an assembly of cells connected in series and parallel. To be operated safely, this assembly must be associated with electromechanical components and electronics management (usually called BMS). This BMS (Battery Management System) ensures safety functions, balancing, thermal management of the battery pack, but it is also capable of estimating the status of the battery pack (state of charge, energy state, state of health). Some innovative battery systems can integrate more functions, such as charger/inverter functions or energy distribution functions to an auxiliary network.

14.1 Battery System Architecture

A power requirement for a battery pack usually results into a mission profile expressed as power *vs.* time. This is illustrated by the type of graph shown in figure 14.1, where the charge and discharge phases can be observed. The integral of power *vs.* time gives the energy consumed or delivered by the battery pack. It is usually expressed in kWh or Wh. For a pack with a given voltage level, we will therefore have a number of Ah absorbed and supplied, which will result in a state of charge on the battery pack (SOC pack = State of Charge of the battery pack).

For a lower pack voltage, with fewer cells in series, the same energy requirement could be met with higher currents. This would require more cells in parallel.

There are therefore different types of parallel and series combinations of accumulators meeting the same energy requirement. We then say that we have an $xSyP$ battery pack (figure 14.2) where x is the number of cells connected in series and y is the number of cells connected in parallel. As the number x can be very high if the

DOI: 10.1051/978-2-7598-2555-4.c014

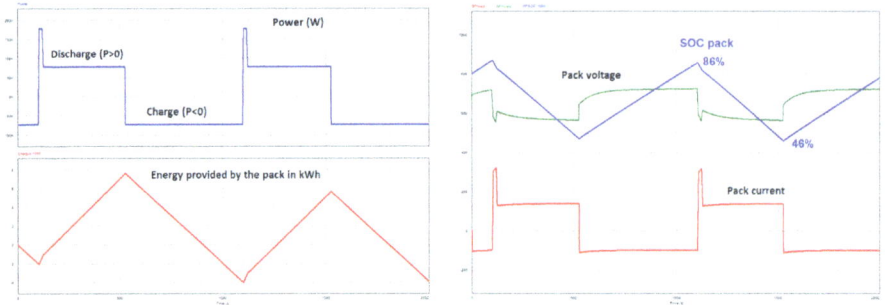

FIG. 14.1 – Evolution of the Energy delivered by the battery pack as a function of the power demand profile-voltage, current and SOC of the battery pack.

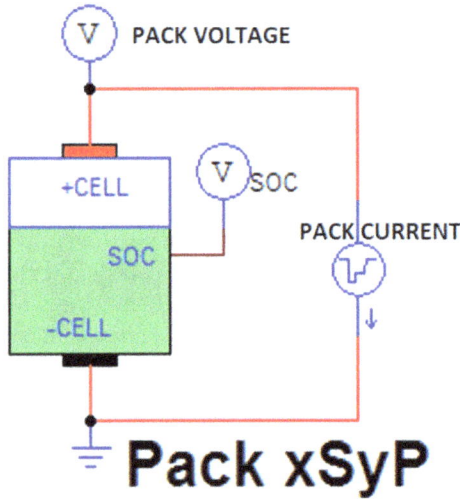

FIG. 14.2 – Series parallel assembly.

voltage is high, the battery pack is often divided into individual modules (figure 14.3) of reduced voltage (<60 V).

The main interest is to work throughout the manufacturing process with Extra Low Voltage Modules (ELV) that do not present any risk of worker's electrification. This limits the duration of the dangerous live phase, which is only linked to the series connection of the modules during final assembly.

Once the assembly of the cells into modules and then the modules into a pack is complete, we do not yet have a battery system. The battery system integrates many other functions such as:

– A BMS (Battery Management System) the main function of which is to manage the safety of the battery pack. The BMS requires the use of different types of sensors (temperature, current, voltage, possibly pressure...).

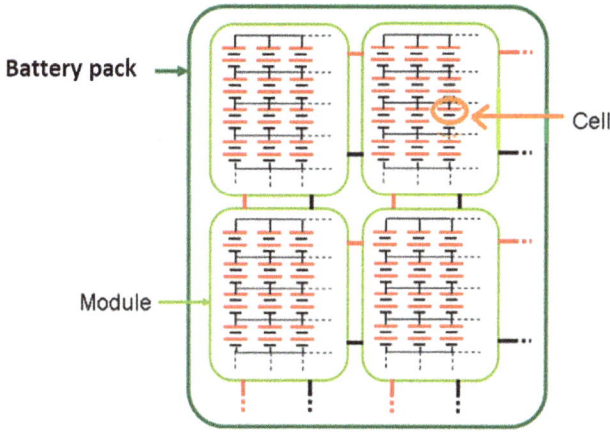

FIG. 14.3 – Assembly of cells, modules and pack.

- A "Junction Box" with contactors, fuses and a preload system to avoid excessive current demand when the contactors are closed.
- A housing or casing to ensure the mechanical maintenance, electrical and sometimes thermal insulation of the battery pack.
- A "service plug" used to electrically disconnect the pack before intervention on the vehicle.
- An IP2X type socket for external connection.

All these components, which are detailed below, constitute a Battery System (figure 14.4).

Here are some examples of modules and battery packs (figure 14.5).

14.2   Battery System in Its Electrical Environment

A battery system can be used for various types of applications (e-mobility, stationary power; emergency units,...). It interfaces with a source (often a charger) and loads which are converters, motors, resistors,... The battery system is therefore not alone in its environment. From an electrical point of view, there may be two types of mechanical ground connections on the battery system.

(a) *Negative terminal of the pack connected to mechanical ground*

The architecture where one of the poles is connected to the mechanics is used for 12 V batteries from the auxiliary network in thermal or electric vehicles. The minus pole is connected to the chassis of the vehicle that ensures the return of the current. With this architecture, an insulation fault on the circuit will then lead directly to a short circuit.

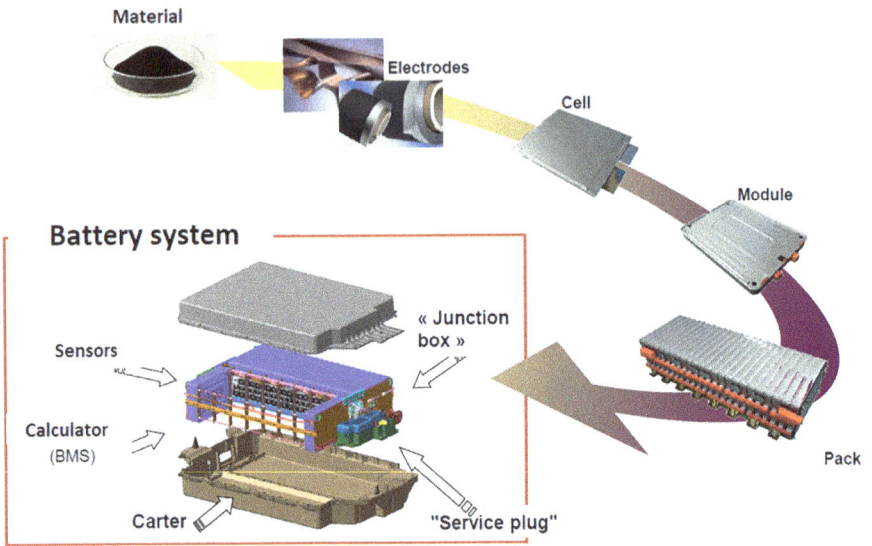

FIG. 14.4 – Whole battery system constitution.

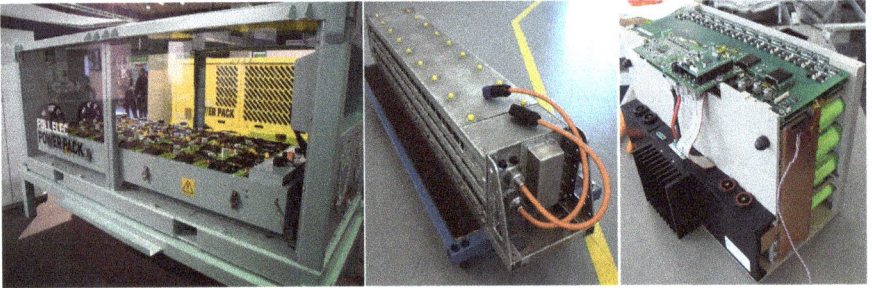

FIG. 14.5 – Examples of modules and battery packs.

(b) *Negative terminal of the pack insulated from mechanical ground*

Traction battery systems (figure 14.6) with higher voltages are usually insulated from the chassis. The difference is significant, since a first insulation fault will not lead to a short circuit. The result will only be a small leakage current to ground with no impact. We can therefore say that there is continuity of service at the first fault: a vehicle in full acceleration, for example, will not be forced to stop immediately, it will be able to continue driving. The fault will be detected by means of an insulation controller and the user will be warned that he will have to take the vehicle to the garage in order to correct the fault.

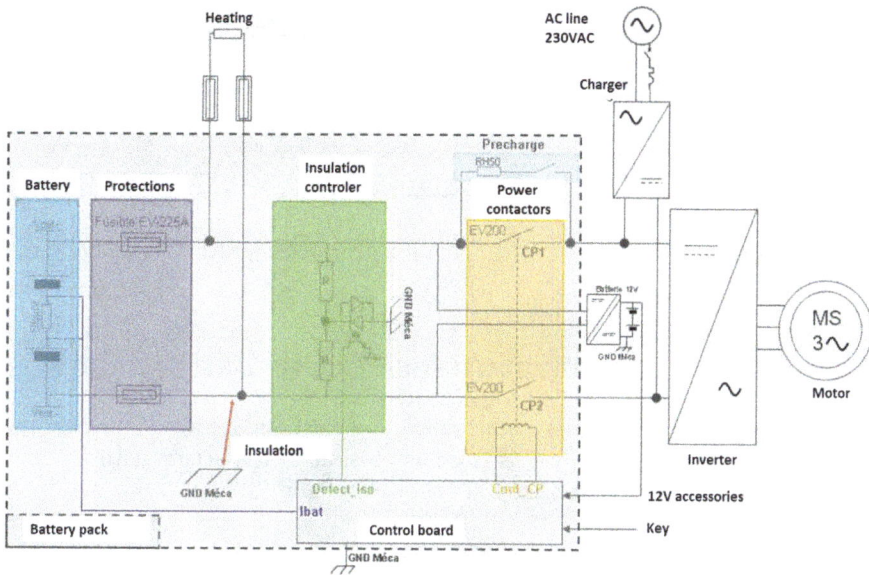

FIG. 14.6 – Battery pack with other elements.

There are various types of insulation monitors. The simplest ones use a high impedance divider bridge whose middle point is connected to the ground *via* a current measurement or detection, which in this case is a photocoupler. As the battery system is isolated from the mechanical ground, the insulation monitor symmetrically references the positive and negative potentials to the mechanical ground. For a 400 V battery pack for example, in normal operation, the polarity is +200 and −200 V with no current flowing through the diode of the photocoupler. If a fault occurs, a leakage current is established. This current flows through the diode of the photocoupler, which lights up, allowing the control board to detect the fault.

Impact of Battery Pack Insulation on Charger Choice

There are two types of chargers on the market:

- Insulated chargers with an insulation barrier of a few kV (2–5 kV in general).
- Non-insulated chargers.

Non-insulated chargers can be cheaper and more compact depending on the voltage range. However, care should be taken before using them as they cannot be used on a battery connected to mechanical ground. On a standard neutral (TT) system, a current to earth will immediately trigger the earth leakage circuit breaker. With a battery pack insulated from mechanical ground, non insulated chargers can be used.

Special cases, depending in particular on neutral regimes exist, but in TT regime, the main information can be summarized as in the table above (figure 14.7).

	Charge mode		Drive mode (charger disconnected)	
	Isolated Charger	Non Isolated Charger	Isolated Charger	Non Isolated Charger
Isolated battery pack	> First insulation fault tolerant > SC(*) at second fault	Possible configuation but : > IC (**) could be perturbated > SC(*) au premier défaut	> First insulation fault tolerant > SC(*) at second insulation fault	> First insulation fault tolerant > SC(*) at second fault
Non isolated battery pack	Short circuit with the first fault	IMPOSSIBLE CONFIGURATION	Short circuit with the first fault	IMPOSSIBLE CONFIGURATION
(*) SC = short-circuit (**) IC = Insulation controller				

FIG. 14.7 – Consequences of an insulation fault depending on whether the battery charger is insulated or not.

14.3   Power Component Associated to Battery Pack

As seen previously, to ensure the management and electrical safety of a battery pack, power elements must be installed as close as possible to the accumulators.

The main components concerned are shown in figure 14.8.

Cable/Wire

Conductors allow electrical interconnection between the battery and the outside of the pack. They must be correctly sized to meet the maximum current and the nominal current demand. The heat generated by the passage of the electric current through the conductors must be correctly dissipated outside the system. In the event of overheating of the conductors, there is a risk of fire, mainly due to the insulation

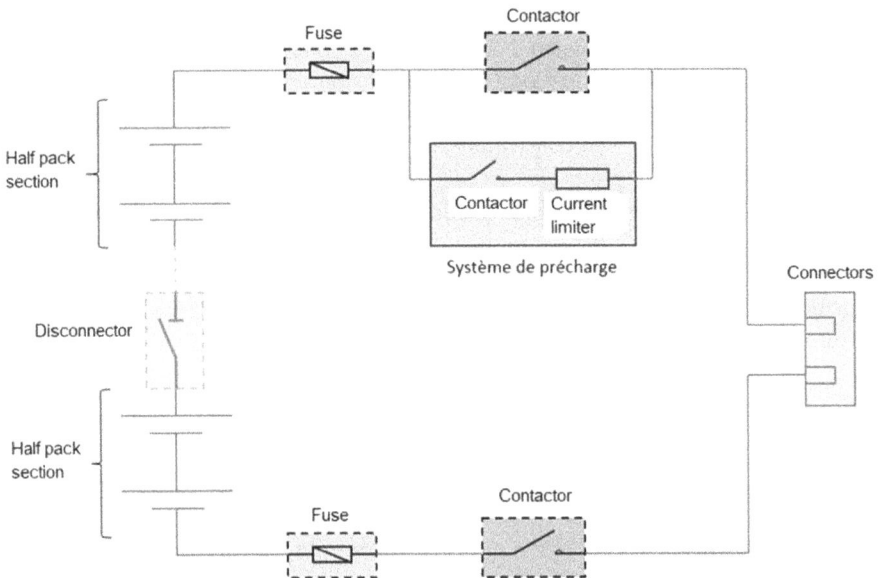

FIG. 14.8 – Electrical diagram of a battery system.

around the conductors, which can ignite and burn. The conductors can be of several shapes: bars, cables, printed circuit boards, flicker,... They can also be of different natures depending on the expected performance: copper, aluminum,...

Fuses

As mentioned earlier, a conductor can overheat if it is no longer able to dissipate the heat generated by the electrical current. A fuse (figure 14.9) is a component designed to melt and physically open the electrical circuit if the current flowing through the conductors exceeds its rating. It is dimensioned to protect the conductors according to several characteristics:

– the rating: corresponds to the current that the conductors can withstand continuously without overheating.
– The breaking capacity: during a short circuit, due to its low internal resistance, a battery can deliver several thousand amperes. The fuse must be able to cut off the electric arc caused and stop the current.
– The voltage at its terminals: once melted, the fuse must support the total voltage of the battery pack at its terminals.

It is preferable to place two fuses at the two polarities of the battery pack. This prevents a short circuit from occurring between one of the internal potentials of the battery pack and one of the two "plus" or "minus" potentials of the battery pack within the pack.

FIG. 14.9 – Fuse.

Contactors

Contactors (figure 14.10) are controlled components that can open and close the electrical circuit. In most battery packs, two power contactors are placed in series with the "plus" and "minus" polarities of the battery pack. This disconnects the battery from the system and thus decouples the power source from the system. In addition, when the system is in sleep mode, it is advantageous, with these components, to be able to completely disconnect the battery to ensure that there is no abnormal discharge of the battery. The contactors are usually controlled by the vehicle's on-board computer or directly by the "battery management system"

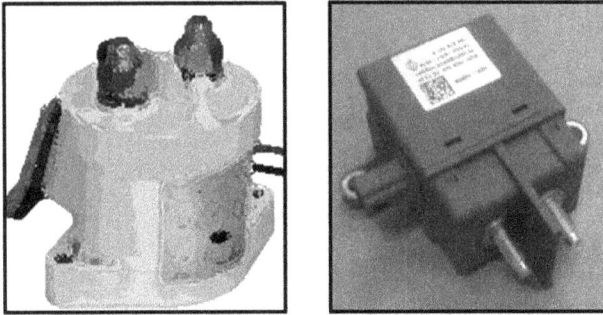

FIG. 14.10 – Power contactors.

(presented elsewhere). The contactors must be dimensioned to support the full voltage of the battery pack and to be able to cut off the maximum current flowing through the battery. Due to the difficulty of extinguishing DC arcs, contactors are typically sealed, with the cut-off taking place in a vacuum or in a gas under controlled pressure. One should here be careful, because some contactors are polarized: they can only cut the specified current in one direction. In the other direction, the breaking capacity is much lower.

Pre-Load

The pre-loading system is mandatory when electric capacitors of high capacitance are used in the equipments embedded in the vehicle, such as the filtering capacitors used *e.g.* for the motor inverter, the charger, or the high-voltage to low-voltage converter of the accessory grid. A direct connection of a battery to a highly discharged capacity is equivalent to creating a short circuit in the battery for a short period. Thousands of amperes then flow through conductors and contactors that are not able to withstand them. This type of fault usually results in the destruction of components, especially the contacts of the contactors, which are soldered together.

In order to limit the current when connecting the battery to the capacitors on the circuit, a current limiting circuit is used to control the charging current of these capacitances. Once the capacitors are correctly charged, the pre-load circuit is shunted by the associated power contactor. The current limiting device is often a power resistor because it is a simple and inexpensive component. It is advisable to associate monitors to the pre-load circuit to close the power contactor only if the residual voltage is small in order to avoid destroying the resistor. Converter-based current limiters are also available.

Connectors

The power connector (figure 14.11) is the electrical interface between the battery and the system. On a vehicle the power connector is orange-colored to be recognizable by the safety services. The connector must be sized to withstand the

FIG. 14.11 – IP2X Connectors.

FIG. 14.12 – Disconnector.

maximum current delivered or absorbed by the battery. In addition, it must withstand the total voltage of the battery between its terminals, taking into account all the safety margins imposed by the electrical safety standards. To guarantee the safety of people, it must have a protection rating of "2X", *i.e.* the electrical contacts of the connector can not be touched with the fingers.

Manual service Disconnector

Whenever possible, it is preferable to integrate a disconnector in the battery pack. The disconnector (figure 14.12) is a device that physically opens the electrical circuit. The opening is visible and thus ensures that there is no voltage on the battery connector and the external circuit. In this way, a maintenance operation can be carried out safely. As with the connector, it must be easily accessible and orange in color.

14.4 Multiples Functions of BMS

The BMS (Battery Management System) is the intelligence associated with the battery pack. The BMS is an electronic system that is in charge of several functions, the main one being to ensure the safety of the battery pack. After having presented the electronic architecture of the BMS, we will detail one by one its functions namely:

- The management of safety during the different states of the battery pack (charge and discharge).
- Battery pack status management (SOC, SOE, SOH).
- Balancing the cells with each other to compensate dispersion.
- Communication with the environment.
- If necessary, thermal management of the battery pack.

Different Architectures of BMS Electronics

There are two types of BMS architectures (figure 14.13): a first type presents a centralized architecture: all the measurements (voltages, temperatures, etc.) are cabled together and connected on the same electronic board, a second type presents a decentralized architecture organized in a modular way.

The centralized BMS has the advantage of simplicity from an electronic point of view and the disadvantage of complexity at the wiring level. In the case of a battery pack with a large number of cells, all the measurements at high voltage are centralized on the same board. This results in a complex wiring harness and the associated high risk of short circuit between these wires.

The modular BMS is organized in a decentralized manner. Measurements are made at the level of each module (MMU = Module Management Unit); the wiring is then local and the information is then sent back *via* a galvanically isolated communication bus to a PMU (Pack Management Unit). The CMU corresponds to the Cell Management Unit. The risk of short circuit between the cables is limited, as it only exist within the modules. Due to the galvanic insulation, there is no risk of short-circuit in the communication cables between modules.

There is no rule but in practice, centralized BMS are rather used on low voltage battery packs (12, 24 or 48 V), low power while modular BMS are found on high voltage battery packs (400 or 700 V) and higher power. In the remainder of the document, we will only mention the case of the modular BMS, a little more complex from an electronic architecture point of view.

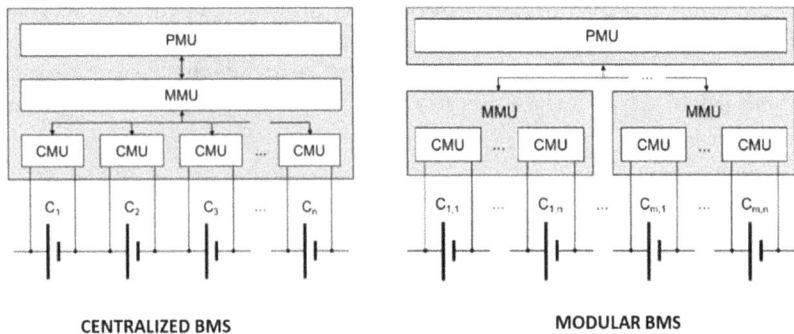

FIG. 14.13 – Electronic architecture of a BMS.

With a decentralized architecture (figure 14.14), there is a MMU board associated with each module and a central PMU board associated with the battery pack. Usually, several communication networks coexist: a first internal CAN network that allows communication between the MMU cards and the PMU cards and a second external CAN network that establishes communication between the battery pack (PMU) and the central electronics of the vehicle (ECU). To avoid potential communication faults, a redundant hardware signal called "Battery Enable/Disable" is frequently implemented on the different electronic boards. This signal alone triggers the opening of the contactors in the event of a critical fault that is not taken into account by the communication bus.

The PMU and MMU cards share the battery pack management functions. The PMU card, generally referenced to the potential of the mechanical ground, usually performs the following functions:

- Measurement of pack voltages and currents.
- Calculation of the states of the pack (SOC, SOE, SOH that we will see later).
- Calculation of the maximum admissible current in charge and discharge according to the state of the pack and transfer of this information to the central electronics.
- Management of load and balancing sequences.
- And sometimes insulation control, contactor control and thermal management pack control.

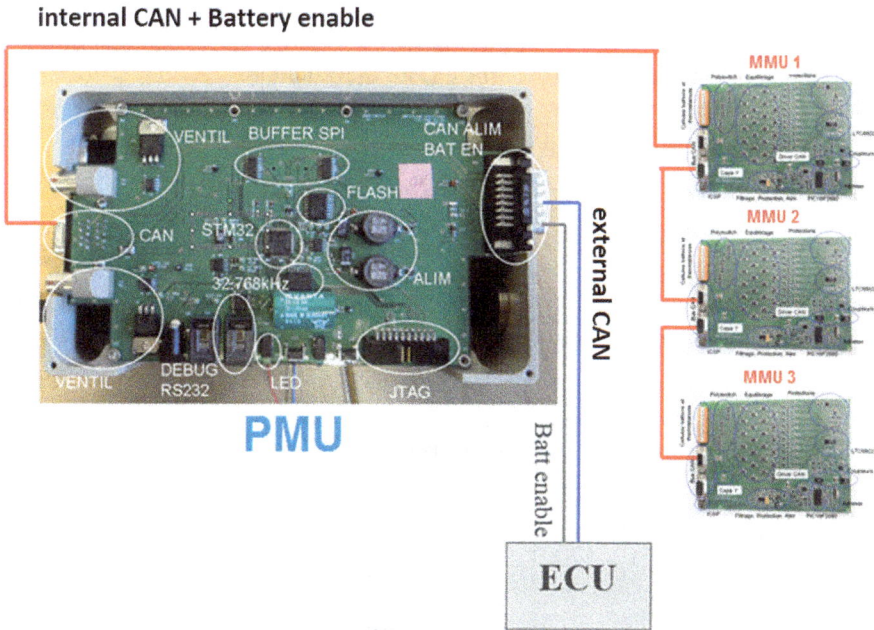

FIG. 14.14 – Interfaces ECU PMU and MMU.

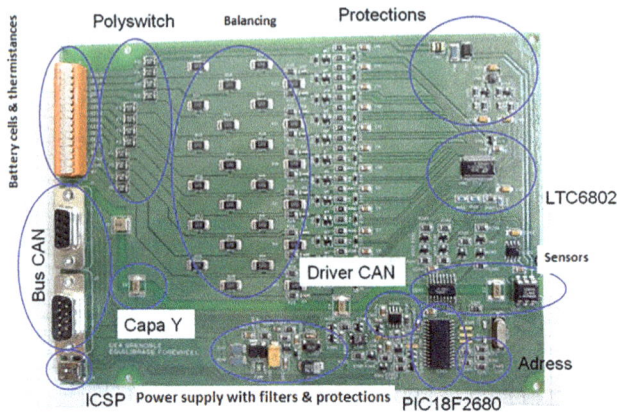

FIG. 14.15 – MMU board representation.

The MMU board (figure 14.15) associated with each module performs the functions:

– Insulation between the module potentials and mechanical ground.
– Measurement of cell voltages.
– Measurement of certain temperatures at module level.
– Cell voltage balancing.

To meet these functions, electronic component manufacturers offer components dedicated to the management of Li-ion batteries. Among these manufacturers are Linear Technology, Texas, Analog device, Intersil, Toshiba,... that all offer components allowing monitoring between 10 and 16 cells, which constitute a module. The modular approach has been extended to the electronic monitoring component.

Battery Pack Safety Management

The main function of the BMS is the safety management of the battery pack. Several dreaded events [iii], classified ASIL (Automotive Safety Insurance Level) C or D according to ISO 26,262 are to be avoided. They can lead to fire starts. In particular, the following ones can be noted:

– Overcharging of one or more cells of the battery pack-rated ASIL D.
– Overheating of one or more cells of the battery pack-rated ASIL D.
– The deep discharge of one or more cells of the battery pack-rated ASIL C.

On this last defined event, it is mainly the recharge that will follow a deep discharge that may be critical.

Management of the Battery States

Another important function is the management of the battery pack status. A full chapter (chapter 15) is dedicated to the definition of the states and the associated algorithms. The best known are the state of charge (SOC), the energy state (SOE) and the state of health (SOH).

Balancing of the Battery Pack

In order to be able to restore/deliver all its nominal energy, a battery pack must be balanced regularly. The faradic efficiency of all cells in a pack is close to 100%, but depending on the quality of the manufacturing process and local ageing phenomena, there may be some slight dispersions. Without proper and regular balancing, the usable energy of the battery pack will gradually decrease. To compensate for this phenomenon, a balancing system for the different stages of the battery pack is put in place. Balancing consists of placing all the stages of the series connection in the same state of charge. This function is mainly carried out at the end of the charge, because during this phase the differences in state of charge give rise to accentuated voltage differences, thanks to discrete measurements of the cell voltages, it is at this phase easier to align the state of charge of the stages to the same value.

The graph in figure 14.16 shows the charging and balancing phases of a 320 V battery pack (96 stages). In pink and blue are the voltage curves of the stages with the lowest and highest voltages respectively. These are the two stages that will theoretically take the longest to balance. The sequencing is as follows:

Phase 1: The battery pack is charged until one of the voltages of the stages reaches a maximum value set at 3.6 V in this case.

Phase 2: The stages with the highest voltage levels are discharged *via* resistors at a current of a few tens of mA until the voltage difference between the most and least charged stages is less than a few tens of mV. This stage dissipates energy. It is then called dissipative balancing unlike other types of balancing called non-dissipative balancing.

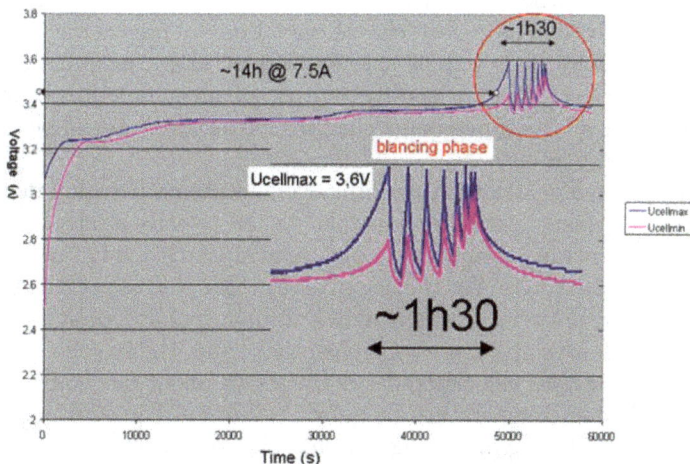

FIG. 14.16 – Balancing process during the end of charge.

Phase 3: Recharge again until 3.6 V is reached.

Phase 4: Repeat phase 2.

Then phases 3 and 4 are repeated until the stage with the lowest voltage has a voltage higher than 3.5 V at the end of the charge.

The battery pack is then balanced with cell voltages at the end of charge all between 3.5 V and 3.6 V.

The balancing time depends mainly on the unbalance, which itself depends on the production quality of the cells and the temperature homogeneity (therefore ageing) in the battery pack.

As previously mentioned, such balancing is dissipative, but due to the very low levels of unbalance to be compensated, the dissipated energies are negligible. Dissipative balancing is the standard solution used because of its simplicity, lower cost and greater reliability than the power electronics-based energy transfer solution.

Thermal Management of the Battery Pack

A final function assigned to the BMS is the thermal management of the battery pack, discussed in more detail in the next chapter.

Compared to the storage of electricity in capacitors, whose operating temperature range extends from − 40 to 85 °C, 105 °C or 125 °C depending on the technology, or compared to ultracapacitors, whose operating range is typically − 40 − 65 °C, lithium ion batteries have a narrower temperature range typically from − 20 to 60 °C in discharge and 0 to 45 °C in charge.

High temperatures have a significant impact on battery life that can be reduced to a few months. At negative temperatures, the internal resistance increases sharply, the capacity of the cells and the energy returned on discharge drop sharply.

When charging at negative temperatures, the electrochemical reaction of lithium metal deposition competes with the main reaction of inserting lithium ions into the negative graphite electrode. Due to this non-reversible reaction creating losses of Lithium in ionic form, the capacity of the battery decreases. Charging at low temperatures, especially if it is a fast charge, can lead to premature ageing of the battery, which is why it is out of specification of many manufacturers. Other manufacturers are beginning to specify the use of low-temperature batteries on charge and defining a level of maximum charging current that should be respected at different temperature levels.

The impact of temperature on performance and ageing is highly dependent on the lithium-ion chemistry used. For example, for the positive electrode, manganese oxide is the most sensitive chemistry for high temperatures due to the dissolution of manganese in the electrolyte. For the negative electrode, the use of Titanate instead of a graphite-based material allows avoiding lithium metal deposition even under rapid charging at low temperatures. On the other hand, the stored energy is lower since the voltage between the electrodes decreases by about one volt.

Due to the limited temperature range of lithium ion batteries and the strong impact on lifetime when approaching extreme high or low temperatures, the batteries are kept in a reduced temperature range by the system in which they are used. For example, by temperature conditioning of the room for stationary applications or very large storage, or by temperature conditioning of the battery pack for vehicle applications.

In thermal terms, a battery pack is characterized by a low level of loss, for example, for an efficiency of 95%, the losses represent only 5% of the stored energy, and a very high thermal inertia value, the heat capacity being the product $m.Cp$ where m is the mass and Cp is the mass heat of the components of the cells and the mechanics of the pack.

The high value of the product m.Cp provides a high pack inertia, which is able to store the losses of the battery pack with a temperature rise limited to a few degrees for a slow charge or discharge rate, *i.e.* typically less than 1 C. Thus, as was the case for the first electric vehicles based on lithium ion batteries, the battery pack may have no cooling if the charge of the vehicle is only a slow one.

During driving, the battery pack rises in temperature by a few degrees and then cools down slowly during slow charging phases.

However, such a battery pack is subject to all weather-related temperature fluctuations.

Currently, battery packs for electric vehicles are increasingly equipped with temperature conditioning to allow rapid charging and increase service life.

The following different technologies are used:

− Air conditioning circulation, so battery conditioning is shared with passenger compartment air conditioning.
− Circulation of glycol water in fine exchangers that meander between cylindrical cells or which are in contact with rigid prismatic cells or pouch cells.
− Refrigerant circulation in fine exchangers in thermal contact with cells.

The heat pump of the air temperature conditioning system, glycol water or by contact with the evaporator can also ensure a low temperature pack heating by cycle inversion, or this function can be performed by an electric resistance to simplify the system and to reduce its cost.

The temperature management system must meet many constraints because the battery pack is an electrical equipment with all the accumulators at different potentials.

14.5   Design and Manufacture of Battery Packs

The pack must be composed of electrical safety requirements, available volume and mechanical stresses that he will receive in his lifetime.

In the design of a battery pack, two requirements are important:

− Mass energy density: amount of energy per kg of battery (expressed in $Wh.kg^{-1}$).
− Energy density by volume: amount of energy per liter of battery (expressed in $Wh.L^{-1}$).

The higher these values, the better the performance of the battery pack.

Mechanical Design

As seen previously, the division of a pack into modules facilitates management:

- Modularity, by allowing, with a single type of "standard" module, creating several packs of variable voltage and capacity.
- Maintenance, because it is possible to replace a defective module quickly.
- Safety, by limiting the risks of incident propagation.

  However, it also has disadvantages:

- The addition of extra parts, and therefore mass and volume...
- Additional electrical connections, which involve more hot spots, *i.e.* connected to a dangerous potential, and the risk of disconnections and failures.

The choice of a division into modules in a pack is therefore the result of a compromise between safety and ease of maintenance on the one hand, and performance on the other.

During the mechanical design, other parameters are important:

Mechanical cell variations while using the pack deserves a particular attention. Indeed, some cell chemistries deform more than others cells, and must sometimes be mechanically stressed.

The operating temperature is also a critical parameter. Only correct sizing can ensure the best possible performance and avoid premature ageing of the battery.

The design of a module or battery pack shall also take into account the manufacturing requirements. For example, it shall take into account the existence of live parts, which may cause a high electrical risk of short circuit. Electrically insulated tools must be used for assembly, making them more cumbersome than those normally used in mechanical engineering. In some cases, specific tools must be designed to make up for the lack of space.

The insulation distances between all conductive parts connected to different potentials are of prime importance. In most cases, double insulation is recommended.

Modules' Assembly Because of the work under voltage, any assembly must be carried out in accordance with the rules defined by the regulations (electrical approvals, PPE (Individual Protective Equipment), EPC (Collective Protective Equipment). During the entire assembly sequence of the module and the pack, it is essential to prevent any risk of short-circuit of the electrical circuit. Voltage and insulation resistance measurements are carried out in real time to guarantee the safety of people and the product.

Module assembly begins with mechanical assembly. The basic rule is that no part can act as both a mechanical support and as an electrical conductor at the same time.

There are two commonly used techniques for the mechanical assembly of cells. By gluing:

- Either the cells are glued together.

– Either they are glued together by means of honeycomb parts, for example.

By mechanical holding:

– Either the cells are clamped between two mechanical parts.
– Either the cells are circled.

Other mechanical holding techniques can be used, while aiming to limit mass and volume.

Once the cells have been held together, the electrical connection can be made:

With commutators soldered on cells with flat terminals (see figure 14.17).

With "busbars" screwed directly onto the terminals of the cells with a threaded hole or stud.

The current collectors consist of 0.2–0.8 mm thick metal strips, cut to match the surface of the terminals to be connected. The materials they are made of make it easy to solder to the cell terminals. In order to increase the current flow cross-section, the collectors can be lined with a thicker copper foil, which is perforated at the solder joints.

At present, three types of solders can be used:

– Resistance welding, the most common.
– Laser welding.
– Ultrasonic welding.

Soldering with a soldering iron directly on the terminals is not recommended. Heat is a source of degradation and danger for the cells, which can even lead to thermal runaway if this heat input is poorly controlled.

"Busbars" (figure 14.17) are conductor bars, sometimes electrically insulated. They offer a large cross-section, facilitating the passage of current.

Once the cells connected, the balancing board can be mounted and then the module can be closed (figure 14.17) to provide electrical insulation or to ensure that it is sealed from the outside.

Final Assembly of the Pack

A pack consists mainly of modules, a BMS and also a casing (packaging), or a structure to hold the whole (figure 14.2). This casing or structure is the backbone of

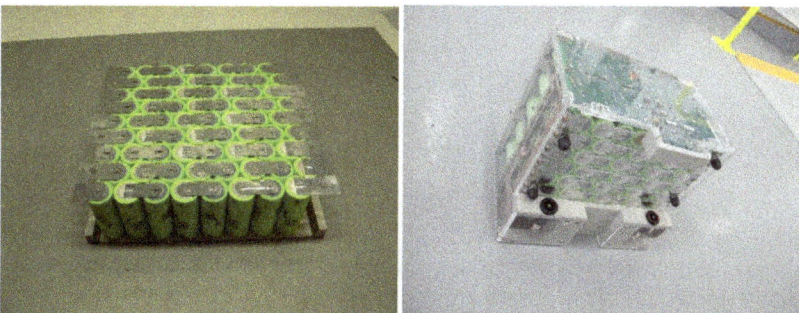

FIG. 14.17 – From the cell assembly to the module realization.

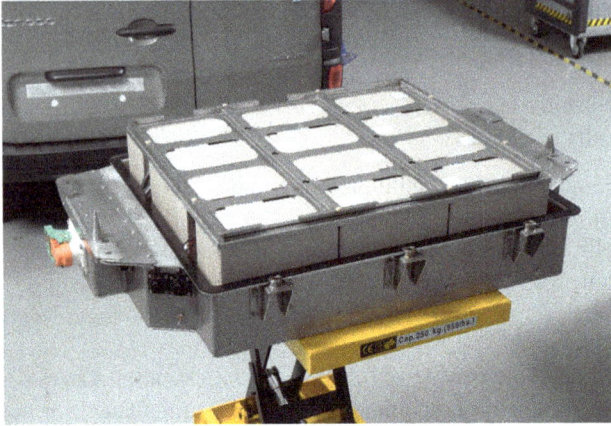

FIG. 14.18 – Final assembly of a battery pack.

the pack, the assembly of this part must be controlled during the completely manufacturing process.

Another essential point, which guarantees the correct operation of the pack, is the wiring and the installation of the BMS, the safety and control devices. Depending on the pack and its assembly sequence, this wiring can be done before or after the modules are installed. The assembly upstream of this cabling allows limiting the working phases under tension. It is therefore preferred.

The installation of the modules is very specific to the packs. If the modules are closed and therefore insulated, this work can be very simple. On the other hand, in the case of open modules, the work can become very complex. The regulations do not allow having several bare parts under high voltage. It is then necessary to protect the working areas as the work progresses.

The electrical connections (power cables and communication circuits) are then installed between each module. The package is then often integrated into a housing (figure 14.18) and tested.

The design and assembly of a battery pack requires special expertise, the mastery of which determines the safety of use by the end customer, as well as the lifespan of the complete system.

14.6   Examples of Innovation on Battery Systems

The CEA has supported numerous partners in the design and production of battery systems, always with the aim of trying to bring a share of innovation and added value to conventional battery systems. The following two examples illustrate this. They have led to the filing of numerous patents and publications.

Example 1: The Switched Battery

The principle of the switched battery [i] is to associate each battery stage or each group of battery stages with a power electronics function that enables the stage in question to be switched on or "bypassed". Then, thanks to an intelligent control system, we get a variable-voltage battery. As the switched voltage levels are low (from a few volts for one stage to a few tens of volts for several stages), it is possible to use transistors with very low losses. The architecture of such a system is shown in figure 14.19. An alternating voltage (figure 14.20) is generated between the two poles, here noted $\phi 1$ and neutral.

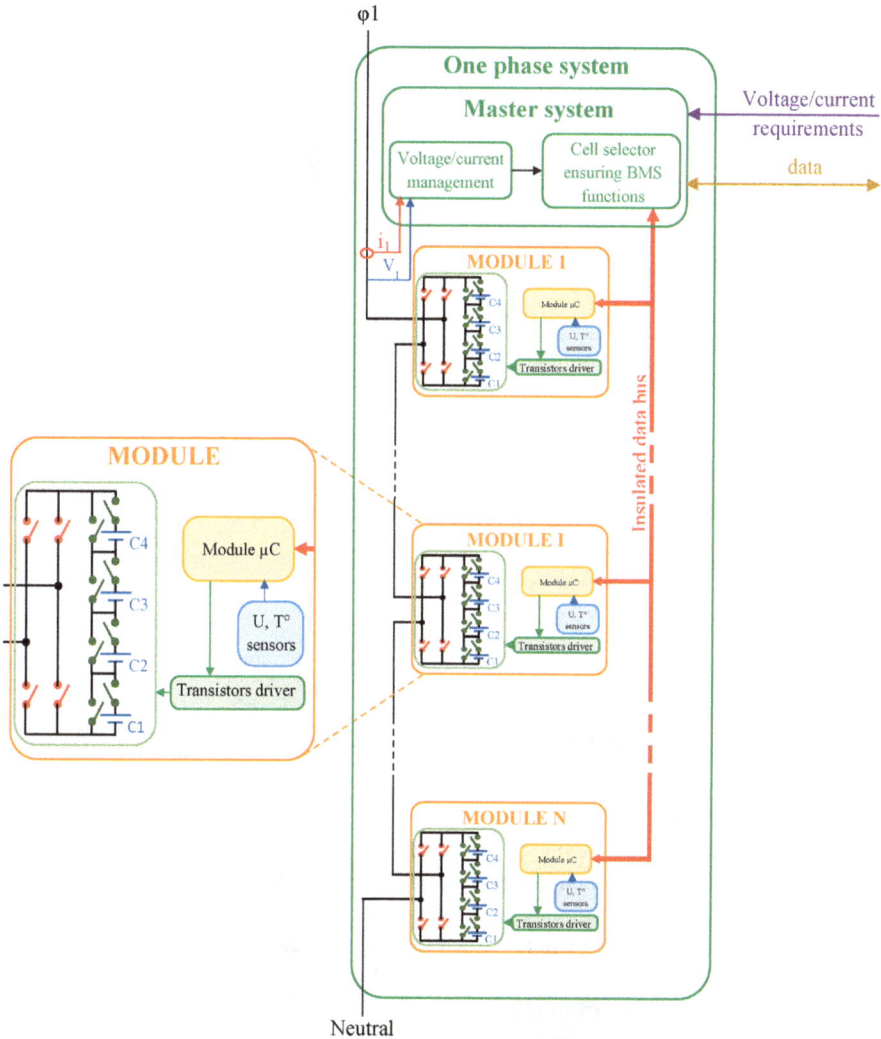

FIG. 14.19 – Electronic architecture of the switched battery.

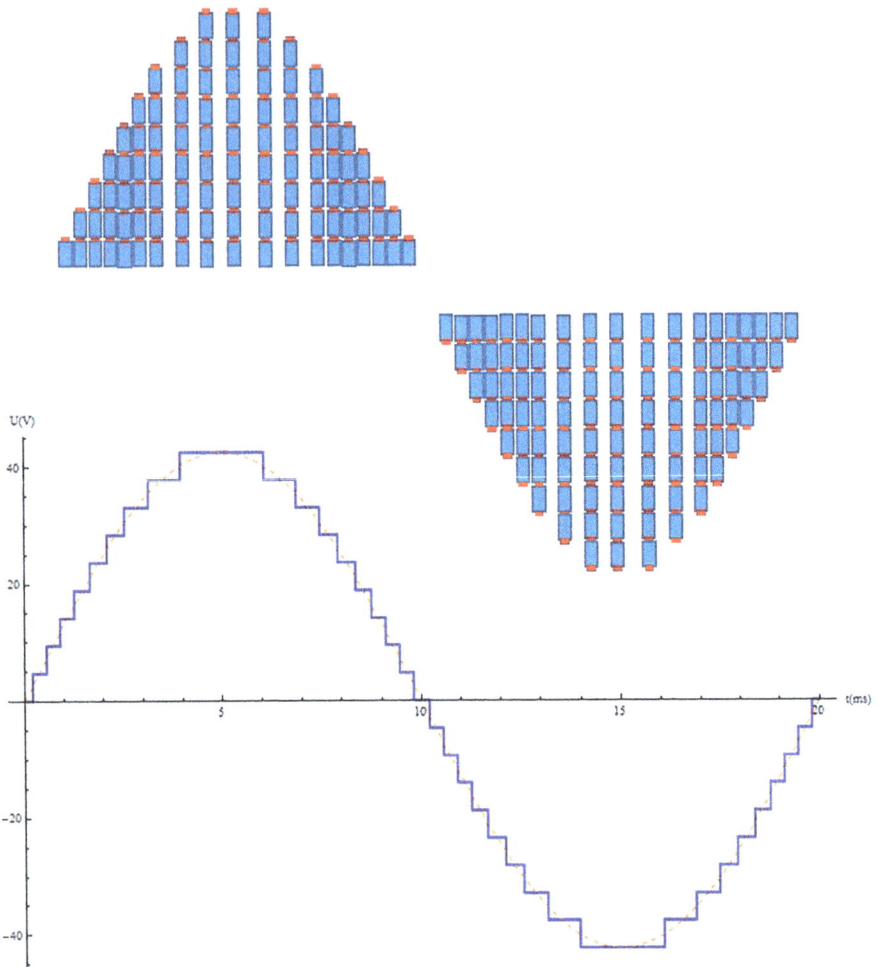

FIG. 14.20 – Reconstituted sinusoid with switched battery.

Compared to a conventional system, the switched battery solution allows:

- Removing the inverter.
- Removing the charger.
- Ensuring continuity of service in the event of a cell failure.
- Making the best use of each cell (no more limitation by the weakest cell): gain in autonomy.
- Increasing conversion efficiencies, both during charge and discharge: less heat to be removed.
- Drastically reducing or even eliminating filter elements (inductor).
- Drastically reducing electromagnetic emissions (EMC).
- Ensuring fast recharging.

- Managing both AC and DC voltage: for example, direct recharging by solar power in DC with MPPT (Maximum Power Point Tracking) type operation and AC restitution of the energy to the grid.

Example 2: System Architecture Without Auxiliary Network Battery

In most electric transport applications, two networks are used simultaneously. A 400 or 700 V high power network is used for electric traction or propulsion and a low voltage network (12 V, 24 or 28 V) supplies the auxiliaries. This second network is

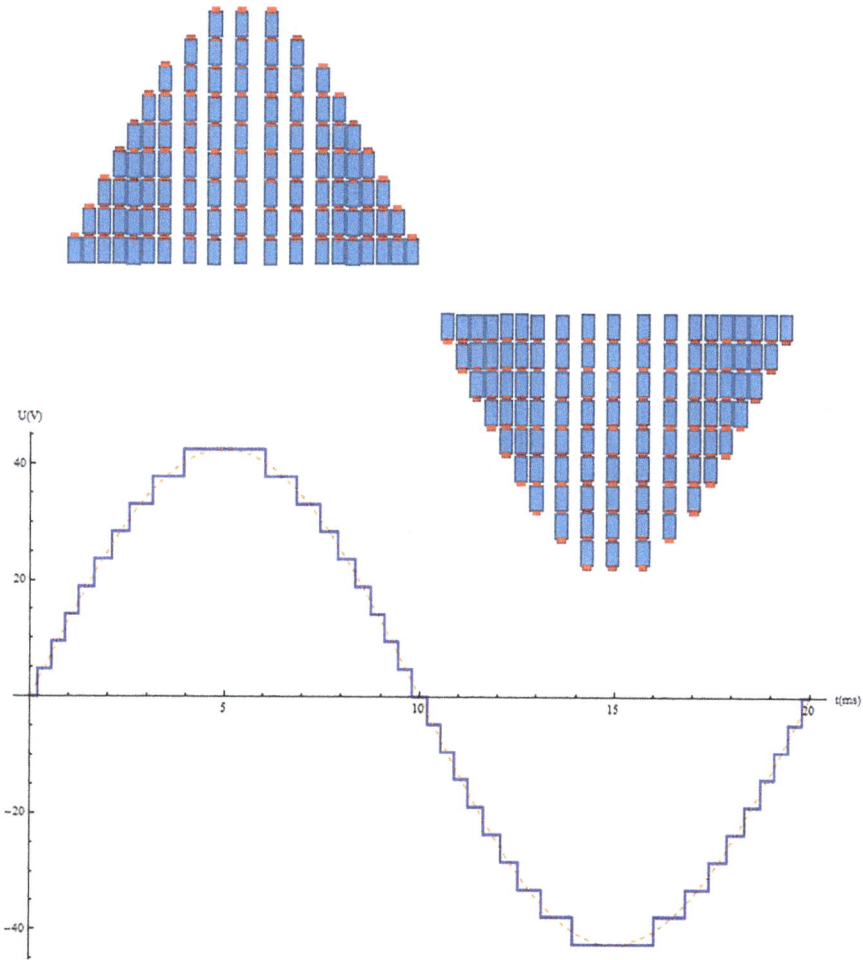

FIG. 14.21 – Battery modules interfacing with auxiliary network.

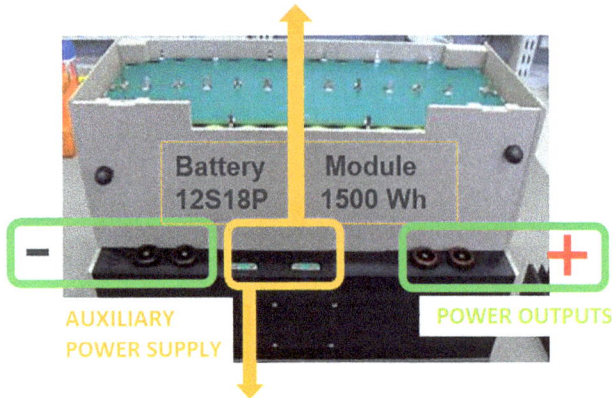

FIG. 14.22 – Battery module with integrated converter (power and auxiliary).

also used to start the system. A low-voltage battery is therefore not only necessary for starting but also for compensating in case of a potential failure of the DC/DC converter supplying the 12 V from the 400 or 700 V.

The battery system shown in figure 14.21 uses an advanced modular battery concept that integrates a low voltage DC/DC converter in each module [ii]. Several smaller converters connected in parallel have replaced the main DC/DC converter.

The three main advantages of this solution are:

Advantage 1: It is possible to remove the auxiliary battery since several small DC/DC converters provide redundancy for the 12 V mains supply.

Advantage 2: The number of DC/DC converters used to supply the auxiliary grid can vary depending on the power used. As each small converter works at nominal power, the overall efficiency of the solution is improved.

Advantage 3: It is possible to ensure a powerful inter-module balancing by prioritizing the module supplying power to the auxiliary network. In this way, disparities in capacity between modules are balanced, which is very interesting when replacing a aged module with a new one.

The standard battery module becomes an advanced module (figure 14.22) integrating two power distribution systems: one for power (*e.g.* traction) and one other for auxiliaries.

In the two examples presented, the objective was to carry out a reflection at the application level in order to pool certain functions and bring added value to the battery subsystem.

References

[i] Despesse G., Sanjuan S., Gery S. (2012) *Battery monitoring system using switching battery cells*, Innovation for Transport Systems of the Future, Nov 2012, Paris, France, pp.12–15.

[ii] Garnier L. (2014) An innovative balancing solution used to supply the 12 V auxiliary network of an electric vehicle, PCIM 2014.

[iii] Everlasting H2020 project No. 713771 Electric vehicle enhanced range, LIfetime and safety through INGenious battery management, Deliverable 6.1.

Chapter 15

Definition of the State Estimation Algorithms of a Battery System and Associated Calculation Methods

Vincent Heiries, P.-H. Michel, Arnaud Delaille and Fathia Karoui

The purpose of this chapter is to introduce the various indicators related to the battery state and the monitoring algorithms used to estimate these indicators. All the indicators characterizing the state of a battery are described, and more specifically, the State of Charge (SOC) and State of Health (SOH) estimators are presented here in detail. These algorithms are based jointly on equivalent electrical circuit scheme modeling of Li-ion cells and on adaptive filtering techniques (Kalman filter).

We first detail how adaptive filtering techniques represent interesting signal processing tools to assess battery cell SOC and SOH. An overview of the Kalman filter is presented to understand its operating principle applied to the estimation of the SOC. Then an algorithm for estimating the cell capacity is presented.

15.1 Battery State Indicator Definition

15.1.1 State of Charge

The SOC (state-of-charge) of a battery is the amount of stored electrical charge $q(t)$ relative to the current capacity $C(t)$:

$$SOC(t) = \frac{\text{stored electrical charge}}{\text{current capacity}} = \frac{q(t)}{C(t)}$$

The state of charge is expressed as a ratio and usually varies between 0% (fully discharged cell) and 100% (fully charged cell), although it can take values outside

DOI: 10.1051/978-2-7598-2555-4.c015

this usual range depending on how it is estimated and how it is defined by convention.

The stored electrical charge of a battery can thus be expressed using the SOC:

$$q(\text{SOC}) = \text{SOC} \times C$$

As the capacity value changes over the life of the battery with aging, an index can advantageously specify the capacity C taken as a reference for the SOC value. As an example, SOC_{BOL} means that the capacity C at the beginning of life is used as a reference value for the SOC calculation; this definition is taken from the reference [i].

15.1.2 State of Energy

The SOE (State-of-Energy) differs from the SOC state of charge by referring to a quantity of energy rather than a quantity of charge. It indicates the amount of electrical energy stored relative to the current energy storage capacity EC:

$$\text{SOE} = \frac{E_{\text{stored}}(t)}{\text{EC}}$$

$$\text{SOE}(\text{SOC}) = \frac{E_{\text{stored}}(\text{SOC})}{\text{EC}} = \frac{\int_{q(\text{SOC}=0\%)}^{q(\text{SOC})} V_{\text{Bat,OCV}}(q).dq}{\int_{q(\text{SOC}=0\%)}^{q(\text{SOC}=100\%)} V_{\text{Bat,OCV}}(q).dq}$$

with $V_{\text{Bat,OCV}}(q)$ the open circuit voltage of the battery.

The energy state is expressed as a percentage and varies in the same way as the usual state of charge between 0% and 100%.

According to the following equation, the evolution of the state of stored energy E_{stored} can be calculated using the SOE at the beginning and the SOE at the end of a charging or discharging process.

$$\Delta E_{\text{stored}} = \int_{q(\text{SOE}_{\text{beginning}})}^{q(\text{SOE}_{\text{end}})} V_{\text{Bat,OCV}}(q).dq = \int_{E_{\text{stored,beginning}}}^{E_{\text{stored,end}}} dE_{\text{stored}}$$

Finally, similarly to the stored electrical charge and SOC, stored energy can be expressed using the value of the SOE that relates to E_{stored}, the stored energy, to the current energy storage EC capacity as per:

$$E_{\text{stored}}(\text{SOE}) = \text{SOE} \times \text{EC}$$

15.1.3 State of Health

The concept for representing the performance and health condition of a battery compared to a new battery is called State-of-Health (SOH). The SOH indicator is defined as the current capacity of a fully charged battery as a percentage of its initial beginning of life fully charged capacity. The definition adopted under reference [ii] is expressed as follows:

$$\text{SOH} = \frac{\text{Current stored capacity(SOC} = 100\%)}{\text{Initial BOL capacity(SOC} = 100\%)}$$

In a similar way to the SOC estimate, an index can advantageously specify the total capacity taken as a reference, which can sometimes correspond not to the total capacity of the battery under given conditions but to the nominal capacity given by the manufacturer, or even correspond to the total available capacity potentially limited by the electronic management board known as BMS (Battery Management System).

It also sometimes happens that the state of health refers to energy rather than capacity or even to a resistance value when the available power is an important factor linked to the use of the battery.

15.1.4   *State of Function*

The SOF state indicator is used to define the performance of a battery during an operation, which is specific to each application. SOF takes into consideration the influence of the state of charge range, charge and discharge rate, ambient temperature and other degradation factors. Indeed, it is used to describe the extent to which the performance of the battery adequately meets the expected requirements during operation.

For example, the SOF of a starter battery is the ability of the battery to deliver or not deliver the power required (and therefore not directly related to the state of health as defined in the previous section). Similarly, the SOF of a backup battery corresponds to the ability of the battery to deliver a certain amount of power over a predetermined time.

15.1.5   *State of Safety*

The state-of-safety (SOS) indicator is based on the concept that safety is inversely proportional to the concept of misuse. This state uses the same range as the other commonly used battery states SOC, SOE, SOH and SOF, with values between 0% (completely unsafe) and 100% (completely safe). It combines the effects of an arbitrary number of sub-functions. Each of these sub-functions describes a particular case of battery abuse using one or more variables such as voltage, temperature, mechanical deformation, etc.... Based on a definition of the SOS as a probability function, this characteristic value can be calculated as the product of the distribution of the individual sub-functions as defined in the reference [iii]:

$$\text{SOS}(x) = f_1(x) \times f_2(x) \times \cdots f_n(x)$$

$$\mathrm{SOS}(x) = \prod_{k=1}^{n} f_k(x)$$

There is no single correlation between a single parameter and a single safety hazard.

For example, thermal runaway is the result of extreme heating, and this heating may come from a variety of sources such as a high current demand related to a short circuit, a local hot spot (internal short circuit related to lithium deposition), or external heating (fire). Therefore, it is not sufficient in most cases to focus on a single variable, *e.g.* current, to prevent thermal runaway.

Engineers designing or implementing battery management systems for an electrical storage system must decide, based on their experience or the priorities they identify, which properties to control in order to correctly describe an abusive condition they want to avoid.

15.2 Battery Diagnosis Methods

Different approaches exist to diagnose the state-of-charge (SOC) and state-of-health (SOH) of batteries that are the most important and most often estimated indicators. A review of these different approaches can be found in the following documents, among others [iv–ix].

For the diagnosis of the SOC, a distinction is usually made between the evaluation of the following parameters and methods, which can be used in a complementary way: Open Circuit Voltage [x], Coulometry [xi], Impedance measurement [xii], Equivalent Electrical Circuit [xiii], Adaptive filtering estimation methods [xiv–xvii], or Machine Learning Methods [xviii, xix].

Figure 15.1 proposes a classification of these different approaches:

We will see, in the following paragraphs, the most classical methods of the state of the art of SOC and SOH estimation.

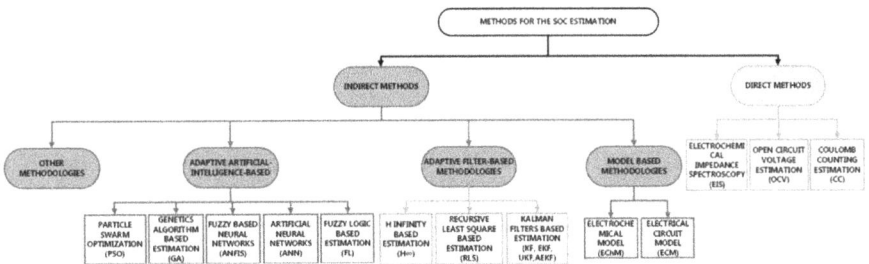

FIG. 15.1 – Classification proposed in reference [ix] of the various SOC estimation methods.

15.2.1 State of Charge Estimation

15.2.1.1 Coulomb Equation – Description of the Conventions

The Coulomb equation defines the SOC (between 0% and 100%) by the following equation (positive current in charge and negative current in discharge by convention):

$$\text{SOC}(t) = \text{SOC}(t_0) + \int_{t_0}^{t} \frac{\eta i dt}{C_t} \text{ (Coulomb equation)}$$

The variable i is the current (A) and C_t is the cell capacity (A.h) at time t. η is the Coulomb efficiency (efficiency of conversion of chemical energy into electrical energy), this value tends towards 1 for Li-ion cells. Over a range of 100% SOC the whole capacity of a cell is theoretically used. However, this equation does not make it possible to determine the SOC at time t precisely, because:

- The SOC at t_0 is not easily known without prior measurement.
- There is a risk of significant bias that increases over time because of the integral term that can accumulate errors (current, capacity).
- The capacity at time t is not easily known, it changes as battery cell ages.
- This current accumulation modeling is not sufficient to describe the evolution of the SOC since in reality other variables are taken into account in its evolution. In particular, the temperature and the charge/discharge rate affect the amount of charge that can actually be used.

15.2.1.2 Battery Cell Voltage Modelling

The terminal voltage of a Li-ion cell is conventionally modelled by a circuit connecting in series a no-load voltage and an impedance based on parallel RC circuits. We use the following model:

The impedance parameters R_0, R_{diff} and C_{diff} are time-dynamic parameters (dependence on SOC and SOH). They are also dependent on the current flow rate as well as on the flow mode (charging or discharging). The open circuit voltage (OCV), on the other hand, is a relatively stable parameter over time. However, it also depends on the mode (charge or discharge) and for this reason the integration of a hysteresis parameter can be useful. An example of OCV mapping as a function of the SOC is given in figure 15.3.

Considering the circuit shown in figure 15.2, the evolution of the cell voltage is as follows:

$$z(t) = \text{OCV}(\text{SOC}(t)) + V_d(t), \text{ with } i < 0 \text{ in discharge mode}$$

From this equation, the SOC is accessible by inversion:

$$\text{SOC}(t) = \text{OCV}^{-1}(z(t) - V_d(t))$$

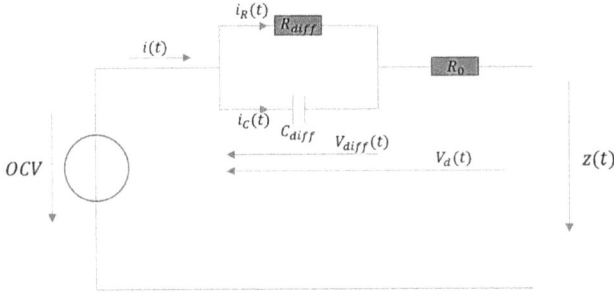

FIG. 15.2 – First-order modeling of a Li-ion cell.

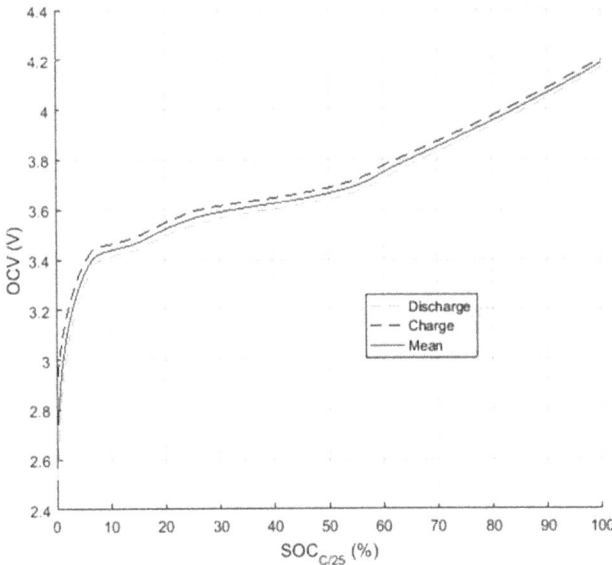

FIG. 15.3 – OCV of a NMC cell.

However, there are limitations. On the one hand, voltage z is not measurable with infinite accuracy, voltage V_d is not measurable and must be estimated efficiently. On the other hand, OCV^{-1} mapping is very sensitive to errors in its center range. Thus, a few mV voltage errors can lead to a SOC error of several percent.

15.2.1.3 *Modelling and Bayesian Inference*

We have seen in the two previous parts that two main models of the evolution of the SOC can be retained. However, each of these models has its own drawbacks, leading to insufficiently accurate estimates of the SOC. The following section introduces a definition of the problem analytically defined through a state observer in order to get closer to a near-optimal estimate of the SOC.

15.2.1.3.1 *State-Space Observer*

- *SOC state variable*

The SOC, because of its integral nature, can be modeled as a state equation that expresses its evolution from time k to time $k+1$. At order 1, the evolution of the SOC state is directly established by the expression:

$$\text{SOC}_{k+1} = \text{SOC}_k + \eta \frac{I_k T_e}{C_k} = f(\text{SOC}_k, u_k, w_k) = \text{SOC}_k + f(u_k, w_k)$$

The term u_k is called command term and is here equal to $\eta \frac{I_k T_e}{C_k}$, with T_e the sampling period. Here a noise term (denoted w_k) is used to model the uncertainty of this transition equation. We do not make any hypothesis for the moment on the nature of the statistic distribution of w_k. Later, in the Kalman filtering formalism, we will indicate them. Figure 15.4 shows what the statistical representation of the evolution of the SOC implies. On the left is the probable distribution of the SOC at instant k and on the right that of the SOC at instant $k+1$.

- *Cell voltage as observation variable*

The measurement (or observation) equation is written as follows:

$$z_k = \text{OCV}(\text{SOC}_k) + V_{\text{diff}_k} + R_0 i_k + v_k$$

Here the term v_k, commonly known as measurement noise, is an additive noise term that mainly represents the noise of the voltage sensor as well as the current sensor, as well as modeling errors. We observe that this measurement equation is not linear due to the OCV term but also due to the term V_{diff_k} that depends in a complex way on the SOC.

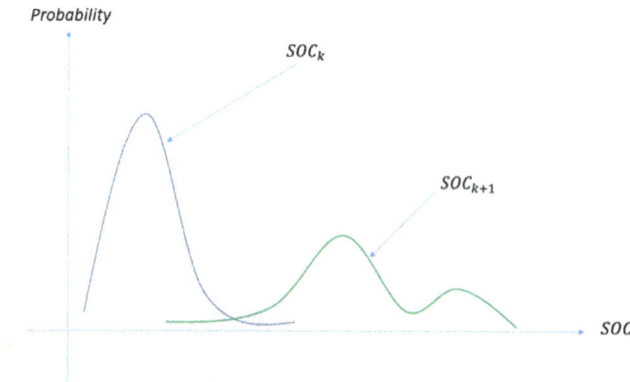

FIG. 15.4 – Pictorial representation of the transition density.

To summarize the following pair of equations can be formed:

$$\begin{cases} X_{k+1|k} = \text{SOC}_{k+1|k} = \text{SOC}_{k|k} + \eta \frac{I_k T_e}{C_k} + w_k \\ z_{k+1} = g\big(\text{SOC}_{k+1}, i_{k+1}, V_{\text{diff}_{k+1}}\big) + v_{k+1} = \text{OCV}(\text{SOC}_{k+1}) + V_{\text{diff}_{k+1}} + R_0 i_{k+1} + v_{k+1} \end{cases}$$

15.2.1.3.2 *Optimal Bayesian Filter*

In order to find the optimal estimate value of X_k it is a matter of finding the most probable value of X_k, knowing the set of previous voltage measurements (hence the Bayesian term of the method, in reference to Bayes' law of conditional probabilities). In other words, we are looking for the most probable mode of appearance of X_k knowing the observations $z_{0:k}$ from the initial instant to instant k, which aim at maximizing the probability density $p(X_k|z_{0:k})$ representing the *a posterior i* density of the state X_k.

$$\hat{X}_{k|k}^{\text{MAP}} = \text{argmax}_{X_k} p\big(X_k|z_{0:k}\big)$$

Equation (15.1): *Optimal Bayesian filtering framework.*

This probability depends on both the evolution described by the state equation and the measurement equation.

In the general case it is complex to express analytically $p(X_k|z_{0:k})$, especially when the state and/or the measurement evolve non-linearly. Moreover, we see that the optimization problem takes an infinite dimension when k tends towards infinity. It is therefore a question of expressing the solution of this problem in a recursive form. Then, under certain hypothesis concerning the state and nature of the perturbations, the Kalman filter can be developed.

15.2.1.3.3 *Kalman Filter Design*

The Kalman filter solves the maximization problem presented in equation 1. The analytical solution is known and optimal (consistent, zero bias and minimum variance estimator) under the following conditions:

- The processes are linear.
- The w_k and v_k noises are white and Gaussian.

  o This hypothesis is difficult to verify, but the Kalman filter remains robust if there is a deviation (content) from the Gaussian character. There are different variants of the Kalman filter. In particular, the Sigma Point Kalman Filter correctly estimates distributions up to 3rd order statistics.

- The w_k and v_k noises are centered.

  o If not, it highlights a modeling error. However, the well-calibrated Kalman filter corrects these bias potentials.

- The w_k and v_k noises are independent of each other.

○ This condition is checked if the noise on the current sensor is independent of the noise on the voltage.

- The X_k state is of Markov 1 type: its evolution at time $k + 1$ depends only on its state at time k.

○ This hypothesis is tested according to our SOC modeling.

- The state and observation equations are linear.

○ This hypothesis is unverified, but the solution can be obtained by linearization (extended Kalman filter) or by estimating the evolution of probability densities (Kalman Sigma Point filter).

Under verification of these hypotheses, the solution takes the very simple form of a recursive filter, where the estimate $\hat{X}_{k+1}$ is a weighted average between the prediction given by the state equation and the measurement error:

$$\hat{X}_{k|k} = \hat{X}_{k|k-1} + K(\text{measured voltage} - \text{computed voltage knowing the prediction})$$

K is the Kalman gain and is derived from the maximization problem posed by equation 1. The higher the relative confidence in the state equation, the lower the Kalman gain, the better the prediction. Conversely, if one has a very high confidence in the measurement equation, then at each observed voltage error, the Kalman filter corrects strongly (high K gain) to catch up with the trajectory of the state variables (here the SOC).

This Kalman filtering process in the SOC estimation can be represented as a comparator type, block diagram, illustrated in figure 15.5, with as input the measured voltage which is compared to the voltage estimated by the observation model applied to the current state prediction.

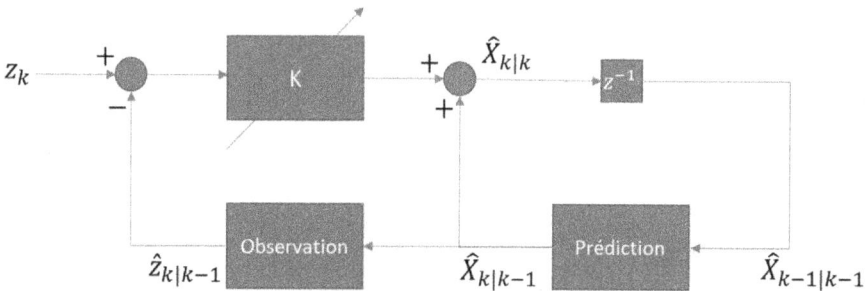

FIG. 15.5 – Block diagram representation of Kalman filtering.

15.2.2   *Kalman Filter Exploitation for State of Charge Estimation*

15.2.2.1   *Improving the Estimation Performance: The Sigma Point Kalman Filter*

Calculation of the Kalman gain requires estimation of the Gaussian error of the estimation and the observation. In the case of non-linear processes (*e.g.* OCV mapping), such an estimate cannot be easily expressed. The extended Kalman filter linearizes these processes at first order, but in the case of too rapid variations the approximation leads to errors that can be significant.

The Sigma Point Kalman filter does not linearize these processes, but estimates *in-situ* the Gaussian evolution relative to the state and observation processes. As the authors point out in the report [xx], it is easier to approximate a Gaussian distribution (*i.e.* to find its mean and variance) than to approximate any non-linear process.

15.2.2.2   *Sigma Point Kalman Filter Application to SOC Estimation*

In order to improve the estimation of the SOC, we present here a possible implementation of the Sigma Point Kalman filter because it is theoretically more robust to cope with non-linearities and is equivalent in computational complexity to the extended Kalman filter (for small dimensions).

The choice of the state – measurement representation is as follows ($N = 2$ states):

$$\begin{cases} X_{k+1} = \begin{bmatrix} \text{SOC} \\ V_{\text{diff}} \end{bmatrix}_{k+1} = \begin{bmatrix} 1 & 0 \\ 0 & 1 - \frac{T_e}{R_{\text{diff}} C_{\text{diff}}} \end{bmatrix} \begin{bmatrix} \text{SOC} \\ V_{\text{diff}} \end{bmatrix}_k + \begin{bmatrix} \eta \frac{T_e}{C_k} \\ \frac{T_e}{C_{\text{diff}}} \end{bmatrix} I_k + w_k \\ z_{k+1} = \text{OCV}(\text{SOC}_{k+1}) + V_{\text{diff}_{k+1}} + R_0 i_{k+1} + v_{k+1} \end{cases}$$

The addition of the V_{diff} state allows filtering potential modeling errors from this voltage and this is particularly beneficial because of the known recursive formulation of its evolution. The SPKF (Sigma Point Kalman Filter) algorithm is formulated very clearly in this publication [i, xx].

An example of SOC estimation performance by Kalman EKF and SPKF filtering is given in figure 15.6. It can be seen that the SOC estimation error is already low for the EKF filter, mostly below 3% error, with a maximum at 5%. The estimation accuracy is even better for the SPKF filter, with an estimation error never exceeding 1.5%.

15.2.3   *Battery Total Capacity Estimation*

15.2.3.1   *Framework*

By estimating the cell capacity and the evolution of this capacity, the state of health (SOH) can be obtained. This battery capacity, which appears in the Coulomb

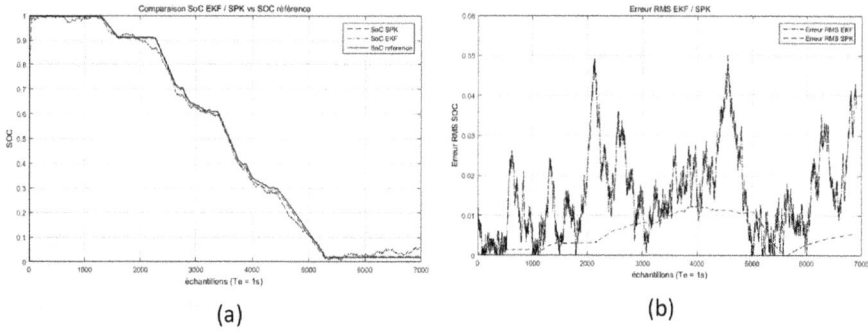

FIG. 15.6 – Example of SOC estimation result by Extended Kalman Filter and Sigma Point Kalman Filter. (a) SOC estimation; (b) estimation error with respect to the reference SOC.

equation, is a parameter that varies particularly with the ageing of the battery, and as a function of the current rate, temperature and storage conditions. However, it turns out that the coupling of the Kalman filter with a well controlled linear regression gives good results to estimate the ageing evolution of this capacity.

15.2.3.2   Linear Regression

Let us take again the Coulomb equation in numerical format ($\widehat{SOC}_k$ is the estimated SOC at instant k):

$$\widehat{SOC}_{k+1} = \widehat{SOC}_k + \frac{\eta I_k T_e}{C_k} \Leftrightarrow \left(\widehat{SOC}_{k+1} - \widehat{SOC}_k\right) C_k = \eta I_k T_e \Leftrightarrow C_k \hat{x}_k = y_k$$

If we consider the dynamics inherent in ageing, then it is assumed that for the duration of a few cycles the capacity is constant, so the previous equation becomes:

$$C \hat{x}_k = y_k$$

This problem can be solved by a linear regression to optimize the least squares minimization criterion:

$$\hat{C}_{RLS} = \operatorname{argmin}_{\hat{C}} \sum_{i=1}^{N} \left(y_i - \hat{C}\hat{x}_i\right)^2$$

The problem thus formulated leads to a biased solution because the inputs $\hat{x} = \widehat{SOC}_{k+1} - \widehat{SOC}_k$ are noisy data. The idea retained in order to reduce the bias of the estimated solution on capacity is to improve the SNR (Signal to Noise Ratio) of the inputs $\hat{x}$. The problem is therefore transformed into the following form:

$$\Leftrightarrow \sum_{j=l}^{l+M} \left(\widehat{SOC}_{j+1} - \widehat{SOC}_j \right) C = \sum_{j=l}^{l+M} \eta I_k T_e$$

$$\Leftrightarrow C \sum_{j=l}^{l+M} \hat{x}_j = \eta T_e \sum_{j=l}^{l+M} I_j$$

$$\Leftrightarrow C \left(\widehat{SOC}_{l+M+1} - \widehat{SOC}_l \right) = \eta T_e \sum_{j=l}^{l+M} I_j$$

$$\Leftrightarrow C \hat{X}_l = Y_l$$

A Recursive Least Square algorithm, or one of its variants, can be used to reach a simple and accurate solution to this problem.

It can be noted that instead of performing the linear regression with points of too low amplitude, one can favor high levels of SOC (and thus current) variation to improve the robustness to the uncertainties of this regression. Figure 15.7 illustrates this principle.

On the right, with a lower derivative value, is the perfect straight line giving access to the C capacity. On the left, of higher derivative value, is the line obtained (and thus the capacity) by regression whose results are affected by an error on the x-axis. In median, the line obtained by regression is in the same way with another set of observation points. We can see that the error on the capacity, with a constant error of abscissa, decreases as we process points spaced from the origin. Note that this drawing represents the worst case since:

- the errors on the x-axis are opposite according to the sign of the ordinates;
- the line is defined with few points, our algorithm is in fact recursive and operates continuously to "adapt" to the best estimate.

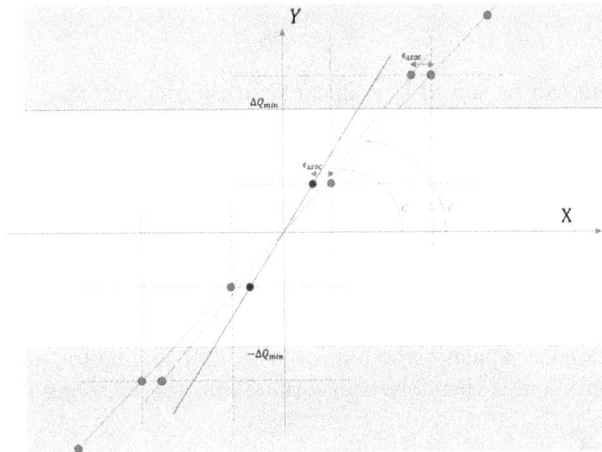

FIG. 15.7 – Principle of linear regression of the cell capacitance.

15.2.4 Alternative Battery State Diagnosis Method

An innovative method that differs from all previous methods is based on the use of ultrasonic measurements. It consists of using two piezoelectric sensors, one to emit a signal and the other one to receive it. The multiple modifications of the waves transmitted through the battery (amplitude, frequency, time of flight, etc.) allow identifying the state of charge after mathematical processing, as shown in figure 15.8, which is the result of work carried out by CEA-Liten. In this figure, the grey level is representative of the characteristics of the received wave, and therefore shows an excellent correlation between these characteristics and the SOC, whatever the discharge regime is here.

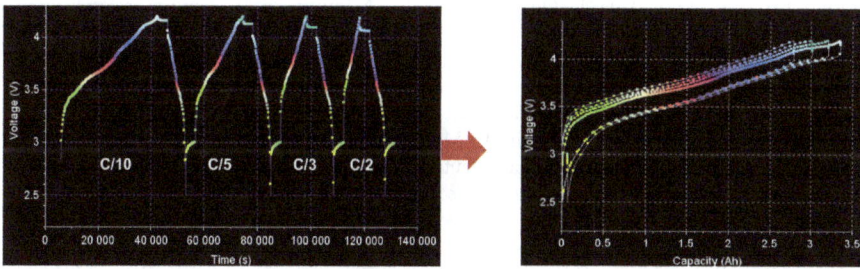

FIG. 15.8 – Evolution of the characteristics of the ultrasonic waves transmitted through a cyclic Li-ion cell, highlighted here by a colorimetric processing method (each unique color corresponding to a unique set of characteristics of the transmitted wave).

Although this ultrasound method is still at the development stage and is therefore not yet embedded in systems in operation, the results obtained are promising to say the least, since they show that it is possible to access to state indicators estimations without using the traditional measurements (current/voltage). This technique is therefore particularly promising in the case of battery chemistries that do not show a significant variation in open circuit voltage with the SOC (*e.g.* Li-ion LFP or LTO chemistries), as well as in the case of primary (non-rechargeable) batteries for which coulometry is not possible.

Li-ion batteries can be characterized by several critical indicators during operation. Accurate and robust knowledge of these indicators allows operating the battery in the best conditions; first of all in terms of safety, but also in terms of performance. Unfortunately, these indicators are not directly accessible, and therefore require the implementation of estimation algorithms with varying levels of precision and robustness.

The implementation of algorithms initially developed in other scientific fields (in particular Bayesian Kalman-type filters) and the increase in the on-board computing capacity in battery management systems are now providing very promising results in terms of on-line estimation of state indicators. This work will allow a fine and real-time characterization of the battery, paving the way for an optimized battery control within the BMS.

Bibliography

[i]   Rubenbauer H., Henninger S. (2017) Definitions and reference values for battery systems in electrical power grids, *J. Eng. Storage* **12**, 87.

[ii]   Rezvanizaniani S.M., Liu Z., Chen Y., Lee J. (2014) Review and recent advances in battery health monitoring and prognostics technologies for electric vehicle (EV) safety and mobility, *J. Power Sources* **256**, 110.

[iii]   Cabrera-Castillo E., Niedermeier F., Jossen A. (2016) Calculation of the state of safety (SOS) for lithium ion batteries, *J. Power Sources* **324**, 509.

[iv]   Piller S., Perrin M., Jossen A. (2001) Methods for state-of-charge determination and their applications, *J. Power Sources* **96**, 113.

[v]   Pop V., Bergveld H.J., Notten P.H.L., Regtien P.P.L. (2006) State-of-the-art of battery state-of-charge determination, *Inst. Phy. Publ. Meas. Sci. Technol.* **16**, 93.

[vi]   Lu L., Han X., Li J., Hua J., Ouyang M. (2013) A review on the key issues for lithium-ion battery management in electric vehicles, *J. Power Sources* **226**, 272.

[vii]   Di Domenico D., Creff Y., Prada E., Duchêne P., Bernard J., Sauvant-Moynot V. (2013) A review of approaches for the design of li-ion BMS estimation functions, *Oil & Gas Sci. Technol. – Revue d'IFP Energies nouvelles* **68**, 127.

[viii] Hannan M.A., Lipu M.S.H., Hussain A., Mohamed A. (2017) A review of lithium-ion battery state of charge estimation and management system in electric vehicle applications: Challenges and recommendations, *Renew. Sustainable Eng. Rev.* **78**, 834.

[ix]   Pablo Rivera-Barrera J., Muñoz-Galeano N., Sarmiento-Maldonado H.O. (2017) SOC estimation for lithium-ion batteries: review and future challenges, *Electron.* **6**, 102.

[x]   Lee S., Kim J., Lee J., Cho B.H. (2008) State-of-charge and capacity estimation of lithium-ion battery using a new open-circuit voltage versus state-of-charge, *J. Power Sources* **185**, 1367.

[xi]   Mingant R., Martinet S., Lefrou C. Method for determining the state of charge of a battery in charging discharging phase at constant current, Patent of United States, US 20100007309A1.

[xii]   Piret H. *et al.* (2016) Tracking of electrochemical impedance of batteries, *J. Power Sources* **312**, 60e69.

[xiii] Chen X., Shen W., Cao Z., Kapoor A. *A Comparative study of observer design techniques for state of charge estimation in electric vehicles*, 2012 7th IEEE Conference on Industrial Electronics and Applications (ICIEA).

[xiv]   Plett G. (2004) Extended Kalman filtering for battery management systems of LiPB-based HEV battery packs - Part 3. State and parameter estimation, *J. Power Sources* **134**, 277.

[xv]   Lee J., Nam O., Cho B.H. (2007) Li-ion battery SOC estimation method based on the reduced order extended kalman filtering, *J. Power Sources* **174**, 9.

[xvi]   He H.W., Xiong R., Guo H.Q. (2012) Online estimation of model parameters and state-of-charge of LiFePO(4) batteries in electric vehicles, *Appl. Eng.* **89**, 413.

[xvii] Li J., Klee Barillas J., Guenther C., Danzer M.A. (2013) A comparative study of state of charge estimation algorithms for LiFePO4 batteries used in electric vehicles, *J. of Power Sources* **230**, 244.

[xviii] Parthiban T., Ravi R., Kalaiselvi N. (2007) Exploration of artificial neural network [ANN] to predict the electrochemical characteristics of lithium-ion cells, *Electrochimica ACTA* **53**, 1877.

[xix]   Charkhgard M., Farrokhi M. (2010) State-of-charge estimation for lithium-ion batteries using neural networks and EKF, *IEEE Trans. Indus. Electron.* **57**, 4178.

[xx]   R.v.d. M. Eric A. Wan (2000) «The unscented Kalman Filter for Nonlinear Estimation,» IEEE.

Chapter 16

Standards and Safety

Philippe Azaïs and Pierre Kuntz

Introduction

The development of a technology is constrained by two types of nomenclatures: regulations (mandatory) and standards, which are not mandatory, but whose observance is strongly advised so that the object can be easily marketed. The analysis of the normative framework is linked to the application and the object in which it will be integrated.

Electrochemical energy storage systems are objects containing energy intrinsically: even in case of complete discharge, the voltage of a battery is usually not zero and the chemical energy contained in a residual way is potentially degradable under certain conditions. For this reason, standards have been developed. These standards distinguish between primary and secondary (rechargeable) systems on the one hand, and single-cell batteries and more complex assemblies (modules, packs) on the other.

The rapid emergence of Li-ion batteries and the multiplicity of applications led to incidents in the years 2005–2007, particularly in the field of "consumer electronic" (mobile phones, laptops, etc.). These incidents (fires, high battery temperatures, etc.) led the organisations dedicated to standards (CEI[1], ISO[2], SAE[3]...) and regulations (UN, UNECE, etc.) to impose a certain number of increasingly restrictive frameworks that are better adapted to market developments. The expected very strong growth of the electrified vehicle market [1] leads car manufacturers to become very involved in the field of standards, not only to limit safety risks when placing batteries on the market but also to influence efforts to improve the safety of these products. What distinguishes electrochemical systems for electrified vehicles

---

[1]CEI: Commission électrotechnique internationale or in English IEC: International Electrotechnical Commission.

[2]ISO: International Organisation for Standardisation.

[3]SAE: Society of Automotive Engineers International.

DOI: 10.1051/978-2-7598-2555-4.c016
© Science Press, EDP Sciences, 2021

from those for portable equipment is the large amount of electrochemical energy on board while a battery of a few Watt-hours (typically less than 30 Wh) is enough to power a mobile phone, the energy contained in an electric vehicle pack is 1000 times higher. In addition, the constraints of use differ: a vehicle is required to be usable between −30 and +60 °C and to have a lifespan of at least 7 years.

16.1   Phenomena Involved in Abusive Conditions

The previous chapters explained how a battery system is made up of many components:

- Cells (=accumulators), that is to say, unit electrochemical storage cells. The heart of these accumulators is the site of the electrochemical reactions allowing the reversible storage of energy. Today there are three standard formats for these accumulators: cylindrical, prismatic or pouch cell. The type of format chosen has an impact both on the choice of module assembly and on safety.
- The modules are made up of electrical and mechanical assemblies of cells in series and/or in parallel. These modules may contain additional electronic, thermal, electrical and mechanical management devices. These devices are essential to guarantee the safety and lifetime of the cells. In some cases, the module alone constitutes the system and has all the safety functions.
- The battery system. This usually consists of the modules with an additional safety layer (also mechanical, electrical, electrotechnical and thermal).

The tests implemented in the regulations and standards aim at verifying that the safety barriers proposed by the manufacturers at the various scales can withstand the stresses applied in an abusive manner.

The stresses which influence safety and which can be combined in the tests are:

- *Temperature*

A distinction is made between high and low values. High values can be generated by an external overtemperature (furnace), by an overload (by electrical means) or an internal physical phenomenon such as a short circuit. The tests are intended to demonstrate that the batteries do not experience any thermal runaway that could lead to a loss of mechanical integrity (*e.g.* material projection). The low values are generated externally, for example when recharging in a cold environment, which may lead to a risk of internal short circuit due to the appearance of lithium dendrites.

- *Voltage*

A distinction is also made here between high values (*e.g.* during an overload under impressed current) and low values (*e.g.* due to an external short-circuit *via* a resistor connected in series across the cell). The aim of the tests is to demonstrate that the products (cells, modules, battery system) behave in a way that is consistent with the intended application. The hypothesis tested is the non-functioning of the electronics and electrical engineering that normally protect the cells.

- *Charge and discharge rate*

These tests aim at increasing the temperature by Joule effect, *via* the relation RI^2.

- *Mechanics*: crushing, shock, vibration, fall.

The aim of the tests carried out is to check the robustness under mechanical stress of the accumulator and the battery module/system, and to demonstrate that the object does not suffer any internal short-circuit, or that a possible internal short-circuit does not lead to thermal runaway.

- *Specific environment*: tests may be carried out under partial vacuum (aeronautics or aerospace), in a humid environment, in salt fog, in total immersion. The idea of these tests is also to demonstrate robustness under external aggression.

Temperature and voltage are the two parameters that are generally the most constraining, as they directly influence the nature of the electrochemical reactions. Above or below certain values, the electrochemical reactions involved become irreversible as they lead to material degradation, which in turn may lead to safety problems. Regulations and standards aim at verifying behaviour under normal and also abusive conditions in order to guarantee, as a priority, the safety of people.

16.1.1 *Phenomena at Cell Level*

To assess the level of safety of batteries, a clear distinction is made between two types of use:

- Under "normal" conditions: these are the conditions of use defined by the manufacturer, they are designed to optimise the operation of the system (voltage, current and temperature).
- Under "abusive" or "accidental" conditions: these are all conditions of use that go beyond the limits defined by the manufacturer.

Apart from abusive tests carried out as part of battery safety studies or during accidental use, batteries are always used in the conditions recommended by the manufacturer.

There are two types of ageing that can affect the proper functioning of batteries:

- Calendar ageing: this is ageing due to pure storage, when the battery is not in use.
- Cycling ageing: this is ageing due to the cycling of the battery, outside of storage.

In reality, in most applications, the two types of ageing (calendar and cycling) cannot be dissociated. A passenger car spends more than 90% of its time parked in a car park (*i.e.* calendar ageing).

Figure 16.1 distinguishes between nominal (green zone) and out-of-specification (red zone) operating limits as a function of temperature and voltage at the cell terminals, and identifies the main causes of degradation.

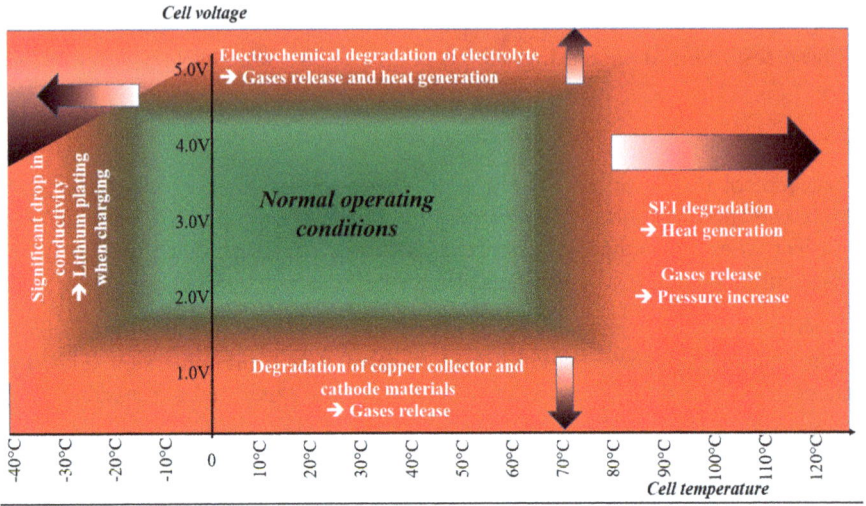

F IG. 16.1 – Diagram of the operating areas under normal and abnormal conditions as a function of temperature and voltage of the electrochemical cell.

Several borderline cases can be distinguished:

- *Under high voltage*:

Carbonate electrolytes degrade from 4.5 V *vs.* Li^+/Li. Above this level, there is a strong degradation which can lead to cycling of the SEI, degradation of the cathode materials, lithium plating if the carbon is overlithiated. This leads to heat and gas generation. These phenomena are irreversible.

- *Under low voltage*:

The copper collector, in case of complete discharge of the battery, can oxidize and cause irreversible degradation of the interface if the voltage is maintained. The cathode materials may also overload in the event of a very deep discharge. This overlithiation leads to irreversible and potentially gas degradation within the battery.

- *At low temperatures*:

The materials become more resistive; the conductivity of the electrolyte becomes very low: as an example, the conductivity of the LP30 electrolyte (Merck®, EC/DMC 1:1 $LiPF_6$ 1M) decreases from 11 mS.cm^{-1} at 20 °C to about 3.5 mS.cm^{-1} at −20 °C. SEI also becomes less conductive to ions. During recharging, the potential of the low-graphite negative electrode reaches a value equal to that of lithium metal due to the multiple contributions to the ohmic drop. This then leads to an irreversible reduction of the lithium ions on the electrode surface: "*lithium plating*". The lithium metal deposit generated is then very difficult to redissolve in solvents: there is then a loss of capacity through reversible lithium loss. As an

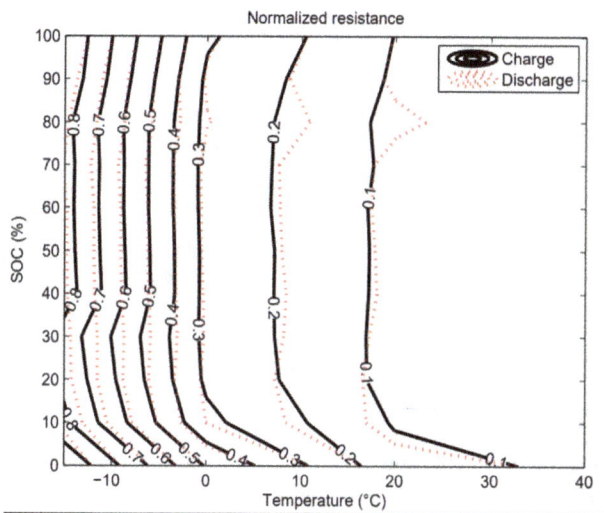

FIG. 16.2 – Influence of temperature and state of charge (SoC) on the resistance value.

example, figure 16.2 illustrates the normalized resistance (10%–90% SoC at 20 °C) as a function of the state of charge and temperature for an electric vehicle cell.

- *At high temperatures*:

LiPF$_6$ salt under ideal conditions (inert atmosphere, no contact with active materials) [2] starts to degrade from 107 °C. However, the slightest impurity (water, alcohols,...) significantly reduces this stability temperature [3].

The separator can also melt, if it is made of polyolefins (PE, PP), from 135 °C. This closes the porosity of the separator and, advantageously, prevents thermal runaway. However, the separator can curl up on itself, which can lead to a short circuit between the electrodes, especially at the outer edges of the electrodes.

The SEI may begin to degrade [4] from 60 °C, but significant degradation is generally observed from 85 °C, whether the anode is graphite or silicon based.

The solvents in the electrolyte are usually mixtures: the phase diagrams of these mixtures generally show a boiling temperature [5, 6] between 130 and 200 °C. In addition, the stability temperature of the electrolyte depends on its composition [7, 8], the active materials [9] and the salt [10].

Finally, from 150 °C onwards, thermal degradation of the cathodes in the presence of the electrolyte is observed. The dissipated energy and the reaction temperature has been measured in several articles [11–22] whose results are compiled in figure 16.3.

On the *x*-axis is represented the specific energy for the cathode material alone (*vs.* Li$^+$/Li). The *y*-axis represents the degradation temperature between the cathode material and an electrolyte based on a mixture of carbonates (usually EC/DMC-based) and 1M LiPF$_6$. The area of each circle represents the energy dissipated during the reaction for a given amount of cathode material (J.g^{-1}).

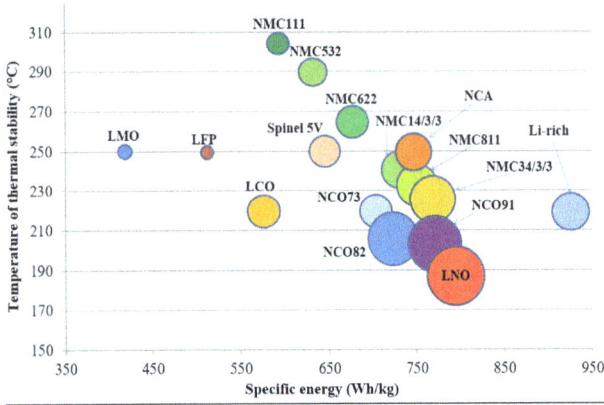

F<small>IG</small>. 16.3 – Schematic representation of the dissipated energy (proportional to the surface area of the circles) as a function of the reaction temperature (y-axis) between the electrolyte (carbonates mixture + LiPF$_6$ 1 M) and the cathode material, for different varieties of cathode materials with different specific energies (the latter being shown on x-axis).

However, this energy (limited to the reaction between the cathode and the electrolyte) represents only a portion of the energy dissipated during thermal runaway. Therefore, it should not be concluded from this figure, for example, that the use of an LFP-based cathode is necessarily safe: this would be tantamount to not taking into account the influence of the design of the cell (case, separator), nor the anode materials (with their SEI), nor the safety measures implemented at the module or battery system levels (electronic, electrotechnical, thermal, mechanical).

Thermal runaway occurs when the energy released by the exothermic reactions occurring inside the battery exceeds the energy that can be dissipated. This thermal imbalance leads to a rise in internal temperature, which activates a cascade of chemical and diffusional reactions, causing the battery to runaway thermally [23]. This runaway phenomenon therefore corresponds to a point of no return between a situation of thermodynamic equilibrium and the loss of control of the phenomena occurring in the cell.

Figure 16.4 shows the various phenomena [24] that can occur in the cell when the temperature of the cell rises, as well as the thermal exchanges involved.

For all these reactions, a point of initiation of the reaction (T_{onset}) is defined. At this temperature, it is still possible for the reaction to stop spontaneously, provided no more energy is supplied to the system and no unexpected chemical reaction is triggered. This is the case, for example, with the decomposition of the SEI.

One of the most common means used by manufacturers to identify potential failures is the FMECA[4]. For example, one can refer to the literature on the

---

[4]FMECA: Failure mode effects and criticality analysis is used to chart the probability of failure modes against the severity of their consequences.

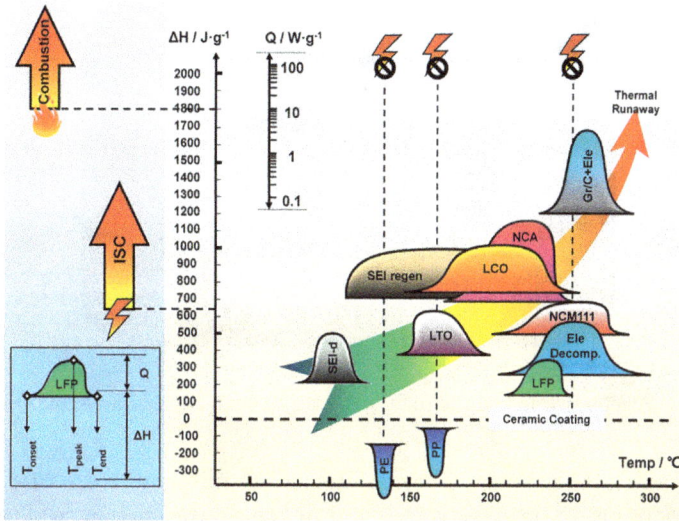

FIG. 16.4 – Diagram of the phenomena encountered during the thermal runaway of a Li-ion cell.

subject [25–28]. The failure modes internal to the cell [29] are, under normal operating conditions (ageing):

- The increase in the thickness of the passivation layer (SEI).
- Fracture of active material particles (mechanical stress).
- Reduction in electrode porosity through electrolyte decomposition.
- The generation of lithium metal deposits ("*lithium plating*") due to increased internal resistance and by depletion of reversible lithium ions.
- Corrosion of current collectors.
- Gas release.

In the case of abusive stress conditions, failure modes will be set out in the paragraph on standards.

16.1.2   Phenomena at Module and Pack Level

Figure 16.5 shows the different levels of protection of a battery system integrated in a vehicle (according to IEC 61511 [30]).

The energy of a battery module is proportional to the number of cells associated in series or in parallel that constitute it.

It should be noted that improper testing at the scale of the elementary battery may fully comply with the requirements of standards or regulations, but may nevertheless lead to safety problems at higher scales (module, pack). A FMECA type analysis is necessary at the module or pack scale in order to identify potential interactions between the elementary cells [31].

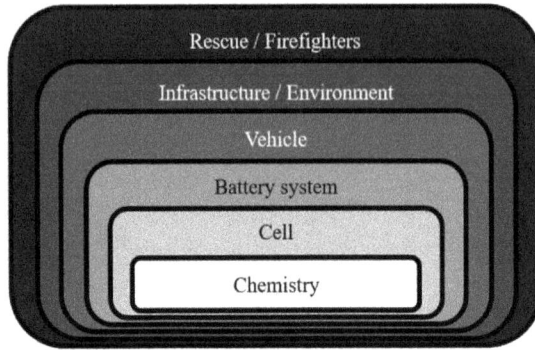

FIG. 16.5 – Schematisation of the different levels of protection in an electrified vehicle.

One of the major risks in the event of thermal runaway of a battery within a module is the propagation of a fault to the other batteries. This can lead to a domino effect on the whole module and then on the battery system. Tests aimed at identifying the impact of the arrangement of the cylindrical accumulators on the propagation have been carried out [32]. Propagation between batteries is related to the electrical architecture [33] and the surfaces facing each other.

The risks associated with these assemblies are essentially:

- *Electrical.* The most common phenomena encountered are problems of dimensioning the whole chain of electrical connections within the pack: cables, busbar, fuses, and contactors. If the electrical protections are not correctly dimensioned, there is an irreversible degradation of the elements to be protected, which generally leads to smoke emissions, then potentially, after degradation of the insulating elements (sheaths, plastic boxes...) short circuits and finally a fire. The housing of the module or pack must also be electrically insulated from the electrical architecture to avoid any electric arc. For example, in an electric vehicle, the voltage reaches about 400 V. In case of electrical contact between the metal case and the positive pole of the pack, the arc is maintained by the pack. These electric arcs generally generate holes in the batteries and molten metal splashes (and thus potentially a battery fire). It should be noted that these arcs can occur even at low voltages (*i.e.* below 48 V) [34].

- *Electronics.* Batteries are not all the same during manufacture (manufacturing dispersion). This dispersion, even with efficient balancing, increases with ageing) [35], as shown in figure 16.6.

- *Chemicals.* In the event of damage to an accumulator (opening, heating), it is possible that certain gases may be released: CO, CO_2, HF, volatile organic compounds (VOC)..., or fumes (electrode materials, VOCs), or that molten materials (metals, plastics) may be projected. These materials can be flammable (case of gases and fumes). In the presence of a hot spot (*e.g.* molten material, electric arc...), these materials ignite as soon as the flammability limit is reached. When an accumulator is opened, the casing, which is potentially conductive, may

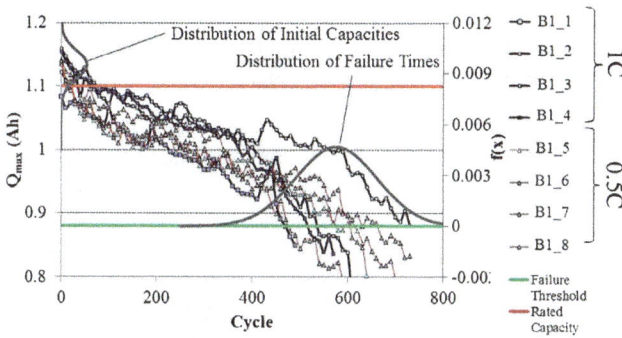

FIG. 16.6 – Evolution of the dispersion of the capacity values of groups of accumulators during ageing.

come into contact with adjacent accumulators whose electrical potential is not necessarily the same. In this case, an electric arc occurs. Combustible materials (solvents, polymers) that become available as a result of the opening of the battery can also ignite and release toxic gases and fumes.

- *Thermal.* The chemical reactions leading to the irreversible degradation of the materials present in the batteries and the other components of the module or pack are generally very exothermic (combustion).
- *Mechanical.* In the case of prismatic cells and pouch cells, the modules are generally made up of cells compressed one on top of the other by holding means (strapping, sheathing, folding, etc.) at a compression value chosen at the beginning of their life. The generation of gas during ageing or abusive conditions leads to mechanical stress in the cell casing and the compression means. It is possible, under certain abusive conditions, to reach the rupture of the compression means and/or to generate an internal short circuit in one or more cells.

16.2 Regulation

The regulations apply to the transport and storage of the battery (before integration), then in its life phase (related to its application) and finally at the end of its life (dismantling, transport, recycling). For these three phases, the following regulatory framework is defined:

- Transport: UN38.3 (in the case of lithium batteries), storage.
- Life phase (electrified vehicle: R100, including post-crash).
- End of life: in Europe [36], the batteries Directive was adopted in 2006 and has been subjected to a number of revisions. Last amendments were incorporated in 2013. The consolidated version of the Directive is presented below:

  o Directive 2006/66/EC of the European Parliament and of the Council of 6 September 2006 on batteries and accumulators and waste batteries and accumulators and repealing Directive 91/157/EEC. Consolidated version.

o Overview of EU Waste Legislation on Batteries and Accumulators.

o Frequently Asked Questions on Directive 2006/66/EU on Batteries and Accumulators and Waste Batteries and Accumulators (May 2014).

The regulations apply to all types of batteries and accumulators, regardless of their shape, volume, weight, constituent materials or use. These provisions do not apply in the case of two types of application: batteries and accumulators intended for military applications (weapons, ammunition, etc.) and those intended for use in space.

The main safety aspects of the R100 are:

1. *Protection against direct contact:*

   - Definition of voltage levels (taken from the ISO 6469-3 standard). It can be noted that these thresholds are different from those used by the UTE 18-510 and 18-550 (low-voltage range at 120 V DC or 50 V AC).
   - Protection against direct electric shocks: IP protection level requirement, depending on the location (inside or outside).
   - Specific marking for class B voltage elements.
   - Class B voltage cables insulated and orange coloured.

2. *Protection against indirect contact:*

   - All visible conductive parts are connected to vehicle ground (vehicle ground must be connected to earth when charging).
   - High voltage circuit isolated from vehicle ground (insulation resistance = 100X operating voltage in DC and 500X in AC).

3. *Functional safety:*

   - It must be impossible to move a vehicle while it is being recharged.
   - Drive mode information must be passed back to the driver.
   - 2 distinct actions must be necessary to switch from stop to drive mode (*e.g.* brake pedal + "start" button for the Prius).
   - 2 separate actions must be required to change from forward to reverse and vice versa.
   - 1 action shall be necessary to change from forward to stationary.
   - Failure protection (also related to ISO 6469): a failure shall put the vehicle in a state that prevents it from moving forward.

4. *Battery protection* (these protections are transpositions of ISO 12405 and ISO 6469):

   - Reliability tests: thermal shock, vibration and humidity.
   - Short-circuit, overload and underload resistance.
   - Protection against accumulation of gases (especially hydrogen).
   - Minimum insulation resistance 100 Ohms/V in DC and 500 Ohms/V in AC.
   - Crash resistance.
   - Protection of the occupants (limited movements allowed). No electrolyte leakage.
   - Protection of third parties (no ejection of the battery) out of the vehicle.

- The power supply system must be protected against short circuits.
- One second after disconnection of the charging plug, the terminal voltage must be below the Class A thresholds.

5. *Impact resistance*

- Following an accident, at least one of the following criteria must be checked:
  - o Between 5 and 60 s after the impact, the voltage of the high voltage bus shall be below the Class A thresholds.
  - o Between 5 and 60 s after the impact, the total energy available on the "high voltage" bus must be less than 2 J (which limits the consequences of a short circuit).
  - o The vehicle wiring shall remain protected to IP XXB level after impact and the protective elements of the high voltage bus shall remain connected to the vehicle ground.
  - o The high voltage components of the vehicle shall remain protected to IP XXB level and the insulation resistance shall be at least 100 Ohms/V in DC (500 if the DC and AC circuits are not galvanically protected) and 500 Ohms/V in AC.

In addition to these regulatory requirements, a number of tests described in Annex 8 of R100 are also required:

A. Vibration test according to a defined template.
B. Thermal cycling and thermal shock test according to a defined procedure (with associated ramp).
C. Mechanical shock test according to a template.
D. Mechanical integrity test (crushing).
E. Fire resistance test. This test is not required if the lowest part of the storage tank system is positioned more than 1.5 m from the ground. This test is quite restrictive (direct exposure to flames for 70 s).
F. Test of external short-circuit protection.
G. Test of protections against overload. If an automatic cut-off function fails (or if there is none), charging must be continued until the battery system under test is charged up to twice its rated charge capacity (which very generally leads to a fire if the protections do not work).
H. Testing of over-discharge protections.
I. Testing the overtemperature protections.

The R100 is also complemented by the R10, which also applies to electric vehicles [37]. R10 concerns the regulations related to electromagnetic compatibility. R10 is based on IEC 61000.

16.3 Standards

Abusive testing is intended to verify that the countermeasures (barriers) put in place by manufacturers respond appropriately to avoid endangering people, first and foremost, and property, to a lesser extent. There are many articles providing results of abusive tests carried out according to standards. A few journal articles are relevant and present in detail the sequence of events taking place during testing at the cell level [38–41], module level [42–45] or battery system level [46].

The aim of the standards is to identify whether the phenomena described at the beginning of the chapter can occur. Several types of aggression are proposed, as shown in figure 16.7:

- Mechanical aggressions. They are intended to deform, crush or penetrate the battery by means of shock, vibration, crushing or drilling. The result of these tests may be a short circuit between the two electrodes. In the best case, this leads to local heating and a rapid drop in voltage without gas, smoke, or open circuit.
- Electrical aggression. The aim is to generate:
  - A low resistive short circuit (in the case of external short-circuit tests) or "soft", *i.e.* much more resistive than the previous one (contamination test by a metal particle positioned between the two electrodes). Finally, it may be the result of a mechanical aggression previously described (internal short-circuit). In this case, no energy is supplied to the system: it is then possible to completely discharge the battery and stop the reactions.

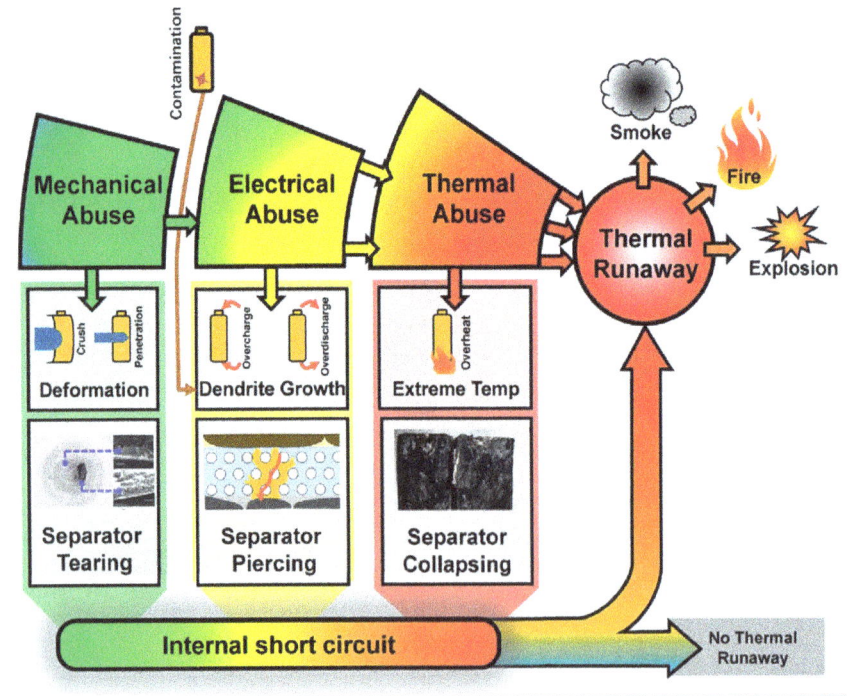

FIG. 16.7 – Diagram of the different types of tests designed to simulate behaviour under abusive conditions and relationships between the tests [46].

– An overload by external means. In this case, energy is supplied to the system. It is then possible to voluntarily reach a much higher voltage than that recommended by the manufacturer. This particularly critical case generally generates undesirable reactions producing gases, fumes... At the end of these tests, it is possible to return to a state of stability, despite the irreversibility of the reactions generated.

- Thermal aggressions. As previously stated, temperature generates irreversible exothermic reactions within the accumulator. It is then possible to degrade the accumulator and reach thermal runaway. Thermal runaway can be avoided by the use of high-performance barriers. The outcome of these thermal tests is usually smoke generation (favourable case), fire or even explosion (in the worst case).

The following tables contain the most commonly used standards for Li-ion batteries. A large bibliographical work is continuously carried out in Europe by the private research centre VITO located in Belgium. As specified throughout this chapter, it is necessary to distinguish the level at which the test is carried out: cell, module, system, vehicle.

The references of the standards are as follows (non-exhaustive list):

– IEC 62660-1:2018: "Secondary lithium-ion cells for the propulsion of electric road vehicles – Part 1: Performance testing".
– IEC 62660-2:2018: "Secondary lithium-ion cells for the propulsion of electric road vehicles – Part 2: Reliability and abuse testing".
– IEC 62660-3:2016, "Secondary lithium-ion cells for the propulsion of electric road vehicles – Part 3: Safety requirements".
– ISO 12405-4:2018, "Electrically propelled road vehicles — Test specification for lithium-ion traction battery packs and systems — Part 4: Performance testing".
– ISO 6469-1:2019: "Electrically propelled road vehicles — Safety specifications — Part 1: Rechargeable energy storage system (RESS)" (*replace ISO 6469-1:2009 and ISO 12405-3:2014*).
– SAE J2929:2013, "Safety standard for electric and hybrid vehicle propulsion battery systems utilizing lithium-based rechargeable cells".
– SAE J2464:2009, "Electric and hybrid electric vehicle rechargeable energy storage system (RESS) safety and abuse testing".
– SAE J1798:2008, "Recommended practice for performance rating of electric vehicle battery modules".
– SAE J2380:2013, "Vibration testing of electric vehicle batteries". *Please refere to J.M. Hooper, J Marco, EVS29 Symposium, World Electric Vehicle Journal 8 (2016) for detailed comments.*
– UL 2580:2013, "Outline of investigation for batteries for use in electric vehicles".
– QC/T 743-2006, "Lithium-ion Batteries for Electric Vehicles".
– IEC 62281:2019, "Safety of primary and secondary lithium cells and batteries during transport".
– UL 1642:2013, "Safety of Lithium-ion Batteries – Testing".

- IEC 62619:2017: "Secondary cells and batteries containing alkaline or other non-acid electrolytes – Safety requirements for secondary lithium cells and batteries, for use in industrial applications".
- IEC 62620:2014: "Secondary cells and batteries containing alkaline or other non-acid electrolytes – Secondary lithium cells and batteries for use in industrial applications".
- IEC 62133-2:2017: "Secondary cells and batteries containing alkaline or other non-acid electrolytes – Safety requirements for portable sealed secondary cells, and for batteries made from them, for use in portable applications – Part 2: Lithium systems".

Standard type	N°	Application	Test number	Type of test	cell	module	system	vehicle
				Performance test / **Ageing test** / **Abusive test** / **Certification**				
IEC	62660-1:2018	BEV & HEV	7.3	Capacity	X			
			7.5	Power	X			
			7.6	Energy	X			
			7.7.2	Charge retention test	X			
			7.7.3	Storage life test	X			
			7.8.2	BEV cycle test	X			
			7.8.3	HEV cycle test	X			
			7.9	Energy efficiency test	X			
IEC	62660-2:2018	BEV & HEV	6.2.1	Vibrations	X			
			6.2.2	Mechanical shock	X			
			6.2.3	Crush	X			
			6.3.1	High temperature endurance	X			
			6.3.2	Temperature cycling	X			
			6.4.1	External short circuit	X			
			6.4.2	Overcharge	X			
			6.4.3	Forced discharge	X			
IEC	62660-3:2016	BEV & HEV	6.2.1	Vibrations	X			
			6.2.2	Mechanical shock	X			
			6.2.3	Crush	X			
			6.3.1	High temperature endurance	X			
			6.3.2	Temperature cycling	X			
			6.4.1	External short circuit	X			
			6.4.2	Overcharge	X			
			6.4.3	Forced discharge	X			
			6.4.4	Internal short circuit test	X			
ISO	12405-4:2018	BEV & HEV	7.1	Energy and capacity at RT		X	X	
			7.2	Energy and capacity at different temperatures and discharge rates		X	X	
			7.3	Power and internal resistance		X	X	
			7.4	No load SOC loss		X	X	
			7.5	SOC loss at storage		X	X	
			7.6	Cranking power at low temperature		X	X	
			7.7	Cranking power at high temperature		X	X	
			7.8	Energy efficiency		X	X	
			7.9	Energy efficiency at fast charging		X	X	
			7.10	Cycle life		X	X	
ISO	6469-1:2019	BEV & HEV	6.2.2	Vibrations			X	
			6.2.3	Mechanical shock			X	
			6.3.1	Thermal shock cycling			X	
			6.4.1.1	Simulated vehicle accident test: RESS level			X	
			6.4.1.2	Simulated vehicle accident test: vehicle level				X
			6.4.2	Immersion into water			X	
			6.4.3	Exposure to fire			X	
			6.5.1	Short circuit			X	
			6.6.2	Overcharge protection			X	
			6.6.3	Overdischarge protection			X	
			6.6.4	Protection against internal overheating			X	

– IEC 61960-3:2017 "Secondary cells and batteries containing alkaline or other non-acid electrolytes – Secondary lithium cells and batteries for portable applications – Part 3: Prismatic and cylindrical lithium secondary cells and batteries made from them".

Standard type	N°	Application	Test number	Type of test	cell	module	system	vehicle
				Performance test (blue) / **Ageing test** (yellow) / **Abusive test** (red) / **Certification** (green)				
SAE	J2929:2013	BEV & HEV	4.2.2.1	Vibration Alternative 1. Complete battery system vibration test		X		
			4.2.2.2	Vibration Alternative 2. Battery Subsystem Vibration test.		X		
			4.2.3	Thermal shock		X		
			4.2.4	Humidity/Moisture Exposure		X		
			4.2.5	Electromagnetic Susceptibility		X		
			4.3	Drop Test		X		
			4.4	Immersion Test		X		
			4.5	Mechanical Shock		X		
			4.6	Battery Enclosure Intergrity		X		
			4.7	Exposure to Simulated Vehicle fire		X		
			4.8	Electrical Short Circuit		X		
			4.9	Single Point Overcharge Protection System Failure		X		
			4.10	Single Point Over Discharge Protection System Failure		X		
			4.11	Single Point Thermal Control System Failure		X		
			4.13	Protection against High Voltage Exposure		X		
SAE	J2464:2009	BEV & HEV	4.3.1	Shock tests	X	X	X	
			4.3.2	Drop test		X		
			4.3.3	Penetration test		X	X	
			4.3.4	Roll-over test		X	X	
			4.3.5	Immersion test		X	X	
			4.3.6	Crush test		X	X	
			4.4.1	High temperature hazard test		X	X	
			4.4.2	Thermal stability test	X			
			4.4.3	Cycling without thermal management		X	X	
			4.4.4	Thermal shock cycling	X	X	X	
			4.4.5	Passive propagation resistance test		X	X	
			4.5.1	Short circuit test	X	X	X	
			4.5.2	Overcharge test	X	X	X	
			4.5.3	Overdischarge (Forced Discharge) test	X			
			4.5.4	Separator shutdown integrity test	X			
SAE	J2380:2013	BEV & HEV		Vibrations			X	
SAE	J1798:2008	BEV		Capacity		X		
				Energy		X		
				Power		X		
				resistance		X		
UL	2580:2013	BEV & HEV	17	Manufacturing and Production Line Testing and Production Quality		X		
			25	Overcharge Test		X		
			26	Short Circuit Test		X		
			27	Overdischarge Protection Test		X		
			28	Temperature Test		X		
			29	Imbalanced Charging Test		X		
			30	Dielectric Voltage Whitstand Test		X		
			31	Isolation Resistance Test		X		
			32	Continuity Test		X		
			33	Failure of Cooling/Thermal Stability System Test		X		
			34	Rotation Test		X		
			35	Vibration Endurance Test		X		
			36	Shock Test		X		
			37	Drop Test		X		
			38	Crush Test		X		
			39	Thermal Cycling		X		
			40	Salt Spray Test		X		
			41	Immersion Test		X		
			42	External Fire Exposure Test		X		
			43	Internal Fire Exposure Test		X		

- IEC 61960-4:2020: "Secondary cells and batteries containing alkaline or other non-acid electrolytes – Secondary lithium cells and batteries for portable applications – Part 4: Coin secondary lithium cells, and batteries made from them".

Standard type	N°	Application	Test number	Type of test	cell	module	system	vehicle
				Performance test / **Ageing test** / **Abusive test** / **Certification**				
QC/T	743-2006	BEV & HEV	6.2.5	Discharge Capacity at 20°C	X			
			6.2.6	Discharge Capacity at -20°C	X			
			6.2.7	Discharge Capacity at 55°C	X			
			6.2.8.1	Rate Discharge Capacity at 20°C, High energy density battery	X			
			6.2.8.2	Rate Discharge Capacity at 20°C, High power density battery	X			
			6.2.9.1	Charge holding and recovery characteristics at normal temperature	X			
			6.2.9.2	Charge holding and recovery characteristics at high temperature	X			
			6.2.10	Storage	X			
			6.2.11	Cycle Life	X			
			6.2.12.1	Overdischarge	X			
			6.2.12.2	Overcharge	X			
			6.2.12.3	Short Circuit	X			
			6.2.12.4	Fall	X			
			6.2.12.5	Heat	X			
			6.2.12.6	Crush	X			
			6.2.12.7	Prick	X			
			6.3.5	Discharge Performance at 20°C		X		
			6.3.6	Simplified loaded mode		X		
			6.3.7	Resistance to vibration		X		
			6.3.8.1	Overdischarge		X		
			6.3.8.2	Overcharge		X		
			6.3.8.3	Short circuit		X		
			6.3.8.4	Heat		X		
			6.3.8.5	Crush		X		
			6.3.8.6	Prick		X		
UN	38.3:2015 (v6)	Transport of dangerous goods	T1	Altitude simulation	X	X	X	
			T2	Thermal test	X	X	X	
			T3	Vibration	X	X	X	
			T4	Shock	X	X	X	
			T5	External short circuit	X	X	X	
			T6	Impact	X			
			T7	Overcharge	X	X	X	
			T8	Forced discharge	X			
IEC	62281:2019	Transport of dangerous goods	T1	Altitude simulation	X	X	X	
			T2	Thermal test	X	X	X	
			T3	Vibration	X	X	X	
			T4	Shock	X	X	X	
			T5	External short circuit	X	X	X	
			T6	Impact	X	X	X	
			T7	Overcharge	X	X	X	
			T8	Forced discharge	X	X	X	
UL	1642:2013	General	10	Short-Circuit Test			X	
			11	Abnormal Charging Test			X	
			12	Forced Discharge Test			X	
			13	Crush Test			X	
			14	Impact Test			X	
			15	Shock Test			X	
			16	Vibration Test			X	
			17	Heating Test			X	
			18	Temperature Cycling Test			X	
			19	Low Pressure (Altitude Simulation)Test			X	
			20	Projectile Test			X	

Standard type	N°	Application	Test number	Type of test	cell	module	system	vehicle
				Performance test (blue) / **Ageing test** (yellow) / **Abusive test** (orange) / **Certification** (green)				
IEC	62619:2017	Industrial	7.2.1	External short circuit	X	X		
			7.2.2	Impact	X	X		
			7.2.3	Drop	X	X	X	
			7.2.4	Thermal abuse	X	X		
			7.2.5	Overcharge	X	X		
			7.2.6	Forced discharge	X	X		
			7.3.2	Internal short-circuit	X			
			7.3.3	Propagation		X		
			8.2.2	Overcharge control of voltage		X		
			8.2.3	Overcharge control of current		X		
			8.2.4	Overheating control		X		
IEC	62620:2014	Industrielle	6.3.1	Discharge performance at +25 °C	X	X		
			6.3.2	Discharge performance at low temperature	X	X		
			6.3.3	High rate permissible current	X	X		
			6.4	Charge (capacity) retention and recovery	X	X		
			6.5.2	Cell and battery internal resistance: measurement of the internal a.c. resistance	X	X		
			6.5.3	Cell and battery internal resistance: measurement of the internal d.c. resistance	X	X		
			6.6.1	Endurance in cycles	X	X		
			6.6.2	Endurance in storage at constant voltage (permanent charge life)	X	X		
IEC	62133-2:2017	Portable	7.2.1	Continuous charging at constant voltage (cells)	X			
			7.2.2	Case stress at high ambient temperature (battery)		X		
			7.3.1	External short-circuit (cell)	X			
			7.3.2	External short-circuit (battery).		X	X	
			7.3.3	Free fall	X			
			7.3.4	Thermal abuse	X			
			7.3.5	Crush	X			
			7.3.6	Over-charging of battery		X	X	
			7.3.7	Forced discharge	X			
			7.3.8	Mechanical tests		X	X	
			7.3.9	Design evaluation – Forced internal short-circuit	X			
IEC	61960-3:2017	Portable	7.3.1	Discharge performance at 20 °C (rated capacity)	X	X	X	
			7.3.2	Discharge performance at –20 °C	X	X	X	
			7.3.3	High rate discharge performance at 20 °C	X	X	X	
			7.4	Charge (capacity) retention and recovery	X	X	X	
			7.5	Charge (capacity) recovery after long term storage	X	X	X	
			7.6.2	Endurance in cycles at a rate of 0,2 It A	X	X	X	
			7.6.3	Endurance in cycles at a rate of 0,5 It A (accelerated test procedure)	X	X	X	
			7.7.2	Measurement of the internal AC resistance	X	X	X	
			7.7.3	Measurement of the internal DC resistance	X	X	X	
			7.8	Electrostatic discharge (ESD)		X	X	
			8	Test protocol and conditions for type approval	X	X	X	

16.4 Tests and Additional Analysis

The tests provided for in the standards consist in being the most representative of the uses. The trigger of the thermal runaway within a battery is the temperature. This is why certain tests are performed in the literature to understand, identify and quantify the heat exchanges that take place during thermal runaway. Three types of tests are performed:

- ARC (Accelerating Rate Calorimetry) tests, at cell level [47, 48], and at the material level [49].
- DSC (Differential Scanning Calorimetry) testing (material level or material mix) [50, 51].
- TGA (Thermo-Gravimetric Analysis, at material level) testing. This method gives only mass loss information and is therefore of little interest.

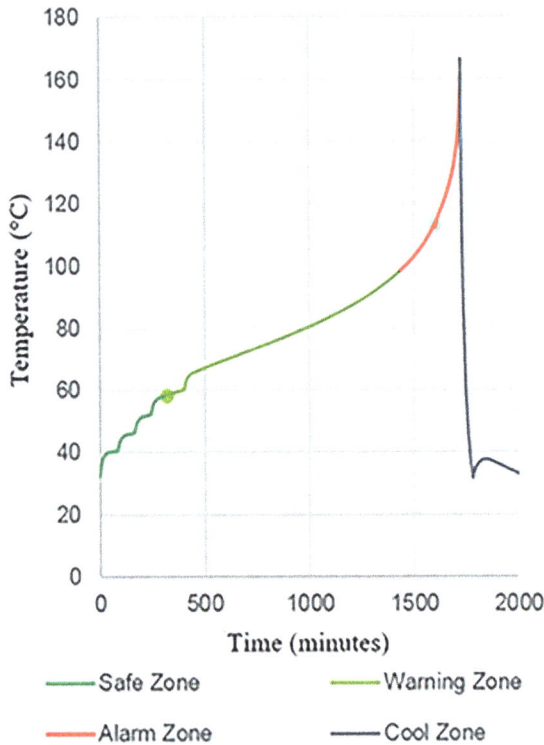

Fig. 16.8 – ARC test: Different steps of the warm-up curve for a Li-ion battery. In green, no reaction is observed. In light green, the cell heats up by itself (self-warming) but with the possibility of a reversible reaction due to the slow kinetics. In red, point of non-return (with opening of the accumulator and gas release). In grey, cooling of the accumulator after evacuation of the energy elements.

ARC consists of applying, in temperature steps (*e.g.* 1 °C/step), in an adiabatic chamber, a controlled heating medium. For each temperature increase, it is examined whether the system observed (a cell, for example) does not heat up by itself (characteristic of a thermal runaway). As the observed phenomena are controlled by thermodynamics and kinetics, it is essential to allow the system a sufficiently long time at each temperature level.

Figure 16.8 shows a standard characteristic curve obtained for a Li-ion battery tested in an ARC [52].

16.5   Solutions to Improve Safety at Different Levels

In order to ensure the safety of battery users, there are several levels of possible actions (figure 16.9). All safety devices developed are designed to reduce the risk of battery failure.

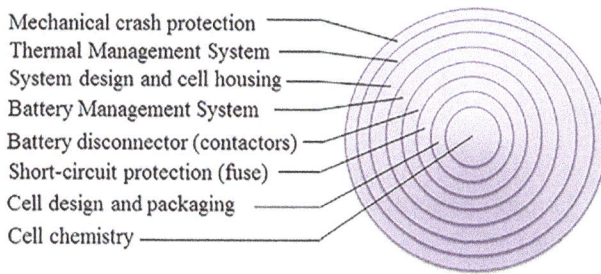

FIG. 16.9 – The "safety onion" showing examples, layer by layer, of different safety actions that can be used to establish a safe battery system in electric vehicles [53].

16.5.1 Improvement of the Components within the Cell

16.5.1.1 Separator

As previously mentioned, the separator ensures the physical separation of the two electrodes; however, when the temperature rises excessively, the separator melts and no longer ensures the physical separation of the electrodes. This causes an internal short circuit.

In the case of a PP/PE/PP three-layer separator, also known as a shutdown separator, when the temperature exceeds 135 °C, the PE melts (figure 16.10). The porous PE film then becomes a solid film, so the separator no longer allows the transport of ions between the two electrodes. From 135 °C, this separator becomes a barrier, which makes electrochemical processes inside the cell difficult [54].

Theoretically, this separator can therefore completely stop the operation of the cell and slow down its thermal runaway when the temperature rises above 135 °C. However, if the temperature rises too quickly and exceeds 165 °C, the PP in turn melts. The polymer film then curls back on itself (in all directions), which brings the electrodes into contact in places (usually at the ends) without any physical

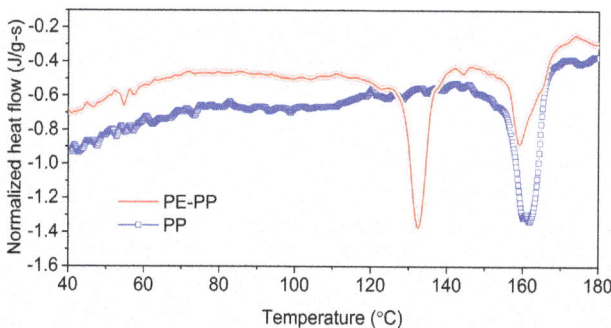

FIG. 16.10 – DSC measurement on two different separators: a PE-PP shutdown separator and a conventional PP shutdown separator [56].

separation. This can lead to a short circuit, especially in the case of dimensional changes of the electrodes (*e.g.* volume expansion due to overload).

16.5.1.2 *Negative Electrode*

The SEI formed on the surface of the negative electrode is critical to the operation of the cell, both in terms of performance and safety [55]. This is why processes have been studied in order to promote its formation, protection and stabilization [51, 56]:

- Moderate oxidation of graphite [53]

This technique consists of slightly oxidizing the surface of the graphite particles to promote the formation of the SEI after the first lithiation and to stabilize the SEI formed.

The oxidation can be done from different solutions:

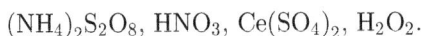

$$(NH_4)_2S_2O_8, \ HNO_3, \ Ce(SO_4)_2, \ H_2O_2.$$

- Deposition of metals and metal oxides

In the case of an electrolyte containing PC, the decomposition of PC is unfavourable to the proper formation of SEI. If the SEI is not sufficiently formed, solvent molecules can intercalate in the graphite and cause exfoliation. The deposition of metals and metal oxides (*e.g.* nickel) on the surface of the graphite creates a layer that is impermeable to solvent molecules while being permeable to lithium ions. This technique can prevent exfoliation of the graphite and promote the formation of SEI.

- Polymer or other carbon-based coatings [53]

Conductive polymer or carbon coatings are also used to protect the graphite, improve the electrochemical performance of the negative electrode and stabilize the SEI. This type of coating is not yet widely used and requires extensive research.

16.5.1.3 *Positive Electrode*

As stated above, the electrochemical and thermal stability of the positive electrode is strongly dependent on the active material of which it is made. A method often used to improve the stability of these materials is to deposit oxides on their surface [52, 53]. For example, an oxide layer of MgO, Al_2O_3, SiO_2, TiO_2, ZnO, SnO_2 or ZrO_2 can be deposited on the active material to prevent direct contact with the electrolyte. This limits reactions with the electrolyte, crystallographic phase transitions on the surface of the particles, the dissolution of transition metals and the disruption of cations in the crystal sites. This type of deposition therefore improves the structural stability of the material.

16.5.1.4   Electrolyte

Conventional liquid electrolytes are composed of one or more lithium salts ($LiPF_6$, LiTFSI...) dissolved in a mixture of solvents (EC, DEC, DMC...). Additives are generally added in small quantities (<10% by mass or volume) and have a significant impact on the performance of the battery. Additives can be divided into six categories depending on the intended purpose [51, 57]:

- To promote the formation of the SEI ("promoters" of the SEI).
- Positive electrode protecting agent.
- Salt stabilizer.
- Overload and flammability protection agent.
- Promote uniform deposition of lithium (avoid growth of dendrites).
- Other agents.

16.5.1.4.1   Fostering the Formation of the SEI

SEI's training and stability are two important factors in the safety, performance and durability of the battery. There are several types of additives that promote SEI:

- Reducing additives

This type of additive generally has a higher reduction potential than solvents. It is preferentially reduced and forms an insoluble solid product on the surface of the graphite of the negative electrode. The deposited layer promotes the formation of SEI [54]. There are two types of reducing additives: polymerizable monomers and reducing agents.

There are a wide variety of polymerizable monomers, one of the most common being vinylene carbonate (VC) [58]. This type of additive reduces gas generation, reduces irreversible capacity and improves the stability of the SEI during cycling [59].

Most reducing agents are sulfur compounds (ethylene sulfite, propylene sulfite or aryl sulfites). The higher the sulphur content of the molecule, the more effective the additive appears to be [60]. However, one of the most widely used reducing agents is fluoroethylene carbonate (FEC) because its reduction product is none other than VC [61]. FEC is particularly useful for high current and low temperature applications as it limits the formation of lithium dendrites [62].

- Reaction additives

Reaction additives are capable of scavenging anion radicals from intermediate solvent reduction compounds or combining with the final solvent reduction product (lithium alkyl dicarbonate and lithium alkyloxide). They participate in the formation of a stable SEI [63].

- Absorption additives

Absorption additives generally have a strong affinity with the graphite surface, they are adsorbed on the active sites of the graphite, which promotes the formation of a layer conducive to the growth of a stable SEI and thus suppresses reactions with the electrolyte. Organic halogen compounds can be used as absorption additives [64].

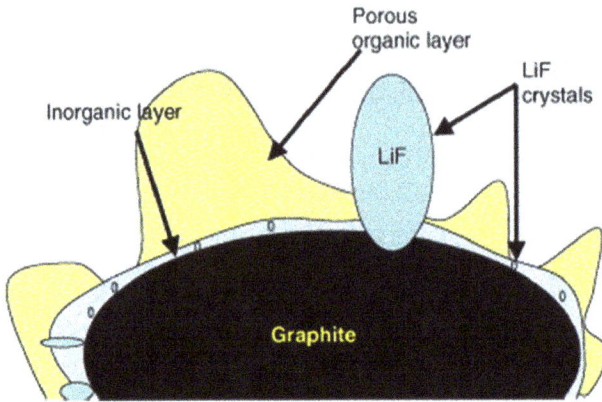

FIG. 16.11 – Schematic representation of the SEI structure on the surface of a graphite particle [65].

- Additives for morphological modification of the SEI

Electrolytes composed of carbonates and $LiPF_6$ form an SEI on the negative electrode containing mainly Li_2CO_3 and LiF (figure 16.11). The presence of large crystals isolated from LiF seems to be responsible for the instability of the SEI.

For this reason, boron-based anions have been developed to dissolve LiF crystals of the SEI [66]. One of the most widely used additives for this purpose is tris-(pentafluorophenyl) borane (TPFPB), which can theoretically dissolve up to 1M LiF in organic solvents forming a 1:1 complex with LiF. However, it is important to note that TPFPB can also react with $LiPF_6$ to form PF_5, a highly reactive gas that is harmful to battery components [67].

16.5.1.4.2 Positive Electrode Protection Agents

The degradation of the performance of the positive electrode by reaction with the electrolyte can have two origins: the presence of HF and H2O acid impurities or the irreversible oxidation of the solvents in the electrolyte. For this reason, agents have been developed to trap acidic impurities and water. In the case of LMO positive electrodes, an organic amine-based compound can be used to trap acidic impurities [68] and a carbodiimide-based molecule is used to consume water impurities [69].

Another approach is to form a protective film on the surface of the positive electrode. The idea is to use additives that combine with the transition metals dissolved in the electrolyte to form an insoluble layer on the positive electrode. Once formed, this layer can limit the dissolution of the transition metals from the electrode and thus limit its degradation [70].

16.5.1.4.3 Salt Stabilizers

The thermal instability of the most common lithium salt ($LiPF_6$) is attributed to two factors: the high equilibrium constant of the salt decomposition reaction and the high reactivity of the product gas PF_5 with organic solvents. In addition, PF_5 can also deteriorate the SEI by reacting avec Li_2CO_3, RCO_2Li, and $ROCO_2Li$. The problem of salt degradation can be significantly reduced [71] by adding LiF dissolved in the electrolyte (0.05 w%). Thus the excess LiF prevents the degradation of $LiPF_6$ by the principle of chemical equilibrium. The reactivity of PF_5 can be reduced by adding a small amount of basic additive (strong Lewis base). The additive generally used for this purpose is TTFP (tris(2,2,2-trifluoroethyl)phosphite), as it forms a stable complex [72] with PF_5.

16.5.1.4.4 Protection Agents

- Overcharge protection:

Two types of additives protect against overload: "*redox shuttles*" and "*shutdown*".

The *redox shuttles* oxidize reversibly at a potential slightly higher than that at the end of the charge (4.3–4.4 V *vs.* Li^+/Li). Thus, if overcharging occurs, this additive oxidizes on the surface of the positive electrode and then migrates to the negative electrode to be reduced there. This additive therefore acts as an internal battery discharger during overcharging. In addition, it prevents the potential of the positive electrode from increasing by maintaining it at the oxidation potential of the additive [73].

Shutdown additives polymerize irreversibly on the surface of the positive electrode when their potential exceeds that at the end of charge [74]. Polymerization insulates the electrode and generates gas. In the event of severe or repeated overcharging, the quantity of gas released is sufficient to trigger the CID (Current Interrupting Device) safety device and thus stop the battery operation. The majority of these additives are aromatic compounds, such as xylene, cyclohexylbenzene or biphenyl.

- Flame retardant additives:

Since some electrolyte solvents are highly flammable (DMC, DEC), flame retardant additives have been developed to reduce the flammability of the electrolyte [75]. Organic phosphorus compounds are commonly used for this application, for example TTFP already used as a salt stabilizer, but also partially fluorinated alkyl phosphates [76] or cyclophosphazenes [77]. However, these compounds are expensive and must be non-toxic.

16.5.1.4.5 Improving Lithium Deposition (Avoid Growth of Dendrites)

As detailed above, *lithium plating* is a major problem that affects both battery performance and safety of use [78]. In order to limit this phenomenon, surfactants are added to the electrolyte. These have the particularity of being adsorbed on the surface of the lithium, especially on dendritic sites, due to their larger specific surface area. By forming a layer on the surface of the lithium, the growth of dendrites can be stopped. Surfactants such as tetraalkylammonium chlorides, cetyltrimethylammonium chlorides and tetraethylammonium perfluorooctanesulfonate salt have been studied and have proven their effectiveness [79].

16.5.1.4.6 Other Agents

• Solvating agents:

The addition of solvating additives increases the solubility of lithium salts in the electrolyte and increases the ionic conductivity of the electrolyte [62].

• Aluminum anticorrosion agents:

Corrosion of aluminum in electrolytes is a well-known problem. The choice of salt has a significant impact on aluminum corrosion, while the solvent plays a minor role. Indeed, it has been shown that 5 mol.% of LiBOB and LiODFB salts are sufficient to suppress aluminium corrosion in a PC-DEC or EC-DMC electrolyte. This protection is due to the fact that the O–B bond(s) of the salt anion (LiBOB or LiODFB) breaks, producing a new anion that combines with Al^{3+} to form a stable passivation layer on the surface of the aluminium [80].

• Wetting agents:

If the electrolyte does not quickly wet the separator, wetting agents can be added. This can happen in batteries sized for high temperature use where the electrolyte contains more PC, EC and GBL. Cyclic alkyls and low molecular weight aromatic compounds can be used to improve the wettability of the electrolyte on the separator [81]. In addition, some *shutdown* additives that protect against over-charging can also act as wetting agents [82].

16.5.2 Safety Devices at Cell Level

In the event of a problem, the safety of a cell can be ensured thanks to safety devices inserted in the cell packaging. However, while the integration of certain devices is easy in cells with rigid casings, it is more difficult to implement in *pouch cells*. The devices generally implemented were initially envisaged to be effective at the cell level, but sometimes prove to be counterproductive in certain situations when using cells in modules or packs (especially PTC) [83].

16.5.2.1 Positive Temperature Coefficient

The "Positive Temperature Coefficient" (PTC) safety device protects the cell when the temperature rises (*e.g.* due to an external short circuit) [84]. This safety element is part of the electrical circuit that the current must flow through when the battery is used (figure 16.12). It has the particularity that its resistance increases abruptly when the temperature rises [85] close to 100 °C. The increase in resistance of the PTC has the effect of considerably reducing the current delivered by the battery. The advantage of this device is that it is completely reversible [86] and does not cut the electrical circuit [87].

There are two types of PTC: those made of ceramics and those made of con-ductive polymers. Ceramic PTCs have the advantage that they can be used in high-voltage applications and are highly reversible. However, they have a somewhat

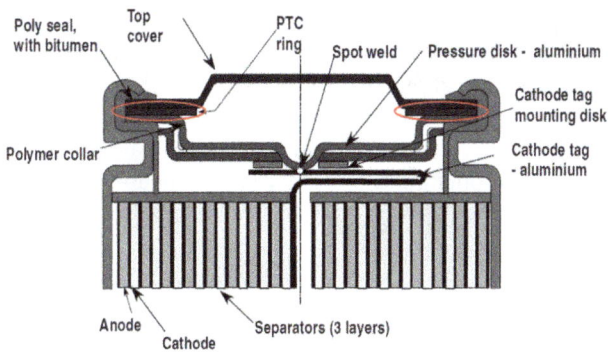

FIG. 16.12 – Cross sectional view of safety devices in a cylindrical cell [80].

long reaction time for moderate current exceedances. PTCs made of conductive polymers react much faster to current overflow, but their use is limited to relatively low currents and voltages [51].

16.5.2.2   Current Interrupter Device

The Current Interrupter Device (CID) is a metal disc on the end of the cylindrical cells (figure 16.13). It is part of the electrical circuit that the current must flow through when the cell is used. In case of internal overpressure, the disc bulges, the electrical circuit of the cell is opened and the cell can no longer deliver current [88]. Actuation of the CID is irreversible and therefore the cell is out of service.

16.5.2.3   Venting

The venting of a cell is achieved by causing a weak point in the cell envelope (it is a thin metal membrane in which a mechanical imprint is made, intended to make it

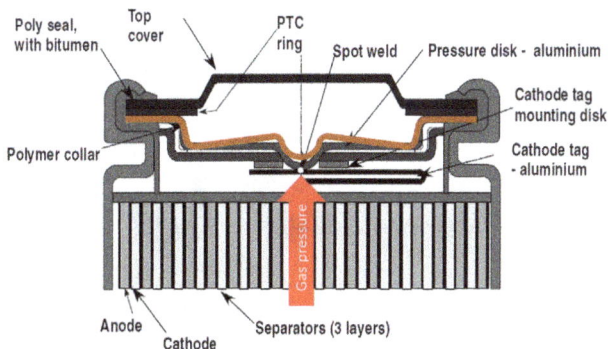

FIG. 16.13 – CID safety device located on the top of a cylindrical cell, before (grey) and after opening (brown) [80].

thinner and more fragile). The vent is designed to open when the pressure rises too high. It is generally placed on the upper end of the cell for cylindrical cells. However, in type 21 700 cells, controlled opening modes have also been added on the opposite end, due to the high energy contained in the cell (compared to type 18 650 cells). They prevent uncontrolled opening by imposing the battery-opening mode [89]. This is not reversible and opening the battery usually causes damage to the internal components of the battery.

16.5.2.4   PCB ("Printed Circuit Board")

It is a purely electronic device consisting of a printed circuit containing diodes limiting a defined current. The role of this PCB is to avoid too high currents. An example of a PCB positioned on a 18 650 cylindrical accumulator is shown in figure 16.14.

16.5.3   Safety Devices at the Module and Battery System Level

Safety and monitoring devices have been developed to address the operational safety risks mentioned at the beginning of this chapter at module and system level. Of course, these devices have an impact on the cost, energy density and power density of the system. However, they often avoid endangering people and limit damage to objects.

16.5.3.1   Electrical Devices

A battery system is made up of many elements that ensure the chain of electrical connections between the two poles of the system: cables, busbar, fuses, contactors... Each element has degradation characteristics corresponding to a current/time couple. These curves are called "I^2t" curves. The elements arranged hierarchically in the system must have I^2t curves without overlapping, allowing the fuses to act before the degradation of the other elements making up the system.

FIG. 16.14 – PCB disassembled from a 18 650 cell. The two tabs are connected by soldering to the two poles of the cell.

The entire storage tank system is designed for a specific application (and therefore an associated degree of pollution). This allows the calculation of two parameters: leakage lines and air clearances (IEC 60664 standard) [90]. These distances theoretically make it possible to avoid arcing within the module or system.

The first essential device in a battery system to guarantee the safety of people and property is the fuse. A fuse is designed to operate at a given voltage. It melts by Joule effect. It is therefore controlled by a current and a time. An example [91] of an I^2t curve is given in figure 16.15. Due to the low impedance of Li-ion batteries, the breaking capacity that fuses have to cope with is particularly high (usually several kA).

Another important element at the system level is the IMD (Insulation Monitoring Device, please refer to IEC 61557-8). It continuously verifies that the energy storage system (under voltage) is well insulated from external electrical conductors accessible to the passengers (bodywork).

Contactors are controlled components that allow access to the system's energy. One power contactor is positioned per pole. An example of a power contactor is shown in figure 16.16.

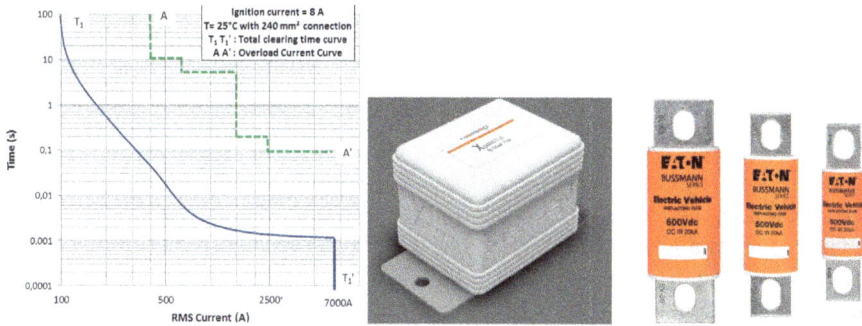

FIG. 16.15 – Example of a time–current curve (left) of a MERSEN XpST series pyrotechnic fuse (middle). On the right, examples of fuses used in electric vehicles (Eaton – Bussmann).

FIG. 16.16 – Example of a power contactor for an electric vehicle (TE Kilovac EV200).

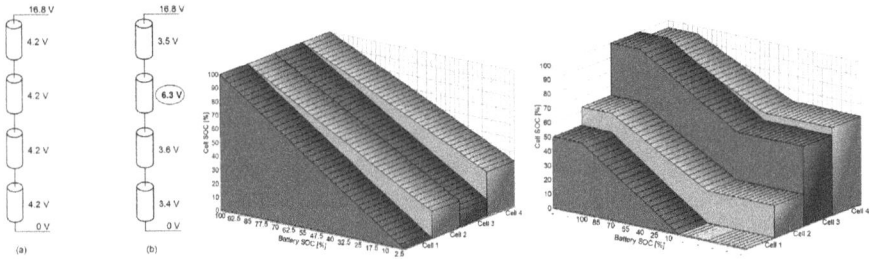

FIG. 16.17 – Schematization of cell performance dispersion in the presence or absence of intercell balancing.

16.5.3.2 Electronic Devices

Electronic balancing balances the voltage values between the batteries so that they continue to operate in the normal voltage range. If this electronic balancing system fails, the batteries will age rapidly (favourable case) and the charging and discharging voltage may even increase significantly, as shown in figure 16.17 [92]. On the left are the diagrams for four batteries in series with a maximum charging voltage of 16.8 V (a: with balancing electronics, b: without). In the event of a high unbalance between the batteries and for a given capacity range, it is possible, without balancing, to reach a battery voltage which leads to thermal runaway and thus destruction of the module. In the centre is shown a balancing made of 100% SoC. On the right, a module without balancing by cycling over a given capacity range: cell 1 potentially reaches a voltage that is too low. In case of failure of the electronics, it is therefore possible to see a fire starting.

The electronics also plays a role in controlling the cells, in particular by feeding back to the electronics the information on current, temperature, etc. In the event of information that does not conform to a nominal operation, the electronics can avoid the use of the storage module to avoid any risk during charging or discharging. These electronic systems are not safety devices in the sense of the standard: they monitor and control the energy storage system. However, the level of failure must be adapted to the application in question. The safety analysis determines a so-called SIL (Safety Integrity Level) [93], which is classified from SIL 1 to SIL 4, with SIL 4 being the highest risk. This analysis is based on IEC 61508 [94].

16.5.3.3 Chemical Devices

In the same way as for electronics, chemical control devices allow feeding back information to the management electronics. These sensors are designed to identify as quickly as possible any leaks that may occur in the accumulators and propagate into the module or storage system [95]. Many chemical sensors exist [96]. However, the cost and time of detection are high and these sensors are sometimes difficult to implement efficiently in an energy storage system.

16.5.3.4 Thermal Devices

Most of the current electronic boards have the possibility of integrating thermo-couples. These thermocouples are positioned within the module and the battery system in such a way that any abnormal temperature rise is identified as early as possible. An example of an electronic control board (balancing, temperature,...) is shown in figure 16.18.

Feedback is necessary but totally insufficient to guarantee safety at the module or battery system level. Thermal management between cells and between modules is essential to guarantee a long service life and the closest possible safety management. As we have seen in the section dedicated to standards, the thermal runaway of a cell leads to its destruction. Therefore, the idea of the different thermal management systems is to avoid thermal runaway (ideal case), or at least to avoid the propa-gation of this thermal runaway to other batteries (realistic case). It is on this real-istic case that the latest standards or recommendations [97, 98] in force are based: it is assumed that runaway, generated by internal or external means, can occur. The safety barriers implemented at module or system level must prevent the fault from spreading to other batteries (or even to modules if the system has several modules).

The thermal management system is essential to help prevent this spread. Various means of thermal management exist on the market [99]. Articles have recently been published in the field [100, 101]. Cooling can be achieved by:

- Air (case of the Renault Zoe).
- Contact with a liquid circulating in a cold plate (case of the Tesla vehicles, with a coil placed between the accumulators.
- Liquid/gas (phase change) (case of the BMW i3).
- Solid/liquid (phase change). These are phase change materials (PCM). Proto-types exist (*e.g.* by All Cell Technologies [102]) but this solution, although it is relevant to limit the presence of oxygen between the batteries, makes the energy storage system heavier.

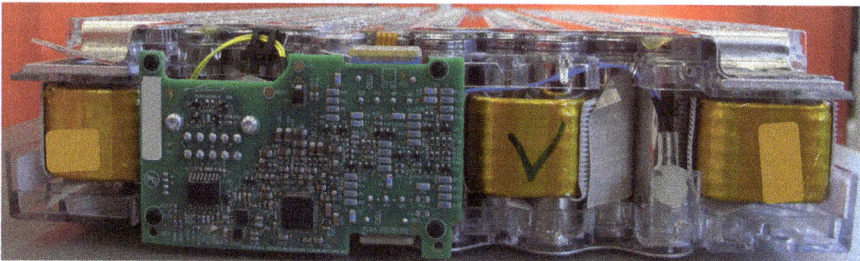

FIG. 16.18 – Front view of module of a Tesla Model S. The module is equipped with control and management electronics. Two thermocouples are connected to the board (2 blue and 2 yellow wires).

16.5.3.5   Mechanical Devices

As mentioned in the paragraph on risks, ageing and improper testing lead to gas generation within the accumulators. These gases, given that the accumulators are leak-proof, lead to internal overpressure and thus to deformation of the casing. Prior to opening the safety vent, abnormal pressure in the accumulator can be detected by means of appropriate sensors (deformation gauges [103–105], for example).

The structure of the pack and the purely mechanical means to maintain the integrity of the system and to dampen vibrations and shocks are clearly conventional mechanics. The only important difference is that it is essential to ensure that during the life of the storage system, leakage lines and air isolation distances are not degraded in order to avoid any arcing phenomena.

16.6   Conclusions and Prospects

The rapid emergence of Li-ion batteries has led all the players in the field (material and battery manufacturers, integrators, car manufacturers and certification and standards bodies) to take into account market developments and to build up standard and regulatory frameworks in order to guarantee the durability of the sector. Safety is therefore at the heart of everyone's concerns.

As we have seen, Li-ion batteries present, at all scales (from material to system), risks inherent to the technology. Safety must therefore be managed at all these scales and the dismantling of safety barriers must also be carried out in order to consider the marketing of a battery system, whether it is small (mobile phone, for example), intended for an electric vehicle, or dedicated to a large stationary system.

The vast majority of tests are nowadays carried out on new objects. However, tests show that the behaviour of batteries and systems changes over time (ageing). The latest developments should therefore focus on these problems. Finally, the emergence of "all-solid" lithium batteries implies rethinking the entire normative framework and studying safety issues in depth. An anode made of metallic lithium can, for example, melt as early as 180 °C (which is not the case with current anode materials). Failure modes can therefore be very strongly modified in the event of metal melting and potentially lead to an internal short circuit releasing a large amount of energy. Solid electrolytes must therefore also be considered and developed taking into account all these new failure modes.

Finally, as has been the trend in other fields, modelling and simulation are rapidly expanding avenues: the main idea is to be able to anticipate abusive behaviour at different scales on the basis of all the physical and chemical parameters. This requires a very good understanding of the phenomena, associated with their energetic and temporal quantification. These activities therefore go well beyond the norms ("trial and error" concept) and aim at developing new characterization methods at different scales.

Bibliography

[1] Since 2018, the major market for Li-ion batteries has become, in value and volume, that of the electrified vehicle (BEV and PHEV): this has led to new constraints on these batteries. The historical market for lead batteries remains the majority market. Lead battery applications remain essentially dedicated to the automotive industry (starter and traction batteries). The same is true for NiMH: more than 75% of the NiMH battery market is dedicated to hybrid vehicles.

[2] Yang H., Zhuang G.V., Ross P.N. (2006) *J. Power Sources* **161**, 573.

[3] Campion C.L., Li W., Lucht B.L. (2005) *J. Electrochem. Soc.* **152**, A2327.

[4] Kim J., Lee J.G., Kim H., Lee T.J., Park H., Ryu J.H., Oh S.M. (2017) *J. Electrochem. Soc.* **164**, A2418.

[5] Ding M.S., Xu K., Jow T.R. (2005) *J. Electrochem. Soc.* **152**, A132.

[6] Tarascon J.-M., Guyomard D. (1994) *Solid State Ionics* **69**, 293.

[7] Botte G.G., White R.E., Zhang Z. (2001) *J. Power Sources* **97–98**, 570.

[8] Ding M.S., Xu K., Jow T.R. (2000) *J. Electrochem. Soc.* **147**, 1688.

[9] Jiang J., Dahn J.R. (2004) *Electrochem. Comm.* **6**, 39.

[10] Chagnes A., Allouchi H., Carré B., Odou G., Willmann P., Lemordant D. (2003) *J. Appl. Electrochem.* **33**, 589.

[11] Liu D., Guerfi A., Hovington P., Trottier J., Dontigny M., Charest P., Mauger A., Julien C. M., Zaghib K. (2011) Abstract #598, 220th ECS Meeting.

[12] Deng H., Belharouak I., Sun Y.-K., Amine K. (2009) *J. Mater. Chem.* **19**, 4510.

[13] Cho J., Jung H.-S., Park Y.-C., Kim G.B., Lim H.S. (2000) *J. Electrochem. Soc.* **147**, 15.

[14] Wang Y., Jiang J., Dahn J.R. (2007) *Electrochem. Comm.* **9**, 2534.

[15] Zhang Z., Fouchard D., Rea J.R. (1998) *J. Power Sources* **70**, 16.

[16] Huang Y.Y., Zeng X.L., Zhou C., Wu P., Tong D.G. (2013) *J. Mater. Sci.* **48**, 625.

[17] Deng H., Belharouak I., Cook R.E., Wu H., Sun Y.-K., Amine K. (2012) *Nat. Mat.* **11**, 942.

[18] Deng H., Belharouak I., Cook R.E., Wu H., Sun Y.-K., Amine K. (2010) *J. Electrochem. Soc.* **157**, A447.

[19] Noh H.-J., Youn S., Yoon C.S., Sun Y.-K. (2013) *J. Power Sources* **233**, 121.

[20] Lee B.-R., Noh H.-J., Myung S.-T., Amine K., Sun Y.-K. (2011) *J. Electrochem. Soc.* **158**, A180.

[21] Irii Y., Kobayashi G., Kataoka T., Ikehara T., Matsumoto F., Ito A., Ohsawa Y., Hatano M., Sato Y., Abstract #189, Honolulu PRiME 2012.

[22] Arai H., Okada S., Sakurai Y., Yamaki J. (1998) *Solid State Ionics* **109**, 295.

[23] Abada S., Compréhension et modélisation de l'emballement thermique de batteries Li-ion neuves et vieillies, Ph-D thesis, University Pierre & Marie Curie, Paris, December 14th 2016.

[24] Feng X., Ouyang M., Liu X., Lu L., Xia Y., He X. (2018) *Energ. Stor. Mat.* **10**, 246.

[25] Hendricks C., Williard N., Mathew S., Pecht M. (2015) *J. Power Sources* **297**, 113.

[26] Lyu D., Ren B., Li S. (2019) *Acta Mech.* **230**, 701.

[27] Ganesan S., Eveloy V., Das D., Pecht M., Proceedings of IEEE Workshop "Accelerated Stress Testing & Reliability" (ASTR), Austin, Texas, October 3–5, 2005.

[28] Williard N., Sood B., Osterman M., Pecht M. (2011) *J. Mater. Sci. Mater. Electron.* **22**, 1616.

[29] Lee S.-M., Kim J.-Y., Byeon J.-W. (2018) *J. Nanosci. Nanotech.* **18**, 6427.

[30] IEC 61511: functional safety.

[31] Williard N., He W., Osterman M., Pecht M., 13th International Conference on Electronic Packaging Technology & High Density Packaging, IEEE, 2012.

[32] Ouyang D., Liu J., Chen M., Wenig J., Wang J. (2018) *Appl. Sci.* **8**, 1263.

[33] Lamb J., Orendorff C.J., Steele L.A., Spangler S. (2015) *J. Power Sources* **283**, 517.

[34] Augeard A., Singo T., Desprez P., Abbaoui M. (2016) *IEEE Trans. Components, Packaging and Manuf. Technol.* **6**, 1066.

[35] Williard N., He W., Osterman M., Pecht M., 13th International Conference on Electronic Packaging Technology & High Density Packaging, 1051–1055, IEEE, 2012.

[36] Please refer to European Commission for more details: https://ec.europa.eu/environment/waste/batteries/legislation.htm.

[37] Regulation No. 10, Revision 5, Uniform provisions concerning the approval of vehicles with regard to electromagnetic compatibility, E/ECE/324/Add.9/Rev.5 & E/ECE/TRANS/505/Add.9/Rev.5, 16 October 2014.

[38] Mikolajczak C., Kahn M., White K., Long R.T., Lithium-ion batteries hazard and use assessment, Exponent Failure Analy. Assoc., juillet 2011.

[39] Maleki H., Howard J.N. (2009) *J. Power Sources* **191**, 568.

[40] Xu J., Liu B., Hu D. (2016) *Sci. Rep.* **6**, 21829.

[41] Lamb J., Orendorff C.J. (2014) *J. Power Sources* **247**, 189.

[42] Feng X., Sun J., Ouyang M., Wang F., He X., Lu L., Peng H. (2014) *J. Power Sources* **275**, 261.

[43] Xia Y., Chen G., Zhou Q., Shi X., Shi F. (2017) *Eng. Failure Analy.* **82**, 149.

[44] Feng X., Fang M., He X., Ouyang M., Lu L., Wang H., Zhang M. (2014) *J. Power Sources* **255**, 294.

[45] Feng X., He X., Ouyang M., Lu L., Wu P., Kulp C., Prasser S. (2015) *Appl. Energ.* **154**, 74.

[46] Feng X., Ouyang M., Liu X., Lu L., Xia Y., He X. (2018) *Energy Storage Mater.* **10**, 246.

[47] Ishikawa H., Mendoza O., Sone Y., Umeda M. (2012) *J. Power Sources* **198**, 236.

[48] Mendoza-Hernandez O.S., Taniguchi S., Ishikawa H., Tanaka K., Fukuda S., Sone Y., Umeda M. (2017) E3S Web of Conferences **16**, 07001.

[49] *J. Power Sources* (2003) **119–121**, 794.

[50] Wu T., Chen H., Wang Q., Sun J. (2018) *J. Hazardous Mater.* **344**, 733.

[51] Zheng S., Wang L., Feng X., He X. (2018) *J. Power Sources* **378**, 527.

[52] Kuntz P., Thèse de Doctorat.

[53] Larsonn F., Andersson P., Mellander B.-E. (2016) Batteries **2**, 9.

[54] Balakrishnan P.G., Ramesh R., Prem Kumar T. (2006) *J. Power Sources* **155**, 401.

[55] Wang Q., Ping P., Zhao X., Chu G., Sun J., Chen C. (2012) *J. Power Sources* **208**, 210.

[56] Fu L.J., Liu H., Li C., Wu Y.P., Rahm E., Holze R., Wu H.Q. (2006) *Solid State Sci.* **8**, 113.

[57] Zhang S.S. (2006) *J. Power Sources* **162**, 1379.

[58] Simon B., Boeuve J.P., U.S. Patent 5,626,981 (1997).

[59] Aurbach D., Gamolsky K., Markovsky B., Gofer Y., Schmidt M., Heider U. (2002) *Electrochim. Acta* **47**, 1423.

[60] Wrodnigg G.H., Besenhard J.O., Winter M. (2001) *J. Power Sources* **97–98**, 592.

[61] Mogi R., Inaba M., Jeong S.-K., Iriyama Y., Takeshi A., Ogumi Z. (2002) *J. Electrochem. Soc.* **149**, A1578.

[62] Jaumann T., Balach J., Langklotz U., Sauchuk V., Fritsch M., Michaelis A., Teltevskij V., Mikhailova D., Oswald S., Klose M., Stephani G., Hauser R., Eckert J., Giebeler L. (2017) *Energ. Stor. Mat.* **6**, 26.

[63] Zhuang G.V., Yang H., Blizanac B., Ross P.N. Jr. (2005) *Electrochem. Solid State Lett.* **8**, A441.

[64] Zhang S.S., Xu K., Jow T.R. (2006) *J. Power Sources* **160**, 1349.

[65] Peled E., Menkin S. (2017), *J. Electrochem. Soc.* **164**, A1703.

[66] Sun X., Lee H.S., Yang X.Q., McBreen J. (2003) *Electrochem. Solid-State Lett.* **6**, A43.

[67] Sloop S.E., Pugh J.K., Wang S., Kerr J.B., Kinoshita K. (2001) *Electrochem. Solid-State Lett.* **4**, A42.

[68] Saidi M.Y., Gao F., Barker J., Scordilis-Kelley C. (1998) U.S. Patent 5,846,673.

[69] Takechi K., Koiwai A., Shiga T. (2000) U.S. Patent 6,077,628.

[70] Amine K., Liu J., Kang S., Belharouak I., Hyung Y., Vissers D., Henriksen G. (2004) *J. Power Sources* **129**, 14.

[71] Hiroi O., Hamano K., Yoshida Y., Yoshioka S., Shiota H., Aragane J., Aihara S., Takemura D., Nishimura T., Kise M., Urushibata H., Adachi H. (2001) U.S. Patent 6,305,540.

[72] Zhang S.S., Xu K., Jow T.R. (2003) *J. Power Sources* **113**, 166.

[73] Buhrmester C., Chen J., Moshurchak L., Jiang J., Wang R.L., Dahn J.R. (2005) *J. Electrochem. Soc.* **152**, A2390.

[74] Reimers J.N., Way B.M. (2000) U.S. Patent 6,074,777.
[75] Mandal B.K., Padhi A.K., Shi Z., Chakraborty S., Filler R. (2006) *J. Power Sources* **161**, 1341.
[76] Wang X., Yasukawa E., Kasuya S. (2001) *J. Electrochem. Soc.* **148**, A1058.
[77] Dagger T., Rad B.R., Schappacher F.M., Winter M. (2018) *Energy Technol.* **6**, 2011.
[78] Waldmann T., Hogg B.-I., Wohlfahrt-Mehrens M. (2018) *J. Power Sources* **384**, 107.
[79] Lia Z., Huang J., Liaw B.Y., Metzler V., Zhang J. (2014) *J. Power Sources* **254**, 168.
[80] Zhang S.S. (2006) *Electrochem. Commun.* **8**, 1423.
[81] Hamamoto T., Abe K., Ushigoe Y., Matsumori Y. (2005) U.S. Patent 6,881,522.
[82] Wang X., Naito H., Sone Y., Segami G., Kuwajima S. (2005) *J. Electrochem. Soc.* **152**, A1996.
[83] Darcy E., Davies F., Jeevarajan J., Cowles P., Lithium-ion cell PTC limitations and solutions for high voltage battery applications, (n.d.).
[84] Hagart-Alexander C. (2010) Instrumentation Reference Book (4th Edition), Chapter 21 - Temperature Measurement, pp. 269–326.
[85] Zhong H., Kong C., Zhan H., Zhan C., Zhou Y. (2012) *J. Power Sources* **216**, 273.
[86] Lisbona D., Snee T. (2011) *Process Saf. Environ. Prot.* **89**, 434.
[87] Kise M., Yoshioka S., Hamano K., Kuriki H., Nishimura T., Urushibata H. (2006) *J. Electrochem. Soc.* **153**, A1004.
[88] Brand M., Gläser S., Geder J., Menacher S., Obpacher S., Jossen A., Quinger D. (2013) *Hybrid Fuel Cell Electr. Veh. Symp.*
[89] Kong L., Li C., Jiang J., Pech M.G. (2018) *Energies* **11**, 2191.
[90] IEC 60664-1:2007, Insulation coordination for equipment within low-voltage systems - Part 1: Principles, requirements and tests.
[91] Ouaida R., de Palma J.-F., Gonthier G., Hybrid protection based on pyroswitch and fuse technologies for DC applications, Symposium de Génie Electrique, June 2016, Grenoble, France, Hal-01361696.
[92] Davide Andrea (2010) Battery management systems for large lithium-ion battery packs, ISBN 13-978-1-60807-104-3, Artech House.
[93] Please refer to application note by MTL Instruments Group AN9025-3, March 2002, https://www.mtl-inst.com/images/uploads/datasheets/App_Notes/AN9025.pdf.
[94] IEC 61508-1:2010, Functional safety of electrical/electronic/programmable electronic safety-related systems - Part 1: General requirements.
[95] Tobias Vossmeyer, Yvonne Joseph, Akio Yasuda, Kenji Ogisu, Yoshio Nishi (2010) Patent application US20100102975A1.
[96] Szulczyński B., Gębicki J. (2017) *Environments* **4**, 21.
[97] Norwegian Maritime Authority - Guidelines for chemical energy storage - maritime battery systems N° RSV 12-2016.
[98] NAVSEA (2011) High-energy storage system safety manual, SG270-BV-SAF-010.
[99] Liu H., Wei Z., He W., Zhao J. (2017) *Energy Convers. Manage.* **150**, 304.
[100] Chen D., Jiang J., Kim G.-H., Yang C., Pesaran A. (2016) *Appl. Thermal Eng.* **94**, 846.
[101] Deng Y., Feng C., J.E, Zhu H., Chen J., Wen M., Yin H. (2018) *Appl. Thermal Eng.* **142**, 10.
[102] https://www.allcelltech.com, Phase Change Composite (PCC™) Thermal Management Material.
[103] Farmer J., Chang J., Zumstein J., Kotovsky J., Dobley A., Puglia F., Osswald S., Wolf K., Kaschmitter J., Eaves S., Bandhauer T., 2014 MRS Spring Meeting – San Francisco, California – LLNL-PROC-644557.
[104] Zhou S., Wang G., Xiao Y., Li Q., Yang D., Yan K. (2016) *RSC Adv.* **6**, 63378.
[105] Shaikh S. (2016) Wireless approach for determining stress and temperature in lithium batteries with SAW sensors, Master of Science, University of Pune, 2016.

Chapter 17

Li-ion Battery Recycling

Emmanuel Billy, Daniel Meyer and M. Chapuis

17.1 Contextual Elements

This section looks at the fate of Li-ion batteries at the end of their life, after their last use as an energy storage device. Generally, there are two main categories of Li-ion batteries: those used in consumer, portable applications (laptops, cameras, smartphones, tablets...), and those used in transport, essentially the electric vehicle with specific characteristics (table 17.1).

More than two-thirds of the 190 GWh of Li-Ion batteries produced in 2019 are powering electrified vehicles (EVs; PHEVs; HEVs). This ratio will dramatically increase in the next coming years, with the expected boom of the electrified vehicle's market. In 2030, 1000–3000 GWh of Li-Ion batteries produced for transport applications will outnumber by far those meeting the needs of all other applications (mainly portable consumer and stationary applications). This increase in sales of electric vehicles will have a very significant impact on the tonnage of material to be recycled. It is important to consider that the life expectancy of Li-ion batteries for consumer applications and electric vehicles being very different (respectively 3–4 years against more than 10 years), recycling issues of huge quantities of lithium batteries will not become critical before 2030, which lets some time for the industry to prepare for this moment.

By regulation, batteries used in Europe must be recycled up to 50% by mass (directives 91/157/ECC and 2006/66/EC). The development of recycling processes for Li-ion batteries is mainly due to this regulation. Batteries are implemented either as single cells (*e.g.* cell phones), as an assembly of a few cells (laptops, portable tools...) or as an assembly of modules containing a lot of unit cells (electric vehicles). A unit cell schematically consists of a casing, of collectors usually made of aluminum and copper, of electrode materials, electrolyte, binder and of a membrane system. R&D on recycling is stimulated at the laboratory level by the presence in batteries of critical materials such as cobalt, nickel or lithium. The recovery of these sensitive or even critical materials will soon become mandatory to save

DOI: 10.1051/978-2-7598-2555-4.c017
© Science Press, EDP Sciences, 2021

TAB. 17.1 – Li-ion battery sales in 2019 by energy (GWh). Since 2017 the EV sales represent by far the largest part of the Li-Ion batteries market that will sooner or later have to be recycled, and EV batteries sales figures are expected to boom to reach 1–3 TWh in 2030 Source: Avicenne [1].

Application	Li-ion battery Energy 2019 (GWh)	Li-ion battery expected Energy in 2030 (GWh)
Portable Electronic Devices (consumer applications)	32,3 GWh	≈ 50 GWh
Automotive	127,3 GWh	1000–3000 GWh
Industrial	9,5 GWh	≈ 50 GWh
Others	20,9 GWh	≈ 70–80 GWh
Total Energy (GWh)	**190 GWh**	1000–3000 GWh

scarce resources, required to manufacture next generations of batteries. Although profits made with recovered metals do not so far cover industrial recycling corresponding expenses, this activity may become economically profitable in the coming years, depending on multiple factors such as the recycling process efficiency, the level of the cost of the carbon price or the fluctuations of the raw materials cost.

Table 17.2 ([1] and associated cited references) illustrates the diversity in chemical composition of the majority of Li-ion batteries.

The cathode materials represent 25–30% of the total mass of the battery including the current collector and the formulation of the electro-active materials, mainly Li metal-oxide (LMO) and Li Fe-phosphate (LFP). An efficient recycling process deals with with the reuse of as many as possible of the physical and chemical components of the battery.

Although the compositions used to manufacture Li-Ion batteries gradually converge towards "standard" NMC, NCA or LFP compositions, Tables 17.1 and 17.2 show significant variations in presently commercialized battery technologies and chemistries. This situation generates important bottlenecks for the development of a generic recycling process capable of addressing all chemistries.

After collecting, sorting the batteries, possibly securing and dismantling the containers, the main steps of a recycling process are made of (see figure 17.1):

(a) A process head, comprising essentially physical processing steps such as grinding of the components and physical sorting of the battery components.
(b) A process core allowing the recovery of species (metals, electrolytes,...) which generally consists in a separation and generation of reusable end products.
(c) A purpose that addresses the ultimate waste produced by the process.

While points (a) and (b) are widely described in the literature, point (c) is relatively poorly discussed. It is important in the development of a process to smartly interface the different steps in order to address various types of batteries.

TAB. 17.2 – Typical LIB cell composition with NMC: lithium nickel manganese cobalt oxide; LCO: lithium cobalt oxide; NCA: lithium nickel cobalt aluminum oxide; LFP: lithium ferrous phosphate.

		NMC (111)	NMC (622)	NMC (811)	LCO	NCA	LFP
Active Cathode	Li	7.86%	7.82%	7.79%	7.09%	7.22%	4.40%
Material	Co	20.21%	12.07%	6.02%	60.21%	9.20%	–
	Ni	20.13%	36.07%	47.93%	–	48.87%	–
	Mn	18.84%	11.26%	5.61%	–	–	–
	Al	–	–	–	–	1.40%	–
	Fe	–	–	–	–	–	35.40%
	P	–	–	–	–	–	19.63%
	O	32.95%	32.78%	32.66%	32.69%	33.30%	40.57%
Graphite		19.00%	20.70%	20.60%	18.50%	22.00%	16.60%
Carbon black		2.30%	2.10%	1.70%	2.40%	2.10%	2.20%
Binder: PVDF		2.90%	2.90%	3.60%	3.00%	2.90%	2.70%
Copper		16.40%	16.80%	15.70%	16.10%	16.90%	14.50%
Aluminum		8.20%	8.40%	8.00%	8.10%	8.40%	7.50%
$LiPF_6$		2.20%	2.20%	2.60%	2.20%	2.30%	3.30%
EC		6.20%	6.30%	7.20%	6.00%	6.30%	9.40%
DMC		6.20%	6.30%	7.20%	6.00%	6.30%	9.30%
Polypropylene		1.90%	1.90%	1.80%	1.80%	1.90%	1.70%
Polyethylene		0.30%	0.30%	0.30%	0.30%	0.30%	0.20%
Polyethylene terephthalate		0.30%	0.30%	0.40%	0.30%	0.30%	0.40%

17.2 Process Head

This stage generally corresponds to pre-treatment operations of physical nature, such as physical crushing or sorting. During this operation, it is easy to recover materials such as the aluminum current collectors, the organic binder and sometimes the electrode materials, exploiting for that purpose their very different physical properties. It is, for example, possible to grind and to magnetically separate metals [3]. The organic binder can be removed by a thermal approach (see pyrometallurgy section) [4] or by ultrasound [5]. Heat treatment of some organics such as PVDF requires management of the gases produced, because of their toxicity. Mechano-chemical treatments by ball milling are investigated in order to modify the chemical properties of some components, facilitating subsequent processing operations, generally of hydrometallurgical nature [6–9].

The electrolytes of the carbonate type, can be recovered during the grinding process, the volatile species can be recovered by condensation of a circulating gas stream, and the other species may be collected in the aqueous stream of the grinding process (grinding with water spraying is carried out for safety reasons).

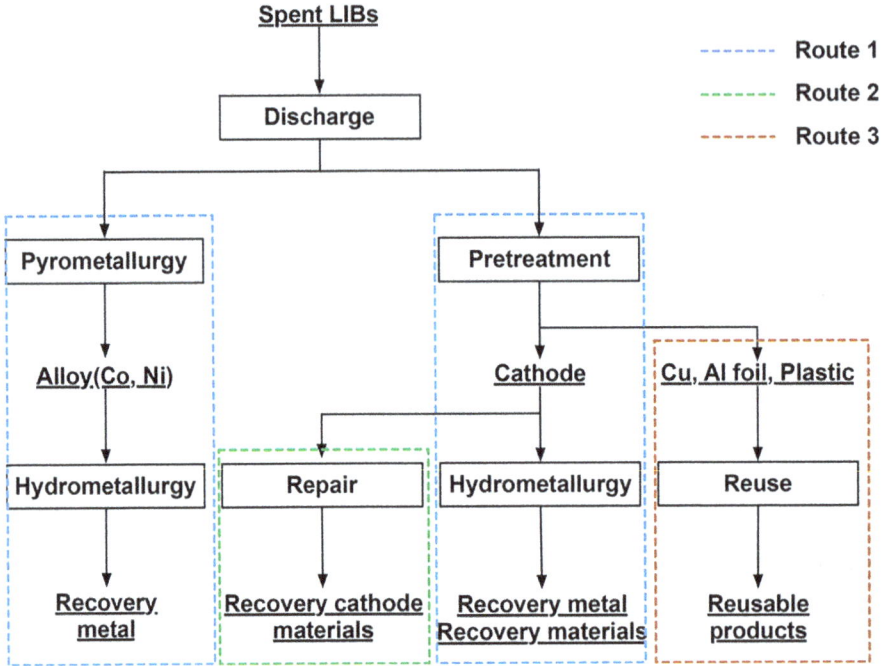

FIG. 17.1 – General flow sheet of spent LIB treatment processes [2].

17.3 Process Core (Separation – Valorization)

The recoverable products are generally collected during this process step. Their nature determines the type of process to set up. Pyrometallurgy and hydrometallurgy are the two approaches generally used, separately or jointly, to recycle Li-ion batteries. Considering the multiple metals or organic species to handle, it is difficult to imagine a generic process capable of treating them all with the same efficiency: compromises are necessary.

17.3.1 Pyrometallurgy

Pyrometallurgy is a branch of extractive metallurgy that is often associated as a thermal operation for extracting metals and removing solvents from batteries prior to wet processing. As it does not finely separate components embedded in a complex matrix, thermal treatment is a very efficient way of transforming and inerting Li-ion battery waste of various natures, for large volumes, with reduced residence times. Heat treatment is effective and practical for preparing and concentrating material for wet processing (cobalt and nickel enriched fraction). Pyrolysis removes the organic binder at 300 °C, promoting the separation of the cathode materials from the aluminum collector [4, 10, 11]. This step is a prerequisite for the separation of the current collectors from the electrode materials.

Many studies refer to pyrometallurgy as a pre-treatment process for battery waste, but few describe the mechanisms involved: rather, the treatment conditions favorable to recycling are described. Heat treatment conditions are typically adapted to separate the slag from the metals of interest that are concentrated in the polymetallic alloy. The separation step requires sometimes the addition of specific species to promote the formation of a high purity alloy. One study reports the addition of $CaO + SiO_2$, pyrolusite, and aluminum at 1475 °C for 30 min to obtain an alloy composed of 99 % copper, cobalt, and nickel [12]. The slag concentrates manganese and lithium, whose composition is 47 % MnO and 2.6 % Li_2O. The most frequent treatment is a vacuum carbothermal reduction treatment, advantageously using the presence of carbon in the batteries to produce lithium carbonate, which is easily leachable in solution [13, 14]. Examples include industrial processes from SNAM, UMICORE, Accurec GmbH, Inmetco or Sony/Sumitomo [15]. The head of the process is typically a pyrolysis step which ensures the safe elimination of resins and polymers contained in the modules [16]. This step is followed by a lithium extraction step by vacuum evaporation, followed by a distillation step for the recovery of metallic lithium.

Pyrolysis is a process adapted to the growing volumes of waste and to the technological constraints for the safety and pre-treatment of lithium-ion battery waste [17, 18]. However, heat treatments require high energy consumption, destroy solvents and polymers, and generate waste that is toxic to humans and the environment, such as gas, dust and slag. The waste treatment step has an impact on the economic model and makes it necessary to recover some of the materials, either by recycling inside the pyrometallurgical apparatus, or by processes generally based on separation techniques. Nevertheless, gaseous emissions and energy consumption are the major current constraints.

17.3.2 Hydrometallurgy

Hydrometallurgy is a succession of chemical steps characterized by the dissolution of a metal and its recovery from this solution; it mainly comprises dissolution or leaching steps, followed by the separation of spare products, and a refining step for packaging in a marketable form. Hydrometallurgy is complementary to pyrometallurgy for the recycling of Li-ion batteries. It has undeniable advantages: (i) as a low temperature step, it decreases energy costs, (ii) it uses small process units that can be designed at reduced costs, and (iii) it allows the fine separation and recovery of metals of higher purity.

17.3.2.1 Leaching of Waste

(a) Complete dissolution

The battery waste leaching step is the first step in the hydrometallurgical process and is downstream of the pre-treatment steps. Thus, the nature of the pre-treatment conditions the dissolution stage, which in turn conditions the metal separation stages. At the end of pre-treatment, the waste or "black mass", in reference to the dark matter loaded with carbon, can vary in nature, structure and composition. Currently, most of studies do not take into account the impact of pretreatment operating conditions on leaching. The battery waste is often dismantled, before undergoing thermal pretreatment, and then the active material of the positive electrode is separated before being dissolved in acid solution. Similarly, the variability of mixtures and chemistries of Li-ion batteries (LCO, NCA, LMO, NMC or LFP), as well as the presence of metallic impurities (collectors, packaging) are rarely considered. However, the composition and the structure of the waste are crucial factors in the implementation of an effective chemical treatment. The simultaneous consideration of all these parameters remains complex, due to the diversity of methods, but also to the pretreatment conditions. Although not exhaustive and not totally representative, current studies give an overview of the composition of baths and the treatment conditions required for a complete dissolution of the black mass. From the Chemical point of view, the solutions are essentially acidic and coupled with a reducing species, such as hydrogen peroxide (*cf.* table 17.3).

The vast majority of approaches aim at a complete and rapid dissolution of the waste. To achieve this goal, conditions are optimized according to the following parameters: acid and reducing agent concentration, solid/liquid ratio, temperature and stirring speed. The dissolution of active materials contained in positive electrodes requires conditions that can generally be described as "harsh". The stability of transition metal oxides implies solutions concentrated between 1 and 6 M, at a temperature of 40–100 °C, and the presence of a reducing agent. The wish to move towards a more "sustainable" chemistry has led to the substitution of conventional mineral acids (HCl, HNO_3 and H_2SO_4) by organic acids (citric, oxalic, malic, ascorbic) [19–22]. However, the prohibitive price of these acids and their environmental impact raises questions about their use for the recycling of lithium-ion batteries [2, 23]. Other studies have shown that the treatment conditions for mineral acids can be softened by the use of metallic copper and aluminum, already present in the black mass [24, 25]. The inherent presence of aluminum and copper is advantageously used to allow oxide reduction by galvanic corrosion [25]. However, copper and aluminum grades are dependent on pretreatment conditions which can lead to incomplete dissolution.

(b) Partial or selective dissolution

Some studies mention conditions for the selective dissolution of Li-ion battery waste, either with respect to lithium or manganese. It is possible to selectively separate manganese from lithium, cobalt and nickel in an ammoniacal medium and in the presence of a sodium sulphite reducer [26]. Strictly speaking, this is not a "selective" dissolution, as the manganese is first dissolved, before precipitating as $(NH_4)_2Mn(SO_3)_2 \cdot H_2O$. However, this approach allows introducing a separation step as early as the dissolution phase. Several studies are interested in the preferential dissolution of lithium by adapting the chemical nature of the solution and/or the chemical state of the waste. A heat treatment adapted according to the temperature and carbon content favours a transformation of the waste in the form of Li_2CO_3, Ni, Co and MnO, so that lithium is easily leachable in a carbonated aqueous medium

TAB. 17.3 – Cases of leaching spent LIBs using different leaching reagents [2].

Raw material	Reagent	Temp (c)	Time (min)	Leaching efficiency (%)		Ref
				Co	Li	
Inorganic acid leaching						
spent LIBs	1.75 mol/L HCl	50	90	99.0	100.0	78
spent LIBs (LiCoO$_2$)	4 mol/L HCl	80	30	90.6	93.1	79
LiFePO$_4$ and LiMn$_2$O$_4$	6.5 mol/L HCl + 5 vol% H$_2$O$_2$	30	60		74.1	80
LIB industry waste (LiCoO$_2$)	2 mol/L H$_2$SO$_4$ + 5 vol% H$_2$O$_2$	75	30	94.0	95.0	81
LiNi$_x$Mn$_y$Co$_z$O compounds	4 mol/L H$_2$SO$_4$ + 5 vol% H$_2$O$_2$	65–70	120	96.0		82
Spent LIBs (mixture)	1 mol/L H$_2$SO$_4$ + 0.075 M NaHSO$_3$	95	240	91.6	96.7	83
Spent LIBs (LiCoO$_2$) (from laptop computers)	2 mol/L H$_2$SO$_4$ + 5 vol% H$_2$O$_2$	75	60	70.0	99.1	84
Spent LIBs (LiCoO$_2$) (from mobile phones)	2% H$_3$PO$_4$ + 2 vol% H$_2$O$_2$	90	60	99.0	88.0	85
Spent LIBs (LiCoO$_2$)	0.7 mol/L H$_3$PO$_4$ + 4 vol% H$_2$O$_2$	40	60	99.0	100.0	86
Spent LIBs (LiCoO$_2$)	1 mol/L HNO$_3$ + 1.7 vol% H$_2$O$_2$	75	60	95.0	95.0	87
Alkaline leaching						
Spent LIBs (Li(Ni$_{1/3}$Co$_{1/3}$Mn$_{1/3}$)O$_2$)	4 mol/L NH$_3$.1.5 mol/L (NH$_4$)2SO$_4$ + 0.5 M Na$_2$SO$_4$	80	300	80.7	95.3	73
Organic acid leaching						
Spent LIBs (LiCoO$_2$)	0.4 mol/L tartaric acid + 0.02 mol/L ascorbic acid	80	60	93.0	95.0	69
Spent LIBs LiCoO$_2$ and CoO	1 mol/L oxalate + 5 vol% H$_2$O$_2$	80	120	96.7		88
Spent LIBs (LiCoO$_2$)	2 mol/L citric acid + 0.6 g/g H$_2$O$_2$ (H$_2$O$_2$/spent LIBs)	70	80	96.0	98.0	89
Spent LIBs (LiCoO$_2$)	1 mol/L oxalic acid	95	150	97.0	98.0	68

TAB. 17.3 – (continued).

Raw material	Reagent	Temp (c)	Time (min)	Leaching efficiency (%)		Ref
				Co	Li	
Spent LIBs (LiCoO$_2$)	1 mol/L iminodiacetic acid + 0.02 M ascorbic acid	80	120	99.0	90.0	66
Spent LIBs (LiCoO$_2$)	1 mol/L maleic acid + 0.02 M ascorbic acid	80	120	99.0	96.0	66
Spent LIBs (LiCoO$_2$)	0.5 mol/L glycine + 0.02 M ascorbic acid	80	120	91.0		70
Spent LIBs (LiCoO$_2$)	1.5 mol/L succinic acid + 4 vol% H$_2$O$_2$	70	40	100.0	96.0	90
Spent LIBs (LiCoO$_2$ and LiNi$_{0.5}$Co$_{0.2}$Mn$_{0.3}$O$_2$)	2 mol/L L-tartaric acid + 4 vol% H$_2$O$_2$	70	30	98.6	99.1	37

[27]. It has also been shown that the use of sodium persulphate oxidants allows the selective extraction of lithium [28]. Other studies have demonstrated the possibility of this extraction by oxalic species [29, 30]. These treatments are not "selective", in the meaning of dissolution, but allow the separation of lithium in solution by the precipitation of transition metals. Finally, a patent reports the separation of manganese from nickel, cobalt and lithium ions. Manganese ions are used as a reducing agent for the oxide. Their oxidation in the form of manganese oxide allows the complete dissolution of the waste. The result is a solution rich in Ni, Co and Li ions that can be separated from manganese oxide particles free of impurities [24].

(c) Dissolution mechanism

Although significant efforts have been made to identify effective dissolution solutions, studies usually do not address dissolution physical and chemical mechanisms. Only one study has so far settled a complete dissolution mechanism for NMC-type materials that can be generalized to other electrode materials [31]. This work is based upon earlier work on the development of electrode materials by acid delithiation [32, 33]. The dissolution mechanism is described in two distinct steps. The first step is characterized by the extraction of lithium, which is the initiation step for the release of electrons required for the reduction of oxides. The electrons are generated by charge compensation of the transition metals and oxygen atoms of the material. The combined presence of electrons and protons leads to dissolution. Thus, the first phase of dissolution is "self-regulating": the phenomenon of delithiation of the material provokes the dissolution. As the reaction takes place, the delithiation slows down and the internal potential of the material increases, reducing the driving force of the reaction until the dissolution process stops (*cf.* figure 17.2).

This explains why, in the absence of reducing species, a plateau in dissolution efficiency is observed in the presence of various acids and for various Li-ion battery chemistries [21, 22, 25, 34–36]. Lessons can be learned from this situation: (i) the quantity of lithium determines the dissolution efficiency, which in turn implies a variability of the dissolution efficiency according to the chemistry of the electrode, (ii) no chemical reducing agent is required during the first phase, allowing the quantity of reagent to be reduced, (iii) a complete dissolution requires the presence of reducing species during the second phase of dissolution. This explains why leaching baths require the presence of reducing species, either by the use of a reducing acid or by the addition of a reducing agent.

In industrial processes, the nature of the leaching baths used is little or poorly described, but the use of mineral acids such as H_2SO_4 or HCl is preferred, in order to reduce treatment costs. In the light of the various works, mineral and organic acids allow the complete dissolution of Li-ion battery waste. In an approach that is both economical and environmental, it is certainly desirable to favor a partial or selective dissolution to reduce the number of steps, and to facilitate the downstream stages of the treatment process.

(d) Bioleaching

The first results of bioleaching tests on Ni–Cd batteries were published in 1998 [37]. This technology has mainly developed over the last ten years on various types of batteries [38]. It is the RedOx properties of metals and enzyme systems generally based on Fe and S that are the driving force behind leaching, with a generally slow kinetics as a major limitation [39, 40].

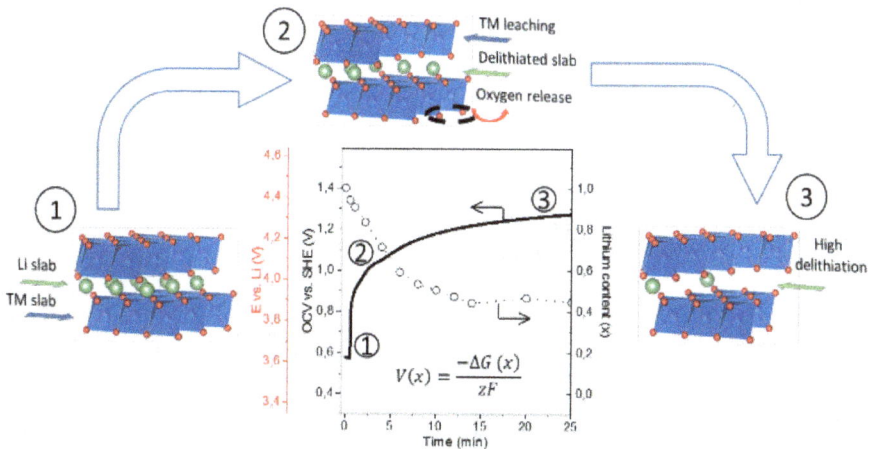

$$V(x) = \frac{-\Delta G(x)}{zF}$$

FIG. 17.2 – Evolution of the open potential of the NMC material during the first phase of dissolution [31].

17.3.2.2   *Treatment to Recover and Minimise Ultimate Wastes*

After a dissolution process or pyrometallurgical treatment, it is generally necessary to add a separation and/or shaping step to recover the products. The technology used depends essentially on the nature of the products and the recovery objectives. The range of treatments available to date makes it possible to draw up a general diagram of the possibilities for the recovery of a Li-ion battery (*cf.* figure 17.1) [2, 15, 41]. It is possible to implement approaches used in mining for the recycling of Li-ion batteries, in particular for Li recovery [15, 41].

The diversity of metals and their physico-chemical properties often require the implementation of several separation techniques in order to obtain a recoverable recycled product [15, 42–44]. The most common techniques are precipitation, solvent separation (liquid–liquid) and, more rarely, the electrochemical approach.

(e) Solvent extraction

Solvent extraction is a liquid–liquid separation technique often used in mineral processing [15, 41, 42] or in the processing of nuclear fuel [43]. A solvent is an industrial name, it is the result of a formulation between a diluent (hydrocarbon, a petrochemical fraction...), one or more extractants (molecules that give solubilization properties of a metal in an organic phase) and one or more modifiers. Extraction is generally used to produce the purest possible metals from concentrated solutions. The implementation of this technology requires the use of several de-extraction separation and washing stages (*cf.* figure 17.3).

The number of solvent separation-extraction systems studied is relatively large, and recent and very comprehensive reviews report on them in detail [2, 4, 41, 42, 44, 45]. For example, the most commonly used solvents for recycling Li-ion batteries are acids derived from phosphorus (D3EHPA, cyanex 272, PC 88A).

As an example:

– Di(2-ethylhexyl) phosphoric acid (D2EHPA) is used to extract manganese and copper with high acidity and cobalt with lower acidity [45–48].

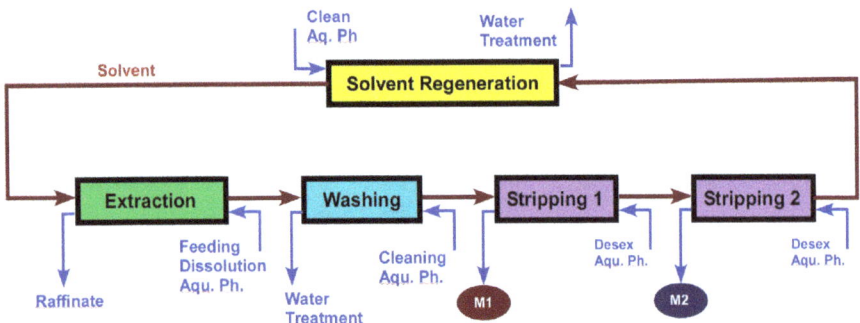

FIG. 17.3 – Solvent extraction steps.

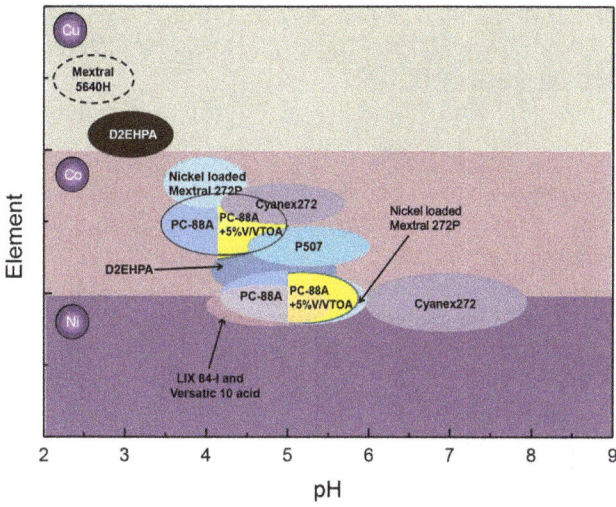

FIG. 17.4 – Extraction of Cu Co and Ni depending on the pH and the solvent (from the ref [2] and ref [46, 49, 52–54]).

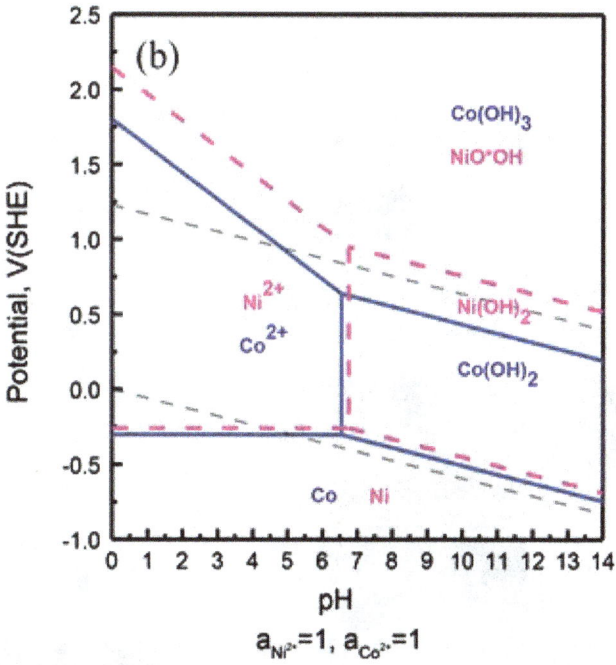

FIG. 17.5 – E-pH diagram Co/Ni from [2].

– PC-88A, also known as P507, separates cobalt from nickel from other metals at pH 4.5 but is not very effective at a pH below 3 [45]. It can be used for cobalt/lithium separation [41, 49].

– Cyanex 272 is generally used to treat cobalt, lithium, aluminum, etc. [35, 41, 50, 51].

A variety of solvent systems and formulations are available for recovering the metals of interest from electrode material dissolution solutions (figure 17.4). Separation is usually controlled by pH.

At the laboratory level, the current trend is the integration of several solvents and the development of new molecules, in particular ionic liquids.

(f) Precipitation

The simplicity of the precipitation method and the low associated costs make it a technique of choice for battery recycling [2, 37]. The simplest approach is to hydrolyze the metals [3, 55]. One problem remains, which is the superposition of cobalt and nickel speciation regions in a potential-pH diagram (*cf.* figure 17.5).

In a separation or conversion framework, it is thus possible to obtain various types of solids such as:

$LiCoO_2$ and $LiMnO_2$ [19];
$LiCoO_2$ [51, 56, 57];
$LiCo_xNi_{(1-x)}O_2$ [58];

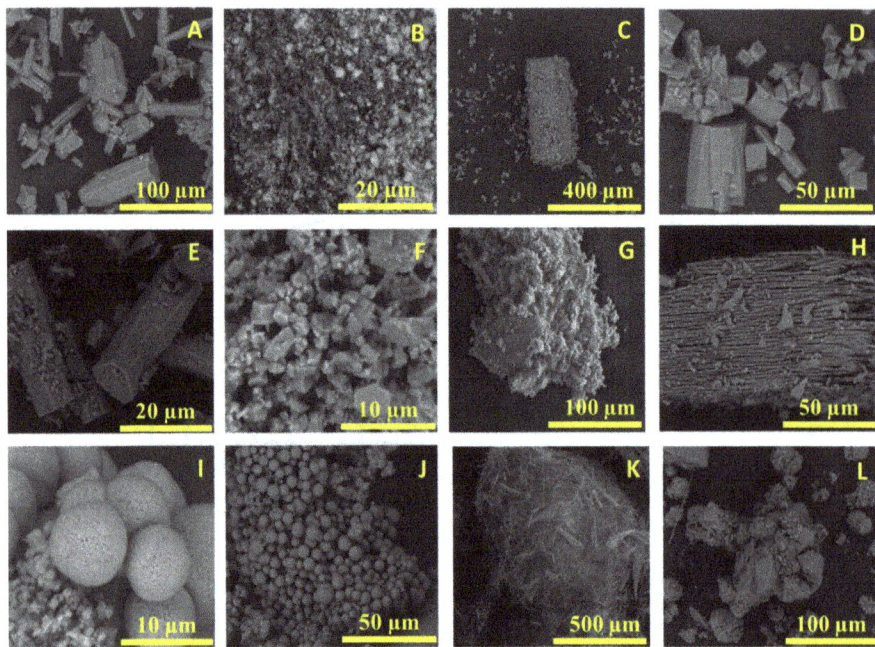

FIG. 17.6 – Various single and mixed MOF based on Mn, Co, Ni [62].

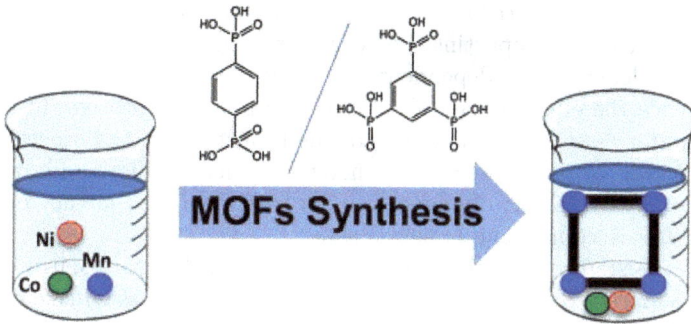

F<small>IG</small>. 17.7 – Selective precipitation of MOF [62].

$LiCoO_2$, $LiMnO_2$ & $LiNiO_2$ from dimethylglyoxime [59] and;
Cobalt oxalate [60].

Precipitation is also a method for producing mixtures of metal oxides (Mn, Co, Ni) that can be directly reused in the manufacture of electrode materials without extensive metal purification [61].

Recent studies have shown that it is possible to recover Mn, Co and Ni type metals by MOF (Metal–organic-framewoks) precipitation using di- or tri-carboxylic acid type molecules (*cf.* figure 17.6) [62].

The substitution of the carboxylic acid function by phosphonate functions also allows a separation of metals (*cf.* figure 17.7) [63].

Precipitation remains mandatory after any separation process (by solvent, column…) to produce a recoverable or reusable solid.

(g) Electrochemical method

Electrochemical processes remain more confidential and are mainly used to produce purified metals [38]. For example, it has been possible to produce Co oxide from Co(III) from the reprocessing of spent batteries [64].

17.4 Conclusion

A recycling process for Li-ion batteries must simultaneously meet economic, environmental and legislative requirements. These three, sometimes contradictory parameters, must be made coherent to lead to an efficient industrial process. Additionally, the nature and composition of the waste is changing, so that the "ideal" process does not exist. It is therefore important to smartly control all the pretreatment and chemical stages in order to address large volumes, with technological and chemical variability. The notions of waste quantity and variability justify the use of pyrometallurgical treatment. Pyrometallurgy meets these constraints, but at the expense of energy costs and the generation of toxic gases. One of the major challenges undoubtedly lies in sorting and pretreatment, in order on one hand, to be able to modulate or adapt the recycling process to changes in battery waste, and on the other hand, to avoid impacting the hydrometallurgical stages. Both leaching and separation of elements are dependent on variations in nature and composition, which influence the efficiency and cost-effectiveness of the process. In fact, modularity is a major asset and a guarantee of the robustness of the process. The indicators of a "good" or "bad" process are therefore difficult to determine *a priori*. The mere notions of yield, purity of a precipitate or treatment time are not sufficient and do not take into account environmental aspects.

Finally, consistency must also apply downstream of the process, in order to make it possible to market the products generated. Chemical processes make it possible to imagine all kinds of purity grades of recoverable metal products. Elemental separations of the elements lithium, manganese, cobalt and nickel or closed-loop recovery are not the only alternatives.

The equation is complex because the quality of a process involves cost, environment, safety and adaptability throughout the treatment chain.

References

[1]   Avicenne Energy – The Rechargeable Battery Market: Value Chain and Main Trends 2019-2030-IBS – 11 March, 2021.

[2]   Lv W., Wang Z., Cao H., Sun Y., Zhang Y., Sun Z. (2018) A critical review and analysis on the recycling of spent lithium-ion batteries, *ACS Sustainable Chem. Eng.* **6**, 1504. DOI: https://doi.org/10.1021/acssuschemeng.7b03811.

[3]   Shin S.M., Kim N.H., Sohn J.S., Yang D.H., Kim Y.H. (2005) Development of a metal recovery process from Li-ion battery wastes, *Hydrometallurgy* **79**, 172. DOI: https://doi.org/10.1016/j.hydromet.2005.06.004.

[4]   Granata G., Pagnanelli F., Moscardini E., Takacova Z., Havlik T., Toro L. (2012) Simultaneous recycling of nickel metal hydride, lithium ion and primary lithium batteries: accomplishment of european guidelines by optimizing mechanical pre-treatment and solvent extraction operations, *J. Power Sources* **212**, 205. DOI: https://doi.org/10.1016/j.jpowsour.2012.04.016.

[5]   He L.-P., Sun S.-Y., Mu Y.-Y., Song X.-F., Yu J.-G. (2017) Recovery of lithium, nickel, cobalt, and manganese from spent lithium-ion batteries using l-tartaric acid as a leachant, *ACS Sustainable Chem. Eng.* **5**, 714. DOI: https://doi.org/10.1021/acssuschemeng.6b02056.

[6]   Yang Y., Zheng X., Cao H., Zhao C., Lin X., Ning P., Zhang Y., Jin W., Sun Z. (2017) A closed-loop process for selective metal recovery from spent lithium iron phosphate batteries through mechanochemical activation, *ACS Sustainable Chem. Eng.* **5**, 9972. DOI: https://doi.org/10.1021/acssuschemeng.7b01914.

[7]   Wang M.-M., Zhang C.-C., Zhang F.-S. (2016) An environmental benign process for cobalt and lithium recovery from spent lithium-ion batteries by mechanochemical approach, *Waste Manage.* **51**, 239. DOI: https://doi.org/10.1016/j.wasman.2016.03.006.

[8]   Guan J., Li Y., Guo Y., Su R., Gao G., Song H., Yuan H., Liang B., Guo Z. (2017) Mechanochemical process enhanced cobalt and lithium recycling from wasted lithium-ion batteries, *ACS Sustainable Chem. Eng.* **5**, 1026. DOI: https://doi.org/10.1021/acssuschemeng.6b02337.

[9]   Ou Z., Li J., Wang Z. (2015) Application of mechanochemistry to metal recovery from second-hand resources: a technical overview, *Environ. Sci.: Processes Impacts* **17**, 1522. DOI: https://doi.org/10.1039/C5EM00211G.

[10] Paulino, J.F., Busnardo N.G., Afonso J.C. (2008) Recovery of valuable elements from spent Li-batteries, *J. Hazard. Mater.* **150**, 843. DOI: https://doi.org/10.1016/j.jhazmat.2007.10.048.

[11] Yang Y., Huang G., Xu S., He Y., Liu X. (2016) Thermal treatment process for the recovery of valuable metals from spent lithium-ion batteries, *Hydrometallurgy* **165**, 390. DOI: https://doi.org/10.1016/j.hydromet.2015.09.025.

[12] Xiao S., Ren G., Xie M., Pan B., Fan Y., Wang F., Xia X. (2017) Recovery of valuable metals from spent lithium-ion batteries by smelting reduction process based on $MnO–SiO_2–Al_2O_3$ slag system, *J. Sustainable Metall.* **3**, 703. DOI: https://doi.org/10.1007/s40831-017-0131-7.

[13] Xiao J., Li J., Xu Z. (2017) Recycling metals from lithium ion battery by mechanical separation and vacuum metallurgy, *J. Hazard. Mater.* **338**, 124. DOI: https://doi.org/10.1016/j.jhazmat.2017.05.024.

[14] Xiao J.F., Li J., Xu Z.M. (2017) Novel approach for in situ recovery of lithium carbonate from spent lithium ion batteries using vacuum metallurgy, *Environ. Sci. Technol.* **51**, 11960. DOI: https://doi.org/10.1021/acs.est.7b02561.

[15] Meshram P., Pandey B.D., Mankhand T.R. (2014) Extraction of lithium from primary and secondary sources by pre-treatment, leaching and separation: a comprehensive review, *Hydrometallurgy* **150**, 192. DOI: https://doi.org/10.1016/j.hydromet.2014.10.012.

[16] Trager T. Friedrich B. Weyhe R. (2015) Recovery concept of value metals from automotive lithium-ion batteries, *Chem. Ing. Tech.* **87**, 1550. DOI: https://doi.org/10.1002/cite.201500066.

[17] Georgi-Maschler T., Friedrich B., Weyhe R., Heegn H., Rutz M. (2012) Development of a recycling process for Li-ion batteries, *J. Power Sources* **207**, 173. DOI: https://doi.org/10.1016/j.jpowsour.2012.01.152.

[18] Wang X., Gaustad G., Babbitt C.W., Bailey C., Ganter M.J., Landi B.J. (2014) Economic and environmental characterization of an evolving Li-ion battery waste stream, *J. Environ. Manage.* **135**, 126. DOI: https://doi.org/10.1016/j.jenvman.2014.01.021.

[19] Castillo S., Ansart F., Laberty-Robert C., Portal J. (2002) Advances in the recovering of spent lithium battery compounds, *J. Power Sources* **112**, 247. DOI: https://doi.org/10.1016/S0378-7753(02)00361-0.

[20] Sun L., Qiu K. (2011) Vacuum pyrolysis and hydrometallurgical process for the recovery of valuable metals from spent lithium-ion batteries, *J. Hazard. Mater.* **194**, 378. DOI: https://doi.org/10.1016/j.jhazmat.2011.07.114.

[21] Lee C.K., Rhee K.I. (2003) Reductive leaching of cathodic active materials from lithium ion battery wastes, *Hydrometallurgy* **68**, 5. DOI: https://doi.org/10.1016/s0304-386x(02)00167-6.

[22] Joulié M., Laucournet R., Billy E. (2014) Hydrometallurgical process for the recovery of high value metals from spent lithium nickel cobalt aluminum oxide based lithium-ion batteries, *J. Power Sources* **247**, 551. DOI: https://doi.org/10.1016/j.jpowsour.2013.08.128.

[23] Li L., Dunn J.B., Zhang X.X., Gaines L., Chen R.J. Wu F. Amine K. (2013) Recovery of metals from spent lithium-ion batteries with organic acids as leaching reagents and environmental assessment, *J. Power Sources* **233**, 180. DOI: https://doi.org/10.1016/j.jpowsour.2012.12.089.

[24] Joulié M., Billy E., Laucournet R., Meyer D. (2015) Procédé de dissolution d'un oxyde métallique en présence d'un métal réducteur. FR3034104 A1 2016-09-30 [FR3034104].

[25] Joulié M., Billy E., Laucournet R. Meyer D. (2017) Current collectors as reducing agent to dissolve active materials of positive electrodes from Li-ion battery wastes, *Hydrometallurgy* **169**, 426. DOI: https://doi.org/10.1016/j.hydromet.2017.02.010.

[26] Zheng X., Gao W., Zhang X., He M., Lin X., Cao H., Zhang Y., Sun Z. (2017) Spent lithium-ion battery recycling – reductive ammonia leaching of metals from cathode scrap by sodium sulphite, *Waste Manage.* **60**, 680. DOI: https://doi.org/10.1016/j.wasman.2016.12.007.

[27] Hu J., Zhang J., Li H., Chen Y., Wang C. (2017) A promising approach for the recovery of high value-added metals from spent lithium-ion batteries, *J. Power Sources* **351**, 192. DOI: https://doi.org/10.1016/j.jpowsour.2017.03.093.

[28] Higuchi A., Ankei N., Nishihama S., Yoshizuka K. (2016) Selective recovery of lithium from cathode materials of spent lithium ion battery, *JOM* **68**, 2624. DOI: https://doi.org/10.1007/s11837-016-2027-6.

[29] Sun L., Qiu K. (2012) Organic oxalate as leachant and precipitant for the recovery of valuable metals from spent lithium-ion batteries, *Waste Manage.* **32**, 1575. DOI: https://doi.org/10.1016/j.wasman.2012.03.027.

[30] Zeng X., Li J., Shen B. (2015) Novel approach to recover cobalt and lithium from spent lithium-ion battery using oxalic acid, *J. Hazard. Mater.* **295**, 112. DOI: https://doi.org/10.1016/j.jhazmat.2015.02.064.

[31] Billy E., Joulié M., Laucournet R., Boulineau A., De Vito E., Meyer D. (2018) Dissolution mechanisms of $LiNi_{1/3}Mn_{1/3}Co_{1/3}O_2$ positive electrode material from lithium-ion batteries in acid solution, *ACS Appl. Mater. Interfaces* **10**. DOI: https://doi.org/10.1021/acsami.8b01352.

[32] Hunter J.C. (1981) Preparation of a new crystal form of manganese dioxide: $\lambda\text{-}MnO_2$. *J. Solid State Chem.* **39**, 142. DOI: https://doi.org/10.1016/0022-4596(81)90323-6.

[33] Thackeray M.M., Johnson P.J., de Picciotto L.A., Bruce P.G., Goodenough J.B. (1984) Electrochemical extraction of lithium from $LiMn_2O_4$, *Mater. Res. Bull.* **19**, 179. DOI: https://doi.org/10.1016/0025-5408(84)90088-6.

[34] Li L., Lu J., Ren Y., Zhang X.X., Chen R.J., Wu F., Amine K. (2012) Ascorbic-acid-assisted recovery of cobalt and lithium from spent Li-ion batteries, *J. Power Sources* **218**, 21. DOI: https://doi.org/10.1016/j.jpowsour.2012.06.068.

[35] Swain B., Jeong J., Lee J.-c., Lee G.-H., Sohn J.-S. (2007) Hydrometallurgical process for recovery of cobalt from waste cathodic active material generated during manufacturing of lithium ion batteries, *J. Power Sources* **167**, 536. DOI: https://doi.org/10.1016/j.jpowsour.2007.02.046.

[36] Stoyanova R., Zhecheva E., Zarkova L. (1994) Effect of Mn-substitution for Co on the crystal structure and acid delithiation of $LiMnyCo1-yO_2$ solid solutions, *Solid State Ionics* **73**, 233. DOI: https://doi.org/10.1016/0167-2738(94)90039-6.

[37] Cerruti C., Curutchet G., Donati E. (1998) Bio-dissolution of spent nickel–cadmium batteries using thiobacillus ferrooxidans, *J. Biotechnol.* **62**, 209. DOI: https://doi.org/10.1016/S0168-1656(98)00065-0.

[38] Ordoñez J., Gago E.J., Girard A. (2016) Processes and technologies for the recycling and recovery of spent lithium-ion batteries, *Renew. Sustainable Eng. Rev.* **60**, 195. DOI: https://doi.org/10.1016/j.rser.2015.12.363.

[39] Mishra D., Kim D.-J., Ralph D.E., Ahn J.-G., Rhee Y.-H. (2008) Bioleaching of metals from spent lithium ion secondary batteries using acidithiobacillus ferrooxidans, *Waste Manage.* **28**, 333. DOI: https://doi.org/10.1016/j.wasman.2007.01.010.

[40] Xin B., Zhang D., Zhang X., Xia Y., Wu F., Chen S., Li L. (2009) Bioleaching mechanism of Co and Li from spent lithium-ion battery by the mixed culture of acidophilic sulfur-oxidizing and iron-oxidizing bacteria, *Bioresour. Technol.* **100**, 6163. DOI: https://doi.org/10.1016/j.biortech.2009.06.086.

[41] Chagnes A., Pospiech B. (2013) A brief review on hydrometallurgical technologies for recycling spent lithium-ion batteries, *J. Chem. Technol. Biotechnol.* **88**, 1191. DOI: https://doi.org/10.1002/jctb.4053.

[42] Swain B. (2017) Recovery and recycling of lithium: a review, *Sep. Purif. Technol.* **172**, 388. DOI: https://doi.org/10.1016/j.seppur.2016.08.031.

[43] Lecomte M., Bonin B. (2008) *Treatment and recycling of spent nuclear fuel*. CEA Saclay, Groupe Moniteur.

[44] Chen X., Chen Y., Zhou T., Liu D., Hu H., Fan S. (2015) Hydrometallurgical recovery of metal values from sulfuric acid leaching liquor of spent lithium-ion batteries. *Waste Manage,* **38**, 349. DOI: https://doi.org/10.1016/j.wasman.2014.12.023.

[45] Wang F., Sun R., Xu J., Chen Z., Kang M. (2016) Recovery of cobalt from spent lithium ion batteries using sulphuric acid leaching followed by solid–liquid separation and solvent extraction, *RSC Adv.* **6**, 85303. DOI: https://doi.org/10.1039/C6RA16801A.

[46] Joo S.-H., Shin D., Oh C., Wang J.-P., Shin S.M. (2016) Extraction of manganese by alkyl monocarboxylic acid in a mixed extractant from a leaching solution of spent lithium-ion battery ternary cathodic material, *J. Power Sources* **305**, 175. DOI: https://doi.org/10.1016/j. jpowsour.2015.11.039.

[47] Zhao J.M., Shen X.Y., Deng F.L., Wang F.C., Wu Y. Liu H.Z. (2011) Synergistic extraction and separation of valuable metals from waste cathodic material of lithium ion batteries using cyanex 272 and PC-88A, *Sep. Purif Technol.* **78**, 345. DOI: https://doi.org/10.1016/j.seppur. 2010.12.024.

[48] Granata G., Moscardini E., Pagnanelli F., Trabucco F., Toro L. (2012) Product recovery from Li-ion battery wastes coming from an industrial pre-treatment plant: lab scale tests and process simulations, *J. Power Sources* **206**, 393. DOI: https://doi.org/10.1016/j.jpowsour. 2012.01.115.

[49] Zhang P., Yokoyama T., Itabashi O., Suzuki T.M., Inoue K. (1998) Hydrometallurgical process for recovery of metal values from spent lithium-ion secondary batteries, *Hydrometallurgy* **47**, 259. DOI: https://doi.org/10.1016/S0304-386X(97)00050-9.

[50] Mantuano D.P., Dorella G., Elias R.C.A., Mansur M.B. (2006) Analysis of a hydrometallurgical route to recover base metals from spent rechargeable batteries by liquid–liquid extraction with cyanex 272, *J. Power Sources* **159**, 1510. DOI: https://doi.org/10.1016/ j.jpowsour.2005.12.056.

[51] Dorella G., Mansur M.B. (2007) A study of the separation of cobalt from spent Li-ion battery residues, *J. Power Sources* **170**, 210. DOI: https://doi.org/10.1016/j.jpowsour.2007.04.025.

[52] Darvishi D., Haghshenas D.F., Alamdari E.K., Sadrnezhaad S.K., Halali M. (2005) Synergistic effect of cyanex 272 and cyanex 302 on separation of cobalt and nickel by D2EHPA, *Hydrometallurgy* **77**, 227. DOI: https://doi.org/10.1016/j.hydromet.2005.02.002.

[53] Joo S.-H., Shin D.j., Oh C., Wang J.-P., Senanayake G., Shin S.M. (2016) Selective extraction and separation of nickel from cobalt, manganese and lithium in pre-treated leach liquors of ternary cathode material of spent lithium-ion batteries using synergism caused by Versatic 10 acid and LIX 84-I, *Hydrometallurgy* **159**, 65. DOI: https://doi.org/10.1016/j.hydromet.2015. 10.012.

[54] Virolainen S., Fallah Fini M., Laitinen A., Sainio T. (2017) Solvent extraction fractionation of Li-ion battery leachate containing Li, Ni, and Co, *Sep. Purif. Technol.* **179**, 274 DOI: https://doi.org/10.1016/j.seppur.2017.02.010.

[55] Xu J., Thomas H.R., Francis R.W., Lum K.R., Wang J. Liang B. (2008) A review of processes and technologies for the recycling of lithium-ion secondary batteries, *J. Power Sources* **177**, 512. DOI: https://doi.org/10.1016/j.jpowsour.2007.11.074.

[56] Contestabile M., Panero S., Scrosati B. (2001) A laboratory-scale lithium-ion battery recycling process, *J. Power Sources* **92**, 65. DOI: https://doi.org/10.1016/S0378-7753(00)00523-1.

[57] Li J., Shi P., Wang Z., Chen Y., Chang C.-C. (2009) A combined recovery process of metals in spent lithium-ion batteries, *Chemosphere* **77**, 1132. DOI: https://doi.org/10.1016/j. chemosphere.2009.08.040.

[58] Kang J., Senanayake G., Sohn J., Shin S.M. (2010) Recovery of cobalt sulfate from spent lithium ion batteries by reductive leaching and solvent extraction with cyanex 272, *Hydrometallurgy* **100**, 168. DOI: https://doi.org/10.1016/j.hydromet.2009.10.010.

[59] Wang R.-C., Lin Y.-C., Wu S.-H. (2009) A novel recovery process of metal values from the cathode active materials of the lithium-ion secondary batteries, *Hydrometallurgy* **99**, 194. DOI: https://doi.org/10.1016/j.hydromet.2009.08.005.

[60] Chen L., Tang X.C., Zhang Y., Li L.X., Zeng Z.W., Zhang Y. (2011) Process for the recovery of cobalt oxalate from spent lithium-ion batteries, *Hydrometallurgy* **108**, 80. DOI: https://doi. org/10.1016/j.hydromet.2011.02.010.

[61] Dewulf J., Van der Vorst G. Denturck K. Van Langenhove H. Ghyoot W. Tytgat J. Vandeputte K. (2010) Recycling rechargeable lithium ion batteries: Critical analysis of natural resource savings, *Res. Conserv. Recycl.* **54**, 229. DOI: https://doi.org/10.1016/j.resconrec. 2009.08.004.

[62] Perez E. Navarro Amador R. Carboni M. Meyer D. (2016) In-situ precipitation of metal–organic frameworks from a simulant battery waste solution, *Mater. Lett.* **167**, 188. DOI: https://doi.org/10.1016/j.matlet.2015.12.129.

[63] Perez E. Andre M.-L., Navarro Amador R., Hyvrard F., Borrini J., Carboni M., Meyer D. (2016) Recovery of metals from simulant spent lithium-ion battery as organophosphonate coordination polymers in aqueous media, *J. Hazard. Mater.* **317**, 617. DOI: https://doi.org/10.1016/j.jhazmat.2016.06.032.

[64] Myoung J., Jung Y., Lee J., Tak Y. (2002) Cobalt oxide preparation from waste $LiCoO_2$ by electrochemical–hydrothermal method. *J. Power Sources* **112**, 639. DOI: https://doi.org/10.1016/S0378-7753(02)00459-7.

Chapter 18

Li-ion Batteries Environmental Impacts and Life Cycle Assessment (LCA)

Élise Monnier, Fabien Perdu and Stéphanie Desrousseaux

Li-ion batteries were first developed to power portable consumer devices (camcorder, walkman, mobile phones…). In the recent years, they also have been used for electric mobility and stationary storage systems, notably to store electricity produced by renewable energies (photovoltaics, wind energy...). The expected widespread use in these two application fields should lead in the coming years to an extra benefit. Indeed, it will contribute to reduce the greenhouse gas emissions and thus, it should positively act on the global warming issue. The present chapter deals with the environmental impacts of Li-ion batteries. A scientific method to quantify these impacts, along with recent results, is presented. Guidelines for Li-ion battery eco-design are also introduced.

18.1   Why a Focus on Battery Environmental Impacts?

The use of lithium-ion batteries in electric mobility and energy stationary storage systems seems to be one of the solutions to tackle the global warming issue. A massive increase in electric vehicle usage could allow for the transportation field the partial substitution of fossil fuel consumption by electricity consumption. In addition, such storage opportunities for renewable energies would help with the deployment of these less impacting energy sources.

As an example, transport was contributing up to 30% (132 out of 445 Mt Co_{2equ}) of the French greenhouse gas emissions in 2018, half of it being generated by passenger vehicles (SOeS (Service de l'Observation et des Statistiques), 2021) (figure 18.1). Variable with time, the oil bill cost 27–30 billion € to France only for transportation, among which ~ 14 billion € for passenger vehicles.

DOI: 10.1051/978-2-7598-2555-4.c018
© Science Press, EDP Sciences, 2021

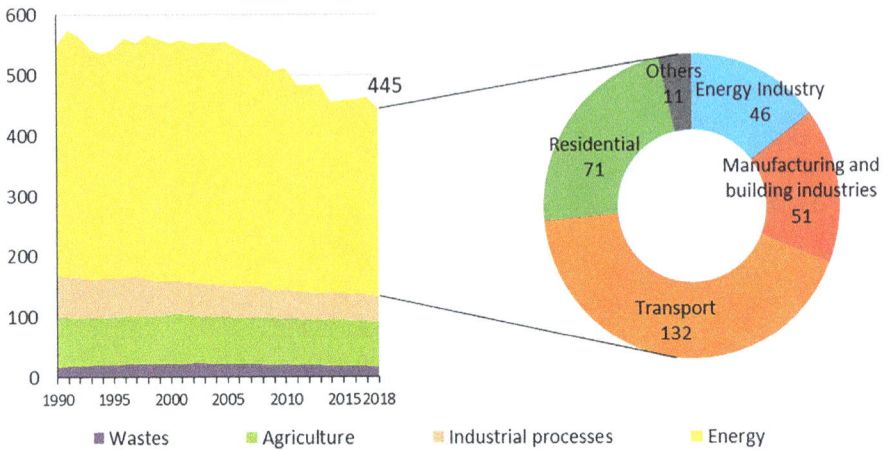

FIG. 18.1 – Greenhouse gas emissions in France by sector between 1990 and 2018, in Mt $CO2_{eq}$.

It is as important to check the ecological footprint of batteries as it is to validate their technical and economical feasibility.

Embodied energy for vehicles ranges from 10 to 100 kWh, depending on vehicle models. This is significantly higher than the quantities required for powering mobile devices (typically in the magnitude of 10 Wh for a smartphone or 100 Wh for a laptop). A very fast market growth is expected, along with a quick and large-scale deployment of the Li-ion battery technology. As for other technologies, Li-ion batteries require the use of raw materials, some of them acknowledged as scarce. The consideration of this material-related environmental issue is critical to allow a sustainable implementation of batteries at foreseen scale of deployment.

To conclude this section on the need for an environmental impact assessment, the context is nowadays very different from when the fossil fuel energies (oil, gas and coal) started to be used as energy sources. Indeed, the current knowledge in the ecology field is now mature enough to support technological developments. Consequently, it is advisable to include the environmental aspects in early stage developments in order to avoid as much as possible the design of irrelevant solutions from an environmental perspective.

18.2   How to Quantify Batteries Environmental Impacts?

Life Cycle Assessment (LCA) is a method to assess the environmental impacts for a product, a service or a system all along the lifecycle stages. International standards ISO 14040 and ISO 14044 (International Organization for Standardization (ISO), 2006) frame this analysis tool.

A product or system lifecycle may include, among others, the following stages:

- Raw material supply.
- Manufacturing.
- Assembling and fitting.
- Use phase.
- Service and repair.
- Transportation.
- End of life (disposal, reuse, recycling, incineration with or without energy recovery).

A schematic example for an electric accumulator lifecycle is given in figure 18.2.

LCA usually provides results for various impacts indicators associated with impact categories on resource consumption, on pollution assessment and on human health. It also presents sensitivity analysis regarding the assumptions used for modelling the system to assess.

In order to compare the environmental impacts of various products with similar functions, it is rather the service provided by a product, also named "functional unit", which is assessed for the quantification of indicators. In this case, functional units are defined in different ways to assess the lifecycle environmental impacts of batteries, depending on the study perimeter selected, as can be seen the examples below.

- Raw materials and manufacturing processes.

   ○ Producing 1 Wh or 1kWh electric storage capacity [22, 24].
   ○ Manufacturing of one battery vehicle pack [5].
   ○ Manufacturing of one battery energy storage (5 MW/5 MWh) for primary control provision [13].

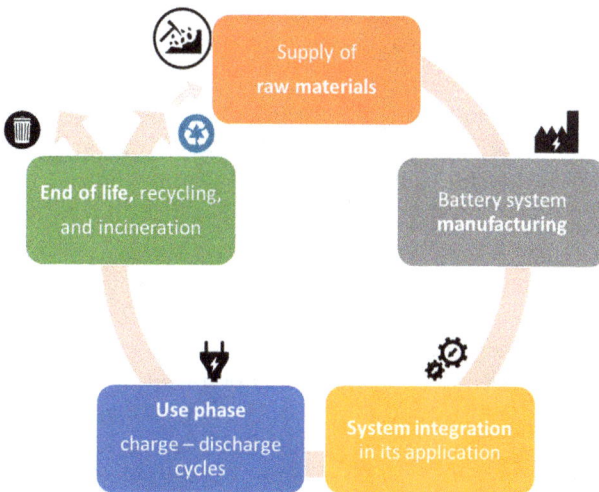

Fig. 18.2 – Example of Li-ion battery lifecycle.

– Lifecycle for electric mobility applications.

    ○ Traveling a 1 km distance with a passenger vehicle [28].
    ○ Delivering goods in a urban area on a 200 km daily average mileage [3].
    ○ Traveling a 1 km distance for one person in Wellington, New Zealand [7].

– Lifecycle for energy stationary storage systems applications.

    ○ Delivering 1 MWh electricity to the electricity grid network [8, 26].
    ○ Supplying a primary control provision of 551 MW in a continuous way over a 20-year period [14].

Thus, LCA results allow dealing with the environmental performances of batteries, while considering several lifecycle stages and the services they provide.

18.3   What are the Main Impacts of Lithium-ion Batteries?

The Li-ion based technology requires the use of critical materials and energy for the manufacturing of electrochemical cells. Upon usage, these electric batteries produce waste that might be difficult to manage. The life cycle of such products has impacts on the consumption of mineral and energy resources, as well as impacts on global warming. Given the processes and chemicals involved in batteries' life cycle stages, impacts are also observed on acidification, water eutrophication and on toxicity, both environmental and human.

(a) Regarding global warming

According to several literature reviews such as ICCT (2021), Romare and Dahllöf [23], Ellingsen (2017), Peters (2017) and Messagie [18], raw material supply and Li-ion battery manufacturing would produce between 30 and 400 kg CO_2 eq. greenhouse gas per produced kWh electric storage capacity. The variability in technologies, chemical processes and battery types involved, and also the quality of data used in the assessment explain the discrepancies in values.

Acknowledged for using good quality data, studies by Ellingsen *et al.* (2014) and Kim *et al.* (2016) quantify greenhouse gas emissions of, respectively, 173 and 140 kg CO_2 eq.kWh$_{\text{storage capacity}}^{-1}$ or 18 and 11 kg CO_2 eq.kg$_{\text{battery}}^{-1}$, for the manufacturing (including raw material supplying) of several Li-ion battery solutions used in the electric mobility application field. 2019 values (Dai *et al.*) assess 72 kg CO_2 eq.kWh-1. More recent 2019-2021 data are around 100 kg CO_2 eq.kWh^{-1}. (ref Dai *et al.*, (2019) or ICCT(2021). Discrepancies may be explained by the rapid high volume production ramp up in Gigafactories, which allow decreasing electrodes losses and optimizing the energy consumption.

Figure 18.3 shows that assessed GHG emissions of batteries decrease with time related to technology maturity improvement and production volumes related to plants' maturity improvements. This trend enables to estimate a value of GHG emissions for manufacturing in 2021 under 100 kg CO_2 eq.kWh$_{\text{storage capacity}}^{-1}$ when batteries are produced in Giga-factories.

For a battery used for stationary energy storage, greenhouse gas emissions of the storage system are calculated for 1 MWh of electric energy exchanged with the grid network upon charging and/or discharging. Koj *et al.* [14] calculated the environmental impacts for a system allowing the full primary control of frequency for the German grid network. This study concluded that over the whole lifecycle, the system produces around 120 kg CO_2 eq./MWh$_{\text{charge or discharge}}$, with approximately 220 kg CO_2 eq./kWh$_{\text{storage capacity}}$ considering only the manufacturing stage (including raw material supplying). It is noteworthy that contributors to GHG emissions, are related, for $\sim$ 100 kg CO_2 eq. per MWh of charged or discharged electricity, to the energy consumption of the battery to provide the service (auxiliaries, yield related losses...) and for only $\sim$ 20 kg CO_2 eq. to the battery manufacturing. Which means that the carbon content of the electricity mix determines in the first order the final CO_2 content of the MWh exchanged. New scenarios envisage the use of electric vehicle as stationary energy storage systems while providing low-emission mobility. Sharing use phases is often a way to reduce drastically environmental impacts.

Regarding batteries for passenger vehicles, greenhouse gas emissions are calculated per km travelled. With a Li-ion battery powered vehicles, literature describes performances ranging from 11 to 290 g CO_2 eq./km [17]. Most recent publication from ICCT (2021) place BEV emissions in 2021 under 100 g CO_2 eq./km, with battery production emissions corresponding with Dai (2019). Comparatively, models of thermal vehicles manufactured after 2010 emit between 124 and 205 g CO_2 eq./km according to Ellingsen (2016).

The Ellingsen *et al.* [5] study highlighted above provides figures between 11 and 50 g CO_2 eq. /km, depending on the electric vehicle and the battery lifespan, the vehicle size and the efficiency of the power train. These values are also highly dependent on the carbon footprint of the electricity mix used to charge the vehicle (see section 18.4 for more details on this topic).

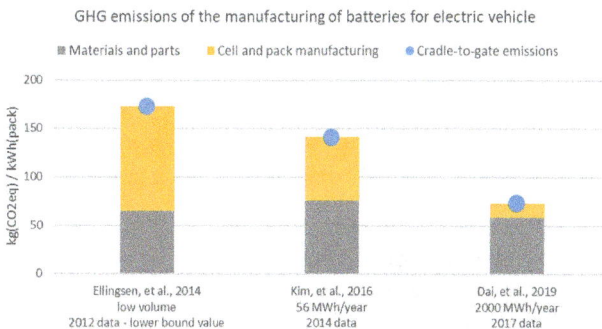

GHG emissions of the manufacturing of batteries for electric vehicle

■ Materials and parts   ■ Cell and pack manufacturing   ● Cradle-to-gate emissions

FIG. 18.3 – GHG emissions of three LCA based on industrial data with production volume detail and date of data.

(b) Regarding energy consumption

A battery consumes energy when it is manufactured, when it provides the service it was designed for and when it reaches end of life for recycling operations.

Energy consumption for raw material supply and for battery manufacturing shows large variations depending on the study parameters, on the production sites and their scale and on the chemical technologies involved. According to the Romare and Dahllöf [23] literature review, the amount of energy specifically needed for the manufacturing of a battery pack (supply and transformation of raw materials excluded) ranges from 350 to 650 $MJ_{elec}.kWh_{storage\ capacity}^{-1}$.

Based on good quality data, studies by Ellingsen *et al.* [5] and Kim *et al.* [12] respectively value to 586 and 530 $MJ_{elec}.kWh_{storage\ capacity}^{-1}$ the energy required for the manufacturing stage of various Li-ion battery technologies in the electric mobility application field. Based on more recent publications (Dai *et al.* 2019), manufacturing energy for a battery pack is around 300 $MJ_{elec}.kWh_{storage\ capacity}^{-1}$.

This amount of energy required for the manufacturing stage is equivalent to the amount stored by the battery while in operation over 150 charge discharge cycles, corresponding to:

- Either to a 2-year usage phase for an 100% electric vehicle battery, with one full charge discharge cycle every 10 days,
- or to a 3-month usage phase for stationary energy storage or for a PHEV type hybrid vehicle, with one full charge discharge cycle per day.

Considering that raw materials supply and transformation major the energy need by 30%–50% [6], the amounts calculated above are increased, respectively to 4–6 years and to 5–9 months.

Generally speaking, heavy charge/discharge cycling activities enable to recover the environmental cost of battery manufacturing, with one full cycle per day or every other day to reach thousands of cycles over the lifespan.

Upon usage (figure 18.4), a battery exhibits a global yield over charge and discharge between 80 and 90%, making it a highly performant energy storage system. This yield corresponds to the electrochemical conversion and to thermal and electric management. The remaining 10%–20% are energy released as heat and thus, not useful for the usage purpose. This energy loss upon usage affects the total energy consumption along the lifecycle of the battery.

FIG. 18.4 – Electric vehicle charging.

(c) Regarding abiotic resource consumption

If technological solutions involving Li-ion batteries are to be widely implemented, it is necessary to address the issue related to material availabilities and assess the opportunities to substitute critical materials.

> Materials used in batteries and classified as critical materials are: natural graphite, copper, and metals for electrodes (Lithium, Cobalt, Nickel, Titanium...).

The European Union considers that natural graphite used by the negative electrodes is a critical raw material. Indeed, it is a major contributor to the European economy and it is subject to supply shortage risks. However, it can be substituted either by synthetic graphite (with same performances at a slightly higher cost) or by other carbon based materials. For instance, hard carbon is suitable for high power batteries but less appropriate for high energy batteries.

Copper, used as current collector at the negative electrode, will probably meet supply issues in the coming years [27]. However, this is the result from other factors rather than its use in batteries.

Metals used as active materials in the positive electrodes, but also lithium, are battery technologies specific and they can hardly be replaced.

Figure 18.5 shows for several Li-ion battery technologies and for various critical raw materials, the total amount of batteries (expressed in energy) which could be produced with all available reserves or resources.

In order to figure what 50 TWh energy represents, one can consider the following two reference situations:

– Converting the 2017 entire automotive fleet (1 billion vehicle) to electric mobility using 50 kWh batteries (2017 median figure for electric vehicles) would require a 50 TWh energy need.

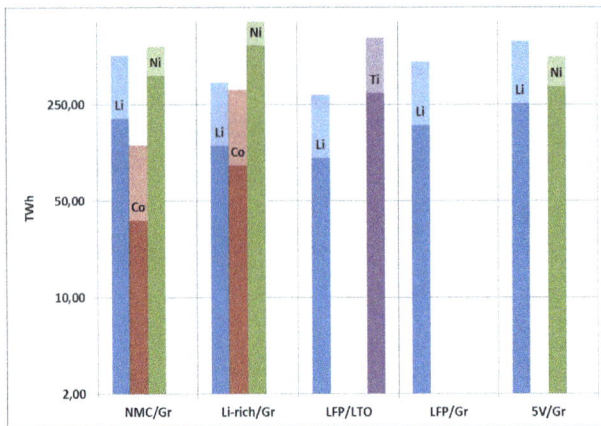

F<small>IG</small>. 18.5 – Theoretical values for energy stored in batteries if using either all the reserves (dark colours) or all the resources (light colours) for materials. Reserves and resources' figures are provided by USGC 2017 data (Log Y-axis).

– The worldwide need for energy storage for smoothing variable energy sources, both solar and wind generated, could reach 50 TWh. This would actually store 20 h of the 2017 global electricity production and thus level over a few days period[1] the intermittency of an energy mix with high contributions from solar and wind powering.

The above two scenarios would possibly use the same batteries. On the one hand, electric vehicles, by being connected to the grid while charging, would provide the levelling service for non-dispatchable energy sources. On the other hand, end of life vehicle batteries would then be used in a second life as stationary storage, provided that the usage conditions are less intensive.

Obviously, batteries have a limited lifespan (10–20 years, depending on the usage). The battery pool would then need to be renewed on a regular basis. Although recycling can reduce the material demand for end of life battery replacement, this option is limited and cannot be self-sustaining.

Considering reserves (known amounts available at current prices) and resource (available if technology improves or if prices increase) data provided in 2017 by USGS (US Geological Survey), the most limiting materials are cobalt (used in "NMC" positive electrode), lithium then nickel. With these figures, it is difficult to provide enough batteries for a 50 TWh storage capacity and replacement could be done at best only a few times. In the case of a 50 kWh battery based on the "NMC" technology, the material requirements are approximately 10 kg cobalt, 20–30 kg nickel and 6 kg lithium.

---

[1]Note that this does not solve the intermittency issue over longer periods.

To mitigate these values, one needs to consider that reserves and resources are often underestimated, particularly for materials that are not yet under strain. For instance, Gavin Mudd provides values almost twice higher for cobalt and nickel [19, 20]. However, the material amounts used for calculations are theoretical values, *i.e.* considering that the full amount of material is involved in the charge/discharge reactions. Actually, for lithium, only 60% of the theoretical capacity is used at the cell level and it is even less when considering a battery pack. This leads to overestimate the potential available battery amounts, while the reserve and resource assessment underestimates it.

Furthermore, regarding material availabilities, available stocks are a higher limit but issues appear long before depletion. Indeed, prices quickly increase when the material extraction no longer meets material demands. For that matter, cobalt exhibits high risks. 60% of cobalt mines are located in the Democratic Republic of the Congo and 90% of refining is done in China. In addition, cobalt is mainly a by-product of copper and nickel extraction. This means that cobalt extraction rate depends on nickel and copper demand, rather than on cobalt demand. Therefore, the price of cobalt is very fickle. The basic value is close to 10 à 20 $/lb (2010–2016 period), but high fluctuations were observed when demand increases: more than 50 $/lb in 2008 prior to the economic crisis and then above 40$/lb in 2018.

(d) Regarding acidification, eutrophication and toxicity

Materials and processes involved in battery manufacturing and usage affect other impact indicators such as terrestrial and water acidification[2], water eutrophication[3] and human toxicity.[4]

A substance, a product or a process contributes towards acidification when their lifecycle produces large amounts of nitrogen and sulfur containing by-products. Emissions can be released in water, in air or in the ground to contribute to the phenomenon.

A substance, a product or a process contributes towards eutrophication when their lifecycle produces large amounts of phosphorus and nitrogen containing by-products. Emissions are released in water or in the ground to contribute (figures 18.5 and 18.6).

Batteries lifecycle affects these impact categories mainly in three ways.

– Either by the use of substances directly affecting the impact:

One can mention specific requirements for the syntheses of active materials, such as the use of ammonia (NH_3), or the need for a calcination step, which can affect

---

[2]Global pollution coming from localized gas emissions that create precipitation of high pH water by combination with atmospheric water (acid rain). When these local emissions are released in the ground or on water, they can contribute to grounds or oceans acidification.

[3]Localized pollution coming from local effluent discharge containing large amounts of nutrients. These nutrients feed fast growing plants (algae) resulting in the asphyxiation of the aquatic environment, threatening other plant and animal species.

[4]Indicator measuring local pollution resulting from the emission of substances, in air, in water or in the ground, which can be toxic for human beings.

FIG. 18.6 – Environmental contamination (eutrophication) by green algae in Brittany (2006).

both acidification and eutrophication. In addition, when the CMR solvent such n-methyl-2-pyrrolidone (NMP) is used for the coating of active materials on electrodes for Li-ion cell, issues regarding eutrophication and human toxicity arise.

– Or by the use of energy consuming materials:

Energy resources and metal extraction and purification produce impacts in toxicity and acidification categories.

Copper and aluminum are the main contributors for batteries as they are used in electrodes, in the pack electronic architecture (cables, Battery Management System or BMS...) and even in packaging. When manufactured, Li-ion batteries also require chemicals with very high purity grades and carbonaceous additives, such as carbon fibers. This requires high amounts of energy, often produced by fuel combustion. Large quantities of nitrogen and sulfur oxides are emitted, thus contributing to acidification and to eutrophication (figure 18.7).

– Or by the need of electric energy:

Manufacturing and using a Li-ion battery require a large amount of electric energy. The more the involved electricity mix contains energy coming from fuel combustion, the larger the contributions to acidification, due to the release of sulfur and nitrogen oxides, and to eutrophication for nitrogen emission.

In addition, the production of copper that is used in the supply infrastructure emits large amounts of nitrogen and sulfur derivatives. This is due to the combustion of fossil fuel to run furnaces that transform sulfur rich ore to produce copper.

Anyway, it is still hard to conclude on the share of the battery manufacturing, compared with other activity sector contributions to these environmental impacts.

FIG. 18.7 – Copper casting, Glencore Fonderie Horne, Québec, Canada.

In addition, these kinds of pollution are highly related to the ability of the environment where manufacturing sites are located, to deal with the level of contaminations. For instance, if a manufacturing plant produces a phosphorous-based active material for batteries and is located on an eutrophication-sensitive coastline, it may aggravate the eutrophication issue already associated with intensive agricultural activities, for example. On the contrary, the localization of this plant, another area not yet impacted by eutrophication issue, may not have significant impacts on the equilibrium of the aquatic environment.

18.4   What are the Impact Sources?

Environmental impacts for electric powered vehicles generally come from:

– The vehicle and battery manufacturing stage.

   o With materials for components, such as cells.
   o With processes and their associated energy consumption.

– The usage stage with the environmental performance of the electricity mix invloved.

For impact indicators, such as human toxicity, ecotoxicity or non-energetic mineral resource consumption, the manufacturing stage generates the largest part of the environmental impact.

For other indicators, electricity required for the battery manufacturing or consumed upon usage in various applications has an impact highly dependent on the electricity mix composition. In the case of an electricity mix strongly contributing to a specific impact indicator, the usage phase will be more impacting than the manufacturing phase. On the contrary, with a lower impacting electricity mix, the manufacturing stage will be become the most impacting stage for most indicators.

As an example, several LCA studies on electric vehicles assessed the greenhouse gas (GHG) emissions for the various lifecycle stages, drawing a distinction between emissions in the manufacturing step, including the battery manufacturing, and emissions during the usage phase, namely the GHG emissions from electricity used for driving. Figure 18.8 drawn from Marmiroli *et al.* [17] shows the impact of each of these two lifecycle stages as a function of CO_2 emitted by the electricity mix used for mileage.

When compared to thermal vehicles, electric powered vehicles emit less GHG, whatever the composition of the electric mix, unless it is mainly based of coal. Indeed, the environmental performances in terms of emissions from electric vehicle are closely related to the performance of the electric mix involved in both manufacturing and usage. For the average European mix, for which CO_2 content is similar to the German mix content, GHG emissions of both manufacturing (vehicle and battery) and usage (consumed electricity) phases compare well. However, a low carbon energy mix, such as in France, is very advantageous for electric vehicles. In this specific case, the manufacturing phase becomes the main contributor to the global warming impact category.

In other words, an efficient transition to lower carbon mobility can only be associated to a transition to low carbon energy sources. This will also depend on the size and weight reduction in vehicles: a vehicle powered with a small battery will be significantly more environmentally virtuous than with a larger one, unless the vehicle is heavily used with a daily mileage corresponding to the battery capacity.

According to most of the studies, when manufacturing an electric vehicle pack, the main source for environmental impacts is the manufacturing of the cell (elementary accumulator), specifically the cathode preparation, whereas the pack integration and the electronic components are only minor contributors.

As an example, regarding the environmental impact related to mineral resource depletion, variations in the composition of active materials for cathode induce strong effects. Figure 18.4 shows, if considering the maximal theoretical capacity of NMC materials, a limitation at 36 TWh due to cobalt reserves, as assessed by USGS. However, as the actual capacity for the material is around 40% below the theoretical value, the potential value is reduced to 20 TWh for batteries (figure 18.9). Initially using materials with $33\%_{stoichiometry}$ of cobalt amount (NMC111), current industrial developments aim at progressively reducing the cobalt amount from $20\%_{stoichiometry}$ (NMC622) to $10\%_{stoichiometry}$ (NMC811). Figure 18.9 exhibits the impact of these composition variations on the potential technology deployment, using the actual performances obtained with the material alternatives. For NMC811 material, the composition of which in cobalt and nickel is close to the NCA composition (technology used in Tesla vehicles), an 80 TWh target is a reachable with the same mineral reserves. In this case, one can observe that the limitations due to cobalt are not anymore far from the limitations from nickel.

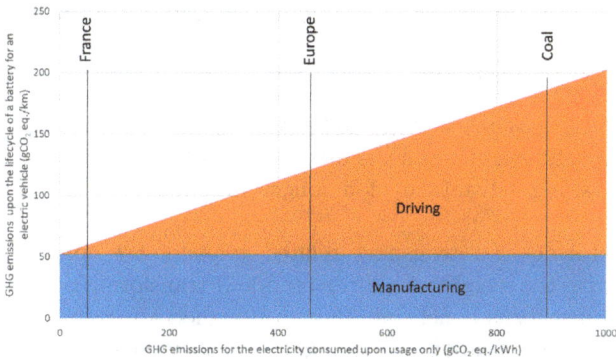

FIG. 18.8 – GHG emission contributions for an electric vehicle to manufacturing and usage lifecycle stages, according to Marmiroli *et al.* [17]. As a comparison, a thermal vehicle emits between 150 and 200 gCO_2eq./km. The electric mix used for manufacturing is close to 500 gCO_2eq./kWh.

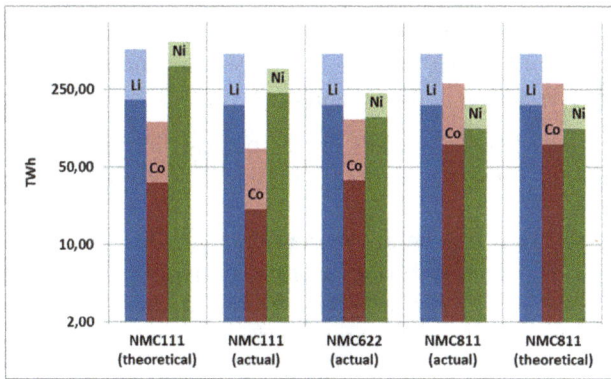

FIG. 18.9 – Actual total energy storable in batteries using the all the reserves (dark colours) or resources (light colours) for materials composing various NMC technologies. Values for reserves and for resources are provided from USGS 2017 data (Log Y-axis).

Regarding battery manufacturing for application related to stationary storage [14], three components are equally important: cells, electronics (such as converter) and buildings.

18.5 Guidelines for Ecodesign

This section on guidelines for ecodesign is based on literature papers by Larcher and Tarascon [15], Arbabzadeh *et al.* [2], Ellingsen *et al.* [4], Romare and Dahllöf [23], and Peters *et al.* [22]. It mainly deals with the impact reduction of the battery manufacturing stage and gives some useful directions to decrease impacts upon usage.

Regarding battery pack,

- A long lifespan, mutualised usage and a good cycling resistance allow limiting the negative effects associated to battery manufacturing, taking advantage of a longer and more intense usage.

  Lifespan

- As manufacturing and transport related impacts are commensurate with the weight of matter used and transformed, a good energy density is beneficial in an ecodesigned approach.

  Energy density

- A good energy yield allows reducing the energy loss upon battery usage. This is all the more relevant when lifespan is increased and with a high carbon content electric mix.

  Energy yield

Moreover, if considering that the above parameters are set, two approaches are to be equally favored to reduce the impacts related to battery fabrication.

- The use of abundant materials and recycling allows the reduction in energy consumption and externalities associated with ore extraction and transformation.

  Materials & recycling

- The use of energy-efficient processes reduces the impact of materials transformation and cell manufacturing.

  Low energy processes

To have an actual beneficial effect on the battery lifecycle, these materials and process substitutions must not damage the lifespan, the energy density and the energy yield.

It is noteworthy that all the above developments lead to a lower energy consumption for battery manufacturing. Consequently, they could also imply cost reductions and make them more attractive from an industrial perspective.

Regarding processes, two main steps need optimization: on the one hand, the synthesis of active materials and on the other hand, the cell assembly.

The development axis to follow for material synthesis is the search for lower temperature processes. Larcher and Tarascon [15] reviewed among other synthetic processes, sorted by decreasing temperatures: solvothermal, hydrothermal, ionothermal syntheses, and ideally biomineralisation.

The cell fabrication is nowadays affected by the electrode coating and drying step. Indeed, coating involves a toxic solvent (NMP[5]) for compatibility with the polymeric binder (PVdF[6]) used in cells. Besides the intrinsic impacts from the solvent, the coating step is followed by drying, which requires large amounts of heat. It is thus desirable to replace NMP by a harmless solvent such as water (already used in negative electrode manufacturing). Ideally, low solvent content or even dry processes should be developed. Two other areas for improvement are the reduction of the need for anhydrous environment in the last manufacturing steps (currently

---

[5]N-Methyl-2-pyrrolidone.

[6]Polyfluorure de vinylidene.

related to the cell electrolyte) and the identification of defects at very early stages of the fabrication chain to reduce production waste (enabled at large scale production).

Due to energy consumption, it should be preferred to locate plants for material synthesis and for cell manufacturing in countries where the energy mix has a lower carbon content.

Research for abundant materials needs to focus not only on cobalt replacement, but also on nickel, for the positive electrode. The most promising candidates are sulfur and ambient oxygen, which could enable to reach high energy densities (practically around 500 Wh.kg^{-1} at cell level). However, important developments are necessary for cyclability and power performances and for energy yield (see chapter 7).

For the negative electrode, although it is a less urgent matter, it is desired to replace lithium. Other cations are investigated such as sodium, for which research is more advanced (see chapter 6), and also potassium, divalent cations (calcium, magnesium) and anions (chloride), for which result are only nascent.

An improvement in energy density can be obtained by reducing the negative electrode weight (silicon addition in graphite, see chapter 3), or by replacing the anode with metal lithium. This latter option requires the development of solid electrolytes able to prevent the formation of dendrites, but these materials should not involve the use of new critical raw materials (see chapter 8).

Finally, batteries could be designed by only using organic materials (see chapter 4). One has to pay attention to the use of additives that could have a significant impact on the cell global footprint (for instance carbon fibers or nanotubes).

Bibliography

[1] Arbabzadeh M., Johnson J.X., Keoleian G. (2017) Parameters driving environmental performance of energy storage systems across grid applications, *J. Energy Storage* **12**, 11.

[2] Arbabzadeh M., Johnson J.X., Keoleian G.A., Rasmussen P.G., Thompson L.T. (2015) Twelve principles for green energy storage in grid applications, *Environ. Sci. Technol.* 1046, doi: https://doi.org/10.1021/acs.est.5b03867.

[3] Bartolozzi I., Rizzi F., Frey M. (2013) Comparison between hydrogen and electric vehicles by life cycle assessment: a case study in Tuscany, Italy, *Applied Energy* **101**, 103, doi: https://doi.org/10.1016/j.apenergy.2012.03.021.

[4] Ellingsen L.A.-W., Hung C.R., Strømman A.H. (2017) Identifying key assumptions and differences in life cycle assessment studies of lithium-ion traction batteries with focus on greenhouse gas emissions, *Transp. Res. Part D*, 82.

[5] Ellingsen L.A.-W., Majeau-Bettez G., Singh B., Srivastava A.K., Valøen L.O., Strømman A.H. (2014) Life cycle assessment of a lithium-ionbattery vehicle pack, *J. Indus. Ecology* **18**, 113, doi: https://doi.org/10.1111/jiec.12072.

[6] Ellingsen L.A.-W., Singh B., Strømman A.H. (2016) The size and rage effect: lifecycle greenhouse gas emissions of electric vehicles. (I. Publishing, Éd.), *Environ. Res. Lett.* **11**, 054010, doi: https://doi.org/10.1088/1748-9326/11/5/054010.

[7] Elliot T., McLaren S.J., Sims R. (2018) Potential environmental impacts of electric bicycles replacing other transport modes in Wellington, New Zealand, *Sustain. Prod. Consump.* **16**, 227, doi: https://doi.org/10.1016/j.spc.2018.08.007.

[8] Hiremath M., Derendorf K., Vogt T. (2015) Comparative life cycle assessment of battery storage systems for stationary applications, *Environ. Sci. Technol.* **49**, 4825.

[9]   A global Comparison of the Life-Cycle Greenhouse Gas Emission of Combustion Engine and Electric Passenger Cars, White Paper, The International Council on Clean Transportation (ICCT), July, 2021.

[10] International Organization for Standardization (ISO) (2006) ISO 14040, *Environmental Management - Life Cycle Assessment - Principles and Framework.*

[11] International Organization for Standardization (ISO) (2006) ISO 14044, *Enivronmental Management - Life Cycle Assessment - Requirement and Guidelines.*

[12] Kim H.C., Wallington T.J., Arsenault R., Bae C., Ahn S., Lee J. (2016) Cradle-to-Gate Emissions from a commercial electric vehicle Li-ion battery: a comparative analysis, *Environ. Sci. Technol.* 7715, doi: https://doi.org/10.1021/acs.est.6b00830.

[13] Koj J.C., Schreiber A., Stenzel P., Zapp P., Fleer J., Hahndorf I. (2014) Life cycle assessment of a large-scale battery system for primary control provision, *6th Advanced Battery Power Conference,* Münster. Consulté le Mars 2018, sur https://core.ac.uk/download/pdf/35009465.pdf.

[14] Koj J.C., Stenzel P., Schreiber A., Hennings W., Zapp P., Wrede G., Hahndorf I. (2015) Life cycle assessment of primary control provision by battery storage systems and fossil power plants, *Energy Procedia* **73**, 69, Elsevier.

[15] Larcher D., Tarascon J.M. (2015) Towards greener and more sustainable batteries for electrical energy storage, *Nature Chem.* **7**, 19, doi: https://doi.org/10.1038/nchem.2085.

[16] Lin Y., Johnson J.X., Mathieu J.L. (2016) Emissions impacts of using energy storage for power system reserves, *Applied Energy* **168**, 444.

[17] Marmiroli B., Messagie M., Dotelli G., Van Mierlo J. (2018), Electricity generation in LCA of electric vehicles: a review, *Appl. Sci.* **8**, 1384, doi: https://doi.org/10.3390/app8081384.

[18] Messagie M. (2016) *Life Cycle Analysis of the Climate Impact of Electric Vehicles,* Vrijet Unicersiteit Brussel: Transport & Environment.

[19] Mudd G.M., Jowitt S.M. (2014) A detailed assessment of global nickel resource trends and endowments, *Economic Geology* **109**, 1813, doi: https://doi.org/10.2113/econgeo.109.7.1813.

[20] Mudd G.M., Weng Z., Jowitt S., Turnbull I., Graedel T. (2013) Quantifying the recoverable resources of by-product metals: the case of cobalt, *Ore Geology Rev.* **55**, 87, doi: https://doi.org/10.1016/j.oregeorev.2013.04.010.

[21] Nordelöf A., Messagie M., Tillman A.-M., Söderman M.L., Joeri V. (2014) Environmental impacts of hybrid, plug-in hybrid, and battery electric vehicles - what can we learn from life cycle assessment? *Int. J. Life Cycle Assessment* **19**, 1866, doi: https://doi.org/10.1007/s11367-014-0788-0.

[22] Peters J.F., Baumann M., Zimmermann B., Braun J. (2017) The environmental impact of Li-ion batteries and the role of key parameters - a review, *Renew. Sustain. Energy Rev.* 491.

[23] Romare M., Dahllöf L. (2017) *The Life Cycle Energy Consumption and Greenhouse gas Emissions from Lithium-ion Batteries,* Stockholm, Sweden: IVL Swedish Environmental Reserach Institute, Récupéré sur https://www.ivl.se/download/18.5922281715bdaebede9559/1496046218976/C243+The+life+cycle+energy+consumption+and+CO2+emissions+from+lithium+ion+batteries+.pdf.

[24] Dai Q., Kelly J.C., Gaines L., Wang, M. (2019) Life cycle analysis of lithium-ion batteries for automotive applications, *Batteries* **5**, 48, doi: https://doi.org/10.3390/batteries5020048.

[25] SOeS (Service de l'Observation et des Statistiques) (2021) Chiffres clés du climat France et Monde - Édition 2021 (DATA LAB), ISSN : 2557-8138, www.statistiques.developpement-durable.gouv.fr.

[26] Vandepaer L., Cloutier J., Amor B. (2017) Environmental impacts of lithium metal polymer and lithium-in stationary batteries, *Renew. Sustain. Energy Rev.* 46.

[27] Vidal O., Rostom F., François C., Giraud G. (2017) Global trends in metal consumption and supply: the raw material–energy nexus, *Elements* **13**, 319, doi: https://doi.org/10.2138/gselements.13.5.319.

[28] Zackrisson M., Fransson K., Hildenbrand J., Lampic G., O'Dwyer C. (2016) Life cycle assessment of lithium-air battery cells, *J. Cleaner Prod.* 299, doi: https://doi.org/10.1016/j.jclepro.2016.06.104.

Chapter 19

Applications and Markets – User Cost

Laurent Garnier, Daniel Chatroux, Fabien Perdu, Bruno Béranger, Frédéric Le Cras, Didier Bloch and Sébastien Martinet

This chapter presents some elements of market analysis with a focus on the electrified vehicle market and on the issue of user cost. It is largely based on the results of AVICENNE ENERGY work, which proved to be realistic all along the past few years, *e.g.* [1] "Christophe PILLOT; The Rechargeable Battery Market : Value Chain and Main Trends 2019-2030-IBS -11 March, 2021".

19.1 General Elements of Market Analysis – Focus on the Electrified Vehicle Market

Figure 19.1 shows that lead-acid batteries, mainly used as starter batteries in the automotive industry, dominates the market in terms of MWh produced. Since 2010, the share of accumulators Li-ion grows year after year, driven by the take-off of the electric and hybrid vehicle market. With the gradual ban on the use of lead, and the booming of electric vehicles, the lead-acid batteries market should, within 2025–2030, become smaller than the lithium batteries market.

Figure 19.2a illustrates the very strong growth, in terms of value, observed for the Li-ion battery market. Almost entirely due to the development of electric mobility. The market is presently with totally dominated by Asian players (Korea, China and Japan) (figure 19.2b) which supply almost 100% of demand. The main manufacturers of lithium batteries are Samsung SDI, LG Chem, Panasonic, Lishen, CATL, BYD (which are integrated player up to the battery pack).

DOI: 10.1051/978-2-7598-2555-4.c019

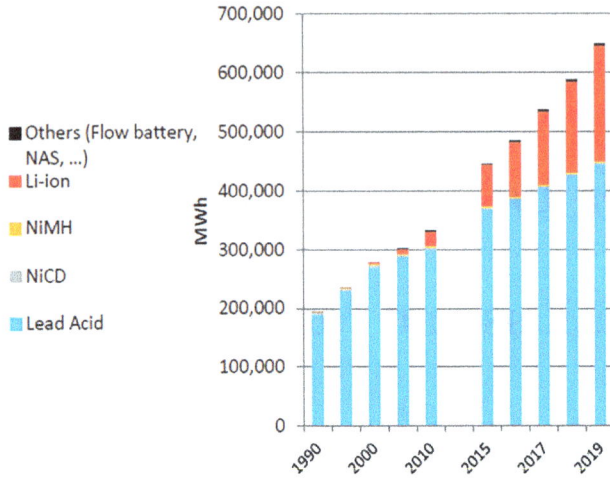

FIG. 19.1 – Worldwide battery market 1990–2019 in MWh energy content of batteries sales [1]. The rapid development of the electric vehicle market should lead Li-Ion batteries to dominate the world's market in the coming years (>3 TWh expected as soon as 2030), and take precedence over all other technologies, including the previously dominating Lead Acid batteries, so far mainly used as starter batteries in internal combustion engine vehicles.

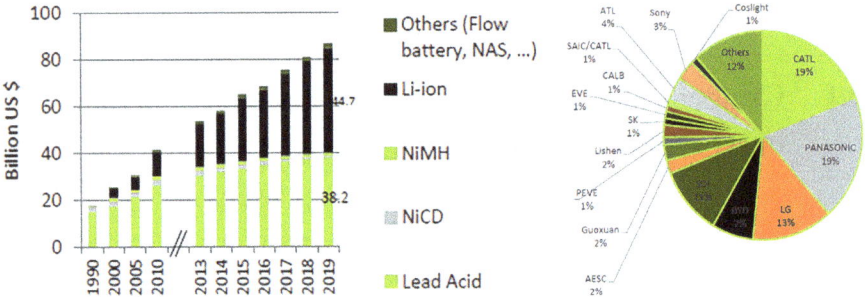

FIG. 19.2 – (a) Worldwide battery market 1990–2019 (B$) [1]. Since 2019 Li-Ion batteries dominate the world's market in value. (b) Lithium Ion batteries industrial players in 2019.

Figure 19.3 forecasts the EV's battery demand growth: ≈ 900 GWh is expected in 2030. However, CO_2 emissions reduction policies may well lead to much stronger national policies in favour of vehicle electrification. Other players [2, 3] speculate on a much larger demand, as high as 2 TWh, or even 3 TWh as soon as 2030. These last figures seem more in line with the expected production, in 2030, of the gigafactories being currently built around the world.

This very rapid development is made possible by a significant reduction in the production costs of Li-ion accumulators (figure 19.4). While the cost of Li-ion cells

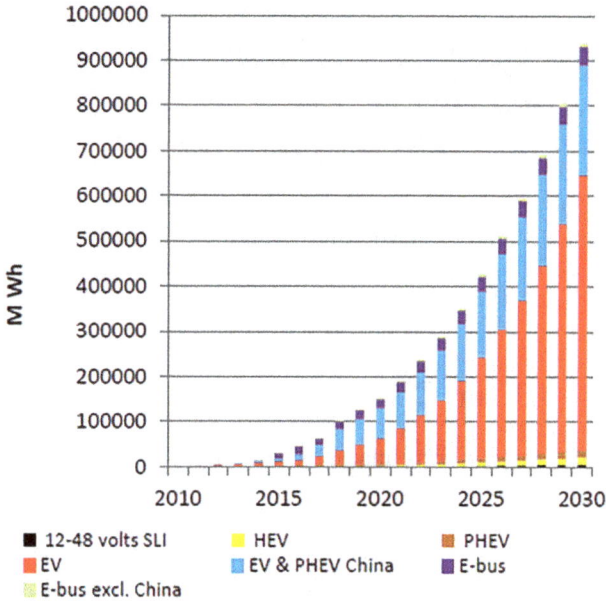

FIG. 19.3 – World battery demand forecast for electrified vehicles (BEV; HEV; PHEV) [1]. Some other players such as Tesla expect a much larger demand, as high as 2 or even 3 TWh in 2030.

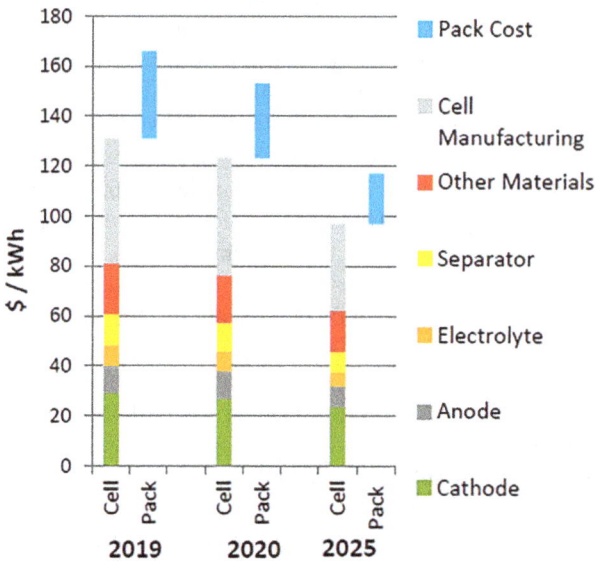

* For Production > 100 000 packs/year

FIG. 19.4 – Lithium Battery cost at pack level for electric vehicle [1]. Some other players speculate on a cost as low as \$80–100 kWh^{-1} at pack level.

was over \$2000 kWh^{-1} in 1991, it fell to just under \$160 kWh^{-1} in 2020. The forecast for 2025 is estimated around \$120 kWh^{-1} [1], or even lower than \$100 kWh^{-1} by some other major players such as Tesla [2].

A large part of the observed cost reduction is due to the production ramp up of very large production plants. Nevertheless, the cost structure of the cells includes a very large share of raw materials (usually higher than 60%), with in particular the contribution of the positive electrode, most often based on cobalt. An important research direction is therefore devoted to reducing this contribution by optimizing the composition of the electrode material, or even eliminating it by switching to new technologies such as Lithium–Sulfur or Sodium-ion.

19.2   Issue of User Cost

Although lithium batteries were marketed as early as 1991, their possible use for electric mobility did not begin to be considered until a few years later, around 1997. One may even remind that when, in 1995, PSA and Renault, in France, ahead of most of their competitors, marketed electric vehicles models based on concepts designed in 1990, they used a "Nickel Cadmium" battery pack, manufactured by the SAFT Company. The price of an electric Citroën Saxo 4-seater sedan was, in 1996, about €14 000 (not corrrected for inflation), to which €100 had to be monthly added for the battery rental.

The performance of vehicles such as the electric Citroën Saxo or its Peugeot 106 twin), however, were quite poor: despite their interesting cranking torque value, their relatively high weight (1100 kg – including 260 kg of the Nickel-Cadmium battery pack) and the poor continuous discharge power capability of the battery limited the performances (difficulties going up the slopes, maximum speed limited to 90 km/h). The 12 kWh battery energy was in practice never available. Measurements showed that around 8.5–10 kWh only were actually usable, and these cars only covered 80 km in the best conditions, 60 km more generally. In addition, batteries had extremely variable lifetimes, which depended on the vehicle's use (state of charge of the batteries during storage, temperature, frequency of recharging, etc.). Some use modes lead to battery failure after 5 years and 30 000 km, while others allowed 100 000 km and a life expectancy of around 10 years.

All of these drawbacks constituted a considerable obstacle to the purchase of such vehicles by individuals. As a result, only around 10 000 electric vehicles were manufactured between 1995 and 2006. The decision to stop their marketing in 2005 concluded this experience with an observation of failure, painfully felt by many car manufacturers, especially European and North American. This situation led, in these countries, to a considerable redution of the effort on electric mobility. This was the very moment when the Asian industry, anticipating the considerable improvements that could be achieved with lithium batteries, put its efforts on their development. Lithium ion batteries soon revealed their very high capacity associated with high continuous power performances behavior. The coupling of these two characteristics allowed a revolution in electric vehicles performances.

From 2011 on, new electric vehicle models, such as the Mitsubishi "i-Miev" and the Nissan "Leaf" were marketed in Europe. The initial high prices of these two vehicles, powered by Lithium-ion accumulators, decreased rapidly, whereas a national financial incentive helped users to purchase new electric vehicles and scrap of old diesel cars.

Since 2013, the offer of new electrified vehicles has accelerated, with models such as: the Renault Zoe, the Volkswagen E-UP and E-Golf, the BMW i3, the Kia Soul EV, the Hyundai Electric Ioniq, the Tesla S and X...

Field results, mainly published by Tesla, are now showing that the battery life expectancy may allow mileage (much) higher than 200 000 km. The idea to change the battery twice during the car's life vanished, and car manufacturers have abandoned the idea to rent the battery.

Electric cars become day after day more affordable: with batteries included, their cost usually ranging in 2021 between €26 000 and €40 000 excluding state incentive. Tesla's top-of-the-range cars stand out with prices ranging from around €45 000 to €95 000, but Elon Musk announced early 2021 the future commercialization of a more affordable Tesla model at € 25 000.

To conclude this chapter, old batteries technologies such as Cadmium Nickel are not adapted to electric vehicle. The transition to Lithium-ion is a major step ahead. It is now possible to consider that thanks to the considerable improvements achieved, Lithium ion batteries fit now with the electric vehicle technical specifications, paving the way for a revolution in the automotive industry. Moreover, the 10-fold battery cost decrease between 2010 and 2020 allows now the TCO (Total Cost of Ownership) of an electric vehicle to be lower than its internal combustion engine counterpart.

References

[1] Avicenne Energy -The Rechargeable Battery Market : Value Chain and Main Trends 2019-2030-IBS -11 March, 2021.
[2] Tesla Batteries Day, September 2020. https://www.rev.com/blog/transcripts/tesla-2020-battery-day-transcript-september-22.
[3] Benchmark Mineral Intelligence "Lithium Ion Battery Megafactory Assessment" July 2020, https://www.benchmarkminerals.com/.

Conclusion

**Didier Bloch, Sébastien Martinet, Thierry Priem
and Frédéric Le Cras**

To efficiently combat climate change and reduce GHG emissions, many countries in the world decided to ban thermal engine powered passenger vehicles before 2040, and accelerate the development of electromobility. The expected tsunami, allowed by the performances improvements and the cost decrase of Lithium-Ion batteries, drives the entire automotive industry to reconsider and completely overhaul its model. In the revolution which is taking place, the mastering of the Li-ion technology, which has become key for the future, may condition the survival of many car manufacturers. Thanks to their intrinsic decisive advantages (energy and power density performances; very high energy efficiency; continuously decreasing cost; acceptable environmental impact...), Li-ion batteries are becoming a standard technology to store the on-board energy required to power vehicles. Their booming world's production is expected to increase from ca. 200 GWh in 2020 to more than 3–4 TWh in 2030 (expressed in battery energy content). During 20 years until the end 2016, Europe considered as a non-priority the fabrication of batteries, mainly used to power portable equipments. With the notable exception of a few players -such as Blue Solutions in France, which developed its proprietary solid-state battery technology, the market was abandoned to Asia. Industrials players such as PANASONIC; LG; SAMSUNG or BYD soon anticipated the full potential of Li-ion batteries for electromobility, and invested huge amounts of money and worked hard to improve batteries performances. Sometimes as part of collaborations with visionary people like Elon Musk, or financially backed by national authorities, they managed to penetrate the automotive market with high added value products. In 2021, the Chinese, Korean and Japanese technologies dominate outrageously the market, and Li-ion batteries appear, as a vehicle's key component, to perfectly meet not only passenger vehicle and commercial delivery vehicles specifications, but also, under certain conditions, meet the needs of heavier transport vehicles as well. Battery fabrication becomes all of a sudden a strategic issue, and automotive companies, such as, in Europe, Volkswagen or Stellantis to name a few, decide to build consortia and invest tens of billions of euros in a no-return strategic move in

DOI: 10.1051/978-2-7598-2555-4.c902

the perspective of catching up with Asian manufacturers and master the skills required. In Europe only, the building of more than 20 Gigafactories – each of them representing an investment of several billions of euros-is proving necessary to meet the demand of the vehicle's electrification market. The European Commission decided to encourage these initiatives : following a first December, 2019 decision to fund with a €3.2 billion grant a first French-led Important Project of Common Interest (IPCEI) involving seven Member States and 17 private companies covering all segments of the battery value chain, it approved in November, 2020 to fund a second, German-led, IPCEI involving twelve Member States with € 2.9 billion public funding. To meet the massive expected demand, The European Commission, sometimes helped by local authorities, even backs Chinese (CATL, ENVISION...); Korean (LG, SAMSUNG...) and Japanese (PANASONIC...) industrial players to lay a foot in Europe and build Gigafactories in Hungary; Great Britain; France and Poland. The cumulated investment to be consented in Europe by European as well as Asian players in Li-ion battery development within 2030 could be as high as € 300 billion.

During the past ten years since the first Gigafactory commissioned in Nevada by Panasonic for Tesla's vehicles, Li-ion batteries improved very quickly. Their market price has decreased by a factor of ten at pack level. In 2021, it fluctuates around € 100–120/kWh. Driven by the scale effect induced by the automotive market, it should pursue its decrease, but as the cost of materials used for fabrication already represents more than 70% of the battery's cost, it is more than likely that it should reach its lower limit in the coming years around € 60–70/kWh, depending on materials cost fluctuations. The major part of the cost decrease is already currently driven by improvements of the battery pack design and its integration in the vehicle as a structural component rather than the material's cost decrease.

This situation does not mean the end of R&D activities. Although mature, Li-ion battery technology can still be improved in terms of battery performance and environmental impact. Research and development efforts are still required to ensure that their massive deployment can simultaneously reveal to be environmentally sustainable. In this respect, the importance of *e.g.* the battery life expectancy can be highlighted: doubling it from 1000 to 2000 charge/discharge cycles thanks to materials stability improvements will not only strongly reduce the battery intrinsic environmental impact and its use cost: it will allow using the vehicles, during parking times, to other purposes than mobility. A very simple calculation is here possible: 2000 cycles represent, for a 50 kWh battery, the equivalent of 800 000 km, or more than 60 years of use for an average driver (13 000 km/year). But no driver will keep his vehicle for such a long time, and battery calendar ageing will anyway not allow a 60 year lifetime. Considering, on the other hand, that the same driver will decide using its vehicle both for mobility (250 000–300 000 km) and for power grid support (the equivalent residual 500 000 km represent 60 MWh all along the vehicle's lifetime), one sees that at system level, beyond their use as massive Renewable Electricity storage buffers, electrified vehicles will offer a lot of interesting additional features compared to thermal engines vehicles: demand peak shaving; grid frequency regulation, self consumption...not considering that the vehicle's owner will be paid back for the given service. All these features, which will avoid building and operating additional fossil fuels powered plants, have to be closely assessed by LCA and business model studies to comply with the reduction of the world's CO_2 emissions targets. As another example, battery recycling at its end of life is also a critical step to recover materials for next generations of batteries and limit the dependance on strategic resources.

Current and future developments of Li-ion batteries may also lead to incremental changes, perhaps even to real technological breakthroughs.

Incremental developments include improved "LFP" (Lithium Phosphate) cathode material compositions using much less sensitive materials or improved "NMC" type cathode materials with significantly reduced cobalt content, thanks to the development of "nickel rich" compositions such as "NMC811", or NMC9/0.5/0.5 (figures indicating the stoichiometric proportions of the materials present: nickel, manganese, cobalt). It is also worth mentioning the use of gelled and/or "solid" electrolytes, which are expected to improve the safety of use and simplify the manufacturing processes by avoiding a costly solvent recovery process step.

Breakthrough developments may concern the so-called "post-lithium-ion" battery families, such as, for example:

- "Solid State" lithium Batteries, which are likely to offer increased energy density and safety thanks to the use of a solid, polymer or ceramic electrolyte, making it possible to dispense with a graphite anode that is replaced by a thinner metallic lithium anode.
- New generations of sodium-ion batteries using vanadium-free cathode active material.
- Lithium–sulphur batteries.
- ...

Finally, the aspects linked to standardisation and regulation may also reveal important: it will be useful, in order to allow a multi-source supply, to standardise to a certain extent the batteries available on the market in terms of geometry, capacity or performance and thus facilitate their wide distribution.

As we can see, the future of lithium-ion or post-lithium-ion batteries, while promising, is far from being definitively written: many efforts still need to be made to include these technologies in a sustainable development approach. We hope that reading this book will have given the reader a better idea of the developments underway.

Glossary

Term	Unit	Meaning	Definition
Accumulator or Cell or Unit Cell			Electrochemical accumulator: reversible electrochemical energy storage device contrarily to primary cells that are non-rechargeable
AFM		Atomic Force Microscopy	Atomic force microscopy (AFM) is a very-high-resolution type of scanning probe microscopy (SPM), with demonstrated resolution on the order of fractions of a nanometer, more than 1000 times better than the optical diffraction-limit.
Ampere.hour	Ah	Capacity Unit. 1 Ah = 3600 Coulombs	An ampere hour or amp hour (symbol: A·h) is a unit of electric charge, having dimensions of electric current multiplied by time, equal to the charge transferred by a steady current of one ampere flowing for one hour, or 3600 Coulombs
Anode			Electrode where an oxidation reaction takes place, designates by misuse of language the negative electrode (where this reaction occurs during discharge)

Term	Unit	Meaning	Definition
Battery			Assembly in series and/or parallel of primary or rechargeable cells of same type
Battery Pack			Assembly of cells modules; electronic; packaging and thermal management unit ready to be integrated into the final application
BET			Specific Surface Area Measurement Method (m^2/g)
BMS		Battery Management System	Electrical and Thermal Management System of a battery pack that controls for instance some safety functions: cell balancing, thermal management, state of charge and state of health measurements of the battery pack...
BOL		Beginning of Life	Performances relative to beginning of life
CAN Network			A Controller Area Network (CAN bus) is a vehicle bus standard designed to allow microcontrollers and devices to communicate with each other's applications without a host computer.
Capacity	Ah		The Battery nominal capacity is the amount of Ampere.hour that can be withdrawn from the battery at a given constant current, starting from a fully charged state. the 1C rate drains the battery after 1 h; the 10-h rate (C/10) drains the Battery after 10 h.

Term	Unit	Meaning	Definition
Cathode			Electrode where a reduction reaction takes place, designates by misuse of language the positive electrode (where this reaction occurs during discharge)
Charge Rate (or Discharge Rate)			Rate at which a cell is charged or discharged. A 1C rate corresponds to a charge in 1 h. 2C in 1/2 h, C/2 in 2 h, 10 C corresponds to a charge in 10 min.
CMU		Cell Management Unit	Unit cell Management System linked with the BMS
Discharging Rate			Rate at which a battery cell is discharged. A 1C rate corresponds to a discharge in 1 h. 2C in 1/2 h and C/2 in 2 h
DSC		Differential Scanning Calorimetry	Differential scanning calorimetry (DSC) is a thermoanalytical technique in which the difference in the amount of heat required to increase the temperature of a sample and reference is measured as a function of temperature
DVA		Differential Voltage Analysis	A differential voltage (dV/dQ) curve, obtained by differentiating the charge/discharge voltage with respect to capacity, is as an indicator of the degradation of electrodes. The peak shift of dV/dQ and the change in the peak-to-peak capacity are usable indicators to understand the capacity fade of electrodes inside the cell.

Term	Unit	Meaning	Definition
ECU		Electronic Control Unit	Central Electronics of an Electric Vehicle
EDLC		Electric Double Layer Capacitor	An Electric double layer capacitor is an electric energy storage system based on charge–discharge process (electrosorption) in an electric double layer on porous electrodes
EDR		Equivalent Distributed Resistance	Equivalent Resistance to a Capacitor
EDX		Energy Dispersive X-ray	Energy-dispersive X-ray spectroscopy or energy is an analytical technique used for the elemental analysis or chemical characterization of a sample. It relies on an interaction of some source of X-ray excitation and a sample.
Electrochemical Cell or Unit Cell			See accumulator
Electrolyte			An Electrolyte is a substance conducting charged particles called ions, which migrate towards electrodes (cathode and anode) of an electric circuit submitted at its terminals to a voltage difference
ESR		Equivalent Series Resistance	Equivalent Series Resistance of a Capacitor
FIB		Focused Ion Beam milling	FIB is a technique used in materials science for site-specific analysis, deposition, and ablation of materials. A FIB setup uses a focused beam of ions to image the samples.

Term	Unit	Meaning	Definition
Floating			End of Charge Voltage Holding until reaching a specified current limit
FT-IR		Fourier-Transform InfraRed Spectroscopy	Fourier-transform infrared spectroscopy (FTIR) is a technique used to obtain an infrared spectrum of absorption or emission of a solid, liquid or gas. An FTIR spectrometer simultaneously collects high-resolution spectral data over a wide spectral range. The term Fourier-transform infrared spectroscopy originates from the fact that a Fourier transform (a mathematical process) is required to convert the raw data into the actual spectrum.
GC-MS		Gas Chromatography – Mass Spectroscopy	Gas chromatography–mass spectrometry is an analytical method that combines the features of gas-chromatography and mass spectrometry to identify different substances within a test sample. Applications of GC-MS. It allows analysis and detection even of tiny amounts of a substance
ICA		Incremental Capacity Analysis	Incremental Capacity Analysis is a method used to investigate the capacity state of health of batteries by tracking the electrochemical properties of the cell. It is based on the differentiation of the battery capacity over the battery voltage, for a full or a partial cycle regarding the experimental conditions.

Term	Unit	Meaning	Definition
ICP-OES		Inductively Coupled Plasma Optical Emission Spectrometry	Inductively coupled plasma optical emission spectrometry, is an analytical technique used for the detection of chemical elements. It uses a plasma maintained by inductive coupling to produce excited atoms and ions that emit electromagnetic radiation at wavelengths characteristic of a particular element. The intensity of the emissions from various wavelengths of light are proportional to the concentrations of the elements within the sample.
Intercalation			Intercalation is the reversible insertion of ions into materials with host structures.
LCO		$LiCoO_2$	Positive Electrode Material for a Li-ion cell based on lithiated cobalt oxide
LFP		$LiFePO_4$	Positive Electrode Material for a Li-ion cell based on lithiated iron phosphate
LIC		Lithium-ion capacitor	A lithium-ion capacitor (LIC) is a hybrid type of capacitor where the anode is the same as those used in lithium-ion batteries and the cathode usually made of activated carbon is the same as those used in supercapacitors. The anode consists of carbon material which is often pre-doped with lithium ions, which lowers the potential of the anode and allows a

Term	Unit	Meaning	Definition
			relatively high output voltage compared to other supercapacitors.
LiPF$_6$		Hexafluorophosphate de lithium	Lithium salt commonly used in Li-ion cell electrolyte
LISICON		LIthium Super Ionic CONductor	Ceramic Material Family that can be used for solid electrolytes of all solid-state batteries
LiTFSI		Bis (trifluoroethanesulfonyl) imide de lithium	CF$_3$SO$_2$NLiSO$_2$CF$_3$: lithium salt used in Li-ion cell electrolyte
LMO		LiMn$_2$O$_4$	Positive Electrode Material for a Li-ion cell based on lithiated manganes oxide
LNMO		LiNi$_{0.5}$Mn$_{1.5}$O$_4$	Positive Electrode Material so-called 5 V spinel
LTO		Li$_4$Ti$_5$O$_{12}$	Negative Electrode Material for a Li-ion cell based on lithium titanate oxide
MMU		Module Management Unit	BMS component at the level of each module of a battery pack
NASICON		Sodium (Na) Super Ionic CONductor	Ceramic Material Family that can be used for solid electrolytes of all solid-state batteries
NCA		LiNi$_{0.8}$Co$_{0.15}$Al$_{0.05}$O$_2$	Positive Electrode Material for a Li-ion cell based on lithiated nickel-cobalt-aluminum oxide
NMC		LiNi$_x$Mn$_y$Co$_z$O$_2$	Positive Electrode Material for a Li-ion cell based on lithiated nickel-manganese-cobalt oxide
NMP		N-Methyl-2-pyrrolidone	C$_5$H$_9$NO: organic solvent used the manufacturing of inks for battery electrodes

Term	Unit	Meaning	Definition
NMR		Nuclear Magnetic Resonance	Nuclear magnetic resonance is a physical phenomenon in which nuclei in a strong constant magnetic field are perturbed by a weak oscillating magnetic field and respond by producing an electromagnetic signal with a frequency characteristic of the magnetic field at the nucleus. NMR is widely used to determine the structure of organic molecules in solution and study molecular physics and crystals as well as non-crystalline materials.
NVPF		$Na_3V_2(PO_4)_2F_3$	Positive Electrode Material for Na-ion Batteries
OCV		Open Circuit Voltage	Open Circuit Voltage (without applied current) of an electrochemical cell
Oxydation			Oxydoreduction reaction that leads to the production of electrons
PMU		Pack Management Unit	Pack-level Battery Management Systeme
Primary Battery			Non rechargeable battery
PTFE		Polytetrafluoroethylene	$(C_2F_4)_n$
PVDF		Polyfluorure de vinylidène	$(C_2H_2F_2)_n$
Reduction			Oxidoreduction reaction that leads to the comsumption of electrons
Secondary Battery			Rechargeable battery

Term	Unit	Meaning	Definition
SEI		Solid Electrolyte Interphase	Solid products generated by the electrochemical decomposition of electrolyte (liquid) at the surface of the electrode. To allow good operating conditions for Li-ion cells, the SEI formed on the graphite negative electrode must be both passivating (dense, insoluble) and ionic conductor of Li^+ ions
SEM		Scanning Electron Microscopy	A scanning electron microscope (SEM) is a type of electron microscope that produces images of a sample by scanning the surface with a focused beam of electrons. The electrons interact with atoms in the sample, producing various signals that contain information about the surface topography and composition of the sample.
Separator			Electronic Insulator Component that contains electrolyte and is place between positive and negative electrodes
SHE		Standard Hydrogen Electrode	Standard Hydrogen Electrode used as reference electrode
SIMS		Secondary Ion Mass Spectrometry	Secondary-ion mass spectrometry is a technique used to analyze the composition of solid surfaces and thin films by sputtering the surface of the specimen with a focused primary ion beam and collecting and analyzing ejected secondary

Term	Unit	Meaning	Definition
			ions. The mass/charge ratios of these secondary ions are measured to determine the elemental, isotopic, or molecular composition of the surface to a depth of 1–2 nm
SOC	% of nominal capacity	State of Charge	State of Charge of the Battery
SOE		State of Energy	State of Energy of the Battery
SOH		State of Health	State of Health of the Battery, often defined as $SOC_{actuel}/SOC_{initial}$
Specific Capacity	mAh/g		Capacity related to the mass of the active material
TEABF$_4$		Tetraethylammonium tetrafluoroborate	$(C_2H_5)_4N(BF_4)$: electrolyte used for supercapacitors
TEM		Transmission Electron Microscopy	Transmission electron microscopy is a microscopy technique in which a beam of electrons is transmitted through a ultrathin less than 100 nm thick specimen to form an image. The image is formed from the interaction of the electrons with the sample as the beam is transmitted through the specimen. TEM microscopes are capable of imaging at a significantly higher resolution than light and SEM microscopes.

Term	Unit	Meaning	Definition
TOF-SIMS		Time-of-Flight Secondary Ion Mass Spectrometry	Time of Flight Secondary Ion Mass Spectrometry is a highly sensitive analytical technique that describes the chemical composition and distribution of a sample surface. It uses a range of incident ion sources to impact on solid surfaces and generate secondary ions that can be analysed by a time of flight mass spectrometer to determine the surface chemistry of that surface or layer.
Watt.hour		Energy Unit. 1 Wh = 3600 Joules	
xC	h^{-1}	Charging/Discharging Rate of a cell	x = number of charges or discharges in 1 h
XPS		X-ray Photoelectron Spectroscopy	X-ray photoelectron spectroscopy (XPS) is a surface-sensitive quantitative spectroscopic technique based on the photoelectric effect that can identify the elements that exist within a material (elemental composition) or are covering its surface, as well as their chemical state, and the overall electronic structure and density of the electronic states in the material. It not only shows what elements are present, but also what other elements they are bonded to...
XRD		X-Ray Diffraction	X-ray Diffraction is the experimental science determining the atomic and molecular structure of a crystal, in which the crystalline structure causes a beam of incident X-rays to diffract into many specific

Term	Unit	Meaning	Definition
			directions. By measuring the angles and intensities of these diffracted beams, a three-dimensional picture of the density of electrons within the crystal is produced. From this electron density, the mean positions of the atoms in the crystal can be determined, as well as their chemical bonds, their crystallographic disorder, and various other information.

The Authors

The Coordinators

Didier BLOCH is a graduate engineer from the Institut National Polytechnique de Grenoble. He worked for several years in industry (Schneider Electric, VARTA GmbH), then was recruited by the CEA in 1992. In 1996, he became head of the "Electrochemical Energy Storage" Laboratory (Lithium Batteries and Fuel Cells), then Program Manager "New Technologies for Energy". He is an expert for OPECST, the EC, the ANR. Since 2016, he has been in charge of the "Battery Materials" laboratory at CEA-LITEN. In 2017–2018, he is an expert on "batteries and electric vehicles" within the "Coexistence of Nuclear and Renewable Energies" Working Group led by the High Commissioner for Atomic Energy.

Dr. Sébastien MARTINET has more than 25 years of experience in the battery field. After a Ph.D related to NiMH batteries with SAFT, he was involved on Li-ion safety improvement in SAFT research center. He then joined CEA in 1999 where he was first in charge of developing the Li-ion process line. He was in charge of the battery laboratory from 2006 to 2009 and then managed scientific activities related to batteries, both Li-ion and post Li-ion.

Dr. Thierry PRIEM, Storage and Flexibility Solution Program Manager at CEA, has an engineering degree from the elite French "École Polytechnique" and "École Nationale Superieure des Mines de Paris". He also has a Ph.D in solid-state physics. Thierry Priem has a broad technical and scientific background at CEA (French Alternative Energies and Atomic Energy Commission) in different research fields: material science, new energies, etc. Thierry Priem teaches in several engineering schools and is regularly involved in national and European expert missions.

For 4 years, Thierry Priem has been at the head of a service of 60 people involved in new energies (hydrogen and fuel cells, photovoltaics, etc.). He has also an experience in technology transfer and negotiation with industrial partners. Mid-2007, he has been the Hydrogen & Fuel Cell Program Manager within the Direction of the New Technologies for Energy at CEA in Grenoble. In 2014, he joined the Scientific Direction at CEA/LITEN. Since 2020, he is Storage and Flexibility Solution Program Manager at CEA. He is also member of the Board of the FCH-JU2 Research Grouping "Hydrogen Europe Research".

Christian NGÔ has published more than a dozen books, alone or in collaboration with another author, on several subjects ranging from basic physics (statistical physics, quantum mechanics, nuclear physics, semiconductor physics) to more applied fields such as energy, nanotechnologies, waste and pollution, the sun, information theory, etc. He did fundamental research for twenty years and published around 200 papers before moving to applied research where he held several management operational and functional positions. In 2008, he created the consulting company EDMONIUM.

The Contributors

Dr. Philippe AZAÏS, after his Ph.D thesis on the aging of supercapacitors in collaboration with SAFT Company, Philippe Azaïs joined the Bollore Group (Batscap/Blue Solutions) in 2006 as Head of Innovation and Research for the Supercapacitors activity. At the end of 2011, he joins the CEA to coordinate multidisciplinary research and development projects in the field of energy storage (in particular lithium-ion batteries) with industrial and academic partners. Philippe Azaïs is a specialist in carbon and energy storage systems. He also supervises thesis in these fields.

Dr. Céline BARCHASZ is a research scientist at CEA LITEN, working on the development of post lithium-ion technologies for more than 10 years. Her research activities mainly concern lithium metal batteries, metal/sulphur systems and all solid-state technologies, both at the material and cell levels.

Dr. Michel BARDET has been working at CEA since 1982. Its main fields of interest are high-resolution solid and liquid state NMR. For more than 15 years he has been involved in NMR studies of organic and inorganic materials of interest for harvesting and storage of energy.

Dr. Anass BENAYAD is Senior Researcher at CEA since 2013 at the department of nanomaterial and nanotechnology. He is developing new activities around operando studies of redox process and electrolyte degradation mechanism in lithium ion batteries. From 2006 to 2013, he was Senior Researcher at Samsung Advanced Institute of technology (SAIT), where he was in charge of different projects related to band gap engineering of the surface and interface electronic structure for battery, 2D-materials (structured carbon and MoS2), and semiconductor materials for thin film transistors for LCD technology. Particularly, he developed a large expertise in Photoemission spectroscopy (XPS, UPS) supported by ab-initio calculations applied to a large variety of inorganic and organic materials. He completed his Ph.D in physics and chemistry of condensed matter from university of Bordeaux at the Institute of Chemistry and Condensed Matter of Bordeaux (ICMCB). He is author and co-author of ~ 100 review papers and 7 patents.

Bruno BERANGER, is an electronics engineer; he is at the origin of several conversions of electric vehicles with lithium ion batteries. He has extensive knowledge of these technologies and has owned several electric cars for the past 18 years.

Dr. Emmanuel BILLY holds a Doctoral degree in electrochemistry, Materials and Process Engineering, France. During his Ph.D, he developed a HydroMetallurgical process to recover precious metals from electronic swarf (WEEE). Since 2012, he is in charge of a recycling team and recycling program on new components for energy

(Li-ion and NiMH batteries, fuel cells, PV cells, permanent magnets) by HydroMetallurgical processes.

Pr. Renaud BOUCHET is Professor in Electrochemistry and materials Science and makes his lecture at the engineering school PHELMA at Grenoble Alpes University since 2012. After his diploma of electrochemical engineering (ENSEEG, 1996), he obtained his Ph.D degree in Electrochemistry from Grenoble INP in 1999. After one year postdoc at Ecole Polytechnique (Palaiseau), he was recruited as Assistant Professor at Aix-Marseille University (AMU) in 2000 and obtained his HDR diploma from AMU in 2006. His main field of interest concerns the ionic charge transport through divided materials (nanocomposite, nanostructured materials...) in applications linked to electrochemical storage. The aim is to analyze the relationships – composition, structure, microstructure and the electrical properties, to design efficient multifunctionnal materials based on polymers and/or ceramics for the applications. He received in 1999, the Ph.D first prize of Grenoble INP, in 2005 the Electrochemistry Prize from the French Chemical Society, and in 2014 the international prize EDF Pulse Science and Electricity for his work on electrochemical storage. He co-authored 76 publications, 3 book chapters, 9 patents, and 82 oral communications (18 invited).

Dr. Adrien BOULINEAU is a solid-state chemist who completed in 2008 his Ph.D focused on the understanding of the layered oxides for Li-ion batteries at the Institute of Chemistry and Condensed Matter of Bordeaux (ICMCB) under the supervision of Dr. Laurence Croguennec and Dr. François Weill. He moved in 2009 to the Laboratoire de Reactivite et Chimie des Solides (LRCS) where he specialized in electron microscopy where he revealed the polymorphism within the silicates family of materials for Li-ion batteries. Since 2010, he is scientist at CEA-Liten where he is developing activities around the application of electron microscopy for characterizing the wide variety of materials used in Li-ion batteries. He is author or co-author of more than 30 articles in the field.

Carole BOURBON, R&D Chemical Technician, is in charge of the synthesis and characterizations of active material for batteries. She is involved since more than 20 years on projects relative to battery materials and she is co-author or author of more than 20 patents and 13 publications in that field.

David BRUN-BUISSON is a research engineer and battery expert. He has been working at CEA for more than 13 years in the field of battery tests and post-mortem analysis. He is now in charge of the CEA safety tests platform, supervising and conducting abuse tests on all kinds of Li-ion batteries and analogues, from cell to pack level. He is the author or co-author of more than 4 publications and the main inventor or co-inventor of 13 patents.

Dr. Claude CHABROL, R&D engineer, worked on Surface Treatments and forging tools in USINOR-SACILOR steel making group during 9 years. In 1994, she joined CEA for projects about CVD and PVD coatings. She is now in charge of characterization of internal components of Li-ion batteries and electrode materials using X-ray diffraction (ante & post-mortem studies).

Marlène CHAPUIS is graduated of Electrochemistry Engineering school, with honors (INP Grenoble – 2008). She spent 2 years in RIO TINTO ALCAN industry Research center dedicated to aluminum production by electrolysis process-Activities on New materials development for R&D potlining (R&D programs). Then moved in GRENOBLE CEA-LITEN institute to manage during 6 years a team dedicated to Lithium ion Batteries components development. Since 2018, is in charge of a recycling laboratory.

Daniel CHATROUX is a power electronic expert. After six years in industry for specific power converters development, he joined CEA in 1992. First, he developed innovative power electronic for high frequency high voltage generators for pulsed laser (cost effective and very reliable high voltage switches with thousands of MOSFETs). Now, his research domain is power electronic for new energies: photovoltaic, fuel cell systems and batteries for automotive applications. He participates at 80 patents on these technologies.

Guillaume CLAUDE, following 5 years of experience in semiconductor activities, he joined CEA/LITEN teams in 2010 to work on the setting up of a pilot line for lithium ion batteries. After installing the equipment and starting it up, he was in charge of maintenance of equipment and dry room. Since 2017, he has been a project manager on "process" themes.

Dr. Jean-François COLIN is a CEA senior expert in the field of materials for batteries. He has 15 years of experience in the development of electrode materials for Li-ion batteries. He is focusing on the development and the characterisation of cathodes for high energy density applications.

Lise DANIEL is engineer in electrochemistry (Grenoble INP, France) and works in CEA-LITEN for 12 years. From 2012 to 2018, she was in charge of a laboratory dedicated to batteries and PEMFC characterization. Her work is focused on ante/post-mortem/operando studies on batteries to increase knowledge on reaction mechanisms in normal, severe and abuse conditions.

Julien DAUCHY received a master's degree in Electronics Engineering from University of Grenoble Alpes/Polytech (Grenoble, France) in 2009. He joined CEA in 2009 where he is a system electrical architect. He first worked in fuel cell system during 2 years. He developed a power DC-DC converter for the interface between fuel cell voltage and battery system. He has worked for 9 years on battery system for electrical vehicles (car, bus, boat, train, plane). During this period at CEA, he contributed to 21 active patents and presented several papers at the conferences PCIM (Power Conversion Intelligent Motion), EPF (Electronique de puissance du future) and EVS (Electrical Vehicle Symposium). He developed battery systems for many partners like Renault, Airbus DS, Thales, Michelin, Toyota and worked on two European projects (ESPRIT and NENUFAR).

Pr. Rémi DEDRYVERE is professor at the University of Pau and Pays Adour, working in the Institute of Analytical Sciences and Physical Chemistry for Environment and Materials (IPREM). His research activities deal with phenomena occurring at surfaces and interfaces in Li-ion and post Li-ion batteries, with a special focus on ageing processes at electrode/electrolyte interfaces.

Dr. Arnaud DELAILLE holds a Ph.D from the University of Paris 6, specializing in electrochemistry. For 4 years, he managed the CEA-Liten LSEC laboratory in charge of the development of battery management indicators and laws, before co-founding the PowerUp startup in which he now holds the position of Director of Operations.

Ghislain DESPESSE graduated (MS degree) in electrical engineering from Ecole Normale Superieure de Cachan (ENS) in 2002 and he received his Ph.D. Degree in microelectronics in 2005. He joined the CEA Laboratory in 2005. He was involved in the development of energy harvesters and he is implicated for nine years in the development of Battery Monitoring Systems (BMS) including inverter functions. He has published 65 papers in refereed journals and conferences and he holds 70 patents.

Stéphanie DESROUSSEAUX is graduated in chemistry and physics and has a Ph.D in Organic Chemistry. After working in R&D in the industry sector for 10 years, she joined CEA in 2009, where she carried applied research in material science to support industrial partners. She is co-inventor in about 20 patents. She is now involved in Life Cycle Assessment, eco-design and circular economy projects related to new material, components and process research for chemistry processes, additive manufacturing and printed electronics.

Didier DEVAUX is a CNRS scientist since 2016 working at the LEPMI laboratory in Grenoble. His researches are focusing on operando electrochemical characterizations of materials and interfaces within accumulators. He graduated in 2012 from Aix-Marseille University in materials science applied to lithium batteries. After his Ph.D, He joined the Lawrence Berkeley National Laboratory and the University of California at Berkeley (USA) as a post-doctoral fellow studying polymer electrolytes by synchrotron X-ray characterization technics.

Dr. Eric DE VITO is senior expert at CEA. He received his Ph.D in 1992 at Universite Pierre et Marie Curie (Paris VI). In 2016, he obtained his "Habilitation to supervise research" (HDR) at Universite Grenoble Alpes. During the last ten years, he has been involved in several research projects related to the study of materials for energy, and more specifically to the development and optimization of electrode materials for Li-ion batteries. He is co-author of more than 40 publications and supervises several Ph.D projects in relation with these studies.

Dr. Jean-Baptiste DUCROS received his Ph.D in electrochemistry/material science in 2006, from the University of Paris XII. He has more than 10 years experience in the field of Li-ion batteries and supercaps. He is now a senior researcher in CEA Grenoble since 7 years.

Dr. Xavier FLEURY has a Ph.D in Materials Science from the University of Grenoble Alpes. He is currently a research engineer in the R&D department of Carbone Savoie, where he works on the development of a new grade of artificial graphite for lithium-ion batteries. Throughout his thesis, which took place at CEA-Grenoble (under the supervision of Pierre-Xavier Thivel, Senior Lecturer at the University of Grenoble-Alpes), he studied the correlation between the aging of internal components within lithium-ion cell and the safety impact. He was particularly interested in the behavior of the separator at the interface with the negative electrode towards the growth of SEI.

Dr. Cédric HAON, Engineer from INPG-ENSEEG (2003) and Ph.D. in Materials Science (INSA Lyon – 2006), was R&D Engineer at Praxair MRC (3 years), before joining CEA in 2010. He is specialized in the synthesis and electrochemical characterization of inorganic electrode materials for lithium-ion batteries. He works mainly on the development of silicon-based anode materials. He is project manager for several European and national projects. He is co-author of 7 patents, 12 publications and co-supervision of 5 theses as well as numerous master students.

Laurent GARNIER is power electronics expert. He graduated in electrical engineering from Ecole Centrale Marseille in 1991. After spending 18 years in industry as an R&D engineer, project manager and then R&D manager in the field of energy conversion, Laurent Garnier joined the CEA in 2010. Since then, he has been leading various research projects in the field of battery and fuel cell systems for industrial partners such as Renault, Thales, etc. He is also scientist correspondent for the LITEN management on these same topics.

Dr. Sylvie GENIES has a Ph.D in electrochemistry from the University of Grenoble I. She joined CEA in 2005, first at the National Solar Energy Institute located in Le Bourget du Lac as a researcher and she characterized the behavior in cycling of different energy storage technologies with a comprehensive post-mortem analysis approach. Since 2011, at the Grenoble site, she has been studying the degradation mechanisms in lithium-ion accumulators by developing or implementing operational diagnostic tools or techniques. This activity is associated with comparative ante/post mortem studies consisting in carrying out in-depth physicochemical and electrochemical characterizations of the internal components recovered after dismantling (electrodes, separator and electrolyte).

Dimitri GEVET holds a license degree in electrical distribution and renewable energies, Dimitri GEVET joined the CEA in 2012 as a technician of batteries assembly for the electrical transport within the LITEN. The core of its functions is the development of new assembly processes and techniques while improving assembly safety.

Dr. Nicolas GUILLET, after a Ph.D in Process Engineering at the Ecole des Mines de Saint Etienne (France) and two years of post-doctoral training in electrocatalysis at INRS-Energie et Materiaux de Varennes (Canada), Nicolas Guillet joined CEA-Liten in Grenoble (France) in 2005 in the team dedicated to the development of PEM fuel cells. Since then, he has been involved in numerous projects related to different energy conversion systems (fuel cells, electrolyzers, batteries...).

Dr. Thibaut GUTEL was graduated from Ecole Superieure de Chimie Organique et Minerale (ESCOM) in 2004 and obtained his Ph.D from Claude Bernard University (Lyon I) under the supervision of C. Santini et Y. Chauvin. His Ph.D focused on the specific behavior of ionic liquid as solvent for catalysis and synthesis of metal nanoparticles. In 2008, he was recruited by HUTCHINSON R&D center (branch of TOTAL group) in order to study the synthesis of porous carbon and their use for supercapacitor electrodes. Since 2010, he is research scientist at CEA-LITEN. His work is dedicated to the synthesis and the characterization of new inorganic and organic electrode materials but also to the study of emerging electrolytes for Lithium battery. In 2019, he obtained his "Habilitation to supervise research" (HDR) at Grenoble Alpes university.

Vincent HEIRIES obtained an engineering degree from ENAC (Ecole Nationale de l'Aviation Civile) in 2003 and a Ph.D in signal processing and digital communications from ISAE (Supaero) in 2007. He worked several years at THALES Space in the field of satellite navigation systems (GPS, GALILEO). Since 2012, he has been working at CEA and his research activities are mainly focused on signal processing applied to fault detection in electrical systems and battery management system. He co-authored 30 publications and is the main inventor or co-inventor of 17 patents.

Dr. Severine JOUANNEAU SI LARBI has been Director of the Department of Electricity and Hydrogen for Transport for 6 years within CEA/LITEN. The main mission of this research unit is to develop solutions for the storage of energy and its components, in particular with innovative batteries and fuel cells. Previously, Severine JOUANNEAU SI LARBI carried out, for about fifteen years, research activities on components (materials, electrodes, cells) and manufacturing processes for improvement of Li-ion batteries performance. Ph.D in material sciences, she is the author of 45 publications and more than 20 patents in the field of batteries.

Dr. Fathia KAROUI is a research engineer at the Laboratory of Electrochemical Storage at CEA-LITEN. She holds a Ph.D, a master's degree in electrochemistry and an engineering degree in decentralized production and energy storage from Grenoble INP. She has more than 15 years of experience in battery modeling, management, diagnosis and sizing.

Dr. Pierre KUNTZ obtained his Ph.D in December 2020 (University of Grenoble-Alpes). He has worked at CEA on the evolution of the safety behavior of Li-ion batteries during their ageing. He studies and characterizes the degradation mechanisms that appear during the use of Li-ion 18,650 commercial cells and determines their impact on safety. His research activities are mainly aging performances analysis, electrochemical and physico-chemical material characterization and safety tests on Li-ion cells.

Dr. Frédéric LE CRAS is a senior researcher at CEA LITEN with expertise in the field of Li-ion battery materials and electrochemistry. His work has been focusing for many years on all-solid-state devices, including microbatteries, with the development of thin film materials, inorganic ionic conductors and the study solid/solid interfaces.

Dr. Florence LEFEBVRE-JOUD is senior scientist in the field of Materials for Energy and holds the position of deputy director of the Institute CEA-LITEN in charge of scientific activities. Graduated for Phelma Engineer School, with a Ph.D and Habilitation from "Grenoble Institut National Polytechnique" she has supervised around 10 master thesis and 11 Ph.D theses and has authored and co-authored more than 50 publications in refereed scientific journals (H-index 20). She is also member of several national (HCERES) and international (ERC) panels in the field of materials and processes for renewable energies.

Dr. Sandrine LYONNARD holds a Ph.D in Solid State Physics. She is a senior researcher at CEA-IRIG, specialized in the investigation of nanomaterials by means of ex situ and operando synchrotron and neutron techniques. Recently, her interests focused on understanding basic mechanisms in battery electrodes and electrolytes, *e.g.* lithiation process, structural/morphological ageing and ionic transport.

Nicolas MARIAGE is a graduate engineer in Material Science from Grenoble POLYTECH. At CEA LITEN, he works as a specialist of electrodes for Li-batteries manufacturing.

Dr. Jean-Frédéric MARTIN is a research engineer at CEA LITEN. After a thesis defended in 2008 dealing with surface phenomena in Li-ion batteries, he worked on the subject of liquid electrolytes (high voltages, low temperatures...) in the framework of several industrial and academic projects. Attached to the study of aging and degradation in these systems, he is now interested in the electrochemistry of post-lithium-ion systems.

Dr. Daniel MEYER is Director of Research at the CEA, he leads a research team at the Institute of Separative Chemistry in Marcoule. He is interested in fundamental and applied studies related to the development of material cycles in a hydro-metallurgical waste recycling process.

Laure MONCONDUIT is director of research at the CNRS at the ICGM, in Montpellier. She has led the "Batteries" research group since 2011. Her research focuses on the synthesis and characterization of new electrode materials for Li-ion and post-Li (Na, K, Mg, Ca-ion) with particular attention focused on the redox mechanisms of electrode materials and those involved at the electrode/electrolyte interface, by operando techniques (DRX, Mössbauer, IR-ATR, Raman, XAS). She coordinated (2 ANR) or participated in numerous French or international academic projects (H2020), of collaboration with industrials (Total, SAFT, Renault, Nanomakers ...).

Elise MONNIER is graduated in processing, material and electrochemistry engineering and in Ecodesign (joint degree) in France. She joined CEA in 2010. From then on, she works on Life Cycle Analysis (LCA), environmental studies, ecodesign of New Energy Technologies and eco-innovation. She is currently conducting LCA, eco-design or eco-innovation projects on electrical mobility, on new energy production and storage means, on circular economy and on innovative electronics.

Djamel MOURZAGH is an R&D engineer at CEA Tech, Grenoble. He developed Li-ion prototypes for various applications (medical, automotive, space, PV, etc....); he then focused on the development of battery components and more specifically on battery separators. He notably participated in the development of functionalized separators (Li salt rich separator, ceramic-coated separator). He is currently involved in the development of a hybrid polymer technology Solgain™ in collaboration with Solvay.

Sébastien PATOUX is manager of the battery technologies division at CEA for more than 10 years. He is also an international expert for his company. He has a background on material sciences, involved on battery activities for more than 20 years. He defended a Ph.D on Li-ion cathode materials in 2003 at the University of Amiens (France), before joining the Lawrence Berkeley National Laboratory (US) for a postdoctorate position. He has been going back to France to work for CEA. He holds 20 patents and more than 40 publications in the domain.

Dr. David PERALTA received his Ph.D in materials chemistry in 2012, from the IFP energies nouvelles. He joined CEA in 2012 as chemical researcher. His research focuses on the synthesis and optimization of new materials for Li-ion applications.

Dr. Fabien PERDU is engineer and researcher at CEA LITEN. He worked for eight years on the neutronics and thermal hydraulics simulations of innovative nuclear reactors, followed by ten years on electrochemical energy storage. His current research activities cover mainly battery technologies performances, how they match mobility or stationary applications, and the associated environmental impacts linked to the materials and energy used.

Dr. Bramy PILIPILI MATADI holds a Ph.D from the University of Grenoble Alpes, specialized in electrochemistry. He conducted his research work based on understanding and modeling the aging of Li-ion batteries at the LSEC laboratory of CEA-Liten, before joining the PowerUp startup where he now holds the position of Battery Expert.

Pr. Philippe POIZOT was appointed as full Professor at the University of Nantes (Institut des Materiaux Jean Rouxel, IMN-CNRS, Nantes, France) in 2012. After a Master of Science in analytical chemistry and electrochemistry (University of Paris VI, 1998), he obtained his Ph.D. degree in Materials Science (2001) focused on "conversion reactions" at the University of Picardy Jules Verne (UPJV-LRCS) in Amiens, France. Following a postdoctoral training with J.A. Switzer at the University of Missouri-Rolla (USA) to develop the electrodeposition of nanostructured materials, he came back to UPJV-LRCS as Associate Professor in 2002. He proposed the concept of "renewable" batteries in 2008 by promoting novel electrode materials based on redox-active organic compounds deriving from biomass. His current research interests are mainly focused on rechargeable batteries, and the development of organic batteries. A recipient of the Bronze Medal of the French Society for Encouragement and Progress (2001), Junior fellow of the Institut Universitaire de France (2012–2017), Prof. Poizot has published more than 90 peer-reviewed articles and 10 patents.

Dr. Willy PORCHER, graduated from Electrochemistry Engineering school (Grenoble INP, France), he obtained his Ph.D. degree in Materials Science (2008) on "aqueous formulation of positive electrode based on LiFePO4" at the University of Nantes. He is an expert at CEA LITEN in Li-ion technology and was coordinator of the H2020 SPICY project. He has twenty publications and ten patents relating to the aqueous formulation of LFP and NMC cathodes, cells with a silicon-based anode, ultra-thick electrodes or the prelithiation process.

Dr. Yvan REYNIER graduated from Electrochemistry Engineering school (Grenoble INP, France), then studied Li-ion battery electrode thermodynamics and kinetics to obtain his Ph.D in materials science in 2005. He is now battery expert specializing in battery design and prototyping. He is an author of 9 patents, 19 publications and co-advisor of 4 Ph.Ds and 7 master trainees.

Dr. Hélène ROUAULT, R&D engineer at CEA, is involved in the development of innovative lithium batteries for all types of applications since 1998. Today, her R&D activity is mainly dedicated to a new hybrid polymer technology developed with SOLVAY, the SOLGAIN™ technology.

Florence ROUILLON is in charge of the Battery Prototyping and Processes Laboratory at CEA since 2011. She is responsible for the platform including dry rooms and pilot line for cells assembly. After graduating in 1993 as Master's Degree in Engineering from National Polytechnic Institute of Grenoble (INPG-ENSEEG), France, in process engineering and electro-chemistry, she pursued her studies by specializing in industrial engineering during 1 year at the National Polytechnic Institute of Lorraine (INPL-ENSGSI), France. She first worked as organizational consulting engineer in different fields (automotive, packaging, luxury, food...) in France and Spain.

Pr. Saïd SADKI was appointed as full Professor at University Grenoble Alpes in 2004. He develops his research activity in the joint laboratory SyMMES (CEA, CNRS, University grenoble Alpes) based at CEA Grenoble. Pr. Said Sadki holds a Ph.D in Electrochemistry and polymer science (1992) from Univ. Paris Sorbonne (France) and a research habilitation (HDR) in Electrochemistry of Soft Matter (2003) from Univ. Cergy Pointoise (France). Dr. Sadki seeks to optimize the properties of polymers for color changing electrochromics, visible and near-infrared light emission, and flexible organic solar cells. Since 2010, Dr. Sadki is active in developing research project dealing with nanostructured silicon electrodes for (μ)-supercapacitors and battery applications. Pr Sadki has supervised 12 Ph.D students; he co_leads 3 Europeans Projects and was the leader of 3 ANRs in the field of Supercapacitors and Batteries.

Dr. Virginie SIMONE is a research engineer in materials chemistry. Holder of an engineering degree from ENSCBP and a Ph.D from CEA-Liten, she focused on energy storage and more particularly on the development of Na-ion batteries during her Ph.D. Her thesis work develops the study of a complete Na-ion system, from the synthesis of the electrode materials to their electrochemical performances and their detailed characterization.

Dr. Loïc SIMONIN has been a Research Engineer at CEA-LITEN for 10 years. He holds a Ph.D in Materials Chemistry from the Technical University of Delft (Netherlands) and has been working for 15 years on electrode materials for batteries, such as transition metal oxides, polyanions, alloys and carbonaceous materials. He is an expert in Na-ion batteries, from materials to system integration. He coordinates numerous R&D projects with academic and industrial partners.

Dr. Dane SOTTA received his Ph.D from Universite de Picardie (Amiens, France) in 2011. The topic of his Ph.D thesis was to develop new gelled electrolytes based on ionic liquids for lithium-ion batteries. Then, he joined CEA and now works as a research engineer within a laboratory dedicated to development and manufacturing of components for energy storage devices (batteries, supercapacitors). He was the coordinator of the FP7 project MAT4BAT and he is now in charge of electrode formulations and new process developments for battery technologies in industrial and European projects.

Samuel TARDIF is a researcher at the CEA-IRIG (Institut de Recherche Interdisciplinaire de Grenoble) and beamline scientist on the French beamline InterFace-BM32 at the European Synchrotron Radiation Facility (ESRF). His expertise is in the study of crystalline phases and mechanical strain using synchrotron radiation techniques (Laue microdiffraction, micro or nano X-ray diffraction), as well as in the development of synchrotron-based operando characterization techniques for the study of batteries.

Dr. Vasily TARNOPOLSKIY holds a Ph.D in inorganic chemistry. He has been working at SAMSUNG SDI (South Korea) where he developed low temperature electrolytes for Li-ion cells, and then at Münster University where he developed Li-ion cathode materials operating at high voltage. He joined CEA-LITEN (Grenoble, France) in 2011 where his expertise in inorganic chemistry and in

Li-ion cells is focused on the development of All-Solid-State Li-ion Batteries and liquid electrolytes. He is managing several collaborative industrial and European projects.

Romain TESSARD has been a research engineer at the CEA for 10 years, after 10 years as an engineer and R&D project manager for consumer electronic products. He has specialized in the characterization and selection at cell level of storage devices, mainly lithium-ion cells, with a wide knowledge of suppliers and markets.

www.ingramcontent.com/pod-product-compliance
Lightning Source LLC
Chambersburg PA
CBHW060747220326
41598CB00022B/2354